GLOBAL TRENDS:
Fisheries Management

GLOBAL TRENDS: Fisheries Management

Ellen K. Pikitch, Daniel D. Huppert,
and Michael P. Sissenwine, Editors

Marcus Duke, Associate Editor

American Fisheries Society Symposium 20

Proceedings of the Symposium
Global Trends: Fisheries Management

Held at Seattle, Washington, USA
14–16 June 1994

American Fisheries Society
Bethesda, Maryland
1997

The American Fisheries Society Symposium series is a registered serial. Suggested citation formats follow.

Entire book

Pikitch, E. L., D. D. Huppert, and M. P. Sissenwine, editors. 1997. Global trends: fisheries management. American Fisheries Society Symposium 20, Bethesda, Maryland.

Article within the book

Garcia, S., and C. Newton. 1997. Current situation, trends, and prospects in world capture fisheries. Pages 3-27 *in* E. L. Pikitch, D. D. Huppert, and M. P. Sissenwine, editors. Global trends: fisheries management. American Fisheries Society Symposium 20, Bethesda, Maryland.

Cover design by Cathy Schwartz

Library of Congress Catalog Card Number: 97-072126
ISBN 1-888569-03-4
ISSN 0892-2284

Printed in the United States of America on recycled, acid-free paper

American Fisheries Society
5410 Grosvenor Lane
Bethesda, Maryland 20814-2199, USA

Contents

Preface

The idea for a symposium on global trends in fisheries management was stimulated by a series of discussions among the conveners—Ellen Pikitch, Dan Huppert, and Mike Sissenwine—and by observations about the way fisheries were being managed in the United States and elsewhere in the world. Clearly, fisheries are facing some very serious problems, problems that are drawing increased attention, especially from the public, and that are being addressed around the world. Yet these problems are being approached to a large degree independently by each region. The many symposia and conferences held each year enable researchers to exchange information on the scientific aspects of fisheries; unfortunately, few occasions offer people the opportunity to get together to talk about fisheries management. This symposium was convened to provide such a forum, a forum that addresses the issues on a global scale. Moreover, a goal of this forum was to emphasize actual experiences, not just theories of how we might manage fisheries.

An important underlying theme for this conference was, "What can the United States learn?" While we were interested in the global overview, we were also interested in learning from nations that have more experience in certain aspects of management than we do, such as those countries that have employed individual transferable quota systems. Further, there is much we can learn from our colleagues' experiences here in the United States. Thus, this conference—the first major collaborative effort on a national level between the National Marine Fisheries Services and the University of Washington College of Ocean and Fishery Sciences—was an important step toward facilitating such learning and thereby improving our national perspective.

Given the goals of the conference, the University of Washington was an excellent location for bringing together fisheries professionals from around the world to discuss and learn from each other's experiences. The University's reputation as a center of excellence extends to the Schools of Fisheries and Marine Affairs, where fisheries science and management are taught, and where many of our best national and international fisheries managers were educated.

The timing of the symposium also was appropriate because of the many pressing problems fisheries management currently faces. Throughout the world, fisheries are important in terms of commerce and recreation, in providing a source of high-quality protein and other products, and in fulfilling cultural and life-style needs. In the United States, we are particularly fortunate to have large and diverse fishery resources throughout our exclusive economic zone. But these valuable assets come with responsibility for conservation and wise use. As a nation, the United States can and must do better in fulfilling these responsibilities, and we underscore that this will require improved, sound, and comprehensive scientific information. We must not allow scientific uncertainty to fuel controversy and confusion.

Ultimately, the problems facing fisheries management—overfishing, overcapitalization, environmental degradation, habitat loss, bycatch—relate to the conservation of living resources. We need to better assess fish populations; we need to better regulate commercial and recreational fisheries; we need to better formulate policies for aquaculture and its relationship with wild fisheries; and we need ultimately to better maintain the economic contribution of fisheries to society for the long term. Sound public policy in dealing with fisheries requires good science—both natural science and social science. It also requires the logical design of laws, institutions, and industry organizations.

At the time of the symposium, the biological conservation and economic goals described above were the subject of active debate in the United States and were a high priority in considering the reauthorization of the Magnuson Act. Since then, the Act has been reauthorized, and it does place greater emphasis on fisheries habitat protection, fish stock conservation, and avoidance of bycatch waste. Unfortunately, it places constraints on new limited access programs, especially individual quotas, in the United States. The authors of the papers contained in these proceedings, as well as various participants of the symposium, played an important role in contributing to the debate and thoughtful consideration that resulted in the reauthorized Act.

The proceedings of the symposium "Global Trends: Fisheries Management" represent cooperative efforts among the education, research, and management communities worldwide to improve our communication towards the goal of resolving the formidable array of problems that face fisheries management. The participants of the conference came together to communicate with and learn from each other, and it is our sincere hope that some of what we learned will be used in the future to improve the state of fisheries management in the United States and around the world.

Ellen Pikitch, Associate Director, School of Fisheries
Ross Heath, Dean, College of Ocean and Fishery Sciences
Rolland Schmitten, Assistant Director, National Marine Fisheries Service

Acknowledgments

The symposium "Global Trends: Fisheries Management" and the preparation and publication of these proceedings were made possible through the generous support of the National Marine Fisheries Service (NMFS). The joint sponsorship of this symposium by the University of Washington College of Ocean and Fishery Sciences and NMFS reflects a long-standing cooperative arrangement between the University and NMFS. This cooperation involves both scientific research and the development of fisheries management strategies. Within the College, the principal sponsors—the Schools of Marine Affairs and Fisheries and the Washington Sea Grant Program (part of the Office of Marine Environment and Resource programs)—of the Global Trends symposium and proceedings have long histories of research and training in the science and art of fisheries management. The more than 2,000 Fisheries and Marine Affairs graduates—employed as harvesters, processors, managers, and researchers—make up a substantial part of the fisheries community, particularly in the Pacific Northwest. We also appreciate the affiliate sponsorship of the symposium by Wards Cove Packing Company and Ocean Trawl, Inc.

We are particularly grateful for the participation in this symposium of distinguished experts from other regions of the United States and from other countries. Among the nations represented here are the People's Republic of China, Indonesia, Australia, New Zealand, France, the Netherlands, Canada, Norway, Great Britain, Iceland, Mexico, and South Africa—the net has been cast wide. The time and effort expended by the authors of the papers herein, as well as their patience and cooperation throughout the editorial and production process, are also greatly appreciated.

We appreciate the support of and participation in the conference by Rolland Schmitten, Assistant Director, NMFS; Ross Heath, Dean, College of Ocean and Fisheries Sciences; and Marsha Landolt, Director, School of Fisheries. In addition, we are grateful for the support and encouragement of Louie Echols, Director, Washington Sea Grant Program, as well as that of Bill Aron, Director, NMFS Alaska Fisheries Science Center. We acknowledge the time and effort in organizing and chairing the Individual Transferable Quota Forum and two panel discussions by Daniel Huppert, School of Marine Affairs; Edward Wolfe; Clarence Pautzke; R. Bruce Rettig; and Richard Marasco.

We are grateful to the following members of the local fishing industry who helped to organize this symposium: Robert Alverson, Fishing Vessel Owners' Association; Joe Blum, American Factory Trawler Association; Vince Curry, Pacific Seafood Processors Association; Douglas B. (Bart) Eaton, Trident Seafoods Corporation; and Dave Fraser and Mark Lundsten. We also appreciate the following people for their assistance in symposium organization: Richard J. Marasco and William L. Robinson, NMFS; Al Millikan, Washington State Department of Fish and Wildlife; Penny Pagels, Greenpeace; Joseph M. Sullivan, Mundt, MacGregor, Happel, Falconver, Zulauf & Hall.

Special acknowledgment is made to the Food and Agriculture Organization of the United Nations for permission to reproduce in this work text, graphs, and tables previously published under FAO copyright.

We acknowledge Beverly Gonyea and her staff in the Office of Continuing Education, University of Washington College of Forest Resources, for their organizational support and logistics management for the symposium.

We extend our thanks to the people who helped to edit and produce these proceedings: Andrea Jarvela's substantive technical editing is greatly appreciated. At the School of Fisheries, secretary Abby Simpson was invaluable in providing editorial and proofreading assistance. Graphics artist Cathy Schwartz was responsible for ensuring the high quality of the illustrations throughout the book. Faculty member Ted Pietsch provided essential expertise in ensuring correct taxonomic nomenclature. We acknowledge the assistance in the early stages of production by Willis Hobart, Chief, Scientific Publications Office, NMFS. We especially appreciate Robert Kendall, Managing Editor, American Fisheries Society, who encouraged us to have the Society publish the proceedings—a decision we are quite glad we made—and for his outstanding support and endurance throughout the editorial and production process.

Finally, we especially appreciate the School of Fisheries—including Director Ken Chew and the administrative staff—for their administrative, financial, and moral support for the production of these proceedings.

The Editors

Contributors

Alverson, Dayton Lee
Natural Resource Consultants
4055 21st Ave W, Suite 100
Seattle, WA 98199 USA
Tel: 206-285-3480
Fax: 206-283-8263
Email: NRCseattle@aol.com

Allsopp, W. Herbert L.
Smallworld Fishery Consultants, Inc.
2919 Eddystone Crescent
North Vancouver, British Columbia V7H 1B8
Canada
Tel: 604-929-1496
Fax: 604-929-1860
Email: smafi@aol.com

Anderson, James L.
Department of Environmental and Natural
    Resource Economics
University of Rhode Island
5 Lippitt Road
Kingston, RI 02881 USA
Tel: 401-874-4568/2471
Fax: 401-783-8883
Email: jla@uriacc.uri.edu

Anderson, Lee G.
College of Marine Studies
University of Delaware
Newark, DE 19716 USA
Tel: 302-831-2650
Fax: 302-831-6838
Email: Lee.Anderson@MVS.UDEL.EDU

Armstrong, David
Head of Unit Conservation Policy & Environment
Directorate General for Fisheries
European Commission
Rue de la Loi, 200
B-1049 Brussels
Tel: 322-295-3129
Fax: 322-296-6046
E-mail : Francis.Olbrechts@dg14.cec.be

Arnason, Ragnar
Faculty of Economics and Business Administration
University of Iceland
Oddi v/Sturlugotu
101 Reykjavik, Iceland
Tel: 354-525-4539

Fax: 354-552-6806
Email: Ragnara@rhi.hi.is

Beckett, J.S.
RR #2
Mountain, Ontario K0E 1S0
Canada
Tel: 613-989-2860
Fax: 613-989-1644

Branson, Andrew
New Zealand Fish Industry Board
Private Bag 24-901
Wellington, New Zealand
Tel: 644-385-4005
Fax: 644-385-2727
Email: andrew@fib.co.nz

Burke, William T.
University of Washington
School of Law
Box 354600
Seattle, WA 98105 USA
Tel: 206-543-2275
Fax: 206-685-4469
Email: burke@u.washington.edu,
sealaw@marinelaw.com

Butterworth, D.S.
Department of Applied Mathematics
University of Cape Town
Rondebosch 7700
South Africa
Tel: 2721-650-2343
Fax: 2721-650-2334
Email: DLL@maths.uct.ac.za

Campbell, R.A.
CSIRO Division of Fisheries
GPO Box 1538
Hobart, Tasmania, Australia 7001
Tel: 6102 325 368
Fax: 6102 325 199
Email: Rob.Campbell@ml.csiro.au

Chamberlain, George
Ralston Purina International
Checkerboard Square—11T
St. Louis, MO 63164 USA
Tel: 314-982-2402
Fax: 314-982-1613
Email: gchamberlain@ralston.com

Christy, Francis T.
IMARIBA
2853 Ontario Rd NW
Washington, DC 20009 USA
Tel: 202-483-6768
Fax: 202-328-3975
Email: imariba@netrail.net

Cochrane, K.L.
Fishery Resources and Environment Division
 Food and Agriculture Organization of the United
Nations
Via delle Terme di Caracalla
00100 Rome, Italy

Daan, Niels
Netherlands Institute for Fishery Investigations
PO Box 68
1970 AB IJmuiden
The Netherlands
Tel: 31-255-064646
Fax: 31-255-064644
Email: niels@rivo.dlo.nl

De Oliveira, J.A.A.
Sea Fisheries Research Institute
Private Bag X2
Roggebaai 8012
Capetown, South Africa
Tel: 2721-402-3144
Fax: 2721-25-2920
Email: JDOLIVEI@sfri.sfri.ac.za

Duke, Marcus G.
University of Washington
School of Fisheries
Box 357980
Seattle, WA 98195-7980 USA
Tel: 206-543-4678
Fax: 206-685-7471
Email: mduke@fish.washington.edu

Exel, Martin
Australian Fisheries Management Authority
Burns Centre
28 National Circuit
Forrest ACT 2603, Australia
Tel: 616-272-3260
Fax: 616-272-5036

Garcia, Serge M.
Fishery Resources Division
Fisheries Department
Food and Agriculture Organization
 of the United Nations
Viale delle Terme di Caracalla
0100 Rome, Italy

Tel: 396-522-56467
Fax: 396-522-53020
Email: serge.garcia@fao.org

Hannesson, Rögnvaldur
The Norwegian School of Economics and Business
Administration
Helleveien 30
N-5035 Bergen, Sandviken, Norway
Tel: 47-596-9000
Fax: 47-559-59543
Email: sam_rh@debet.nhh.no

Hilborn, Ray
University of Washington
School of Fisheries
Box 357980
Seattle, WA 98195-7980 USA
Tel: 206-543-4650
Fax: 206-685-7471
Email: rayh@pisces.fish.washington.edu

Homans, Frances R.
Department of Applied Economics
University of Minnesota
St. Paul, MN 55108 USA
Tel: 612-625-6220
Fax: 612 625 6245

Huppert, Daniel D.
University of Washington
School of Marine Affairs
PO Box 355685
Seattle, WA 98195-5685 USA
Tel: 206-543-0111
Fax: 206-543-1417
Email: huppert@u.washington.edu

Ísaksson, Arni
Institute of Freshwater Fisheries
Vagnhofda 7
112 Reykjavik, Iceland
Tel: 354-676-6400
Fax: 354-567-4869

Kaufmann, Barry
Australian Fisheries Management Authority
Burns Centre
28 National Circuit
Forrest ACT 2603, Australia
Current address:
11/244 Campbell Parade
Bondi Beach
NSW, 2026 Australia
Tel: 029-130-1542
Fax: 612-913-03549
Email: bkaufman@s054.aone.net.au

Kirkwood, Geoff
Renewable Resources Assessment Group
Centre for Environmental Technology
Imperial College of Science, Technology
    and Medicine
8 Prince's Gardens
London SW7 1NA
United Kingdom
Tel: 44-171-594-9272
Fax: 44-171-589-5319
Email: g.kirkwood@ic.ac.uk

Laurec, Alain
Director for Internal Resources, Conservation Policy
and Environment
Directorate General for Fisheries
European Commission
Rue de la Loi, 200
B-1049 Brussels
Tel: 322-295-9601
Fax: 322-296-6046
E-mail: Francis.Olbrechts@dg14.cec.be

Lindholm, R.
c/o Jeff Dunn
CSIRO Division of Fisheries—Marine Labs
GPO Box 1538
Hobart, Tasmania, Australia 7001

Lu, Xiangke
University of Washington
School of Fisheries
Box 357980
Seattle, WA 98195 USA

Major, Philip
Ministry of Agriculture and Fisheries
101-103 The Terrace
Wellington, New Zealand
Tel: 644-472-0367
Fax: 644-470-2669
Email: Kershawv@fish.govt.nz

Newton, C.
1655 22nd Street
West Vancouver, British Columbia V7V 4EZ
Canada
Tel: 604-925-9121
Fax: 604-926-4854

Parsons, L. Scott
Canada Department of Fisheries and Oceans
200 Kent St.
Ottawa, Ontario K1A 0E6
Canada
Tel: 613-993-0850
Fax: 613-990-2768

Pauly, Daniel
International Center for Living Aquatic Resources
Management (ICLARM)
MC PO Box 2631
0718 Makati
Metro Manila
Philippines
Tel: 632 818 9283
Fax: 632 816 3183
and
Fisheries Centre
University of British Columbia
2204 Main Mall
Vancouver, British Columbia V6T 1Z4 Canada
Tel: 604 822 1201
Fax: 604 822 8934
e-mail: pauly@fisheries.com

Pennoyer, Steve
National Marine Fisheries Service
Alaska Region
PO Box 21668
Juneau, AK 99802-1668 USA
Fax: 907-586-7249
Email: steven.pennoyer@noaa.gov

Pikitch, Ellen
University of Washington
School of Fisheries
Box 357980
Seattle, WA 98195-7980 USA
Tel: 206-543-4650
Fax: 206-685-7471
Email: ellenp@pisces.fish.washington.edu
Current address:
Director, Fisheries Programs
Osborn Laboratories of Marine Sciences
Wildlife Conservation Society
Boardwalk at West 8th St.
Brooklyn, New York 11224
Tel: 718-265-2688

Punt, Andre
University of Washington
School of Fisheries
Box 357980
Seattle, WA 98195-7980 USA
Current address:
Division of Fisheries
CSIRO Marine Laboratories
GPO Box 1538
Hobart, Tasmania, Australia 7001
Tel: 613-6232-5492
Fax: 613-6232-5000
Email: Andre.Punt@ml.csiro.au

Sainsbury, Keith
CSIRO Division of Fisheries
GPO Box 1538
Hobart, Tasmania, Australia 7001
Tel: 613 623 25369
Fax: 613 623 25199
Email: Keith.Sainsbury@ml.csiro.au

Sissenwine, Michael P.
Northeast Fisheries Science Center
166 Water Street
Woods Hole, MA 02543 USA
Tel: 508-495-2000
Fax: 508-548-5124
Email: Michael.Sissenwine@noaa.gov

Whitelaw, A.W.
CSIRO Division of Fisheries
GPO Box 1538
Hobart, Tasmania, Australia 7001
Tel: 6102-325-408
Fax: 6102-325-199
Email: Wade.Whitelaw@marine.csiro.au

Wilen, James
Department of Agricultural and Resource Economics
University of California, Davis
Davis, California 95616 USA
Tel: 916-752-6093
Fax: 916-752-5614
Email: wilen@primal.ucdavis.edu

Reviewers

Each paper in this book was reviewed by at least three individuals. The quality of this book is a direct result of their efforts. We thank the following people and those whose comments reached us anonymously for their time and effort in providing thoughtful and helpful reviews: D. Alverson, L. Anderson, R. Arnason, R. Beverton, D. Butterworth, K. Chew, W. Clark, R. Conser, J. Crutchfield, D. Cushing, S. Edwards, G. Ellis, P. Fricke, D. Gunderson, M. Hall, J. Hastie, R. Hilborn, J. Horwood, D. Huppert, R. Johnston, G. Kirkwood, L. Kochin, G. Lilly, D. Lightner, P. Livingston, A. May, D. McCaughran, W. Michaels, M. Miller, S. Murawski, B. Muse, S. Pooley, J. Pope, A. Punt, V. Restrepo, B. Rettig, K. Ruddle, J. Siber, M. Sinclair, H. Sparholt, D. Squires, R. Stickney, L. Trott, and R. Trumble.

STATUS AND TRENDS IN WORLD FISHERIES

Current Situation, Trends, and Prospects in World Capture Fisheries[1]

S. M. GARCIA AND C. NEWTON

Abstract.—Following an earlier analysis provided by the Food and Agriculture Organization of the United Nations (FAO 1993a), this paper gives an update of the trends and future perspectives of world fisheries. It describes and comments on worldwide trends in landings, trade, prices, and fleet size. It illustrates the decrease in landings in the last 3 years, the relationship between landings and prices, and the large overcapacity in world fishing fleets. It provides a review of the state of world fishery resources, globally, by region, and by species groups, as well as a brief account of environmental impacts on fisheries. It presents an economic perspective for world fisheries that underlines further the overcapacity and subsidy issues that characterize modern fisheries. In conclusion, this paper discusses management issues including the need for fleet reduction policies and the potential combined effect of overcapacity and international trade on resource depletion in developing exporting countries, and on the overall sustainability of the world fishery system.

World fisheries play an important role in development, providing incomes to about 200 million people, directly or indirectly. One-third of the world catches is exchanged through international trade, the volume of which has doubled between 1980 and 1990. Fisheries play a significant role in a number of developing countries where, since 1950, more than 85% of the world demographic growth has been concentrated. The coastal fisheries are potentially threatened by the ongoing progressive migration of people towards coastal areas, particularly coastal urban centers, where 60% of the world population already lives.

The rapid and continuous increase in fishing intensity during the last half of 1900s has had a tremendous impact on the aquatic ecosystem, its resources, and the market. This impact is evident in the depletion of resources, the degradation of the environment, and the evolution of supply, demand, and prices. It is also reflected in the changes in access and property regimes in the ocean, which are still evolving.

The political changes in Eastern Europe are also leading to an important modification of the role of these countries in world fisheries. Between 1961 and 1990, a large part of the catches of small pelagic species was made by fleets from the former USSR, Germany (GDR), Poland, Bulgaria, and Rumania, which specialized in the capture of these abundant low-price species and compensated for natural oscillations in stock abundance by "migrations" between areas of production and by pulse fishing.

The economic consequences of these political changes have led to a curtailment of the activities of these fleets (largely subsidized in the past) and to a shift in their area of operation and target species, with a return to their exclusive economic zones (EEZs) and waters closer to home, and a greater interest in high-value species for the export market. In some developing regions (e.g., in the Gulf of Guinea) that were markets for part of the landings of the Eastern European fleets, the sudden reduction in landings by these foreign fleets has led to shortages in supply and significant increases in prices (e.g., in Guinea Bissau).

All these important changes have affected fisheries and resources, sometimes positively, often negatively. The perspective of the Food and Agriculture Organization of the United Nations (FAO) on fisheries trends and their implications has been presented in many documents prepared for, and following, the UN Conference on Environment and Development (UNCED) in 1992 (Garcia 1992; Garcia and Newton 1994; FAO 1992a, 1993a). This paper presents an update of this perspective, focusing on trends in fisheries landings, trade, prices, and fleet size, and it describes the state of the world fishery resources (globally, by region, and by species groups). It provides a global assessment of the world fishery resources as well as a global economic model for world fisheries, which underscores the huge overcapacity that characterizes modern fisheries. In conclusion, it discusses management issues, briefly addressing environmental impact.

Trends in World Fisheries Production

Total Landings

Marine ecosystems produce 85% of the world fish yields. The process of intensification of marine fisher-

[1]This paper was submitted in 1994. Since then, other more recent analyses with different approaches and figures were published as follows: Grainger, R. and S. M. Garcia. 1996. Chronicles of marine fishery landings (1950–1994). Trends analysis and fisheries potential. Food and Agriculture Organization of the United Nations, Fisheries Technical Paper 359, Rome.

ies, which started just before the World War II, accelerated notably after it and led to an exponential increase in landings (Figure 1). During 1950–92, marine fishery catches increased by 300% from 18.5 to 82.5 million metric tons (mt) (Figure 2). The changes in the rate of increase with time, however, indicate that the upper limit of capture fisheries on conventional species has probably been reached (see later section on global assessment).

From 1950 to 1992, reported landings from marine fisheries increased at an average rate of 6.8%/year in the 1950s (18.5 million mt in 1950 to 31.2 million mt in 1959), 7.4%/year in the 1960s (to 54.5 million mt in 1969), only 1.7% in the 1970s (to 63.7 million mt in 1979), and 3.6% in the 1980s (to 86.4 million mt in 1989). In 1990–92, however, catches decreased at a rate of 1.5%/year (to 82.5 million mt in 1992) for the first time in history, with the exception of the two world wars, despite increased landings of anchoveta (*Engraulis ringens*) of 4.0 million mt in 1991 and 5.5 million mt in 1992.

The first period of low growth, in the 1970s, corresponds with the collapse of the anchoveta resource, possibly aggravated by the first oil price crisis in 1974, which slowed down the activity of long-distance fleets. The second period of low growth, in the early 1990s, corresponds with a decrease of Japanese (*Sardinops melanosticta*) and South American pilchard (*S. sagax*), as well as overfishing of important demersal resources in the northwest Atlantic. Between these two periods, the higher growth rate of the 1980s was mainly due to the simultaneous recovery of the anchoveta and the Japanese pilchard, as well as to the intensification of exploitation of Alaska pollock (*Theragra chalcogramma*) for the surimi industry.

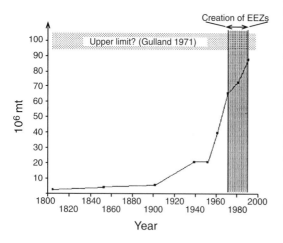

FIGURE 1.—Evolution of fishery production since 1800. Modified from Hilborn (1992).

Species Composition

In 1992, the marine catch consisted of 81.3%, 10.4%, 5.9%, and 2.4% for fish, mollusks, crustaceans, and other species, respectively. An analysis of the trends by species groups indicates that most of the increase in marine catches since the early 1980s (Figures 3 and 4) came from five major pelagic or semi-pelagic species (anchoveta, Alaska pollock, Chilean jack mackerel [*Trachurus murphy*], South American pilchard, and Japanese pilchard), which accounted for 24% of the total marine production (including aquaculture) in 1992 vs. 30% in 1989 (Figure 4; see also FAO 1993a). These species account for about 5% of total value in 1992 vs. 6% in 1989. From 1970 to 1992, the catch of the four major demersal species (silver hake [*Merluccius bilinearis*], haddock [*Melanogrammus aeglefinus*], Cape hake [*Merluccius capensis*], and Atlantic cod [*Gadus morhua*]) decreased by about 67% (5.0–1.6 million mt). Atlantic cod was the second most important marine species in 1970 (after anchoveta) with 3.1 million mt. It was only the sixth most important species in 1989 (after Alaska pollock, anchoveta, Japanese and South American pilchards, and Chilean jack mackerel) with landings of 1.8 million mt, and the tenth most important species in 1992, falling below capelin (*Mallotus villosus*), Atlantic herring (*Clupea harengus*), skipjack tuna (*Katsuwonus pelamis*), and European pilchard (*Sardinops pilchardus*), with landings of 1.2 million mt. The world fish supply is increasingly relying on low-value species, characterized by large fluctuations in year-to-year productivity, concealing the slow but steady degradation of the demersal high-value resources.

Between 1970 and 1992, landings of flatfish, tuna, and shrimp (Figure 5) show that flatfish production has been very stable (around 1.2 million mt yr^{-1}). Tuna and shrimp landings, on the contrary, reflect the large increase in overall pressure. Tuna landings have increased at a rate of 7.4%/year. Total shrimp production increased by 8.3%/year, but a part of that increase came from shrimp culture, which now represents about 25% of the total production (Figure 5). In certain areas, however, shrimp culture expansion seems to be reaching environment-imposed limits; the environmental impacts observed (e.g., in Thailand) and the severe economic losses incurred through diseases (e.g., in China in 1993) indicate that, in some areas, an environmental limit has been reached that cannot be passed without more costly methods of production and more stringent management measures.

Economic Value

Four species or species groupings constitute about half of the value of the world catch (Figure 6). In 1992, by decreasing order of importance, these were shrimp, redfish,

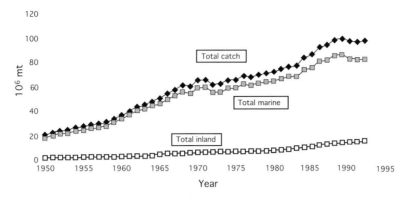

FIGURE 2.—Total reported catches from marine fisheries (1950–92).

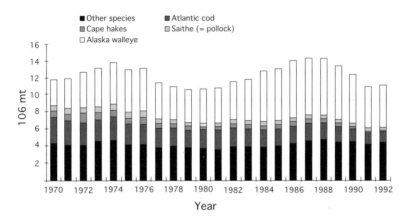

FIGURE 3.—World catch of demersal species (1970–92).

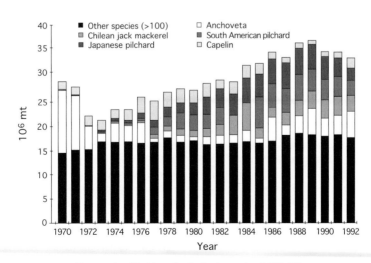

FIGURE 4.—World catch of pelagic species (1970–92).

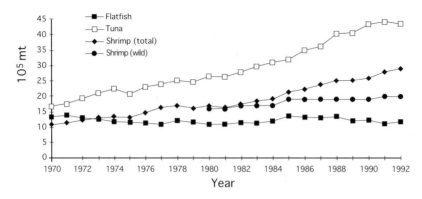

FIGURE 5.—Production of flatfish, tuna, and shrimp (including culture) (1970–92).

miscellaneous marine fish, and tuna. The miscellaneous category occupied the third rank in 1992 (Figure 6A) whereas it did not even appear in the first 13 species in 1970 (Figure 6B). This likely reflects the trends in many fisheries towards landing large quantities of unidentified mixtures of small fish with low economic value ("trash fish") as a result of overfishing and reduction in the size of fish. During the same period, redfish moved from fifth to second rank (reflecting an increase in fishing pressure). Overfishing is reflected in the decreased economic importance of many high-value species such as Atlantic cod (from third to twelfth rank) and hake and haddock (from second to seventh rank). Values are not available for 1992, but data on landings confirm the trends: the miscellaneous category (10 million mt) has doubled between 1970 and 1992 and now occupies the first rank while cod regressed from the third to the tenth rank.

Regional Distribution

Information on the regional distribution of fisheries is extensively documented in FAO (1993c, d, e, f). The data available for 1992 indicate that the Pacific Ocean provides 62.3% of total world landings, followed by the Atlantic (29.2%) and the Indian (8.5%) oceans. The data by FAO Statistical Areas (Figure 7) show that, despite some decrease since 1988 (26.6 million mt), the northwest Pacific continues to have the highest production (24.2 million mt) in terms of landings in 1992, followed by the southeast Pacific (13.9 million mt), northeast Atlantic (11.1 million mt), western central Pacific (7.7 million mt), and western Indian Ocean (3.7 million mt) (Figure 7). The comparison of the productions in 1970 and 1992 (Table 1) shows that in 1990 the largest relative increases have been in the southwest Pacific (+800%) and eastern Indian Ocean (+300%), while the northwest Atlantic decreased (-37%). These differences do not reflect some of the important variations between 1970 and

1992. For instance, southeast Pacific production greatly fluctuated between 5.6 million mt (in 1972) and 15.3 million mt (in 1989) owing to instabilities in small pelagic stocks. A significant decrease (20% or 645,000 mt) was observed between 1990 and 1992 in the eastern central Atlantic mostly because of a decrease in European pilchard and in the fleet activity of the former USSR countries. During the same period, the landings of the northwest Atlantic decreased by 25% (or 650,000 mt) mainly because of continuous declines in Atlantic cod stocks. The greatest increase (1.5 million mt) was observed between 1990 and 1992 in the northeast Atlantic where landings were at their highest level since 1985, mainly as a result of increases in capelin (75% or 0.9 million mt), Atlantic herring (11% or 0.15 million mt), and Norway pout.

Geoeconomic Distribution

In many developing countries, fisheries represent an important source of foreign exchange with a net earning (exports minus imports) of more than US$10 billion (all monetary values in this paper are cited in US$) in 1990, higher than earnings from other selected agricultural commodities such as coffee, tea, or rubber (FAO 1992b). The relative contribution to world production by developed and developing countries has significantly changed since 1970 (Figure 8A, B). In the early 1970s, the developed countries caught 57% of total landings. With the acceleration of the process of extension of the EEZs and the sharp rise in fuel prices (after the 1974 and 1979 oil crises), this share fell progressively to less than 50% in 1985 and less than 40% in 1992. The catches made by long-range fleets in distant fishing areas peaked at 8.9 million mt in 1989 and have decreased since then to 5.7 million mt in 1992. Part of the increase in developing countries' share of the world catches reflects some transfer of foreign fleets' catches under coastal countries' flags

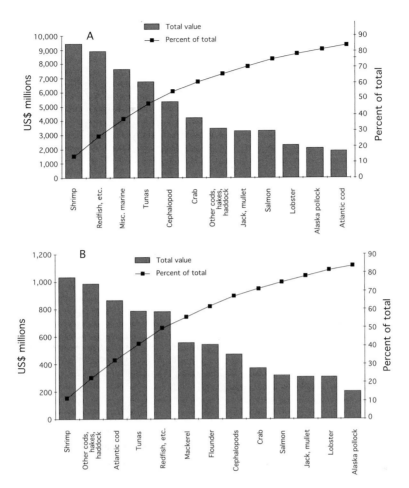

FIGURE 6.—Value of major species and groups of species and cumulative percentage of world total (A) in 1992 and (B) in 1970.

through the establishment of joint ventures, thereby changing the nationality of the catch if not of the real operations. A large part of the increase, however, is due to active fisheries development programs, which are often supported by national credit and subsidy schemes and underwritten by international and regional development banks.

National Distribution

In 1992, the 20 largest fish producers included 11 developing countries, dominated by China and Peru, and 9 developed countries, dominated by the former USSR and Japan. The cumulative curve (Figure 9A) shows that these 20 countries contributed close to 80% of the world production. The first six (China, Japan, Peru, Chile, the Russian Federation, and USA) produced 50% of the world landings whereas in 1970 (Figure 9B), four countries produced 50% (Peru, Japan, USSR, China). China

increased its production from 3.1 million mt in 1970 to more than 15 million mt in 1992, progressing from the fourth to first rank as a result of intensive mariculture expansion, more liberal trade and price policy, and long-range fleet expansion. Chile progressed from the fourteenth to the fourth rank (1.2–6.5 million mt). The Republic of Korea also expanded its distant-water fishing, increasing its production from 750,000 to 1.8 million mt and passing from the eighteenth to the tenth position.

In 1992, Japan's catches increased slightly, from 8.3 to 8.5 million mt. Its catches in distant waters decreased by 41%, from 1.6 million mt in 1982 to 0.9 million mt in 1992. Japanese total production has only been maintained, however, because of the natural increase in its sardine stock between the mid-1970s and the mid-1980s, yielding more than a third of their total production, and a decrease might be expected as sardine stocks return to long-term average levels of abundance. The USA and Canada have increased their catches from the north

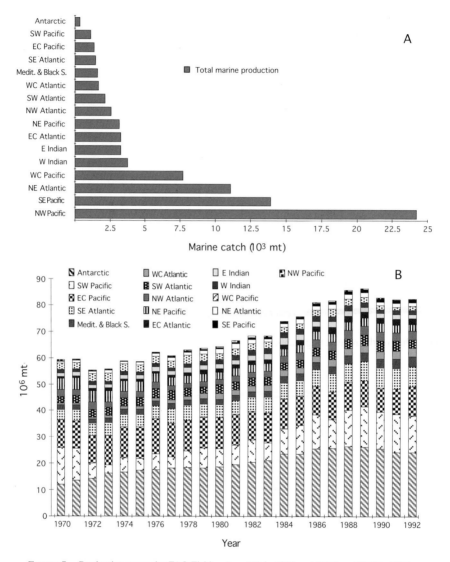

FIGURE 7.—Production per major FAO Fishing Area (A) in 1992 and (B) from 1970 to 1992.

Atlantic and the Pacific during their post-EEZ era (1970s), but the recent collapse of demersal resources in the northwest Atlantic (particularly cod) has severely affected the economic and social conditions of coastal communities.

Landings from distant-water fishing (in the EEZs of other nations and on the high seas) increased from 7.4 million mt in 1982 to a record level of 8.9 million mt in 1989, after which catches decreased to 7.5 million mt in 1991 (-16%) and 5.7 million mt in 1992 (-44%), or a total decrease of about 37% in 3 years, mainly as a result of a sharp decrease in the activities of the former USSR countries. With the formation of the Independent Republics, the shift to market economies has led to a retrench-

ment of the long-range fleets of these countries to less distant waters and in their own EEZs. Between 1991 and 1992, the catch by the distant water fishing fleets of these countries, consisting mainly of small pelagic fish, decreased between 20% and 71% depending on the country (Table 2), with a total decrease for all former USSR countries of about 50% in 2 years. Landings of the other Eastern European countries started to decrease in 1986–87. Between 1986 and 1992, the decrease was 24%, 30%, and 76% for Poland, Rumania, and Bulgaria, respectively, and 41% for these three countries together (from 729,000 to 430,000 mt). The total decrease in activity of the Eastern European fleets has affected fish-food supplies and prices of small pelagic species in West Africa, and it is

TABLE 1.—Total marine landings (metric tons [mt]) in 1970 and 1992 and relative increase.

Fishing area	Landings (10^3 million mt)		
	1970	1992	% increase
Northwest Pacific	12.1	24.2	+100.0
Southeast Pacific	13.8	13.9	+0.7
Northeast Atlantic	10.6	11.1	+4.7
Western central Pacific	3.9	8.2	+110.3
West Indian	1.6	3.8	+137.5
East Indian	0.8	3.3	+312.5
Eastern central Atlantic	2.5	3.3	+32.0
Northeast Pacific	2.6	3.2	+23.1
Northwest Atlantic	4.1	2.6	-36.6
Southwest Atlantic	0.7	2.1	+200.0
Western central Atlantic	1.4	1.7	+92.8
Mediterranean	1.1	1.6	+45.5
Southeast Atlantic	2.5	1.5	-40.0
Eastern central Pacific	0.8	1.4	+75.0
Southwest Pacific	0.1	0.9	+800.0
Antarctic	0.4	0.4	+0.0

not yet clear whether the development of local fleets to harvest these species is an accessible and economic alternative. In addition, there has been an increase in changes in flags to open registers without corresponding reporting on catches by the related flag states.

Trade

Detailed data on international fish trade are available for the period 1960–90 (FAO 1992b). The volume of internationally traded fish has increased from $2.5–$2.8 billion in 1969–71 to $35–40 billion in 1990, an increase from about 5% to 11% of the total trade in agricultural products. This increase indicates that fish trade developed faster than agricultural trade. The growth in fish trade has slowed down, however, from 18%/year in 1969–76 to 8%/year in 1979–90. The trends look similar for developed and developing countries, but the data available for 1979–90 show the following:

- In the developed countries, imports increased faster than exports (8.6% as opposed to 7.4%/year), indicating a net deficit, which increased from $700 million to about $15 billion between 1969–71 and 1990. These countries are the largest importers with more than 85% of the imports in value from 1969 to 1990. Japan's share of world imports tripled during the same period (8–28%), illustrating the impact of the EEZ process. On the contrary, U.S. imports decreased from 25% to 16% of the world imports, indicating an opposite effect.
- In developing countries, high-value species are exported while low-value species find their way into the national and regional markets. These countries are responsible for 70%, 84%, about 66%, and over 80% of the trade in cephalopods, frozen shrimp,

fresh and frozen tuna, and canned tuna, respectively. Their imports increased less than exports (7% as opposed to 8.8%/year). Their share in worldwide exports increased from 32% in 1969–71 (before the establishment of EEZs) to 44% in 1990 while their relative share of the imports increased only from 10.7% to 12.9%. These countries appear, therefore, as net exporters with a positive trade balance that increased from $500 million to $10.6 billion between 1969–71 and 1990, representing a significant source of foreign exchange. Thailand, for instance, multiplied by 6 its share of world exports (from 1% to 6%) while the Republic of Korea and Taiwan (province of China) increased their share from practically nothing in 1970 to 5% in 1990. The countries principally responsible for this net trade balance are China, Chile, and Thailand. In the case of Thailand, expansion of trade is related to the very rapid development of tuna canning and a 400% increase in shrimp culture, although there was a reduction of Thailand's fish meal exports, which were redirected towards its aquaculture industry.

The global balance of fish and fishery products is negative for developed countries, with a deficit of about $15 billion/year, with western Europe (including the European Union) accounting for more than $12 billion. Developing countries have a positive balance of more than $10 billion, with East Asian countries accounting for $7 billion. It is notable that, in a rather grim context of decreasing terms of trade[2] of agricultural products, fish trade in developing countries has progressed and represents a significant opportunity in terms of foreign exchange.

Foreign exchange and national food security objectives may often be conflicting as the incentives to increase exports are reducing the relative availability of food fish for domestic consumption in the developing world. In the medium to long term, demand will continue to grow faster than supply, as a consequence of demography in the developing world and continued increase in demand for food fish in the developed regions. During the 1980s, developed countries started to tighten their controls on levels of effort, decreasing harvest rates, fleet sizes, and access. The recent crisis in Europe, Canada and the USA should accelerate the process. Combining forecasts of population growth with stagnation in fish supplies indicates that the world availability of fish for food, which had increased from 9 to 13 kg per capita (FAO 1992c) between 1961 and 1990, will decrease to 11 kg per capita between 1990 and

[2]The "terms of trade" of a particular product is the ratio between the average unit value of this product and the average unit value of all commercial trade. This ratio reflects the evolution of the relative purchasing power derived from this product.

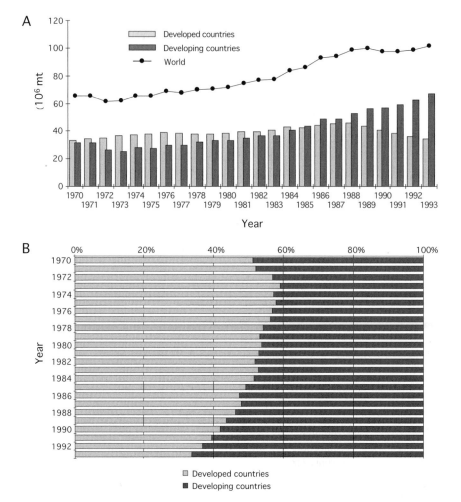

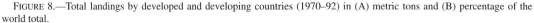

FIGURE 8.—Total landings by developed and developing countries (1970–92) in (A) metric tons and (B) percentage of the world total.

2030 (Brown 1995). Developed countries may have the means to stabilize the availability to them by purchasing the necessary quantities. As a consequence, fish availability in developing countries is likely to decrease further because of increased exports. The overall shortage in supplies will probably further increase the price of fish and fishery products. This should compensate, at least partly, for decreased abundance and quality, particularly as the price of low-value species will be pushed upwards through substitution (e.g., surimi). By stabilizing revenues, however, this increase in price will not provide the necessary incentives to reduce fishing effort as much as is required to rehabilitate fisheries.

Since 1980, the proportion of the total world fish production going to human food has been around 70% without any clear trend, and estimates are that 95 million mt of fish for direct human consumption will be required by

the year 2010 to maintain present per capita consumption. During the same period, about 30% of the world fish production was used essentially for animal feeds in agriculture and aquaculture. The absolute quantities going to fish meal and oils have increased, however. In the developed countries, they increased from 5 million mt to more than 12 million mt between 1961 and 1990. During the same period, in the developing world, the quantities going to fish meal and oils increased from 6 million mt to 16 million mt, with oscillations due to the collapse and recovery of the anchoveta, which provides between 25 and 49% of the world production depending on the years.

Prices

Prices of fish in international trade are essential to the understanding of the evolution of world fisheries. They

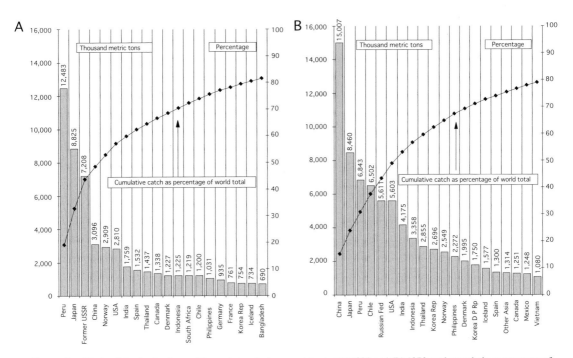

FIGURE 9.—Marine fisheries landings of the main producing countries in (A) 1992 and (B) 1970, and cumulative percentage of the world total.

are affected by various factors including demography, trade circuits (e.g., 90–95% of shrimp imports in Japan come from Asia, and 65–75% of shrimp imports in the USA come from Latin America), competition with local species (e.g., competition between tropical penaeids and cold water shrimp in European markets) and products of substitution (such as surimi), all of which may limit imports and constrain prices. In the long run, the price of fish is affected by its availability; the overall availability of fishery products per capita had started to level off in 1970 (Figure 10), and since then the gap between supply and potential demand has been increasing rapidly.

An analysis was conducted of the relationship between deflated prices (base = 1978) and landings for species used for human consumption (Figure 11) and for species used mainly for industrial reduction (Figure 12). For jacks (*Trachurus* spp.), mullet (*Mugil* spp.), tuna, cephalopods (Figure 11A), and Alaska pollock (Figure 11B), the price has remained fairly constant despite positive changes in landings, indicating that the demand is fairly elastic (i.e., large changes in supply have little effect on prices), the market is demand-driven, and the increased supply has been sufficient to satisfy the increased demand without pushing prices upward. The downward trend for tuna and cephalopods during the last decade may reflect some market saturation and increased production of secondary species, particularly for tuna, as well as cost decreases

resulting from new harvesting technology. The demand appears fairly elastic also for shrimp and salmon (during the last decade at least), indicating probably that the additional production from aquaculture has been able to compensate for the increased demand. The price of crab has increased despite increased production, indicating that the increase in demand has outstepped the supply despite the success of surimi as a substitute. The demand appears very inflexible (supply-driven) for lobster (Figure 11A), flatfish, and redfish (Figure 11B), reflecting their increasing scarcity and the fact that they are difficult to substitute. Notably, the price of cod, hake, and haddock has remained stable despite the significant decrease in landings (mainly as a result of overfishing), probably illustrating the fact that these fish have been substituted on the market by Alaska pollock, whose price remained stable despite large increases in landings.

TABLE 2.—Reported landings from long-range fleets (mt).

Country	1991	1992	% difference
Russian Federation	1,705,870	1,020,876	-40
Ukraine	728,466	396,145	-47
Lithuania	438,515	145,238	-67
Latvia	335,720	94,349	-72
Estonia	286,714	81,998	-71
Georgia	51,109	40,125	-20
Former USSR	3,546,394	1,778,681	-50

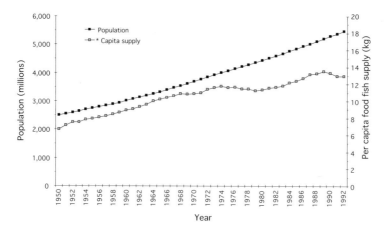

FIGURE 10.—Trends in world populations and fish supply per capita (1970–92).

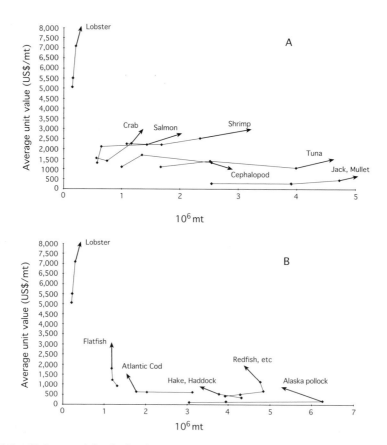

FIGURE 11.—Relationship between deflated unit value and supply for species used for human consumption (1970, 1978, 1989).

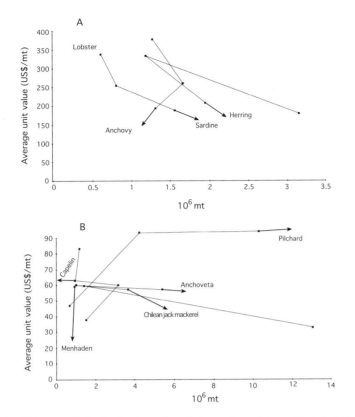

FIGURE 12.—Relationship between deflated unit value and supply for species mainly used for industrial reduction (1970, 1978, 1989).

The picture is less clear for the small pelagic species, partly because of their large fluctuations. Species that are partly used for human consumption (Figure 12A) showED decreasing prices regardless of whether supplies increased during the last decade (sardine and herring) or remained fairly stable, stabilizing at about $200/mt. The price of pilchard, partly used also for human consumption, appears to have increased (Figure 12B) despite very large increases in landings mostly related to environmental fluctuations (Bakun 1995). On the contrary, the price of anchoveta, menhaden (*Brevortia tyrannus*), Chilean jack mackerel, and capelin has fluctuated apparently independently of their landings, converging at a common price around $60/mt. This situation probably reflects the fishes' common destination (fish meal) and that the impact of their price fluctuations is dampened by the much larger production of soya, the main substitute for fish meal.

State of World Fishery Resources and Environment

In 1971, FAO first published (Gulland 1971) a world review of fishery resources, which estimated the world theoretical potential of traditionally exploited species to be around 100 million mt, of which just 80 million mt was probably achievable for practical reasons related to the impossibility to optimize management on every wild stock in a complex multispecies system. Since then, evidence clearly indicates an increase in the number of stocks reported as being under severe fishing pressure and a simultaneous decrease in the number of stocks offering potential for expansion (Figure 13).

The last review made by FAO on the state of world fishery resources (FAO 1994a) has yielded a more detailed picture (Figure 14). On the basis of an extensive analysis of the literature and the work of the FAO regional fishery bodies working groups, this review categorizes, region by region, the stocks for which an assessment exists as underexploited (U), moderately exploited (M), heavily to fully exploited (F), overexploited (O), depleted (D), and recovering (R). The category (F) comprises stocks that are exploited at a level of fishing close to F_{MSY} (the fishing mortality corresponding with the maximum sustainable yield [MSY]) and whose abundance is close to B_{MSY} (the biomass corresponding with

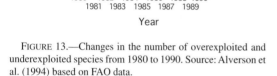

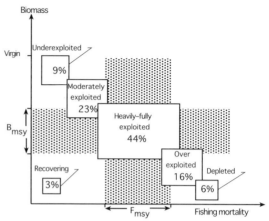

FIGURE 13.—Changes in the number of overexploited and underexploited species from 1980 to 1990. Source: Alverson et al. (1994) based on FAO data.

FIGURE 14.—State of world fishery resources: Proportion of the assessed stocks that are underexploited, moderately or fully exploited, overfished, depleted, or recovering. Source: FAO (1994c).

the maximum sustainable yield).[3] The other categories are exposed to respectively higher (O, D) or lower (U, M) fishing intensity than that corresponding to MSY. When there is uncertainty as to the exact stock status, the stock has been put in the two most likely categories (e.g., F–O) and counted twice. Some stocks or groups of stocks are in an unknown state and have not been taken into account in the following analysis. Figure 14 shows the distribution of all stocks or species aggregated among these categories together with the position of the category in a system of coordinates defined by the fishing mortality (as X axis) and biomass (as Y axis). It shows that 32% of the stocks for which data are available in FAO appear as underexploited or moderately fished and might be able to support some increase in fishing. It also shows that 69% are exploited at or beyond the level corresponding to MSY. This does not imply that 69% of the stocks are improperly utilized. "Full" utilization is generally the goal of fisheries development, and the figure indicates mainly that little scope exists for further development. However, because of (a) the uncertainty in the positions of F_{MSY}, (b) the non-precautionary nature of MSY as a management target for many stocks, and (c) the inertia in fleet dynamics and the fishery development process, "the fully fished" stocks are obvious (and likely) candi-

dates for overfishing in the near future if past behavior persists.

If the situation is examined region by region, the analysis is more difficult because the proportion of stocks and aggregates of stocks for which assessments are not available may sometimes be relatively high, varying from 53% in the northwest Pacific to 7% in the southwest Atlantic. Because they are very aggregated, these values should be taken cautiously, but they are intended to stress that even though the situation appears serious in many respects, the database available to fully assess it is dramatically incomplete. With this caveat, we have calculated for each region the proportion of the assessed stocks that appeared to be exploited beyond F_{MSY} and below B_{MSY}; further, in this last FAO review, the proportion of the assessed stocks and stock aggregates that are either fully exploited, overfished, and depleted or slowly recovering from depletion varies from 100% in the northwest Pacific to 29% in the eastern central Pacific (Figure 15).

In the FAO regular reviews, the situation is also examined stock by stock (FAO 1994a). The state of the stocks varies obviously between species and regions. A statement about the average state of a species or species group, based on many stocks in different regions and receiving different levels of fishing mortality, has little operational and statistical value. However, as similar species tend to have similar market value and are confronted with similar fishing pressures, an attempt has been made to provide for a qualitative classification of the state of stocks as follows. Values from 1 to 5 have been given to stocks considered, respectively, underfished, moderately fished, fully fished (i.e., at MSY or F_{MAX}), overfished, and depleted and recovering (considered as

---

[3]In Figure 14, the two subcategories "heavily" and "fully" exploited have been combined to account for the high level of uncertainty in the estimate of the current fishing level and for the level corresponding with "full" fishing (i.e., conventionally MSY). For stocks in this category, increased effort will not lead to any significant increase in landings.

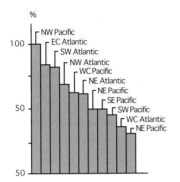

FIGURE 15.—State of regional fishery resources: Proportion of the assessed stocks that are very intensively exploited (i.e., fully exploited + overexploited + depleted + recovering) by major FAO fishing area. Source: FAO (1994c).

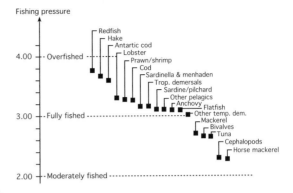

FIGURE 16.—Average state of major fishery stocks and groups of stocks.

depleted).[4] The average value was calculated for each major species as well as for species groups across all regions (Figure 16).

Keeping in mind the caveats about the data, it is interesting to see that redfish, hake, Antarctic cod, lobster, shrimp and prawn, cod, and tropical demersals are, on average, fished beyond full exploitation. In addition, despite the general statement often made that small pelagics are still underfished, sardine, pilchards, menhaden, and anchovy appear fully fished on average. Mackerel, bivalves, and traditional tuna stocks are close to full exploitation (the potential of tropical, small coastal tuna is not well known). The resources that, on average, appear moderately to fully fished are cephalopods (mainly oceanic ones) and horse mackerel (*Trachurus* spp.).

These overall statements on average status of species or regional resource aggregates should be cautiously interpreted. In an aggregate, some stocks are in a much worse state than average and would require more stringent measures while others are in a better state than average and could, in theory, stand higher fishing effort. There are, for example, indications that silver hake in the northwest Atlantic could be further exploited and that anchoveta in the eastern central Pacific is now underfished. Mesopelagic resources are also known to offer a large potential, but regular assessments are not available. Krill is usually also considered as a resource offering a large potential for increased catches although some concern has recently been expressed about the potential impact of the ozone hole and ultraviolet light on these stocks. While the importance of interannual variability is progressively being recognized for most

pelagics and a growing number of demersals, the assessments of many stocks clearly would need to be more frequently revised than they are presently, particularly in the tropics where the research capacity is often deficient. Nonetheless, these overall statistics indicate that the state of world fishery resources should be a subject of major concern and that this global assessment is sufficiently confirmed at regional, country, or stock level to be taken seriously.

Environmental Issues

Reflecting the general pressure exerted on natural systems from development activities, environmental issues—as established after the United Nations Conference on Environment and Development and considered here in the broadest sense to concern living resources and their habitat—have become increasingly significant in fisheries, posing difficult challenges. Some problems are internal to fisheries and concern depletion of the resource base, insufficient selectivity of gear and practices with significant consequences on bycatch and discards, direct damage to the environment by fishing techniques (e.g., trawling, dynamite fishing), and at-sea and onshore processing facilities and fishing ports. Other problems, among the most serious ones, are related to impacts made on the fishery resources and environment by other users.

Discards during fishing operations are a major source of concern. Alverson et al. (1994) have estimated that the annual quantities caught and discarded (probably dead and including unknown large quantities of juveniles) by the world marine fisheries amount to about 27 million mt. The world reported landings being 82.5 million mt, this means that about 25% of the fish caught is discarded and returned to the sea where it is naturally recycled. The distribution of total marine catches (average 1988–92) and estimated discards by major FAO Statistical Area

---

[4]See Figure 14 for a significance of these terms.

(Figure 17) indicates that, in general, the most productive areas are also those where discards are the highest.
Although the technical and economic implications of the
potential solutions to the problem are not easy to address,
this issue is one of the most critical facing fisheries today and the most damaging for their image.

The progressive degradation of the marine environment is another important source of concern. The major
environmental problems come from the coastal zone
degradation; this zone includes the critical habitats, nurs-
eries, and feeding and spawning areas that sustain about
90% of the exploited world fishery resources. Productivity in this area is being affected by an increasing demand for coastal space and resources from a growing
coastal human population. The marine environment is
affected locally by fishing and competing coastal activities but also by inland industrial activities and urban development, the impact of which is transferred to the
coastal zone through rivers and rainfall. According to
the Joint Group of Experts on the Scientific Aspects of
Marine Pollution (GESAMP 1990), 77% of the pollution reaching the coastal areas comes from land-based
sources. The consequences can be particularly acute for
small-scale fishing communities and fish farmers.

An extreme illustration of the problem and its potential consequences is given by the ecological collapse of
the productive system in the Black Sea which, because
of its magnitude and doubtful reversibility, could probably be considered the marine ecological catastrophe of
the century. The fishery resources of this area, which
produced about 1 million mt of landings in the late 1980s,
have collapsed through overfishing and eutrophication
to 100–200,000 mt in 1991 in a degraded ecosystem, 90%
of which is now anoxic. The cost of this ecological disaster has been estimated at hundreds of millions of dol-

lars, leaving more than 150,000 people without a livelihood and an important fishery sector in total disarray
(FAO 1993b; FAO 1994b). In the future, similar problems may also affect other closed or semi-enclosed, low-
energy, and strongly stratified water bodies such as the
Baltic Sea and large lakes. Although the above example
is an extreme one and is not representative of the risks in
an open ocean, it shows that the problem is serious and
that without a change towards integrated management,
the fate of coastal resources may be similar to, and possibly worse than, the fate of wild, freshwater resources.
A reasonable level of organic contamination may, however, have positive effects and indeed increase fish productivity, particularly in shallow and enclosed or semi-
enclosed seas (Caddy 1993); thus, current efforts at
reducing organic pollution from land-based sources may
indeed reduce fisheries potential.

Natural variations in the abundance and resilience of
fishery resources and the potential impact of global
climate change are also a source of uncertainty for fisheries planning and management. A complete analysis
of the trends and future perspectives of fisheries supply
and management should consider the impact of climate
variability and global climate change on fishery systems.
Both phenomena relate to the dynamics of the ocean–
atmosphere coupling and its evolution under global
environmental change. The issue is particularly well
documented for pelagic resources. For example, the devastating effects of El Niño on the pelagic resources of
Peru and Chile (Glantz and Thompson 1981) were described long ago. The impact of less catastrophic but
possibly more frequent environmental oscillations tends
to be blurred by fishing impacts and to remain unde-
tected or difficult to demonstrate. Such oscillation (or
"regime") changes are now being reported for a large

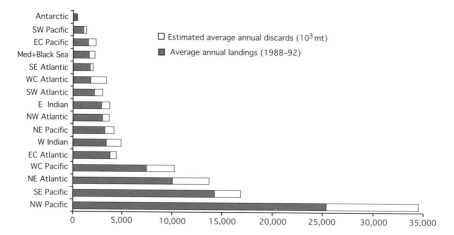

FIGURE 17.—Average annual marine catch (1988–92) and estimates of discards by major FAO statistical division. Source: FAO
and Alverson et al. (1994).

number of stocks, both demersal and pelagic, coastal or offshore, for shallow or deep resources. Coherent oscillations of groups of species are detected (Csirke and Sharp 1984; Lluch-Belda et al. 1992), but the overall trends are hardly predictable. Important resources have undergone increases in potential from the mid-1970s to the mid-1980s, such as sardine in Japan, Peru, Chile, California; anchovy in South Africa–Namibia; north Pacific Alaska pollock and other demersal fish; and lobster and other reef resources in the tropics (Bakun 1994). Some species, such as anchovy and sardine, seem to vary in opposite directions. It is expected that some important pelagic stocks will continue to decrease, but the overall impact on world resource potential and landings remains unpredictable. The impacts of climatic oscillations can be very serious when they effect a series of low recruitment for a fishery where effort is largely in excess of the average F_{MSY}, thereby resulting in sudden recruitment collapses. Exceptionally good recruitment can also be a problem in the sense that, temporarily improving the state of the stocks, it may delay the necessary management measures and allow fishing to grow well beyond sustainable levels.

Global Economic Perspective

The information available at the beginning of the 1970s already indicated that the fishery resources of the world had a limited potential (Gulland 1971) that was being reached rapidly, and the need for improved management, particularly of effort controls, was clearly expressed at the FAO Technical Conference on Fishery Management and Development held in Vancouver, Canada, in December 1973 (Stevenson 1974). As clearly stated in the Chairman's summary (Needler 1974):

> It has been unanimously recognized that the resource is not unlimited, . . . that there is a tendency for prices to rise faster than the general level of commodity prices, . . . that the pressure (on the resource) is already intense but will become more so . . . and that the need for management to sustain the yield is already the rule rather than the exception.

Unfortunately, the process of extending national jurisdictions in the 1970s seems to have turned this central issue into a secondary one for 2 decades, leading to a largely uncontrolled increase in the world fleet size and to the poor situation in which fisheries are today. In the following sections, we examine the trends in world fleet capacity and in its performance in terms of landing rates as well as the relationship between the two, leading to a global bioeconomic assessment of the world fisheries, the limitations of which are discussed in the last section of the paper.

Trends in World Fleet Size

Statistics on world fishing fleet, although not entirely complete, indicate the extent of the size of these fleets. The 1992 Lloyds Register of Shipping lists ships of at least 24 m or, in terms of gross registered tonnage (grt), 100 grt, and fishing vessels (industrial fishing fleet) compose 30% of the total number of all ships in the Register. Although their tonnage is only 3% of the total tonnage of all ships, the replacement value of the fishing fleet is estimated at $173 billion (FAO 1995, Table 1), or almost 45% of the total replacement value of all ships included in the Register (Figure 18A; FAO 1995).[5]

The FAO Bulletin of Fishery Fleet Statistics lists the industrial fishing fleet at 38,400 ships with a tonnage of 16.6 million grt (compared with the 24,400 vessels and 13.0 million grt reflected in the Lloyds Register). The FAO data also indicate that the number of decked vessels less than 100 grt or 24 m is about 1.14 million with a total tonnage of 9.4 million grt. The FAO Bulletin lists 2.3 million undecked vessels in the world, of which only 32% are powered open boats. The tonnage of these powered vessels is not recorded, but it can reasonably be assumed (based on FAO's practical experience) that their tonnage is between 2 and 4 grt. If a rough average of 3 grt per undecked powered vessel is assumed, the fleet would represent about 0.74 million grt or about 3% of the world's total gross registered tonnage. The distribution of the world tonnage in fishing vessels by continent in 1989 (Figure 18B) illustrates the large proportion of vessels from Asia and the former USSR fleets.

A time-series of fishing fleet data is available in the FAO Bulletin of Fleet Statistics 1994 for ships above 100 grt or 24 m (Table 3, column 2). For the purpose of any comparison of catches per grt in a time-series, using these data as a measure of world fleet size for decked vessels would underestimate the actual tonnage of the world fleet by only the tonnage of the undecked vessels (i.e., by about 3%). This time-series indicates that, between 1970 and 1989, the actual tonnage increased at a rate of 4.6%/year, from 13.6 to 25.3 million grt (however, in the latest revisions of the FAO data, which are not taken into account in this paper, this value for 1989 appears to be 26.0 mt).

There are interesting comparisons to be made. During the same period (1970–89), when a large number of EEZs were claimed, the size of coastal developing countries' fleets increased from 26.7% of the total number of fishing vessels to 58%, while the tonnage increased from 12.7% to 28.8%. The developed countries had started their increase much earlier. For example, in Iceland,

---

[5]For further comparisons, see Table 1, page 19, in FAO (1995).

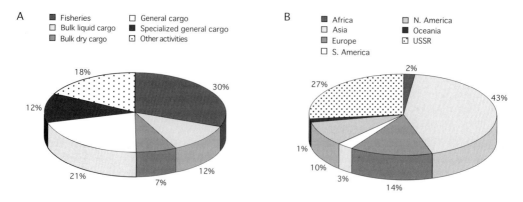

FIGURE 18.—Importance of the world fishing fleet numbers in relationship to (A) other types of ships (>100 mt gross registered tonnage [grt] and 24 m) in the world merchant marine fleet and (B) distribution by continent of the world total fishing fleet in grt (>100 grt and 24 m). Source: World Fleet Statistics, December 1992, Lloyds Register, London.

between 1945 and 1983, the capital employed in fisheries increased by 1,300%, catches increased only by 300%, and the output/capital ratio decreased to less than one-third of what it was in 1945 (Arnason 1994). Many countries (e.g., Europe, Australia, New Zealand) have started programs to control and reduce fishing fleets, sometimes with great opposition. Following economic transition in Eastern Europe, a significant part of the fleet of these countries is to be scrapped.

Trends in Landings per Unit of Capacity

The tonnage data available in the FAO Bulletin on Fleet Statistics, which underestimate the total world fleet size by only about 3%, can be used as an index of world fleet size, assuming that the observed trend is representative of the trend in the overall fleet.

The data summary provided in Table 3 (column 3) indicates that, during 1970–89, the total marine landings increased from 59.2 to 86.4 million mt at an average rate

of only 2.4%/year. An examination of the fleet statistics of countries exploiting the main pelagic species by type of vessels shows that very little of the increase in fleet capacity has been directed to fishing those low-value and pelagic species that have produced most of the increase in world catches during the last 20 years. If the five principal low-value and pelagic species (Alaska pollock, anchoveta, Japanese pilchard, South American pilchard, and Chilean jack mackerel) are subtracted from the total marine catches, the landings of the other species—hereafter called selected landings—have increased from 42.9 to 61.3 million mt (Table 3, column 4) at a rate of 2.3%/year, that is, at half the rate of increase of the world fishing fleet size (4.6%/year). From 1975 to 1989, the total landing rate, which was obtained by simply dividing the total selected landings by the grt index (Table 3, column 6), varied around 3.4 mt grt^{-1} without trend and with the 1970 value of 4.4 appearing as an outlier. The selected landing rate (Table 3, column 7) decreased, however, from 3.2 to 2.4 between 1970 and 1989 with an average

TABLE 3.—Fleet capacity (>100 grt or 24 m), total landings, selected landings (excluding the five main pelagic species), total deflated value (1989 base), and indexes of catch per unit of capacity (mt) and value (10^3 US$/grt).

Year	grt (10^6 mt)	Total landings (10^6 mt)	Selected landings (10^6 mt)	Total value (deflated) (10^9 $ 1978)	Total landing rate (mt/grt)	Selected landing rate (mt/grt)	Value/grt (deflated) (10^3 $ 1978)
1970	13.5	59.2	42.9	28	4.4	3.2	2.1
1975	17.3	58.6	49.2	NA	3.4	2.8	NA
1978	19.3*	63.0	52.7	35	3.3	2.7	1.8
1980	19.8	64.5	52.5	NA	3.3	2.7	NA
1981	20.0	66.5	52.5	NA	3.3	2.6	NA
1982	20.8	68.3	52.3	NA	3.3	2.5	NA
1983	21.2	68.3	53.1	NA	3.2	2.5	NA
1984	21.8	73.9	54.6	NA	3.4	2.5	NA
1985	22.5	75.7	55.2	NA	3.4	2.5	NA
1986	23.5	81.1	57.3	NA	3.5	2.4	NA
1987	24.1	81.7	59.9	NA	3.4	2.5	NA
1988	24.8	85.7	61.4	NA	3.5	2.5	NA
1989	25.3	86.4	61.3	58	3.4	2.4	2.3

of 2.5; in this case, the 1970 value of 3.2 fits perfectly with the rest of the data. It seems, therefore, that during 1970–89, the apparent maintenance of the world fleet productivity in global terms (around 3.4 mt grt^{-1}) conceals the fact that its yield in higher-value species decreased by 25% despite technological progress (spotter planes, factory and mothership, satellite navigation, sounders, wide opening nets, etc.).

One could expect that this relative and progressive degradation of the species composition of the landings would have resulted in decrease in the value of the landings (or revenues) and economic yields, providing the necessary economic signals of overcapacity and of the need to regulate fishing more efficiently. For this assumption to be verified, the deflated value of the total landings (1978 US$), or total revenues, was examined for 1970, 1978, and 1989, the only years for which this type of data was available when preparing this paper,[6] together with the index of revenue per grt. The data given in Table 3 (columns 5 and 8) and Figure 21 show that, while the overall fleet size increased by 87.4%, total landings by only 46%, and selected landings by only 43%, the total value of the landings increased by more than 107%. The same data show that while from 1970 to 1989 the total landings per grt appeared stable around 3.4 mt grt^{-1} and the selected landings per grt decreased by 25%, the revenue per grt increased by 38% (from $2,100/grt to $2,300/grt). This trend indicates that the economic incentive for growth in fleet capacity has been at least maintained and possibly increased over time, despite the repeated signs of overfishing of individual stocks and the repeated warnings of scientists at national, regional, and international levels.

The fact that the total catch and value per grt remained stable from 1970 to 1989 does not mean that fisheries were performing well. The consequences of the expansion have been a drastic reduction in abundance and spawning potential with an increase in resource instability. The phenomenon has been verified at the national level and some examples can be given. In San Miguel Bay (Philippines), for example, available data on species abundance indicate that four species accounted for 75% of the biomass in 1947, against five in 1980–81, and more than seven in 1992–93. In the meantime, stock density decreased by more than 80%, from 10.6 to 2.0 mt km^{-2} (E. Cinco, J. Diaz, R. Gatchalian, G. Silvestre, International Center for Living Resources Management, Manila, unpubl. rep.). In the Philippines' Samar Sea, the resource abundance dropped from 8.0 to 3.5 kg d^{-1} between 1981 and 1990 while the number of commercial

species of major importance dropped from 250 to 10 and the standard of living for 100% of the fishermen dropped below the poverty line and even below the food threshold (Saeger 1993).

Global Biological Assessment

At the beginning of the 1970s, FAO predicted that the potential of the world traditional fish resources (small pelagic, large pelagic, and demersal fish) was close to 100 million mt excluding discards (Gulland 1971). This work stressed, in addition, that "in practice no more than 80% of the potential in an area may be harvestable because of the difficulties of ensuring the best management of each individual stock." For all practical purposes, and because it would not be feasible to extract MSY from every stock (assuming that this would even be an advisable objective), the world potential of traditional species would be close to 80 million mt. The landings from marine fisheries passed this level in the mid-1980s. Adding to these landings the average 27 million mt of fish caught but discarded (Alverson et al. 1994) would bring the present catches above 100 million mt. It seems, therefore, clear that the maximum production of traditional fishery resources is either being approached or already has been reached (see Figure 1); this seems to be confirmed by the fact that the annual rate of increase of the world marine landings is approaching zero (Figure 19).

The data available on world fleet size, selected landings, and landing rates from 1970 to 1989 (Table 3) point to an inverse relationship between fleet capacity and landing rates, which would indicate the possibility to fit a global production model to the data provided that the data on grt represent the trend in fishing mortality and the landing rates represent the trend in global abundance of world resources. During the same period, however,

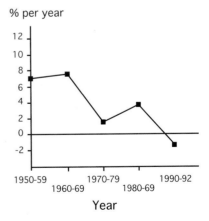

FIGURE 19.—Evolution of the rate of growth of world marine fisheries landings (1950–92).

---

[6]These data (starting from 1990) are now available (FAO 1993g, Table K).

the average fishing power of the components of the world fleet increased owing to technological progress, and the changes in landing rates (quantities landed per grt) may not reflect the true changes in resource abundance and, indeed, may underestimate its decrease.

Fitzpatrick (1996) estimated the relative value of the "technology coefficient" calculated for 13 different types of fishing vessels ranging from super trawlers (of 120 m) to pirogues (of 10 m) in 1965, 1980, and 1995, taking the value of the coefficient in 1980 as a basis. On average, this coefficient has increased from 0.54 ± 0.26 in 1965 to 1.0 in 1980 (the basis) and 2.0 ± 0.9 in 1995. The evolution of this relative coefficient approximates the changes in the efficiency of these vessel types from a technological viewpoint. The coefficient applies to new vessels and not to entire fleets where vessels of various ages and technological levels are mixed. However, new technologies tend to be incorporated rapidly into existing vessels, often with government subsidies. We assumed, therefore, that the trend indicated by Fitzpatrick reflected the trend in efficiency for the world fleet and that these relative efficiency values could be combined with data on world fleet size in grt to better reflect the likely increase in fishing pressure exerted by this fleet. Interpolating between the 1965, 1980, and 1995 values given by Fitzpatrick, we have estimated the relative technology coefficient for the years 1970–89 (Table 4). Multiplying the world fleet capacity in grt (Table 4, column 2) by the relative coefficient of technological efficiency (Table 4, column 3), a corrected index of world fishing fleet capacity in "standard" grt (indicated hereafter as grt*) has been developed (Table 4, column 4). The corrected fleet capacity and index of fishing mortality appears to have increased by 332%, from 9.3 million grt* in 1970 to 40.2 million grt* in 1989. Ex-

cluding the five main low-value and small pelagic species, the selected landings per unit of fleet capacity and abundance index of the selected species decreased from 4.4 to 1.2 mt grt*[-1] (Table 4, column 5). The total landings per unit of capacity, which appeared stable in Table 3, decreased from 6.4 to 2.4 mt grt*[-1] (Table 4, column 6).

The relationships between the fleet corrected capacity, landings, and landing rates for selected and total landings are graphically represented in Figure 20. The limitations of the available data are easily recognized as well as the problems of applying the production model theory to an aggregated world "stock." For want of a better global approach to the dynamics of the world fishery sector, and because the relationships appear coherent, a simple exponential model (Fox model) has been fitted to the data to take into account the nonlinear appearance of the relationship. The results are as follows:

A. For selected landings
   n: 13
   R^2:   0.95
   a: 5.180
   b: -0.033
   MSY: 57.7 million mt
   f_{MSY}:   30.5 million grt*

The results indicate that the MSY of the selected species (excluding the five principal pelagic species) would be at 58 million mt and that the corresponding index of effort would be 30.5 million mt of corrected grt*. The comparison of this estimate with the effort and landings in 1989 (the latest data point available in the analysis) shows that the grt* in 1989 was 132% of the MSY level for selected species and that the landing in 1989 was 106% of the estimated MSY. The results indicate, therefore, that the world resources of selected species is exploited beyond the MSY level with an overcapacity of at least 30%.[7]

B. For total landings
   n: 13
   R^2:   0.95
   a: 5.41
   b: -0.025
   MSY: 82.8 million mt
   f_{MSY}:   42.0 million grt*

The results indicate that the MSY of the total world resource would be at about 83 million mt and that the

TABLE 4.—Fleet capacity (>100 grt or 24 m), technology coefficient, corrected fleet capacity, and landing rates for selected and total landings.

Year	grt (10^6 mt)	Technology coefficient	grt* (10^6 mt)	Selected landing rate (mt/grt*)	Total landing rate (mt/grt)
1970	13.5	0.69	9.3	4.6	6.4
1975	17.3	0.84	14.5	3.4	4.0
1978	19.3[a]	0.93	17.9	2.9	3.5
1980	19.8	1.00	19.8	2.7	3.3
1981	20.0	1.07	21.4	2.5	3.2
1982	20.8	1.13	23.5	2.2	2.9
1983	21.2	1.20	25.4	2.1	2.7
1984	21.8	1.26	27.5	2.0	2.7
1985	22.5	1.33	29.9	1.8	2.5
1986	23.5	1.39	32.7	1.8	2.5
1987	24.1	1.46	35.2	1.7	2.3
1988	24.8	1.53	37.9	1.6	2.3
1989	25.3	1.59	40.2	1.5	2.1

[a]The tonnage for 1978 has been interpolated.

_____
[7]By reference to the capacity required to produce MSY. A more precautionary approach would require larger reductions in capacity.

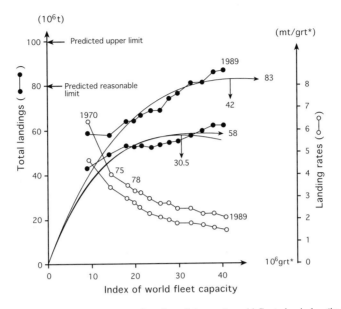

FIGURE 20.—World landings and landing rates as a function of the total world fleet size index (in corrected grt*). Upper curves = all species. Lower curves = selected species (i.e., excluding the five main pelagic species).

corresponding effort would be 42 million grt*. The comparison of this estimate with the effort and landings in 1989 (the latest data point available in the analysis) shows that the grt* in 1989 was 98% of the MSY level for the total world resource and that the landings in 1989 were practically equal to the MSY. The results indicate therefore that, when all species are considered together (including the five main pelagic species), the world resource appears as exploited at MSY level.

The two results would confirm that the species of higher value (the selected species) are more affected by overcapacity and require more drastic management measures than the small pelagic species. The results tend to confirm that the progressive inclusion of fluctuating, small pelagic species in the world landings has concealed the overfishing of the high-value species. The results are also in close agreement with the more detailed resource assessments provided previously in this paper, which showed that 69% of the resources for which data are available are either fully fished or overfished (Figure 14) and that, excepting some pelagic resources and mollusks, most types of resources are fully fished or overfished (Figure 16).

Global Economic Assessment

The production curve obtained previously for the total world fishery resource may be combined with data on value to produce a total world revenue curve, which may be combined with data on the cost of fishing for a

very approximate bioeconomic assessment of world fisheries. On the basis of data from FAO (1993a, page 17, Table 21), which shows the estimated total value of marine landings in 1989, an average price of $862/mt has been calculated for the world landings in 1989. When the production curve for all species is multiplied by their average unit value, a total world revenue curve (US$) is obtained (Figure 21). Rough and conservative estimates of both the total and operating costs for 1989 (not including the opportunity cost of capital and debt servicing) have been calculated by FAO (1993a, page 52, Table 29). These values are, respectively, approximately $3,600/grt and $4,600/grt (uncorrected)[8] leading to total and operating costs of about $91 and $116 billion, respectively, for a fleet size of 25.3 million grt or 40.2 million corrected grt*. These two points have been plotted (Figure 21) and joined to the origin of the graph to represent the relationship between total world fleet capacity and operating or total costs, assuming a simple linear function. For the sake of comparison and validation, the calculated deflated values (base = 1989) of the total catch for 1970, 1978, and 1989 (respectively $34.0, $57.5, and $70.0 billion) have been reported on the graph at their corresponding levels of corrected capacity. Their position in relationship to the calculated revenue curve

---

[8]The present cost for a grt of a fishing vessel is about 10 times the cost for a grt of any other type of vessel excluding military ones (J. Fitzpatrick, FAO, Rome, pers. comm.).

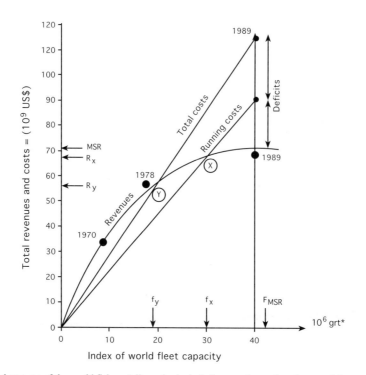

FIGURE 21.—Total revenue of the world fishery (all species included), operating and total costs of the world fleet, in relationship to total world fleet size (in corrected grt*). Based on 1989 prices and costs. ● = the total deflated revenues (base = 1989) for 1970, 1975, and 1989.

(calculated without using them) shows a surprisingly good agreement and indicates that, despite the obvious approximations in the analysis, the results are coherent. The following conclusions might be drawn from the analysis:

- The maximum sustainable revenue (MSR) for the global resource of traditional species (at 1989 prices) is $71 billion, corresponding to a fleet size of 42 million grt*.

- The theoretical value for the revenue at equilibrium corresponding to the fleet capacity available in 1989 (40.2 million grt*) is practically equal to the MSR and very close to the actual value of the landings, estimated at about $70 billion (FAO 1993a, Table 21, page 17). One could, therefore, conclude that the situation prevailing in 1989, both in terms of fleet size and economic yields, corresponded practically with the MSR.

- The total costs (~$116 billion) and running costs ($91 billion) incurred in 1989 are much higher than both the actual and equilibrium revenues for that fleet size (~$70 billion). The deficit (excluding the opportunity cost of capital and debt servicing) is $46 billion (in relationship to total costs) and $21 billion (in relationship to running costs). These levels of deficit have already been emphasized by FAO (1993a).

- To make the world fishery sustainable on an economic basis at 1989 levels of fleet size would, therefore, require lowering the costs per unit grt by about 43%, or increasing ex vessel fish prices by 71%, or a combination of the two. Fish price increases will be limited by the price of substitute products. Production costs could be lowered substantially by making more efficient use of artisanal fisheries and reducing the use of long-range fleets.

- To reduce the deficit by only adjusting the fleet capacity, it would be necessary to reduce the fleet to the point where the cost and the revenue functions intersect. For the revenues to cover operating costs, (point X, Figure 21) the world capacity should be reduced by 25% (from 40.2 to about 30.0 million grt*) with a loss of revenues of about 4% only. For the revenues to cover total costs (point Y, Figure 21), the world fleet should be reduced by 53% (from 40.2 to about 19 million grt*)[9] with a loss of revenues of only 21%. These reductions in fleet capacity would

---

[9]The draft objectives for the European Union fisheries for the period 1994–97 foresee a 30% reduction in effort in most of the important fisheries (Commission of the European Communities 1993).

lead to a significant improvement in catch rates of about 20% in the first case (point X) and 60% in the second case (point Y).

In practice, an overall economic rationalization of the world fishery likely would require a combination of measures related to prices, unit costs, and fleet capacity, particularly when one of the objectives will be to maximize employment.

Conclusions and Discussion

After a long history of fisheries growth, all available data point to the conclusion that the total potential of traditional species has been reached on the average, even though there are differences between species groups and regions. The species composition of landings has changed over time, showing that the world fish supply is increasingly relying on variable, small pelagic and other-low value species, thereby concealing the slow but steady degradation of the demersal, high-value resources.

The problem of overfishing, stressed in the 1946 London Conference on Overfishing, has clearly become general and now concerns all regions of the world. Following the extension of exclusive economic zones, developing countries are progressively taking a "leading role" in the overfishing problem as they develop their own fishing capacity (from 28% to 58% of the world fleet complement), which has been prompted by a high demand on their local markets as well as on the developed countries' markets. More than half of the 20 top producers in the world are developing countries. The trade in fish and fishery products has increased from 5% to 11% of the trade in agricultural products, and the developing countries appear as net exporters while developed countries appear as net importers. Trends in prices depend on species groups and reflect increased scarcity for some high-value species as well as the effect of substitutes (surimi) and aquaculture production. For industrial species used for fish meal, the fluctuations in price are dampened by the much larger production of soya, their main substitute.

The analysis of the state of stocks by species groups and by region shows that about 70 % of the fish resources for which data are available are either heavily or fully fished, overexploited, overfished, depleted, or recovering from depletion. High-value demersal resources (cods, hakes) are the most affected, but many small pelagic stocks are also affected. The analysis also shows that in all regions, the expansion of effort needs to be controlled more strictly and effort needs to be reduced in most cases. Despite the approximations affecting the analysis, we suggest that the results confirm that the state of world fishery resources should be a subject of major concern and taken seriously by all governments with respect to

their EEZs and the high seas. We also stress that the situation created by the world overcapacity is compounded by the progressive degradation of critical environments in the coastal areas and, possibly, by climate change.

The analysis of the trends in the size of the world fleet (in grt), landings, and landings per grt shows that, while the total world fleet size and technological capacity to fish increased, the world fleet landing rates were maintained at about 3.4 mt grt^{-1}, but the landing rates of the higher-value species decreased by 25% despite technological progress. Revenue per grt, however, increased by 38%, providing the incentives for fisheries growth despite the resource decline. This decline, already apparent in the selected species group without correction for the fleet efficiency, becomes even more conspicuous when taking into account the effects of technological improvements, indicating a 62% decrease in the global abundance index and 73% decrease in selected species index between 1970 and 1989.

When the main pelagic and low-value species are excluded, the world resources appear to be overfished with an excess capacity of about 30%. When all species are included in the analysis, the world resource appears to be fished at the level corresponding to MSY. This result hides the fact that many resources are severely overfished and some are still moderately exploited, but it indicates also that there is little or no room for major increases in world catches of traditional species, and that the world priority should be on arresting the growth of fishing fleets and in implementing fleet reduction schemes to return to safer and more economic levels of resource biomass. The analysis confirms that there is little hope that landings of traditional species can be sustainably increased with the current fishing regimes and discard practices.

The economic analysis shows that despite the decrease in the resource base, the incentive to fish and to increase fleet size remained because prices increased, maintaining and even raising the revenue extracted per grt. It also confirms that the present revenues from fisheries at capture level cannot cover the cost of fishing and that a global deficit of $46 billion exists. Eliminating this deficit would require a reduction of fishing costs (-43%), or an increase in price (+71%), or a reduction of the world fleet capacity (-25% to -53%), and probably a combination of all three measures.

Review

In preparing this paper, we had to face the challenge imposed on us by the organizers of the meeting (i.e., to show and explain the global trends in fisheries). We were aware of the dangers of aggregating data to such high levels and of the difficulty to interpret their changes. However, as the data were pooled together and the analysis progressed, a coherent picture emerged despite the sources of potential bias in the data and the analysis, which have probably been pushed to their limits. We are aware that it is perilous to apply non-weighted "averages" of relative levels of exploitation to species groups across discontiguous regions and to regions across species with widely different life cycles and resilience. Using indexes of world fleet capacity (from an incomplete database), global indexes of fishing efficiency, and total landings (of dubious accuracy) to develop a production model for the whole world is also certainly dangerous and may even appear unreasonable to some scientists. However, we argue that, in the absence of better data and alternative analysis, this is the best scientific evidence available of global level at the moment (using the terminology of the 1982 UN Convention for the Law of the Sea) and that it would not be very "precautionary" to totally disregard it because it does not satisfy some of the traditional statistical requirements.

We cannot be sure of the accuracy of the findings, but the emerging picture is so bleak that we believe it is our duty to put out the information and the warning it contains (once again). A global picture is required by governments, by the news media, and by nongovernmental organizations because world fisheries have attracted attention at the global level, in the U.N. General Assembly, at the U.N. Conference on Environment and Development (UNCED), at the U.N. Conference on Straddling Fish Stocks and Highly Migratory Fish Stocks, and at FAO where an International Code of Conduct for Responsible Fishing was adopted in 1995. There is no doubt that better analyses could certainly be undertaken at the regional and national level with less aggregated data. Some have already been done. Too many are lamentably lacking. We hope that this paper will promote more analysis of this kind, particularly at the national level, where the dynamics of the entire fishery system is often yet to be properly understood, if only to disprove locally the global conclusions arrived at in this paper.

The analysis indicates that the world fleet presently available is practically at the f_{MSY} level and that no significant additional landings or revenues can be expected by simply increasing fishing capacity. The present fishing pressure is not evenly distributed, and the conclusion that fish resources are fully fished "globally" conceals the fact that some are already overfished while a few others may be able to produce more. The results obtained in this paper seem to confirm, at the global level, the diagnosis already repeatedly established for many national and international fisheries (for which a lack of economic analysis is obvious). The results imply the following:

- At current levels of costs and prices, global fisheries can only be maintained through significant direct and indirect subsidies to the capture sector, partly externalizing its high costs and dissipating all or most of the economic rent. This conclusion is confirmed by many analyses carried out at the national level.

- At current capacity levels, costs could be reduced by making better use of artisanal fisheries and using fishing techniques that reduce costs such as passive gears (e.g., set or drifting gillnets, longlines) and concentrate fish (e.g., fish aggregating devices and artificial reefs).

- Substantial reductions in effort levels would reduce costs or boost productivity or both. One of the first measures to contain and reduce fleet sizes will be to reduce or suppress subsidies or redirect them towards effort-reducing measures (buy-back schemes, etc.).

- Ex vessel prices may be too low and vertical integration between the capture, processing, and distribution subsectors may also be required to ensure that part of the profits made by processors and retailers is redistributed to better cover the cost of capture. Alternatively, prices could be improved through more competition by increasing the independence of fishers in negotiating prices from buyers.

Stock productivity could be boosted by rehabilitating degraded critical habitats and reducing impacts on juveniles (by closing areas and seasons). It is impossible to assess the potential impact of reducing discards, and improving their marketing would not improve sufficiently the total revenue to be of relevance (even though the problem must be addressed for biological and ethical reasons). All of these measures would move the yield curve upwards.

We are not certain that this global assessment depicts correctly what would happen if capacity was substantially reduced, since the model cannot capture the complex reactions of the fishery sector and the depressed resource base to such a reduction. The abundance and landings of large predators and other preferred species would certainly increase, and this would tend to improve prices. In many areas, protecting juveniles by closing areas and seasons could raise the production curve substantially in tonnage and value, as shown in Cyprus and in the Philippines where biomass and MSY could be doubled in 18 months (Garcia and Demetropoulos 1986). However, if these improvements were obtained too rapidly, the prices could also fall abruptly through market saturation, as shown in Italy where such a closed season was experimentally introduced.

Altogether, it seems difficult to prevent a large part of the presently hidden costs of fisheries from being made more transparent (i.e., known to society) and progressively reduced or passed to the consumer in the future. There are limits to this, however; for instance, Westlund (1995), in a perspective analysis of pelagic fisheries in West Africa, confirmed that "small pelagics are available in the sea but it seems that consumers (i.e. low income groups) cannot pay the price covering the costs of (industrial) production."

The large imbalance between the cost of fishing and its revenues has already been underlined in Garcia (1992) and FAO (1993a). In this last paper, it was roughly estimated that, in 1989, the investment in fishing fleets was about $320 billion. The opportunity cost of this capital, based on a 10% annual return on capital, is $32 billion/year, or 46% of the ex vessel value of the world landings (which amounts to ~$70 billion). Despite the approximations involved, this figure illustrates the disproportionate share of the wealth extracted from the ocean fishery resources being absorbed by a fleet capacity that grossly exceeds what would be required from the economic and biological standpoints. The large deficit observed would indicate that the world capture fisheries are operating under conditions of overinvestment and overcapacity. If the excess of technology is largely imported, as in many developing countries, the wealth generated by fisheries in these countries may be partly transferred abroad (e.g., the acquired excess fleets generate employment and revenues in the developed countries' shipyards).

The progress of world landings in the last 30 years has concealed the worrying and sometimes alarming situation of some of the major fish resources and, in particular, the high-value demersal species. The real situation was, however, very well known. Global assessments such as presented here may not have been so frequently available, but a number of important assessments published at national and regional level have been accessible to managers and policy makers. Put together, they left very little doubt about the relative state of the world resources and fisheries, but for the most part these assessments have been disregarded.

A perspective view of the process of world fisheries development since 1945, the date FAO was established, has been provided (Garcia 1992; Garcia and Newton 1994). These papers described how the heavy fishing rates (and often the overfishing) that characterized the north Atlantic before World War II progressively spread to the north Pacific in the 1950s, to the eastern Atlantic (West Africa) and eastern Pacific (Latin America) in the 1960s, to the Indian Ocean and the Antarctic in the 1970s, and to the south Pacific and southwest Atlantic during the 1980s. This extension of fishing pressure has been supported by remarkable progress in technology (boat design, gear, positioning systems, detection equipment, onboard fish preservation), which has allowed long-range

fleets to stay away from home for longer and longer periods of time.

The increase in fleet size and the development of larger and safer vessels have resulted in significant excess fishing capacity, which can be rapidly transferred from one overfished stock or area to the next. The drastic measures being taken to reduce effort or restructure fisheries in the north Atlantic and Eastern Europe are releasing an important excess effort, which is on the market at very low cost. As a consequence, the full exploitation and depletion of the remaining world resources, which in the 1950s would have taken 10 years or more to reach, can now be reached nearly instantly. In addition, the extension of jurisdiction and restriction of access to resources have led to transfers of at-sea processing capacity from developed to developing countries through barter arrangements (processing capacity against fish). Last, but not least, modernization of artisanal fisheries gear (e.g., introduction of monofilament and multi-monofilament set gillnets, medium-scale driftnets, modern purse seines with outboard engines, portable echosounders, and positioning systems, etc.) can greatly increase fishing capacity and pressure on the resources rapidly, dramatically, and at a relatively low cost.

The process of "colonizing" distant fishing grounds, originally conducted by a limited number of developed countries (USA, Japan, Eastern and western Europe), was accompanied by resource collapses provoked by a combination of exceptional climatic conditions and excessive fishing (e.g., anchoveta, Namibian pilchard [*Sardinops ocellatus*], Atlanto-Scandian herring) or purely through overfishing (e.g., Mauritanian lobster, western Sahara sea bream, Gulf of Thailand demersal fish, Philippines coral reef resources). This colonization has generated conflicts between coastal countries and distant water fishing nations, leading to extension of national jurisdiction to 200 miles through a process that started in 1947 and which eventually was completed with the entry into force of the 1982 United Nations Convention on the Law of the Sea (UNCLOS) at the end of 1994. High-seas resources, which represent about 10% of the world catches, have been progressively deteriorating (FAO 1992b, 1993a; Garcia and Majkowski 1992) because of inadequate or lacking national and international control of high-seas fishing and noncompliance with management measures agreed to, with difficulty, in international or regional fishery management fora.

All these developments, added to the powerful incentive represented by rising prices in a market globally limited by supplies, has led to a very volatile situation for most world fishery resources that have a high risk of overfishing, particularly in developing countries. The impact of mismanagement (including lack of considering natural variability in stock potential and resilience) on northwest Atlantic stocks is costing hundreds of millions of U.S. dollars per year, with a loss of an important number of jobs in the fishery itself plus many more in the related industries and activities. The economic disaster in the Black Sea is of the same order of magnitude, and it is doubtful that developing countries could afford such an economic shock.

It is, therefore, too late to argue about the probability of occurrence of something that has already become a sad reality. It is time to "bite the bullet" and ask the question, "How much should the capacity be reduced?" The UNCLOS requires that stocks be maintained at the level at which they could produce their MSY. Recognizing the scientific uncertainties about this concept and the exact position and value of MSY, and the need for a precautionary approach to fisheries management, the UN Conference on Straddling Fish Stocks and Highly Migratory Fish Stocks (New York, 1993–94), with the advice of FAO, proposed to consider MSY as a minimum international standard, particularly for stock rebuilding strategies and not as a target for catch levels. Under that interpretation, stocks should be exploited, in most cases, at levels of effort below f_{MSY}, and stock biomass should be maintained at levels higher than B_{MSY}; these reference points would be considered as thresholds at which corrective action has to be taken (Garcia 1994; FAO 1994b). Similarly, overfished resources should be rebuilt *at least* to B_{MSY} and preferably at even higher biomass levels.

The large direct and indirect subsidies required to maintain the world fishery indicate that it represents a significant cost to the world society (even when other costs are excluded, such as those related to environmental degradation from fishing and damage to biodiversity). World fisheries may generate social and other benefits, particularly in the coastal areas, that are not reflected in the fisheries revenue curve, thereby justifying the subsidies. But it is not clear whether society, when confronted with an objective choice, would not prefer to see its contribution used differently (e.g., for better schools or health systems). Such an analysis at the global level is impossible and meaningless, but it should be undertaken at national and regional levels (in the case of shared resources).

As a consequence of strengthened management schemes, the real or opportunity price of access to fish stocks in developed countries is likely to increase. On the contrary, in developing countries the need to obtain foreign exchange through fishing agreements and the deficiencies in management schemes (including monitoring, control, and surveillance systems) will put the price of access to their resources at a lower level, particularly if developed countries subsidize the "expatriation" of their excess fleets. As already stressed by Garcia and Newton (1994), the consequence in environmental and economic terms is that the risk of depleting the de-

veloping countries' fish resources to the benefit of the developed markets is increasing through international trade. The trends observed since the early 1970s, in both fleet transfers through joint ventures and international trade, may indicate that the process has already started and that the differential in the price of access to the resource between the two worlds, combined with subsidies, has led to increased use rates in developing countries. The recent examples of economic disasters in the northwest Atlantic demonstrate without any doubt that the risk is not just theoretical for countries that will never have the economic capacity of developed countries to withstand the socioeconomic consequences of such a crisis.

The interaction between environment and trade is one of the most explosive issues following UNCED. It should be obvious that a world fishery system based on active exchange through trade, particularly as a consequence of the Uruguay Round of the General Agreement on Tariffs and Trade, and large exports to the developed world can only be globally sustainable if the resources in the developing exporting world are exploited in a sustainable manner. This is obviously not the case in most areas and, if developed countries continue to export their excess fleet capacity to the developing world, the system can only continue to deteriorate while fisheries will further increase the debt of the developing world.

Acknowledgments

The authors wish to express their most grateful thanks to J. Fitzpatrick, R. Grainger, and the staff of the Fisheries Data and Information Service (FIDI) for the data they provided and their extremely helpful assistance in data processing and preparation of graphics. They also wish to express their grateful thanks to Dr. M. Sissenwine, R. J. H. Beverton, and the two other anonymous reviewers for their very stimulating comments and helpful suggestions. Any misinterpretation or error in the paper, however, remains our sole responsibility.

References

Alverson, D. L., M. H. Freeberg, S. A. Murawsky, and J. G. Pope. 1994. A global assessment of fisheries by-catch and discards. FAO Fisheries Technical Paper 339, Rome

Arnason, R. 1994. Fishery management in Iceland. Pages 29-38 *in* E. A. Loayza, editor. Managing fishery resources. World Bank discussion papers, Fisheries Series 217.

Bakun, A. 1995. Global climate variations and potential impacts on the Gulf of Guinea sardinella fishery. Pages 60-84 *in* F. X. Bard, and K. A. Koranteg, editors. Dynamic and uses of sardinella resources from upwelling off Ghana and Côte d'Ivoire. ORSTOM Editions, Paris.

Brown, L. 1995. State of the world. Report on progress toward a sustainable society, Worldwatch Institute, Washington, D.C., Norton and Co., New York.

Caddy, J. F. 1993. Towards a comparative evaluation of human impacts on fisheries ecosystems of enclosed and semi-enclosed seas. Reviews in Fisheries Science 1:57-95.

Commission of the European Communities. 1993. Proposal for a council regulation (EC) fixing management objectives and strategies for certain fisheries or groups of fisheries for the period 1994 to 1997. Document COM (93) 663 Final. 15.12.93, Brussels.

Csirke, J., and G. D. Sharp, editors. 1984 report of the expert consultation to examine changes in the abundance and species composition of neritic fish resources. FAO Fisheries Report 291 (1), Rome.

Fitzpatrick, J. 1996. Technology and fisheries legislation. Pages 191-200 in Precautionary approach to fisheries, part 2, scientific papers. Food and Agriculture Organization of the United Nations, Fisheries Technical Report 350/2, Rome.

FAO (Food and Agriculture Organization of the United Nations). 1980. The state of food and agriculture (SOFA). World review. Marine fisheries in the new era of national jurisdiction. FAO Agriculture Series 12, Rome.

FAO (Food and Agriculture Organization of the United Nations). 1992a. The state of food and agriculture, 1992. Marine fisheries and the law of the sea. A decade of change. FAO Agriculture Series 25, Rome.

FAO (Food and Agriculture Organization of the United Nations). 1992b. Tableaux par pays. Pages 308-343 in Données de base sur le secteur agricole. FAO Département des Politiques Economiques et Sociales, Rome.

FAO (Food and Agriculture Organization of the United Nations). 1992c. Page 25 in Fish and fishery products: world apparent consumption statistics based on food balance sheets (1961–1990). FAO, Rome.

FAO (Food and Agriculture Organization of the United Nations). 1993a. Marine fisheries and the law of the sea. A decade of change. Special chapter (revised) of the state of food and agriculture (SOFA). FAO Fisheries Circular 853, Rome.

FAO (Food and Agriculture Organization of the United Nations). 1993b. Fisheries and environment studies in the Black Sea system. General Fisheries Council for the Mediterranean, Studies and Reviews 64, Rome.

FAO (Food and Agriculture Organization of the United Nations). 1993c. Trends in catches and landings: Atlantic fisheries: 1970–1991. FAO Fisheries Circular 855.1, Rome.

FAO (Food and Agriculture Organization of the United Nations). 1993d. Trends in catches and landings: Indian Ocean fisheries: 1970-1991. FAO Fisheries Circular 855.2, Rome.

FAO (Food and Agriculture Organization of the United Nations). 1993e. Trends in catches and landings: Pacific fisheries: 1970–1991. FAO Fisheries Circular 855.3, Rome.

FAO (Food and Agriculture Organization of the United Nations). 1993f. Trends in catches and landings: Mediterranean and Black sea fisheries: 1970–1991. FAO Fisheries Circular 855.4, Rome.

FAO (Food and Agriculture Organization of the United Nations). 1993g. FAO yearbook of fishery statistics: commodities, Vol. 77. FAO, Rome.

FAO (Food and Agriculture Organization of the United Nations). 1994a. Reference points for fisheries management: their potential application to straddling and highly migratory resources. FAO Fisheries Circular 864, Rome.

FAO (Food and Agriculture Organization of the United Nations). 1994b. Report of the second technical consultation on the stock assessment in the Black Sea, 15-19 February 1993, Ankara, Turkey. FAO Fisheries Report, 495, Rome.

FAO (Food and Agriculture Organization of the United Nations). 1994c. The state of world marine fishery resources. FAO Fisheries Technical Paper 335, Rome.

FAO (Food and Agriculture Organization of the United Nations). 1995. The state of world fishery resources and aquaculture. A document prepared for the Ministerial Session of the FAO Committee on Fisheries (COFI), FAO, Rome.

Garcia, S. M. 1992. Ocean fisheries management: The FAO programme. Pages 381-418 in P. Fabbri, editor. Ocean management in global change. Elsevier Applied Science.

Garcia, S. M. 1994. The precautionary approach to fisheries with reference to straddling fish stocks and highly migratory fish stocks. FAO Fisheries Circular 871, Rome.

Garcia, S. M., and A. Demetropoulos. 1986. L'aménagement de la pêche à Chypre. FAO Document Technique Pêches (250), Rome.

Garcia, S. M., and J. Majkowski. 1992. State of high seas resources. Pages 173-236 in T. Kuribayashi and E. Miles, editors. The law of the sea in the 1990s: a framework for further international cooperation. Proceedings of the Law of the Sea Institute 24th Annual Conference, 1990, Tokyo, Japan, July 24-27.

Garcia, S. M., and C. Newton. 1994. Responsible fisheries: an overview of FAO policy developments. Marine Pollution Bulletin 29(6):528-536.

GESAMP (FAO/IMO/UNESCO/WMO/WHO/IAEA/UN/UNEP Joint Group of Experts on the Scientific Aspects of Marine Pollution). 1990. The state of the marine environment. Report Studies GESAMP (39).

Glantz, M. H., and D. Thompson. 1991. Resources management and environmental uncertainty. Lessons from coastal upwelling fisheries. Wiley Series in Advances in Environmental Science and Technology 11.

Gulland, J. A., editor. 1971. The fish resources of the ocean. Fishing News (Books) Ltd., London.

Hilborn, R. 1990. Marine biota. Pages 371-386 in B. L. Turner, III, editor. The Earth as transformed by human action. Cambridge University Press.

Lluch-Belda D., R. A. Schwartzlose, R. Serra, R. H. Parrish, T. Kawasaki, D. Hedgecock, and R. J. M. Crawford. 1992. Sardine and anchovy regime fluctuations of abundance in four regions of the world oceans: a workshop report. Fisheries Oceanography 1:339-347.

Needler, A. W. H. 1974. Chairman's summary of the highlights of the conference. In J. C. Stevenson (editor), Technical conference on fishery management and development. Journal of the Fisheries Research Board of Canada 30(12):2508-2511.

Saeger, J. 1993. The Samar Sea, Philippines: A decade of devastation. NAGA, The ICLARM (International Center for Living Aquatic Resources Management) Quarterly, October:4-6

Stevenson, J. C., editor. 1974. Technical conference on fishery management and development. Journal of the Fisheries Research Board of Canada 30(12):1921-2536

Westlund, L. 1995. Report of the study on exploitation and use of small pelagic species in West Africa. FAO Fisheries Circular 880, Rome.

Economic Waste in Fisheries: Impediments to Change and Conditions for Improvement

Francis T. Christy

Abstract.—Extraordinary amounts of economic waste exist in open-access fisheries. Very rough global estimates indicate annual waste may be on the order of $60 billion. For the United States, the estimates indicate waste of $2.9 billion per year. This waste may well become more severe in the future. Demand in the year 2010 is likely to be more than 45% greater than present production. Supplies from capture fisheries are probably now at their maximum limits. Supplies from aquaculture will increase but face significant constraints in view of scarcity of space, clean water, and feed. The result will be continued increases in the real prices of most fish species, placing greater pressures on the stocks and increasing the need for effective management. Means for the prevention of this waste (although imperfect) are generally well known. Yet, with a few exceptions, the dissipation of economic rents continues in both international and national fisheries.

Generally, fisheries management analysts assume that fishery administrators will make the right decisions if there is reliable and credible information about the benefits to be gained (or losses to be avoided) and if there is adequate knowledge about the various techniques for effective management. Thus, considerable effort has been devoted to improving knowledge, reducing uncertainty, and developing refinements in management measures. These investments in research, however, have had little noticeable effect. It is worthwhile to examine the basic assumptions made by fishery analysts and to question why the results have been so meager. Asking this question provides a basis for identifying those conditions and the forces that impede movement to better management as well as those that may contribute to improvements.

Extraordinary amounts of economic waste exist in open-access fisheries. Very rough estimates for the world as a whole indicate that annual waste may be on the order of $60 billion. For the United States, the estimates indicate waste of $2.9 billion per year. Means for preventing this waste, although imperfect, are generally well known. And yet, with a few exceptions, the dissipation of economic rents continues in both international and national fisheries.

Generally, analysts of fisheries management assume that fishery administrators will make the right decisions if there is reliable and credible information about the benefits to be gained (or losses to be avoided) and if there is adequate knowledge about the various techniques for effective management. Thus, considerable effort has been devoted to the improvement of knowledge, the reduction of uncertainty, and the development of refinements in management measures. These investments in research, however, have had little noticeable effect.

It is worthwhile to examine the basic assumptions that have been made by fishery analysts and to raise the question as to why the results have been so meager. Asking this question provides a basis for identifying those conditions and forces that impede movement to better management as well as those that may contribute to improvements.

Fisheries management is defined here as the set of controls and institutions that lead to the production of economic rents from the use of the resources. This subsumes maintenance of the sustainability of the stocks.

The Global Situation

The USA is by no means alone among nations suffering from deficiencies in fisheries management. The Food and Agriculture Organization of the United Nations (FAO) has made some rough estimates of the total economic waste in marine fisheries. These estimates are based on estimates of the total costs of the world's marine fishing fleets and the total gross revenues from marine fisheries in 1989, with some assumptions about the amount of surplus capital and labor.[1]

Total costs were estimated to amount to about US$124 billion in 1989 (Table 1; FAO 1993). The estimates are believed to be generally conservative.[2]

The gross revenues in 1989 were estimated to amount to about US$70 billion[3] for total marine landings of 81 million metric tons (mt). Estimates of average unit values of fish at points of landing are extremely difficult to make. The FAO, the only collector of such information on a glo-

---

[1]This section of the paper borrows heavily from work the I did for FAO in the preparation of a study on changes in global marine fisheries (FAO 1993).

[2]FAO welcomes criticisms and comments on the basic information and calculations and is actively seeking to improve the estimates. Comments should be sent to Dr. Christopher Newton, Chief of the Information, Data and Statistics Service, UN Food and Agriculture Organization, via delle Terme di Caracalla, 00100 Rome Italy (Fax: 011 39 6 5225-3020).

[3]All monetary values in this paper are cited in US$.

TABLE 1.—Estimates of annual total costs of global fishing fleet, 1989.

Item	Estimated costs in US$ billions
Maintenance and repair	30.2
Supplies and gear	18.5
Insurance	7.2
Fuel	13.7
Labor	22.6
Total operating costs	92.2
Capital costs	31.9
Total operating and capital costs	124.1

TABLE 2.—Estimated landed value of major species and species groups, 1989.[a]

Species or species group (US$ millions)	Landed value	% of total
Shrimp	7,370	11
Tuna	6,775	10
Cephalopods	5,344	7
Crab	4,189	6
Salmon	3,278	5
Lobster	2,275	3
Alaska pollock	2,072	3
Atlantic cod	1,904	3
Total	33,207	48

[a]The estimates are for 1989. Interannual variations are significant, so present values may be quite different. Shrimp and salmon prices have recently been affected by supplies from aquaculture and crab prices by production of artificial crab from surimi processing. The estimates of average unit values were prepared by A. Crispoldi-Hotta, FAO Fishery Information, Data and Statistics Service.

bal basis, collects landings data for about 1,000 species from 227 countries and administrative or political entities. A large amount of landings in developing countries occur on isolated beaches along extensive coastlines. The prices among individual species range widely from trash fish at less than $0.05 per pound to luxury fish, such as bluefin tuna (*Thunnus maccoyi*), at $10 per pound. In addition, prices of any individual species vary according to size, quality, and place and time of landing.

Although there are few sources of accurate data on the prices of most species, information on prices for the major species landed is available. About 50% of the estimated value of global marine landings comes from eight individual species or uniform species groups, for which there is relatively good information (Table 2). With exception of Alaska pollock (*Theragra chalcogramma*), these species and species groups are of relatively high unit value. The cephalopods and some of the tuna (skipjack [*Katsuwonus pelamis*] and yellowfin [*Thunnus albacares*]) are not fully exploited, but the rest are heavily fished and, in some cases, the stocks are severely depleted.

There are various possible sources of error in the calculations of costs and revenues. One is under-reporting of landings or gaps in the collection of landings data. Under-reporting is known to be significant in the north Atlantic where fishers catch greater quantities than the quotas allocated to their countries and where fishers flying flags of convenience fail to report their landings. It also tends to occur in some national fisheries that are managed under systems of individual transferable quotas (ITQs). Gaps in data collection occur in some countries owing to the difficulties of monitoring catches that are landed in isolated spots. Thus, total revenues may actually be higher than those estimated.

Errors may also exist in the quantification of total number of fishing vessels. Some of the large vessels listed in various sources may no longer be actively fishing. On the other hand, records of quantity and size of vessels are inadequate in many countries, including certain developed countries, and the amounts may be lower than those estimated.

A particular difficulty in the analysis is that of esti-

mating the capital value of the fishing fleet. In the absence of information on the age of the vessels, deriving estimates of current value is impossible. The use of replacement costs overstates present capital investment. But for the purposes of the exercise, it is not the total amount of capital invested that is important but the annual cost of that capital. This should include sufficient amounts to produce a satisfactory return on the owner's investment as well as the amounts necessary to pay off the costs of the vessel. In the calculations, annual cost is estimated to be 10% of the replacement value. For older vessels that have been amortized, this is clearly too high a figure. But for newer vessels, it may be too low.

Replacement costs are also used as a basis for estimating operating costs. In the case of fuel and labor, the estimates derived as percentages of replacement cost have been double-checked with other approaches and appear to be relatively accurate. For insurance, the percentages of replacement value were modified to make some allowance for presumed present value. The total estimates may, nevertheless, be somewhat higher than they should be. It can be noted that insurance premiums, including those for liability, are considerably higher for fishing vessels than for commercial vessels. Also, during the 1980s there was considerable construction of large new vessels globally. It seems appropriate to use replacement value as a basis for maintenance and repair because the costs of spare parts and labor are at present prices. These possible errors indicate that the results of the calculations must be considered with caution.

On the basis of the estimates of costs and revenues, the operating costs of the global marine fisheries fleet, in 1989, were *$22 billion more than the gross revenues*. If the estimated capital costs are included, the deficit in that year amounted to $54 billion.

Aside from possible errors, the most likely explanation

for the gap between operating costs and gross revenues is the heavy subsidies that are provided to many of the world's fishing fleets, particularly for large-scale operations of many developed countries. Most notable are the subsidies that were provided by the former USSR in 1989 and previous years. On the basis of information in 1989, the operating costs of the former USSR fleet ranged from $10 to $13 billion while gross revenues may have been less than $5 billion. The subsidies are currently believed to be considerably lower due to the retrenchment of the fishing operations and the move to privatization.

Japan also provided significant support for its fleet. According to the Japan Fisheries Association (1991),

> the current credit balance extended to fisheries from both the commercial and government sectors is about $US19 billion . . . In order to support business entities in financial difficulties, the government financing system will assume their liabilities. The amount of liability taken over by the government has been substantial in recent years due to the severe economic status of the fisheries industry.

In addition, European countries provide large subsidies, not only for construction of vessels and fishing operations but also for the purchase of fishing rights in other countries. These purchases provide an inducement for the foreign vessels to maintain excessive effort in the coastal states' zones since the fishers themselves do not incur the costs of access and have no incentive to reduce their amount of effort.

Although overall government support for fishing operations may be declining, a significant deficit in the global fishing economy is still apparent. If the total support by the former USSR is removed, the estimate of the deficit would decline from $54 billion to $41 billion.

In addition to the deficit in operations, large amounts of economic rents are also being dissipated. The actual amounts of this form of waste are not known. A very rough indication can be derived by assuming that, in open-access fisheries, the potential rents may be on the order of 30% of gross revenues. This is based on an analysis of potential or actual rents in Australian fisheries (Campbell and Haynes 1990), which showed that rents ranged from 11% to 60% of gross revenues with a weighted average of 30%.

If this factor is applied to the current global marine gross revenues of $70 billion, the potential rents would be $21 billion per year. However, rehabilitation of depleted stocks could lead to higher global catch levels and increased total revenues. Previous estimates indicate that effective resource management could lead to an additional catch of 20 million mt (FAO 1993), but this is now thought to be unlikely. In addition, it is also believed unlikely that total marine catches will increase in the

future; if they do, the increase will come from fishing further down the food chain for species with low average unit values (see Garcia and Newton 1997). Although some of the rents are currently extracted by coastal states in the form of access fees, they are likely to be less than $1 billion. Thus, the total amount of economic waste in global fisheries might be on the order of $60 billion per year (about $41 billion in actual deficit and another $20 billion in dissipated rents). Moreover, this estimate makes no allowance for the costs associated with the implementation of satisfactory access controls.

In addition to subsidies, another basic reason for the economic waste is the condition of free and open access that marks most national as well as international fisheries. The behavioral pattern of exploiters of an open-access resource is well known. Surplus profits in a fishery appear as a new stock is discovered, as real prices rise, or as costs fall with innovations in technology and techniques. These surplus profits attract new investments, which result in excess fishing effort and in lower average yields and revenues. Although the benefits of controls over access to fish stocks are abundantly clear, the record of successful interventions is sparse.

Future Outlook

Without some significant improvements in fisheries management, the future situation is likely to worsen as increased demand encounters diminished opportunities for increased production. Present global production is about 100 million mt of fish from marine, inland, and aquaculture sources. Of this, about 70 million mt is used for direct human consumption and 30 million mt is reduced to fish meal, mostly as a feed for animals.

With regard to future demand, simply to maintain present levels of per capita consumption, the production of fish for food use will have to increase from 70 to 90 million mt by the year 2010, an increase of 28% (Table 3). It is expected that many countries will increase levels of per capita consumption in response to rising incomes so that total actual demand will be even larger. Estimates cannot be made satisfactorily because of the lack of information on income elasticities of demand.

The demand for non-food use is also expected to increase. The growth rate in the catch of fish for fish meal was about 2.5% per year from 1960 to 1990. If this is rate is extrapolated to the year 2010, it would indicate a demand of about 47 million mt for fish reduced to fish meal for animal feed and other purposes. In fact, this is likely to be greater because of the rapid growth in aquaculture production of shrimp, salmon, and other carnivorous species, which makes heavy use of fish meal as a feed ingredient. Total demand for fish in the year 2010 may thus approximate 138 million mt (Table 4).

TABLE 3.—Projected demand for fish for food, from all sources, based on present levels of per capita consumption (metric tons [mt]).

Country or region	Supplies, 1988/90 (10^3 mt)	Demand, 2010 (10^3 mt)	% change
Japan	9,346	9,949	6.5
Former USSR	8,333	9,491	13.8
Western Europe	8,270	8,586	3.8
Eastern Europe	1,206	1,288	6.8
North America	5,927	6,739	13.7
Other developed	753	1,024	36.0
Total developed	33,835	37,077	9.6
Africa	5,422	9,981	84.1
China	11,875	14,761	24.3
Other Asia	15,857	23,497	48.2
Latin America and Caribbean	3,748	5,274	40.7
Total developing	36,902	53,513	45.0
Grand total	70,737	90,590	28.1

TABLE 4.—Projected total demand (mt) for all fish, 2010.

End use	1988/90 (mt)	2010 (mt)
Food	70.7	90
Feed	28.2	47
Total	98.9	137

These figures are admittedly highly speculative. They are presented, however, to provide a rough indication of the nature of the problems. The most important one is that the calculations indicate a demand for a 39% increase in global production over the present level (without accounting for increased per capita consumption). There are no indications that such an increase is feasible. Estimates of total sustainable supply from capture fisheries are extremely difficult to make, but the leveling off of total catch in the past few years indicates the limit may already have been reached (see Garcia and Newton 1997).

Increased production from aquaculture is quite likely but may be constrained by several factors. Aquaculture production in 1990 was about 12 million mt (excluding seaweed). If capture production remains at the present level of less than 90 million mt, aquaculture production would have to increase about four times to meet the indicated demand.

About 45% of aquaculture production in 1990 was produced by China, mostly from various species of carp grown in conjunction with agriculture. Opportunities exist in other countries for increased production of herbivorous species. Excluding China, increases in Asia may be expected from improvements in culture techniques but are likely to face limitations in space and in supplies of clean water. In Africa and Latin America, culture of herbivorous species has barely taken place in spite of considerable investments. In Africa, investments in aquaculture development amounted to about $150 million between 1975 and 1987 while production from aquaculture declined over that period at a rate of 10% per year (Huismann 1988). There is little likelihood of a significant reversal in the next decade.

Aquaculture production of carnivorous species has increased rapidly in the past few years. Shrimp culture currently accounts for about 25% of total shrimp production from all sources, and salmon culture accounts for about 40% of total salmon supplies. Further increases can certainly be expected, but they face eventual constraints with regard to feed. Alternative sources of feed would have to be found. For example, in China, mussels are grown to use as feed for the culture of shrimp and finfish.

Mollusks (such as oysters, mussels, scallops, and clams) appear to offer opportunities for increased production through cultivation. As plankton feeders, they do not face the same feed constraints as carnivorous fish; and as marine creatures, the need for clean water and space is not so severe as it is for freshwater species (although the 95% decline in oyster production in the Chesapeake Bay since the peak years indicates that production will not be without its problems).

The significant shortfall in supply will lead to increases in real prices for most species of fish. With limited natural supplies and increasing demand, real prices are driven upward and continue to attract more fishing effort even though both average and total yields continue to decline (Figures 1 and 2). Real prices for most fish products are likely to continue to rise in the future, until they reach a point of consumer resistance.

The increase in real prices has two contrary effects. It increases the values that would be created by effective controls over fishing access; on the other hand, it rewards the open-access condition by making it feasible for fishers to continue to earn profits in a depleted fishery.

Impediments to Change

It is frequently assumed that effective fisheries management will be achieved if there is sufficient reliable information, if the participants understand and appreciate the need for management, and if appropriate and acceptable management measures are available. Thus, much of the research and literature focuses on increasing the information, educating the participants, and devising or improving management measures. Although each of these steps may be desirable, a different approach might be taken and used as a means for identifying the impediments to change and the problems of implementing effective management. This approach entails asking the question, "Why have management measures not been adopted more widely?"

A first step in answering this question is to examine some of the basic implicit and explicit assumptions that underlie management attempts:

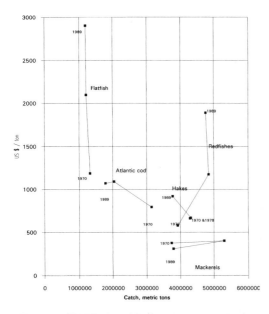

FIGURE 1.—Relationship between catch (mt) and deflated average unit values for selected species.

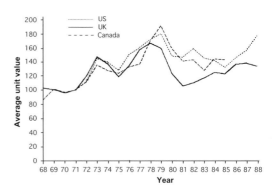

FIGURE 2.—Deflated average unit values for fish in selected countries (US$ 1970 = 100).

1. Administrators have the authority to control access, directly or indirectly. This assumes that there is sufficient legal or institutional authority to restrict entry and that the scope of the authority is sufficient to cover the use of the stocks satisfactorily.

2. Administrators have the mandate to make the necessary decisions. This assumption implies that administrators have the mandate to make decisions to determine who should have access and who should not.

3. The objectives for fisheries management can be clarified and conflicts among objectives can be resolved. This assumption is sometimes derived from the suggestion that management can be achieved if there are clearcut, conforming objectives.

4. Administrators are willing to make management decisions controlling access.

5. Administrators do not perceive the need for controlling access. It is sometimes assumed that educating administrators about management needs will lead them to make the right decisions.

6. Administrators need more information before adopting management measures. Fisheries are marked by a high degree of uncertainty regarding potential yields and stock interrelationships. It is often assumed that such uncertainty must be reduced to an acceptable level before management measures can be introduced.

7. Administrators need to know more about management techniques before they can adopt the most appropriate ones.

8. Participants in large-scale fisheries will be better

off under systems of controlled access. This assumption is derived from the simple logic of the conventional model of an open-access fishery, which demonstrates that considerable economic rents could be gained by limiting capital and effort. On the basis of this assumption, efforts are made to convince the fishers that they should accept access controls.

9. Participants in small-scale fisheries will be better off when fishing is managed. This assumption has the same derivation as the one above but is differentiated on the base of opportunity costs for labor.

10. Society will be better off if access is controlled. It is generally assumed that controlled access will result in major benefits to society by reducing misallocation of capital and labor and by producing economic rents, whether the rents accrue to the fishers or to society, and that there may be a reduction in the costs of management, including those of research and enforcement (depending on the system in effect).

The assumptions discussed above have varying degrees of validity in different situations and different countries. Some are widely held and some not. But, in general, they provide the basis for most of the attempts being made to adopt and implement fisheries management measures. However, since these attempts are not noteworthy for their success, reexamination seems desirable. This will be helpful in identifying and evaluating the conditions that need to be met to achieve effective management and to elicit general principles that have validity, usefulness, and acceptability.

With regard to identifying important conditions and factors, the various assumptions can be divided into four categories. These include assumptions about (1) institutions, (2) wealth distribution, (3) perceptions of problems, and (4) costs and benefits.

Institutions

The first assumption mentioned previously refers generally to national and international institutions and whether these facilitate sufficient authority to make the necessary management decisions. Within national economic zones, there are wide variations in decision-making authority. In some situations, authority is relatively unconstrained. For example, until recently, the State Trading Organization in the Maldives was able to purchase almost all fish produced by domestic fishers at about half the export price. Since almost all of the fish caught are exported, this form of tax led to the production of sizable economic rents. In Saudi Arabia, a concession awarded to a single company has allowed it to exercise full control over capital investment in the shrimp fisheries. In the Mar del Plata groundfish fishery in Argentina, the fishers are organized to control the market and restrict production in order to maintain high prices. Fisher's cooperatives in Japan have been granted exclusive rights to manage certain local fisheries. There are, however, some situations where authority is used to prevent controls over access.

At the other extreme, decision-making authority is diluted. In the USA, responsibility is divided between the states and the federal government. The members of the regional councils come from various and often disparate interest groups within and outside the industry. Their role in preparing management plans is overseen by the federal government. There are, of course, other reasons why management in the USA is far from perfect, but the institutional weakness is important.

The number of national governments that have adopted appropriate institutions is increasing and, as beneficial experience grows, it is likely that more and more states will take the necessary measures. Nevertheless, considerable effort is required to devise institutions that will have authority for management and that will fit into the particular situations of individual countries.

For shared, straddling, and high-seas stocks, there are few institutions, and those that exist have little authority for making decisions that would allow access controls, as indicated by the problems in the European Pond, the "donut hole" in the Bering Sea, and the northwest Atlantic.

An outstanding exception in the past was the Convention for the Preservation and Protection of Fur Seals, created in 1911. Under the agreement, the four parties (USA, Russia, Japan, and the United Kingdom on behalf of Canada) agreed to forgo pelagic sealing and harvest the stocks only on the breeding islands of the USA and Russia. In exchange for giving up their rights, Japan and Canada received rents in the form of a portion of the skins.[4] Today, the only international institution that comes close to this is the South Pacific Forum Fisheries Agency, which fosters cooperation among the member states through facilitating negotiations over access rights with foreigners, and which maintains a Regional Register of Foreign Fishing Vessels.

Clearly, in international waters there is difficulty in creating an institution with sufficient authority to make decisions on controlled access. This is due not only to the reluctance of states to relinquish portions of their sovereignty but also to the fact that, in some situations such as the northeast Atlantic, access agreements require negotiations over the distribution of job opportunities among the concerned countries. Negotiators who agree to the reduction of jobs for their fishers do not last long in their positions. This is a major impediment to the establishment of effective fisheries management.

The problems of misuse are increasing, along with the threat of extensions of national jurisdiction beyond 200 nautical miles. As these costs increase and as the benefits of multinational institutions become more apparent, new forms of international institutions may possibly emerge. Two examples illustrate the contrast between the status quo and the advent of institutional tools that may bring about more effective management: Burke and Christy (1990) suggested new approaches for dealing with the tuna in the Indian Ocean, which were largely ignored in the adoption of the Indian Ocean Tuna Commission. However, FAO initiated and, less than 1 year later, obtained international acceptance of an "Agreement to promote compliance with international conservation and management measures by fishing vessels on the high seas." This agreement provides an important tool in dealing with vessels flying flags of convenience.

Wealth Distribution

The second and third assumptions, dealing with the mandate to make decisions and the establishment of management objectives, are perhaps the most important and difficult since they deal with the distribution of wealth. Access controls, whether ITQs, license limits, or territorial use rights in fisheries (TURFs), create forms of property rights under which some users acquire the rights and others are excluded from free entry. This necessarily constitutes a distribution of wealth, whether it is between sets of present users or between present and future users. Such decisions are essentially political in nature and not generally within the mandate of fishery administrators.

---

[4]An interesting footnote to history is that the Japanese withdrew in 1941, not because they were on the verge of war, but because "both direct and indirect damage had been afflicted on the Japanese fishing industry by the increase in fur seals" (Whiteman 1965). The number of fur seals reportedly increased from 125,000 in 1911 to approximately 2.3 million in 1941. A new agreement was reached in 1957, with the provision that the determination of the total allowable yield would take into account the predation by fur seals on the productivity of other living marine resources.

In most cases, fishery administrators are limited to decisions that equalize the screams of outrage of the contending groups. Rather than control access, they resort to measures that presume to be non-distributional in effect, such as restrictions on gear, time, and season or area of fishing. In some cases, particularly those based on gear restrictions, the presumption of non-distributionality is inaccurate.

This has led to a plethora of controls, greatly complicating the life of the fishers, as well as making them ill disposed to administrators and additional controls, and creating problems of enforcement. Such measures, although they may be desirable in certain situations, do nothing to prevent the economic waste.

For appropriate decisions to be made, it is necessary to involve politicians. The political process can be invoked in various ways. The most frequent is that of crisis, which can occur when average catches of fish fall faster than increases in real price or when different user groups come into conflict over a common resource. The conflict can be between fishers using different gear for the same stock, between fishers from different areas or different countries using the same stock, or in "value-conflicted" fisheries where there is incompatibility among different uses or values (e.g., recreational vs. commercial, commercial fishers vs. animal preservationists).

These kinds of crises tend to lead to decisions affecting the distribution of wealth and sometimes result in access controls within the commercial fisheries. In the case of Indonesia, a

> sudden, growing invasion of 20 to 30 ton gt [sic] trawlers in the traditional grounds was soon felt as a serious unfair competition and threat to the socioeconomic balance among the masses of the fishermen, which led to disturbances like physical clashes between trawlers and traditional fishers, followed by arrests and demonstrations in several fishing villages. This situation attracted very much attention of the press and even the parliament, who were very sympathetic with the big masses, as was the public. (In response to this) the Government reached the political decision to ban the operations of trawlers (around Java and Sumatra) (Sardjono 1980).

In this case, the decision was not followed by direct or indirect controls limiting access in the fishery. In other cases, however, the crises have led to access controls (e.g., the salmon fisheries of Alaska and British Columbia and the northern prawn and southern bluefin tuna fisheries in Australia).

The major problems with crisis decisions are that there is little room for maneuverability by administrators and little time to prepare effective management measures. The resulting decisions are generally imperfect.

Invoking the political process prior to crisis would be desirable, although difficult. The constituencies in favor of effective management tend to be weak while those opposed are generally strong. However, a change in the strength of the interests of the different constituents is possible and is already occurring in some situations. This offers an opportunity for actions and studies that would help to further influence the changes. Some of the major present or potential constituents can be identified and described.

Fishers.—Fishers are clearly the most directly affected by access controls. Generally they tend to be in opposition, but their interests are not clear. Much depends upon their perceptions of the problems and their position in the fishery. Some speculations follow.

The "highliners" (the small proportion of fishers who have high skills and take the largest share of the catch) are able to manipulate the existing regime of regulations in their favor and tend to oppose any change that would diminish their flexibility or require them to operate differently. They are usually the first ones to leave a distressed fishery and develop a new one (if they have the opportunity). Since such freedom is threatened by access controls, they are likely to be in opposition. In cases where they may not have the opportunity to develop alternative fisheries, they may support access controls. They may also choose this option if they feel that access controls can be manipulated to the disadvantage of their competitors. The highliners also tend to be the most vocal fishers, and they exercise a relatively strong influence.

The fishers with less skill and less willingness to take risks may also have less understanding of the issues and less ability to vocalize their interests although they tend to greatly outnumber the highliners and thus may carry more weight in the political process. Part-time fishers are likely to be excluded by access controls and can be expected to be strongly in opposition. Crews on large vessels (and their labor unions) may also tend to oppose access controls because of the likelihood of reduced employment opportunities.

Fishers' positions are also complicated by differences in their views of the status quo. For the highliners, there is a tendency to view the status quo as that of an earlier period when they took larger shares and had fewer competitors. For the other fishers, they tend to believe that next year is the year when they will strike it rich. "For measured access programs like ITQs, where initial rights are granted according to some available suite of criteria, the conflict between those looking to preserve historical harvest levels (backward-looking) and those with high hopes (forward-looking) is a major stumbling block to reaching industry agreement" (J. Hastie, Alaska Fisheries Science Center, National Marine Fisheries Service, Seattle, Washington, pers. comm.).

In some situations, participants in large-scale fisheries have found it advantageous to adopt their own controls on access. In some cases, this emerges from an interest in controlling the market. One example is the groundfish fisheries of Mar del Plata, Argentina, where about 200 vessel owners collaborate in negotiating prices for their products (A. Alberto Gumy, FAO, Rome, pers. comm.). They then agree to limit total production at the level appropriate to the market and allocate shares among themselves. In other cases, the objective of collaboration is to reduce conflict. This approach is found among the shrimp trawler owners in Chilaw, Sri Lanka, where they collaborate to forestall government intervention in resolving conflict with small-scale operations. They have rigid controls on time and place of fishing and limits on the number of vessels. In the Philippines, tuna fishers have agreed among themselves on the placement of "payaos" (fish aggregation devices), and they exercise exclusive use rights within their areas. There are some situations where management, in the sense that economic rents are produced, is achieved illegally. In some countries, fishers must bribe local officials in order to fish. Bribery constitutes a form of tax which serves to restrain entry and reduce overcapitalization (see Christy 1978b).

Small-scale fisheries are generally community-oriented. There are many situations where the community supports controls against access by the large-scale vessels that intrude into their community waters. Within communities that participate in small-scale fisheries, access controls among the fishers may or may not be acceptable. In situations where the community has acquired, by tradition, a de facto TURF, the community generally controls access and limits the number of members of the community who can fish (although some of these systems are now being threatened by population growth within the community). However, for newly developed fisheries, different approaches may be taken.

In situations where no traditional TURF exists, the interests are less clear. Two contrary approaches, for example, were found in the state of Kerala in India (Kurien 1991). In one community, an artificial reef was erected by a group of fishers who then controlled access, limiting it to themselves. In another community, the investment was made by the community as a whole, on the basis of "whatever each one can give happily," and access was open to all members of the community.

Overall, if a group of fishers perceives that the benefits to them of having exclusive use rights are greater than the costs (political, legal, economic), they will accept, or even institute, access controls.

Shipbuilders.—The interests of shipbuilders are generally strongly in support of maintaining open access since this provides opportunities for more vessel construction. Their response to economic hardship in the fishing industry is to encourage the government to provide subsidies.

Consumers.—All consumers are facing rising real prices and diminishing supplies of fish products. Important exceptions are products that are cultivated, such as salmon and shrimp, and products that provide substitutes for luxury commodities, such as crab and lobster analogs produced through protein restructuring. These exceptions, however, provide only a small part of total consumption. Moreover, they also may experience real price increases as feed and the raw materials become scarce.

Improved management, to the extent that it allows rehabilitation of stocks, will alleviate (at least temporarily) some of the price pressures. Overall, consumers would thus tend to support stock conservation measures (though not necessarily access controls). However, the impetus for involvement in management decisions does not appear to be strong at present. The rise in prices has been gradual and consumers have been able to substitute unfamiliar, but satisfactory, fish species for hitherto preferred species.

Taxpayers and finance and planning agencies.—At present, in open-access situations, the losses in terms of dissipated economic rents are not tangible to any particular group of constituents. Although there is some evidence of misallocation of capital and labor (e.g., 2-d seasons for halibut), it apparently is not yet sufficient to raise effective outcry. There may even be a negative effect. Maryland's requirement that oysters can only be dredged by sail boats produces graphic Sunday newspaper supplements of the "romantic" life of the oystermen on the beautiful skipjacks and bugeyes. Economic rationality is not a goal that wins many supporters, other than academics (who wield relatively little influence).

The large potential revenues, however, should be of interest to taxpayers and government agencies concerned with economic revenues and national economic welfare (finance ministries and planning agencies). Recent public attention in the USA to the negligible fees paid for mineral lands and grazing fees may be useful in strengthening the interests of taxpayers and the government.

Environmentalists.—Perhaps the strongest constituents for change are environmentalists and animal preservationists. Although these two interest groups are often lumped together, it is useful to make a distinction between them. Genuine concerns about environmental degradation have attracted increasing public attention, some of which is now focused on overfishing and depletion of fish stocks as well as improving the quality of the coastal zone. Some environmental organizations have played a useful role (in the USA) in influencing the political process in favor of access controls although they tend to oppose ITQs apparently because they believe that these will facilitate domination of the industry by "big business."

The interests of animal preservationists are different in that their concern about access control is that of achieving zero mortality of marine mammals resulting from incidental take. This would be accomplished by precluding commercial fishers from accessing fish stocks, either directly or by controls over gear and fishing techniques. The political constituency of the animal preservationists is largely restricted to the USA, even though it has widespread effects. The opposition of the animal preservationists to ITQs may result from their awareness that the attribution of values to the resources would create a market in which they would have to compete monetarily in order to achieve their goals. At present, it costs them nothing to express their demands while the fishers and society in general bear the costs when preservationists' demands are met.

Although the political constituency of the environmentalists is strongest in the USA and some other developed countries, it is increasing its power in some developing countries. Marine reserves, for example, have recently been adopted in a number of southeast Asian countries.

Perceptions of Problems

Assumptions 4, 5, 6 and 7 (p. 32) are generally related to the question of awareness of the problems of open-access fisheries. These assumptions have driven a great deal of research in the past on the conviction that the provision of information and education of the administrators will lead to their taking the appropriate actions.

A first question to ask is whether administrators are willing to make decisions to control access. Historically, this has not been universally true. The thrust for economic development during the 1950s and 1960s induced many countries to invest heavily in large-scale fishing operations. This was generally supported by development agencies such as the World Bank and regional development banks, which provided investment funds to developing countries for construction of vessels and ports.

In this context, fishery administrators have been rewarded on the basis of the amount of funds they have brought into the country and on growth in capital investment. Staff of development agencies have also been rewarded by the amount of loans they have generated. The attitude that there are large potential resources that can be tapped by increased fishing effort persists today in some countries, and is maintained by agencies created in the past to serve as conduits for development funds. This reward system is not conducive to attempts to control access.

In some cases, the desirability of controlling access is not readily apparent. There may be opportunities for increasing total catches owing to the exclusion of foreign fishers or because of the presence of newly marketable stocks. Invariably, however, there will be other stocks within the country's exclusive economic zone being fished at or beyond maximum sustainable yield. In these cases, rising real prices, due to shortages of supply, may produce high revenues and obscure the fact of declining yields.

It is often argued that deficiencies in information about the status of the stocks preclude decisions on adopting access control measures. This argument may be advanced by administrators unwilling to make the necessary decisions, by fishers unwilling to accept access controls, and by scientists interested in maintaining or increasing their research funds. There is no question but that fisheries science is difficult and that the results are fraught with uncertainty. A big part of the problem is the nature of the beast, which swims in an opaque, three dimensional medium and is subject to extraneous influences and complex interrelationships with its predators, prey, and competitors. These difficulties are compounded by imperfect information and actual disinformation (e.g., when fishers misreport their catches to avoid regulations).

Nevertheless, there is often sufficient biological or economic information, or both, to know that a fishery is in bad shape and that management is essential to prevent it from becoming worse. An example with regard to biological information is the North Sea and the Baltic, which have been investigated for several decades by the International Council for the Exploration of the Sea (ICES). Information on the status of the stocks is about as good as it gets. And yet, in spite of abundant evidence of overfishing, that condition persists. For example, in 1981, ICES recommended that Baltic cod catch be limited to 197,000 mt. However, the International Baltic Sea Fishery Commission could only agree to 272,000 mt and the actual catch was 380,000 mt.

Biological information may be marked by a high degree of uncertainty, but economic information is usually more clear. Most economic analyses of fisheries have amply demonstrated malaise in the particular fishery being studied. They have also frequently indicated the necessary steps to reduce that malaise. For example, a thorough economic analysis was done of the U.S. Pacific halibut fishery by Crutchfield and Zellner (1963). Although access controls are now in the process of being adopted, information on the need for such controls was available 30 years ago, and nothing was done.

Economic and biological information will never be available to determine precisely the point of optimum yield, no matter how that is defined. But in most cases, sufficient information is available to make decisions that will lead to improved benefits to society.

Costs and Benefits

The last three assumptions (p. 32) mentioned relate to the costs and benefits of adopting access controls, the overall assumption being that the fishers and society will be better off under systems of access controls than they are under open access. These assumptions warrant some examination to determine whether the costs of formulating, adopting, and implementing access controls are greater than the benefits. The answers depend upon both the forms of controls adopted and the particular fisheries.

Most systems of access controls create some form of property right in the fishery, whether it is in the form of a license (as in a license limit scheme) or in the form of a share in the allowable catch (as in an ITQ scheme). Automatically, the creation of a property right establishes a value in that right, the amount of which will vary with the degree of exclusion and the economy of the fishery. If the right is transferable, the value is expressed in the sale price. Some countries (e.g., Japan, Namibia, Malaysia) have attempted to make the privileges nontransferable. Such attempts generally do not work and serve to obscure sale prices by driving the negotiations under the table.

The value may accrue entirely to the individual fishers or may be shared with society through the imposition of taxes or user fees. That portion that accrues to the fishers is a "windfall gain" that goes to those who obtained the privileges at the initial allocation. The second-generation fishers who buy the privileges incur the costs of amortizing the purchase price as well as the ordinary costs of fishing.

Whether the net incomes of the second-generation fishers, on the average, are greater or lesser than the incomes of those who were in the fishery prior to the access control depends upon a number of factors. Theoretically, the price that second-generation fishers pay for the privilege would result in income levels sufficient to cover their opportunity costs. If all fishers were the same, then their incomes would be the same in both the controlled and uncontrolled fishery. But it may well be that a fishery subject to effective access controls attracts fishers with higher opportunity costs than an uncontrolled fishery. If the fishery is well managed, it will, presumably, be more stable to the extent that the factor of fishing mortality is controlled. It could be subject to less risk, allowing the fishers to plan their effort and investments more effectively. There could also be a reduction in the complex regulations that mark open-access fisheries and a more uniform and orderly distribution of fishing effort

through the season. However, there likely will be imperfections in the management system. For example, ITQs in fisheries with intraseasonally declining yields may continue to attract excessive effort, or ITQs where landings cannot be monitored satisfactorily may be unenforceable (Christy 1973). The initial allocation may allow excessive effort, either by licensing too many vessels, allowing excessive quotas, or exempting vessels of certain size.

Since such problems may impede participation by fishers with higher opportunity costs or fishers who are more efficient, it is not clear that the controls would result in higher income levels to the fishers (other than those in the first generation receiving windfall gains).

For operators in small-scale fisheries, particularly those in developing countries, the question depends upon whether access controls are already in effect and how they operate. As noted previously, there are many situations where informal, traditional TURFs are in existence (Christy 1993a). These TURFs essentially constitute the exercise of management authority at the local level, and they have been adopted for a variety of reasons. In some cases, TURFs are used to achieve enhancement of the resources through planting of seed stock or artificial reefs. In others, they are used to produce economic rents by the use of fees or auction or to control markets. Rents may also be captured through other means such as propitiatory payments to gods, high interest rates to money lenders, bribery or pay-offs to licensing or enforcement agents (Christy 1982). And some TURFs have been adopted to ensure community stability.

These systems tend to be fragile and are being threatened by intrusion of large-scale operations from outside, by pressures of non-members to enter, and by population growth within the communities. Many TURFs have already been wiped out by national development efforts (often supported by international aid), which failed to recognize the existence of the systems and their potential for management.

Where the TURFs do not exist, or have been destroyed, it would generally be difficult for national or state governments to put access controls into effect other than by allowing the local communities to create or reestablish TURFs. Small-scale fisheries generally take place in isolated communities and frequently operate from beaches. Monitoring and enforcing ITQs or license limits would not be feasible in these fisheries.

The creation or reestablishment of TURFs may or may not improve the welfare of the fishers in small-scale operations. Where the community is suffering from the intrusion of large-scale operations, the exclusion of these operations would benefit the community as a whole. But the provision of property rights to a community is a difficult task and could result in the acquisition of access rights by one or a few individuals within the community. Extraordinary care would be required to prevent the creation of "sea lords" who might worsen the lot of the most impoverished.

An additional difficulty in small-scale fisheries is the general absence of alternative employment opportunities. Indeed, in some developing countries, fisheries are considered to be the "employer of last resort," and evidence indicates that labor displaced from agriculture and other activities enters those fisheries that have no community barriers. Limitations on the amount of fishing effort in these situations may not be appropriate if they stimulate significant unrest.

With regard to society, access controls may generally be beneficial, but this again depends upon the kinds of controls, their effectiveness, and the particular situation. In the short run, in some situations there may be very large transaction costs in adopting and implementing the management measures. Considerable time and effort may be spent on research (to produce information to satisfy opponents) and on negotiation among various competing groups. In order to deal with objections to the proposals, the measures may be compromised to the point where they produce small benefits and incur high continuing costs of research and enforcement. In these situations, society may possibly be better off by benignly neglecting the industry and letting the open-access condition flourish, short of species extinction. If this policy were to be adopted, regulations and controls could be significantly reduced if not entirely abolished. With removal of regulations, there would be no necessity to conduct expensive research or enforcement programs, leading to large savings to the country (see Christy 1978a).

In the long run, closed-access systems may produce several benefits to society including improved allocation of capital and labor; the production of economic rents, previously dissipated, which may be shared with society; reduced costs to the taxpayer in the management of the fisheries; and stable supplies of high-quality products for consumers. Whether these benefits materialize, and the degree to which they do so, depends on the effectiveness of the system and the costs of maintaining it.

Although it is quite likely that there will be positive net benefits, examining the question will help in determining the most effective measures that can be adopted and implemented.

Summary and Conclusions

The condition of open access that is prevalent in marine fisheries is the source of large amounts of economic waste, as well as a cause of depletion and conflict. It has long been known that closing off access to the stocks is necessary to prevent the waste. But there are relatively few examples of success in adopting and implementing such management measures.

There are several reasons why success is so elusive. The most important one is that decisions to close access affect the distribution of wealth and are, therefore, political rather than administrative decisions. Institutional arrangements also appear to be unsatisfactory in many instances, particularly for stocks that are shared or that straddle economic zones and the high seas. Deficiencies in the availability of information do not appear to be a major impediment to decision making, at least for the adoption of measures that will improve the contributions of fisheries to national economies. Overall, it is likely that the adoption of closed-access systems will produce net benefits to fishers and society, if the systems are well designed.

The prevailing approach to fisheries management is to avoid decisions until a crisis emerges. This has significant deficiencies. When the crisis becomes apparent, the problems have already reached a stage of intractability and constrain the range of possible solutions. This generally results in imperfect, if not unsatisfactory, measures.

It is difficult to act before the appearance of crisis, largely because the problems have not entered the consciousness of the politicians. There are, however, several approaches that might be followed by those who seek to promote economic rationality in fisheries. One is to identify, mobilize, and strengthen the constituents who favor closed-access systems. In this regard, more economic analyses of the costs of open access and the benefits of controls might be undertaken and provided to taxpayer groups, planning agencies, and finance ministers. Analyses of biological waste resulting from excess fishing effort could be given increased publicity.

Another approach is to create institutions that will facilitate the necessary decisions and the implementation of the measures. This may require revision of legislation in cases where present legislation impedes the adoption of access controls. The devolution of management authority to local levels of government offers opportunities for local controls in small-scale fisheries. Proposals for new forms of institutional arrangements might be made, including multilateral and international institutions as well as national ones. For example, fishery management districts might be formed with representation from various interest groups.

There are certain conditions that may encourage groups of fishers to create their own access controls, particularly, though not exclusively, in the case of TURFs. Some of these conditions can be influenced by governments (Christy 1993b).

Past approaches have not been noticeably successful in improving fisheries management, and generally for very good reasons. For improvements to be made, we must acknowledge the major impediments and adopt new approaches that will remove or alleviate the constraints.

Acknowledgments

I am indebted to J. Hastie and an anonymous reviewer for their insights and comments on this paper.

References

Burke, W. T., and F. T. Christy, Jr. 1990. Options for the management of tuna fisheries in the Indian Ocean. Food and Agriculture Organization of the United Nations, Fisheries Technical Paper 315, Rome.

Campbell, D., and J. Haynes. 1990. Resource rent in fisheries. Australian Bureau of Agricultural and Resource Economics, Canberra, Australia.

Christy, F. T. 1973. Fisherman quotas: a tentative suggestion for domestic management. Law of the Sea Institute, Occasional Paper 19, University of Rhode Island, Kingston.

Christy, F. T. 1978a. The costs of uncontrolled access in fisheries. *In* R. B. Rettig and J. J. C. Ginter, editors. Limited entry as a fishery management tool. Proceedings of a national conference to consider limited entry as a tool in fishery management, 17-19 July 1978. University of Washington Press, Seattle.

Christy, F. T. 1978b. Fishery problems in Southeast Asia. *In* D. M. Johnston, editor, Regionalization of the law of the sea. Proceedings of the Law of the Sea Institute eleventh annual conference. Ballinger Publishing Co., Cambridge, Massachusetts.

Christy, F. T. 1982. Territorial use rights in marine fisheries: definitions and conditions. Food and Agriculture Organization of the United Nations, Fisheries Technical Paper 227, Rome.

Christy, F. T. 1993a. Enhancement, efficiency and equity TURFs: experiences in management. *In* Papers presented at the FAO/Japan expert consultation on the development of community-based coastal fishery management systems for Asia and the Pacific, Kobe, Japan, 8-12 June 1992. Food and Agriculture Organization of the United Nations, Fisheries Report 474, Supplement, Volume 1, Rome.

Christy, F. T. 1993b. Territorial use rights in fisheries: suggestions for governmental measures. *In* Papers presented at the FAO/Japan expert consultation on the development of community-based coastal fishery management systems for Asia and the Pacific, Kobe, Japan, 8-12 June 1992. Food and Agriculture Organization of the United Nations, Fisheries Report 474, Supplement, Volume 1, Rome.

Crutchfield, J., and A. Zellner. 1963. Economic aspects of the Pacific halibut fishery. Fishery Industrial Research Vol. I(1). U.S. Government Printing Office, Washington, D.C.

FAO (Food and Agriculture Organization). 1993. Marine fisheries and the law of the sea: a decade of change. Special chapter (revised), The state of food and agriculture 1992. FAO Fisheries Circular 853, Rome.

Garcia, S. M., and C. Newton. 1997. Current situation, trends and prospects in world capture fisheries. Pages 3-27 *in* E. K. Pikitch, D. D. Huppert, and M. P. Sissenwine, editors. Global trends: fisheries management. American Fisheries Society Symposium 20, Bethesda, Maryland.

Huismann, E. A. 1988. Note on research priorities in the field of aquaculture. Technical note for the Technical Advisory Committee, Consultative Group on International Agricultural Research, World Bank, Washington, D.C.

Japan Fisheries Association. 1991. Fisheries of Japan, Tokyo.

Kurien, J. 1991. Collective action and common property resources rejuvenation: The case of people's artificial reefs in Kerala State. *In* Indo-Pacific Fisheries Commission, Papers presented at the symposium on artificial reefs and fish aggregation devices as tools for the management and enhancement of marine fishery resources, Colombo, Sri Lanka, 14-17 May 1990. RAPA (Regional Office for Asia and the Pacific) Report 1991/11, Food and Agriculture Organization of the United Nations, Bangkok.

Sardjono, I. 1980. Trawlers banned in Indonesia. *In* ICLARM (International Centre for Living Aquatic Resources Management) Newsletter Vol. 3(4).

Whiteman, M. 1965. Digest of international law. United States Department of State, Washington, D.C.

Small-Scale Fisheries in the Tropics: Marginality, Marginalization, and Some Implications for Fisheries Management

DANIEL PAULY

Abstract.—A brief analysis of tropical small-scale fisheries is presented, structured by two areas of emphasis: marginalization—actual and perceived—and Malthusian overfishing, a concept I proposed previously. It is suggested that marginality is, in part at least, a construction resulting from faulty mental maps, which leads to even more marginalization for small-scale fisher communities. Marginalization is the ultimate cause for Malthusian overfishing, whose identification and prevention, or at least mitigation, should be foremost on the agenda of fisheries scientists and policy makers. Some of the implications of these ideas for multidisciplinary research on coastal fisheries systems are outlined.

In early 1994, a number of articles in major international magazines appeared, which—for a while at least—shifted fisheries from their marginal position in the public discourse to the center of attention. There was no objective reason for this outburst of articles: the methodical grinding down of successive fisheries resources was no worse in 1993–94 than in preceding years (see Garcia and Newton 1997). Yet something of this sort had to happen at some stage, just as it happened in the 1980s for tropical rain forests: the unconstrained and massive destruction of potentially renewable resources by industrial fleets could not go on much longer before the press noticed.

The press, being what it is, could be expected to mix insights with drivel. Thus, for example, the *Economist*, in an anonymous article smartly titled "The catch about fish" (March 1994) correctly identified subsidy-driven overcapitalization as the major culprit for the state of fisheries in developed countries (Garcia and Newton 1997), but also stated that "increasingly, boats will head for third-world waters, where the decline in stocks has not yet started."

This paper is not the place to demonstrate that boats from Europe, North America, and Northeast Asia have been exploiting, for decades, the fisheries resources of developing countries, and that their stocks have long started to decline; these topics have been well covered in recent literature. Neither do I deal explicitly with the development of large industrial fisheries in third-world countries, this topic also having received much attention (see Panayotou and Jetanavanich 1987). Rather, I concentrate on tropical small-scale fisheries. In spite of the important, indeed crucial, role of small-scale fisheries in most developing countries and in many developed countries (Figure 1), they continue to be perceived as marginal to the mainstream of fisheries science as illustrated, for example, by their coverage—or lack thereof—in

major texts. This perception may be strengthened by the emphasis in the next section on features of tropical small-scale fisheries not often considered by fisheries scientists, but which, I believe, explain the dynamics of many of these fisheries better than standard bioeconomic accounts. These features are as follows:

- the interactions between the factual and perceived marginality of these fisheries, and
- their tendency to drift toward what has been called Malthusian overfishing (Pauly 1988, 1990, 1994; Pauly et al. 1989; McManus et al. 1992).

Emphasis is given here to the interrelationships between these features and what they may imply for fisheries management and research in the next decades.

The Marginality of Tropical Small-Scale Fisheries

One of the characteristic features of tropical small-scale fisheries is their marginality, that is, their geographic, socioeconomic, and, ultimately, political remoteness from decision makers in major population centers. This feature is strengthened by mental maps, that is, the mental constructs through which we interrelate facts, experiences, and values (see Hampden-Turner 1982 and Peters 1983 for examples from psychology and cartography, respectively) that fail to account for management implications.

Physical remoteness, wherein "the landings may be dispersed over a great length of shoreline" (Munro 1980), is not only a matter of geographic coordinates. Rather, it is exacerbated by the lack of infrastructure (roads, markets, ice supply, communications) that characterizes most developing countries and by the nature of the gears commonly used for small-scale fishing, which are either fixed (e.g., weirs or traps) or applied from crafts with a small operating radius (Stauch 1966; Smith 1979; Horemans

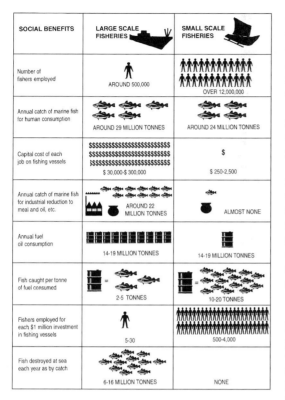

SOCIAL BENEFITS	LARGE SCALE FISHERIES	SMALL SCALE FISHERIES
Number of fishers employed	AROUND 500,000	OVER 12,000,000
Annual catch of marine fish for human consumption	AROUND 29 MILLION TONNES	AROUND 24 MILLION TONNES
Capital cost of each job on fishing vessels	$ 30,000-$ 300,000	$ 250-2,500
Annual catch of marine fish for industrial reduction to meal and oil, etc.	AROUND 22 MILLION TONNES	ALMOST NONE
Annual fuel oil consumption	14-19 MILLION TONNES	14-19 MILLION TONNES
Fish caught per tonne of fuel consumed	2-5 TONNES	10-20 TONNES
Fishers employed for each $1 million investment in fishing vessels	5-30	500-4,000
Fish destroyed at sea each year as by catch	6-16 MILLION TONNES	NONE

FIGURE 1.—How large-scale (industrial) and small-scale (artisanal) fisheries compare globally in terms of catches, ecological impacts and social benefits (from Thomson and Food and Agriculture Organization of the United Nations 1988 [NAGA, the International Centre for Living Aquatic Resources Management Quarterly 11(3):17 with permission]; values are for US$). The balance of social benefits can be expected to tilt even more toward small-scale fisheries when only the tropics are considered.

1993). The latter constraint may lead to localized (or seasonal) resource depletion, leading to influxes into alternative, land-based jobs (Munro 1980), when available, or to alongshore migrations of increasing amplitude (see contributions in Haakonsen and Diaw 1991). The response of government agencies, nongovernment organizations (NGOs), and multilateral or bilateral development agencies to this aspect of marginality has generally been to improve existing infrastructure or create new infrastructure, and especially to implement motorization schemes designed to increase the operational radius of small-scale vessels (see Smith 1979, Pollnac 1981, or Neal 1982 for early critiques of this approach, which largely ignored resource access issues).

Physical remoteness also causes problems in collecting catch and landing statistics (Munro 1980; Vakily 1992), severely hampering management schemes that require

real-time data, that is, data based on transferable or nontransferable quotas. Socioeconomic remoteness from the mainstream of society is related in part to the low incomes of small-scale fishers in most developing countries—even in relative terms—and to the fact that they often belong to ethnic groups (tribes) or social classes (or castes) of low status. This is often compounded by illiteracy or limited formal education (Bailey 1982; Lunianga 1989). This form of remoteness can thus occur in small-scale fisheries immediately adjacent to major cities.

Perceptions of low status—definitely the products of mental maps—have particularly pernicious effects. They often mask—at least to managers and policy makers, usually persons with a high level of formal education— the informal biological and ecological knowledge possessed by successive generations of small-scale fishers (Ruddle and Chesterfield 1977) to catch fish, which also serves as a basis for traditional, community-based fisheries management (Johannes 1981; Ruddle 1988, 1989a, 1989b; Okera 1994).

The geographic and socioeconomic marginality alluded to previously leads inexorably to lack of political power (whether the country has elected officials or not), which itself increases marginality: marginalization becomes systemic. Protest, when it occurs, may take violent form, for example, when industrial trawlers encroach into inshore, traditional fishing grounds (Sarjono 1980).

Further, fisheries, even when industrial or enormously important to food security and foreign exchange earnings, do not usually qualify for a full ministry (Peru is one of a few exceptions). They are usually administered by a department of fisheries (DoF) that is part of a ministry of agriculture, which tends to lack political clout. Indeed, investment decisions directly or indirectly affecting fisheries, such as port development or major fleet expansions funded by international development banks, are usually made through planning or finance ministries, without reference to stock assessment work that might have been done by DoF scientists and without accounting for the ecological costs of such development (see Meltzoff and LiPuma 1986 for a case study).

These various aspects of the marginalization of small-scale tropical fisheries are closely matched by the marginality of the science and scientists studying them. Within developing countries, law, medicine, and even agriculture are far more prestigious disciplines for the sons of the elite to study. The lower status of fisheries science may explain, for some countries at least, the relatively high number of female fisheries scientists (Dizon 1995). Researchers from developed countries who work on tropical small-scale fisheries are frequently resource economists, anthropologists, rural sociologists, or NGO activists, but less commonly fisheries scientists, as evidenced by the dearth of stock assessments in the literature on tropical small-scale fisheries (Roedel and Saila 1980; Platteau 1989a; Pollnac and Morrissey 1989; Agüero 1992).

Common Property, Limits to Entry, and New Entrants

Recent contributions (Aguilera-Klink 1994) have succeeded in overcoming the confusion caused by the application of schemes to tropical small-scale fisheries in which the term common property was thought to necessarily imply open access, and thus resource destruction, as might be assumed after reading Hardin (1968). Indeed, barriers to open access—some subtle, some rather direct—now appear to have been the key characteristics of traditional small-scale fisheries exploiting commons (see contributions in Ruddle and Johannes 1985), as is the case for pastoralists (Behnke 1994). However, this realization may have come too late. Colonial authorities and various development projects have eroded these open-access traditional systems, leaving sociologists and anthropologists only residues to describe, weakened systems that are unable to effectively limit access to commonly owned resources.

This weakening of access limitation benefits particularly the operators of industrial vessels (trawlers, purse seiners, etc.), which can and do force their way onto traditional small-scale fishing grounds. However, this problem is easy to conceive and straightforward to control once the political decision to do so has been made. It may be hard to make, however, given that decision makers, or their political allies, often own stakes in such vessels (Platteau 1989b). Here again, the Indonesian trawling ban of 1980 may serve as an example (Sarjono 1980).

More insidious are developments occurring within the small-scale sector itself, which are more difficult to notice and to conceive as problematic—especially when they occur in response to real or perceived competition from the industrial sector, as in West Africa, Sénégal (Chauveau and Samba 1989), or Ghana (Acquay 1992). Thus, to maintain their catches against pressure from trawlers operating inshore, small-scale fishers might be provided by international aid agencies with more effective gear (e.g., synthetic monofilament gillnets) or subsidized motors, or otherwise enabled to expand their radius of operation, resulting in a massive increase of effective effort, not noticed because of the simultaneous increases of industrial fleets.

What I believe is the most worrisome development within the small-scale fisheries of tropical developing countries in Asia, Africa, and South America is the entry of nontraditional fishers into these fisheries such as Peruvian highlanders, members of traditionally pastoralist groups in Sénégal, or landless rice farmers in the Philippines. In all cases, these people enter fisheries because they have been forced out of their traditional occupations, because there is excessive pressure for land, or because lack of access to grazing range has marginalized livestock production in inland areas. Fisheries have become an occupation of last resort (Neal 1982). The new entrants have been able to become fishers because coastal fisheries resources are vulnerable to simple gear or even to gleaning without gear, and because whatever access limit may have existed was not strong enough to prevent them from fishing.

Recalling Some Basic Principles of Fisheries Science

There are different ways of managing fisheries systems: the most elaborate are probably those that evolved in the Sahel in Africa to regulate access to floodplain resources (Fay 1989a, 1989b) and in the South Pacific, where tradition-based rules still mostly regulate access to nearshore resources, without explicit knowledge of their status (Johannes 1981; Hviding 1991; Ruddle et al. 1992). In developed countries, however, a different tradition evolved, which looked first at the biological status of the fish stocks, then at the fisheries depending on these stocks (Went 1972; Smith 1994). This is well illustrated by the historical sequence of scientific concepts used to define overfishing, viz:

1. growth overfishing, the form of overfishing that was first to be identified and theoretically resolved (Baranov 1918; Hulme et al. 1947; Beverton and Holt 1957; Figure 2A);

2. recruitment overfishing, the second form of overfishing recognized by fisheries scientists, following the seminal work of Ricker (1954; Figure 2B);

3. biological overfishing, the combination of growth and recruitment overfishing leading to catch decline on the right, descending side of surplus-production models (Schaefer 1954, 1957; Fox 1970; Ricker 1975; Figure 2C) and related to ecological overfishing in multispecies fisheries (Pope 1979; Pauly 1979a, 1994);

4. economic overfishing, initially defined in terms of economic theory by Gordon (1953), then combined by various authors with the surplus-production models in (3) to yield the Gordon-Schaefer model (e.g., Anderson 1977; Figure 2D);

These forms of overfishing are well-described in textbooks, and the suggested remedies traditionally involve a mix of management measures aimed at reducing effective fishing effort (such as mesh size regulations, closed sea-

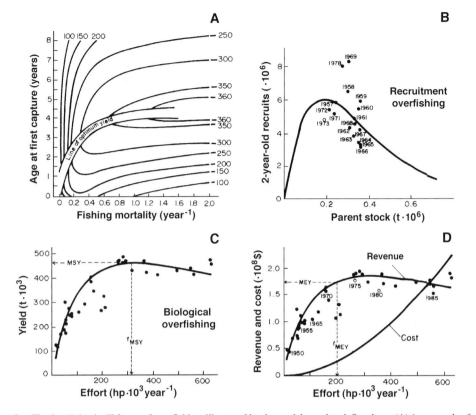

FIGURE 2.—The four "classical" forms of overfishing, illustrated by the models used to define them: (A) An example of a yield per recruit (in g year^{-1}) isopleth diagram for red snapper in the South China Sea, defining growth overfishing (from Pauly 1979b). This model is used for mesh size and related regulation. (B) Example of a stock-recruitment curve for southern bluefin tuna (*Thunnus maccoyi*), defining recruitment overfishing (from Murphy 1982). This model is now used to identify replacement spawning stock levels (Goodyear 1989; Mathews 1991; Mace and Sissenwine 1993). (C) Surplus production model, defining biological overfishing and related parameters (MSY, f_{MSY}) of the small pelagic resources of the Philippines (from Trinidad-Cruz et al. 1993). This model is used for effort regulation. (D) Simple bioeconomic model of a fishery in model C, defining economic overfishing and associated parameters (MEY, f_{MEY}) (from Trinidad-Cruz et al. 1993). Each model implies a certain research program, including field sampling of raw data, collation of secondary data, as well as certain "levers" to implement suggested action.

sons or limits on gear sizes or on craft designs, with individual transferable quotas (ITQs) recently added to the panoply; see Anderson 1997, Hannesson 1997). These measures assume that the fishers concerned are actually in a social and financial position to either implement or comply with those measures. In developed countries, they can, because the textbooks are written in and for such countries, in which fishing is done by corporations (often subsidized by government) or by independent (if small) entrepreneurs who can generate enough political pressure to also obtain governmental subsidies or to take shore-based jobs if all else fails. The situations of various aboriginal groups within developed countries and of small-scale fishers in the Atlantic Provinces of Canada (Ommer 1991, 1995) provide exceptions to this, and resemble more the developing country situation presented in the following text.

Malthusian Overfishing Defined

Small-scale fishers in tropical developing countries are usually poor and lack alternative employment opportunities; that is, once they start fishing, they are forced to continue, even if the resource declines precipitously. The numbers of small-scale fishers tends to increase, both because of internal recruitment and through destitute new entrants. Malthusian overfishing is here defined as what happens when these new fishers, who lack the land-based livelihood of traditional fishers (e.g., a small plot of land or seasonal work on nearby farms or plantations), are faced with declining catches and induce wholesale resource destruction in order to meet their immediate needs.

Overfishing may involve, in order of seriousness and generally in temporal sequence, (1) use of fishing techniques, gears or mesh sizes not sanctioned by government;

(2) use of gears not sanctioned within the fisher communities or catching of fish "reserved" for a certain segment of the community; (3) use of gears that destroy the resource base; and (4) use of destructive techniques such as dynamite and fish poisons that endanger the fishers themselves. (Note that this parallels slash-and-burn agriculture in upland areas, which also leads to environmental degradation and which is also exacerbated by immigration from the lowlands). This sequence, generally misunderstood by administrators and fisheries scientists as based on the ignorance or shortsightedness of fishers, reflects attempts to maintain incomes in the face of declining catches.

The reason I chose the adjective "Malthusian" to characterize this process is not because I wanted to join the chorus lamenting the impacts of rural population growth on natural production systems—these impacts are now obvious (Southgate and Basterrechea 1992; Homer-Dixon et al. 1993). Neither was it because I believe that one should put "population control front and center among the possible ways to confront the problem of overfishing," as suggested by Sunderlin (1994) in an otherwise thoughtful discussion of what he calls "the structural antecedents of poverty and high fertility." Rather, I wanted to emphasize, through this choice of words, what I believe is the key to Malthus' writings—his contention that production (of food) cannot *in the long run* keep up with an ever increasing demand (Malthus 1798).

There are still many people who believe that, globally, terrestrial food production will continue to increase as it has done since 1798 when Malthus published his major essay, despite well-documented, widespread destruction of agricultural production systems as the result of such problems as erosion and salinization (Lightfoot 1990; Southgate and Basterrechea 1992; Mathews 1994; Harris 1996).

Even optimists will have to agree, however, that the biological production of aquatic ecosystems must have an upper limit, and that fish catches will, over time, remain at best constant once a fishery is "developed" (see Pauly 1990). In such situations, catches tend to fluctuate and then gradually decline because of excessive fishing effort (Figure 3) because of the reduction of biodiversity induced by fishing itself and because of impacts from adjacent sectors such as logging-induced siltation of highly productive coral reefs (Hodgson and Dixon 1992). Thus, for capture fisheries at least, Malthus' contention applies: once the "boom" is over, fisheries production will stagnate at best, and certainly not accommodate an ever-growing demand. Further, this ever-growing demand need not be due to local population growth: globally increasing incomes, leading to increased fish consumption and prices, may, through remote markets, affect otherwise isolated fisher communities.

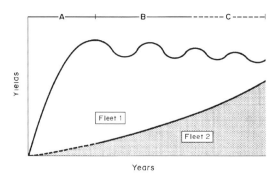

FIGURE 3.—Schematic representation of the evolution of a typical fishery, emphasizing that catches cease to (A) increase after a short development phase, (B) increasingly fluctuate, then (C) head toward a collapse fueled by the competition between groups of fishers or gears or both. Phase C is when fisheries scientists are requested to help while the private fishing costs of Fleet 1 or 2 or both are reduced through subsidies from public funds.

Causal Pattern and Diagnosis of Malthusian Overfishing

Given the prior description and the elements of Figure 4, the following causal pattern is hypothesized to occur in a fishery suffering from Malthusian overfishing:

- Stagnating overall catches and an increasing number of fishers lead to
- decreasing catch per fisher (this may be masked, at the income level, at least for a while by increased value of the catch), which with the first element, jointly lead to
- evidence of (at least localized) biological and ecological overfishing and gradually to
- classical economic overfishing (when the value of the catch does not increase as fast as the costs of fishing), which may coincide with an increased tendency for fishers to undertake seasonal alongshore migrations, a gradual breakdown of traditional management schemes, and non-enforcement of "modern" management regulations.

Important additional symptoms are as follows:
- New fishers are recruited from ethnic groups (e.g., of traditional pastoralists) or regions (e.g., highlands) without a tradition of fishing. The new fisheries will require cheap and easy fishing gears, hence there will be
- increasing use of destructive gears (explosives, poisons). An important, but often neglected corollary of poverty is an
- increasing contribution of women in fisher communities to overall family incomes (i.e., women subsidizing the fisher*men*).

This causal pattern may appear hard to diagnose. However, some fisheries in South and Southeast Asia contain most of the elements of this pattern (see McManus et al. 1992 or Saeger 1993). The last element in this list, which I deduced from observations in several fishing villages (emphasized in Figure 4), still needs empirical verification. However, gender-disaggregated data suitable for verification are rarely collected or analyzed under the prevailing mental map, which tends to relate women to fisheries only when they act as middle*men*.

Malthusian overfishing, which is widespread in Asia, notably South and Southeast Asia, can be expected to spread in the next decades to and within Africa and South America, often in the wake of dynamite fishing. Anecdotal evidence suggests that dynamite fishing is spreading into areas where it was previously unknown, such as the Caribbean or West Africa (see Vakily 1993 for one of the few well-documented cases outside of Southeast Asia). Modern technology will not help in such cases since, as might be seen from Figure 2D, any decrease in fishing costs (such as those induced by economically more efficient gears) tends to further deplete the resource base of the fisheries.

Overcoming Marginalization

Marginalization and Malthusian overfishing, its derived phenomenon, need not occur. A number of remedial actions that would help to alleviate or at least mitigate some of the effects of marginalization are possible. An obvious short-term measure, for which Figure 1 may be seen as providing much of the required justifications, is for central governments to ban commercial fishing on the inshore fishing grounds of small-scale fishers or to enforce existing legislation forbidding such incursions. Such bans have been implemented for the explicit purpose of reallocating coastal resources, and evidence indicates that the intended purposes were achieved, at least in part (Sarjono 1980; Saeger 1981; Martosubroto and Badrudin 1984; Martosubroto and Chong 1987). However, such short-term measures tend to produce only short-term benefits, as are the benefits accruing from enforcing prohibitions on destructive gears such as dynamite or poisons.

In the longer term, dealing with Malthusian overfishing will involve providing the women in fishing communities—obviously in the context of nationwide programs—reasons and the means to limit child-bearing, an option they are largely denied at present by their husbands and such powerful men as conservative politicians and religious leaders, and by economic conditions that make it seem rational to invest in large families (Stevens 1994; Anonymous 1994).

Better educated women are now recognized by development agencies as crucial agents of change in rural settings. Hence, overcoming the marginalization of fisher communities cannot be achieved without empowering women (see contributions in Oestergaard 1992). This may involve the partial devolution of government functions to local fisher communities, leading to arrangements wherein the communities would have the right and the means to establish and enforce exclusive fishing areas, sanctuaries, and gear restriction schemes (Alcala and

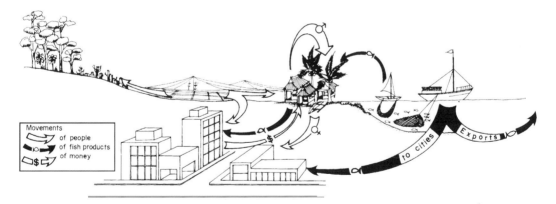

FIGURE 4.—Schematic representation of the processes leading to Malthusian overfishing. A comparatively large agricultural sector releases excess labor, landless farmers who migrate either to urban, upland, or coastal areas. Under this influx, traditional arrangements preventing open access to the fisheries gradually collapse, leading to excessive fishing pressure. This is exacerbated by inshore commercial fishing, by new entrants to fishing as the male children of fishers pick up their fathers' trade, and by the contribution of many young women who leave the communities to work in urban areas, providing a subsidy for men to continue to fish even when resources are depleted. The migrants to upland areas accelerate and complete the deforestation initiated by logging companies, which leads to siltation of rivers and streams, and eventually to smothering of coral reefs and other coastal habitats, thus further reducing coastal fish yields. This model implies a research program and levers to affect events, just as the traditional fisheries models in Figure 2 do.

Russ 1990; Russ and Alcala 1994). Several of the individual ITQ schemes discussed elsewhere in this volume may be useful here as well.

Devolution—to the extent that it does not permit local elites to replace a more distant and perhaps more benign central government and leads to more decentralized modes of governance (described in Putnam 1992)—would represent a lessening of the marginalization presently besetting small-scale fishing communities, which would become the partners, rather than the "target" of government agencies (see contributions in Kooiman 1993). This is particularly important as governments the world over have shown themselves largely unable to manage natural resources without the cooperation of key stakeholders, however large the bureaucracy that is deployed. Also, it is only in the context of devolution that communities can use traditional ("local") knowledge for fisheries management, for example, to establish seasonal or area closures based on empirical knowledge of species life history or to formulate and enforce equitable resource access rules.

For local management to result in increased incomes for fisher families, alternative employment will have to be found for those leaving the fisheries. However, few detailed and realistic plans for phased reduction of fishing effort through alternative livelihood projects exist. The work of McManus et al. (1992) is one of the few exceptions. This work, and related publications on coastal area management, make abundantly clear that eventually reducing the number of small-scale fishers—by providing alternative livelihood opportunities such as mariculture and others—must involve intersectoral arrangements. Hence, fisheries managers must interact with

representatives of sectors operating in coastal areas such as agriculture, tourism, and manufacturing.

Coastal transects, adapted from the transects used in farming systems research, can be used to formalize intersectoral relationships in coastal areas (Pauly and Lightfoot 1992), and thus facilitate the previously mentioned intersectoral conceptualizations and perhaps even consultations. Interestingly, while small-scale fisheries are usually marginal in conventional mental maps, on land's end, they emerge at the core of coastal areas when they are plotted on graphs (Figure 5). Such transects show that biomass exchanges in coastal systems cannot be accurately quantified without accounting for the fish caught by small-scale fishers or gleaned by women and children (Chapman 1987; Talaue-McManus 1989; Wynter 1990). That part of marine production, important in tropical waters, that is derived mainly from terrigenous input, and generated close inshore is recycled back inland. The cycle applies to incomes generated by small-scale fish processing, a major contributor to the coastal economy of communities and one in which women often play the major role (Nauen 1989; Arnal et al. 1992).

For these and related reasons, coastal area management, an important area of systems research, must focus on small-scale fisheries, and thus render them less marginal. Similarly, studies of the world's biodiversity, a new integrative framework for several biological disciplines, cannot but encompass tropical coastal fisheries—coastal systems such as coral reefs being, after tropical forests, the main global sources of biodiversity (see Reid and Miller 1989). For coral reefs, as is the case with the Amazonian rain forest, local knowledge—suitably recorded and validated—will be crucial in complement-

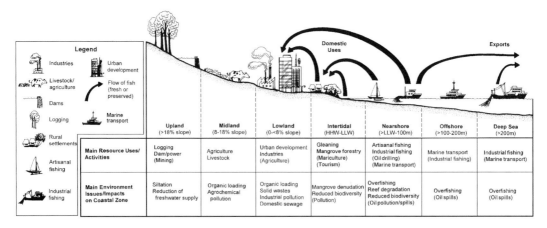

FIGURE 5.—A coastal cross-section, or transect, illustrating the central and structuring role of small-scale fisheries in coastal areas and showing main resource uses and activities along a transect, along with their impact on the coastal zone proper (items in parentheses have no icons). Such graphs, derived from farming systems research (Pauly and Lightfoot 1992) can be used to reconceptualize small-scale fisheries away from their present marginal location on our mental maps toward a central role at the heart of coastal systems.

ing scientific knowledge of endangered species or of unique events or processes, such as spawning aggregations, an area of fisheries biology that is largely driven by local knowledge (Johannes 1981).

A change of mental maps is required for all the factors involved in reducing the marginalization of fishers—changing gender roles, the devolution of political power toward fisher communities, the new governance that devolution implies, and the perception of coastal fisheries as only one sector in coastal areas, albeit an important one. Some of the graphs presented in this paper (Figures 1, 4, and 5) are proposed as elements of new mental maps for fisheries scientists, particularly those working on tropical small-scale fisheries. Small-scale fishers would be at the center of these new maps, not at the edge. Such new maps, I believe, would guide us better than the old maps that I suspect most of us still use to orient ourselves.

Acknowledgments

I thank C. Lightfoot, J. McManus, C. Nauen, R. Pullin, G. Silvestre, K. Ruddle, M. Williams and two anonymous reviewers for extensive comments on earlier drafts of this contribution. I followed most, but not all their suggestions, and hence must accept the blame for whatever may still be unclear, or appear unsubstantiated. This paper is International Centre for Living Aquatic Resources Management (ICLARM) Contribution 1070.

References

Acquay, H. K. 1992. Implications of structural adjustment for Ghana's marine fisheries policy. Fisheries Research 14:59-70.

Agüero, M., editor. 1992. Contribuciones para el estudio de la pesca artesanal en America Latina. ICLARM (International Centre for Living Aquatic Resources Management) Conference Proceedings 35, Manila.

Aguilera-Klink, F. 1994. Some notes on the misuse of classic writings in economics on the subject of common property. Ecological Economics 9:221-228.

Alcala, A., and G. Russ. 1990. A direct test of the effect of protective management on abundance and yield of tropical marine resources. Journal du Conseil International pour l'Exploration de la Mer 46:40-47.

Anderson, L. G. 1977. The economics of fisheries management. The Johns Hopkins University Press, Baltimore.

Anderson, L. 1997. Efficiency and distribution issues during the transition to an ITQ program. Pages 213-224 in E. K. Pikitch, D. D. Huppert, M. P. Sissenwine, editors. Global trends: fisheries management. American Fisheries Society Symposium 20, Bethesda, Maryland.

Anonymous. 1994. Population misconceptions. The Economist, 28 May 1994:81-82.

Arnal, S., A. L. Kodjo-Nyaku, and K. Amegavie. 1992. Les systemes de crédit utilisés par les femmes transformatrices et commerçantes de poisson–Togo, Ghana, République de Guinée, Mali. Programme Régional Afrique de l'Ouest:

valorisation des captures de la pêche artisanale. Dossiers Technico-Economique Bonga, Ser. 4, Abidjan.

Bailey, C. 1982. Small-scale fisheries of San Miguel Bay, Philippines: occupational and geographic mobility. ICLARM (International Centre for Living Aquatic Resources Management) Technical Report 10, Manila.

Baranov, F. I. 1918. On the question of the biological basis of fisheries. Issled. Ikhtiologicheskii Inst. Isv. 1:81-128 (in Russian). Reprinted in Baranov, F. I. 1976. Selected works on fishing gear. Vol. 1. Israel Program for Scientific Translation, Jerusalem.

Behnke, R. 1994. Natural resource management in pastoral Africa. Development Policy Review 12(1):5-27.

Beverton, R. J., and S. J. Holt. 1957. On the dynamics of exploited fish populations. United Kingdom Ministry of Agriculture Fisheries, Investigations Series II:19.

Chapman, M. D. 1987. Women's fishing in Oceania. Human Ecology 15(3):267-288.

Chauveau, J. P., and A. Samba. 1989. Market development, government intervention and the dynamics of the small scale fishing sector: an historical perspective of the Senegalese case. Development and Change 20(4):599-620.

Dizon, L. B. 1995. The impact of publication productivity of scientists in fisheries institutions in the Philippines on the development of fisheries science. Doctoral dissertation. College of Mass Communication, University of the Philippines, Quezon.

Fay, C. 1989a. Sacrifice, prix du sang, "eau du maître": fondations des territoires de pêche dans le delta central du Niger (Mali). Cahiers des Sciences Humaines 25(1/2):159-176.

Fay, C. 1989b. Systèmes halieutiques et espaces de pouvoirs: transformation des droits et des pratiques de pêche dans le delta central du Niger (Mali), 1920–1980. Cahiers des Sciences Humaines 25(1/2):213-236.

Fox, W. W. 1970. An exponential yield model for optimizing exploited fish populations. Transactions of the American Fisheries Society 99:80-88.

Garcia, S. M., and C. Newton. 1997. Current situation, trends and prospects in world capture fisheries. Pages 3-27 in E. K. Pikitch, D. D. Huppert, and M. W. Sissenwine, editors. Global trends: fisheries management. American Fisheries Society Symposium 20, Bethesda, Maryland.

Goodyear, C. P. 1989. Spawning stock biomass per recruit: the biological basis for a fisheries management tool. International Commission for the Conservation of Atlantic Tuna, Working Documents SCRS/89/82, Miami, Florida.

Gordon, H. S. 1953. An economic approach to the optimum utilization of fisheries resources. Journal of the Fisheries Research Board of Canada 10(7):442-457.

Haakonsen, J. M., and M. C. Diaw, editors. 1991. Fishermen's migration in West Africa. Programme for Integrated Development of Artisanal Fisheries, IDAF/WP/36, Cotonou, Benin.

Hampden-Turner, C. 1982. Maps of the mind: charts and concepts of the minds and its labyrinths. Macmilllan Publishing, New York.

Hannesson, R. 1997. The political economy of ITQs. Pages 237-245 in E. K. Pikitch, D. D. Huppert, M. P. Sissenwine, editors. Global trends: fisheries management. American Fisheries Society Symposium 20, Bethesda, Maryland.

Hardin, G. 1968. The tragedy of the commons. Science 162(3859):1243-1248.

Harris, J. 1996. World agricultural futures: regional sustainability and ecological limits. Ecological Economics 17:95-115.

Hodgson, G., and J. A. Dixon. 1992. Sedimentation damage to marine resources: environmental and economic analysis. Pages 421-445 in J. B. Marsh, editor. Resources and environment in Asia's marine sector. Taylor & Francis, New York.

Homer-Dixon, T. F., J. H. Bontwell, and G. W. Kathjens. 1993. Environmental change and violent conflict. Scientific American (February) 268(2):38-45.

Horemans, B. 1993. The situation of artisanal fisheries in West Africa in 1992. Programme for Integrated Development of Artisanal Fisheries in West Africa, IDAF/WP/47, Cotonou, Benin.

Hulme, H. R., R. J. H. Beverton, and S. J. Holt. 1947. Population studies in fisheries biology. Nature 159:714-715.

Hviding, E. 1991. Traditional institutions and their role in contemporary coastal resource management in the Pacific Islands. NAGA, the ICLARM (International Centre for Living Aquatic Resources Management) Quarterly 14(3):3-6.

Kooiman, J., editor. 1993. Modern governance–society interactions. Sage Publications, London.

Johannes, R. E. 1981. Words of the lagoon: fishing and marine lore in the Palau district of Micronesia. University of California Press, Berkeley, California.

Lightfoot, C. 1990. Integration of aquaculture and agriculture: a route to sustainable farming systems. NAGA, the ICLARM (International Centre for Living Aquatic Resources Management) Quarterly 13(1):9-12.

Lunianga, D. A. 1989. Le village de pêche artisanale de Kayar, Sénégal. Canadian International Development Agency, Ottawa.

Mace, P., and M. Sissenwine. 1993. How much spawning is enough? Pages 101-118 in S. Smith, J. J. Hunt, and D. Rivard, editors. Risk evaluation and biological reference points for fisheries management. Canadian Special Publication Fish Aquatic Sciences 120.

Mathews, J. 1994. A small price to pay for proving Malthus wrong. International Herald Tribune (June 9):5.

McManus, J. W., C. L. Nañola, Jr., R. B. Reyes, Jr., and K. N. Kesner. 1992. Resource ecology of the Bolinao coral reef system. ICLARM (International Centre for Living Aquatic Resources Management) Studies and Reviews 22, Manila.

Malthus, T. R. 1798. An essay on the principle of population. Reprinted 1970 by Penguin Books, Harmondsworth, Middlesex, England.

Martosubroto, P., and M. Badrudin. 1984. Notes on the status of the demersal resource off the north coast of Java. Pages 33-36 in Report of the fourth session of the Standing Committee on Resources Research and Development of the Indo-Pacific Fisheries Commission, Jakarta, Indonesia, 23-29 August 1984. Food and Agriculture Organization of the United Nations, Fisheries Reports (318), Rome.

Martosubroto, P., and K.-C. Chong. 1987. An economic analysis of the 1980 Indonesian trawl ban. Indonesian Agriculture and Resources Development Journal 9(1/2):1-12.

Mathews, C. P. 1991. Spawning stock biomass per recruit: a timely substitute for stock recruitment analysis. Fishbyte 9(1):7-11.

Meltzoff, S. K., and E. LiPuma. 1986. The social and political economy of coastal zone management: shrimp mariculture in Ecuador. Coastal Zone Management Journal 14(4):349-380.

Munro, J. L. 1980. Stock assessment models: applicability and utility in tropical small-scale fisheries. Pages 35-47 in P. M.

Roedel and S. B. Saila, editors. Stock assessment for tropical small-scale fisheries. International Center for Marine Resources Development, University of Rhode Island, Kingston.

Murphy, G. I. 1982. Recruitment of tropical fishes. Pages 141-148 in D. Pauly and G. I. Murphy, editors. Theory and management of tropical fisheries. ICLARM (International Centre for Living Aquatic Resources Management) Conference Proceedings 9, Manila.

Nauen, C. 1989. Women in African artisanal fisheries. NAGA, the ICLARM (International Centre for Living Aquatic Resources Management) Quarterly 12(2):14-15.

Neal, R. A. 1982. Dilemma of the small-scale fisherman. ICLARM (International Centre for Living Aquatic Resources Management) Newsletter 5(3):7-9.

Oestergaard, L., editor. 1992. Gender in development. Routledge, London and New York.

Okera, W. 1994. Will traditional African coastal fishermen survive? Ambio 23(2):168-169.

Ommer, R. E. 1991. From outpost to outport: a structural analysis of the Jersey-Gaspé cod fishery, 1767–1886. McGill University Press, Montreal.

Ommer, R. E. 1995. Deep water fisheries, policy and management issues, and the sustainability of fishing communities. Pages 307-322 in A. G. Hopper, editor. Deep water fisheries of the North Atlantic oceanic slope. Kluwer Academic Publishers, Dordrecht, The Netherlands.

Panayotou, T., and S. Jetanavanich. 1987. The economics and management of Thai marine fisheries. ICLARM (International Centre for Living Aquatic Resources Management) Studies and Reviews 14, Manila.

Pauly, D. 1979a. Biological overfishing of tropical stocks. ICLARM (International Centre for Living Aquatic Resources Management) Newsletter 2(3):3-4.

Pauly, D. 1979b. Theory and management of tropical multispecies stocks: a review with emphasis on the Southeast Asian demersal fisheries. ICLARM (International Centre for Living Aquatic Resources Management) Studies and Reviews 1, Manila.

Pauly, D. 1988. Some definitions of overfishing relevant to coastal zone management in Southeast Asia. Tropical Coastal Area Management 3(1):14-15.

Pauly, D. 1990. On Malthusian overfishing. NAGA, the ICLARM (International Centre for Living Aquatic Resources Management) Quarterly 13(1):3-4.

Pauly, D. 1994. On the sex of fish and the gender of scientists: a collection of essays in fisheries science. Chapman and Hall, London.

Pauly, D., and C. Lightfoot. 1992. A new approach for analyzing and comparing coastal resource systems. NAGA, the ICLARM (International Centre for Living Aquatic Resources Management) Quarterly 15(2):7-10.

Pauly, D., G. Silvestre, and I. R. Smith. 1989. On development, fisheries and dynamite: a brief review of tropical fisheries management. Natural Resource Modeling 3(3):307-329.

Peters, A. 1983 The new cartography. Friendship Press, New York.

Platteau, J. P., editor. 1989a. Third world fisheries: the dynamics of incorporation and transformation. Development and Change 20(4).

Platteau, J. P. 1989b. The dynamics of fisheries development in developing countries. Pages 565-597 in J. P. Platteau, editor. Third world fisheries: the dynamics of incorporation and transformation. Development and Change 20(4).

Pollnac, R. B. 1981. Sociocultural aspects of developing small-scale fisheries: delivering services to the poor. World Bank Staff Working Paper 490, Washington, D.C.

Pollnac, R. B., and M. T. Morrissey, editors. 1989. Aspects of small-scale fisheries development. International Center for Marine Resources Development, University of Rhode Island, Kingston.

Pope, J. G. 1979. Stock assessment in multispecies fisheries, with special reference to the trawl fishery in the Gulf of Thailand. South China Sea Fisheries Development and Co-ordinating Programme, SCS/DEV/79/19, Manila.

Putnam, R. D. 1992. Democracy, development and the civic community: evidence from an Italian experiment. Pages 33-73 in I. Serageldin and J. Taboroff, editors. Culture and development in Africa. Environmentally sustainable development proceedings series 1, The World Bank, Washington, D.C.

Reid, W. V., and K. R. Miller. 1989. Keeping options alive: the scientific basis for conserving biodiversity. World Resources Institute, Washington, D.C.

Ricker, W. E. 1954. Stock and recruitment. Journal of the Fisheries Research Board of Canada 11:559-623.

Ricker, W. E. 1975. Computation and interpretation of biological statistics of fish populations. Bulletin of the Fisheries Research Board of Canada 191.

Roedel, P., and S. Saila, editors. 1980. Stock assessment for small-scale fisheries. Proceedings of a workshop, University of Rhode Island, 19–21 September 1979. International Center for Marine Resources Development, University of Rhode Island, Kingston.

Ruddle, K. 1988. Social principles underlying traditional inshore fishery management systems in the Pacific Basin. Marine Resource Economics 5:351-363.

Ruddle, K. 1989a. Traditional sole property rights and modern inshore fisheries management in the Pacific Basin. Pages 68-76 in H. Campbell, K. Menz, and G. Waugh, editors. Economics of fishery management in the Pacific Islands region. Proceedings of an International Conference, Hobart, Tasmania, Australia, 20-22 March 1989. The Australian Centre for International Agricultural Research, Canberra.

Ruddle, K. 1989b. The organization of traditional inshore fishery management systems in the Pacific. Pages 73-85 in P. A. Neher, editor. Rights based fishing. Kluwer Academic Publishers, Hingham, Massachusetts.

Ruddle, K., and R. Chesterfield. 1977. Education for traditional food procurement in the Orinoco Delta. University of California Publications, Ibero-Americana 53:1-172.

Ruddle, K., and R. E. Johannes, editors. 1985. The traditional knowledge and management of coastal systems in Asia and the Pacific. United Nations Education and Culture Organization/Regional Office for Science and Technology, Southeast Asia, Jakarta.

Ruddle, K., E. Hviding, and R. E. Johannes. 1992. Marine resources management in the context of customary tenure. Marine Resource Economics 7:249-273.

Russ, G., and A. Alcala. 1994. Sumilon Island Reserve: 20 years of hopes and frustrations. NAGA, the ICLARM (International Centre for Living Aquatic Resources Management) Quarterly 17(3):8-12.

Saeger, J. 1981. Do trawling bans work in tropical waters? ICLARM (International Centre for Living Aquatic Resources Management) Newsletter 4(1):3-4.

Saeger, J. 1993. The Samar Sea: a decade of devastation. NAGA, the ICLARM (International Centre for Living Aquatic Resources Management) Quarterly 16(4):4-6.

Sarjono, I. 1980. Trawlers banned in Indonesia. ICLARM (International Centre for Living Aquatic Resources Management) Newsletter 3(4):3.

Schaefer, W. E. 1954. Some aspects of the dynamics of populations important to the management of the commercial marine fisheries. Inter-American Tropical Tuna Commission Bulletin 1:27-56.

Schaefer, W. E. 1957. A study of the dynamics of the fishery for yellowfin tuna in the eastern tropical Pacific Ocean. Inter-American Tropical Tuna Commission Bulletin 2:247-268.

Smith, I. R. 1979. A research framework for traditional fisheries. ICLARM (International Centre for Living Aquatic Resources Management) Studies and Reviews 2, Manila.

Smith, T. D. 1994. Scaling fisheries: the science of measuring the effects of fishing, 1855–1955. Cambridge University Press, Cambridge, England.

Southgate, D., and M. Basterrechea. 1992. Population growth, public policy and resource degradation. Ambio 21(7):460-464.

Stauch, A. 1966. Le bassin camerounais et sa pêche. Editions ORSTOM, Paris.

Stevens, W. K. 1994. Poor lands' success in cutting birth rate upsets old theories. The New York Times (January 2):1, 8.

Sunderlin, W. 1994. Beyond Malthusian overfishing: the importance of structural and non-demographic factors. Traditional Marine Resource Management and Knowledge Information Bulletin, South Pacific Commission, Noumea (4):2-5.

Talaue-McManus, L. 1989. The gleaners of Northwestern Lingayen Gulf, Philippines. NAGA, the ICLARM (International Centre for Living Aquatic Resources Management) Quarterly 12(2):13.

Thomson, D., and Food and Agriculture Organization of the United Nations. 1988. The world's two marine fishing industries—how they compare. NAGA, the ICLARM (International Centre for Living Aquatic Resources Management) Quarterly 11(3):17.

Trinidad-Cruz, A. C., R. S. Pomeroy, P. V. Corpuz, and M. Agüero. 1993. Bioeconomics of the Philippine small pelagics fishery. ICLARM (International Centre for Living Aquatic Resources Management) Technical Report 38, Manila.

Vakily, J. M. 1992. Assessing and managing the marine fish resources of Sierra Leone, West Africa. NAGA, the ICLARM (International Centre for Living Aquatic Resources Management) Quarterly 15(1):31-35

Vakily, M. 1993. Dynamite fishing in Sierra Leone. NAGA, the ICLARM (International Centre for Living Aquatic Resources Management) Quarterly 16(4):7-9.

Went, A. E. J. 1972. Seventy years agrowing: a history of the International Council for the Exploration of the Sea. Rapports et Procès-Verbaux du Conseil International pour l'Exploration de la Mer 165.

Wynter, P. 1990. Property, women fishers and struggle for women's right in Mozambique. Sage 7(1):33-37.

Trends in International Law
for High-Seas Fisheries Management

W. T. Burke

Abstract.—This paper discusses trends in international law relating to high-seas fisheries management. After briefly describing the major elements of the current regime for such management, I examine the recent developments intended to improve the international legal framework for high-seas fisheries management. The major efforts to this end consist of the contemporary general international negotiations under the auspices of the United Nations and of the Food and Agriculture Organization (FAO) of the United Nations. The first of these, to which FAO also contributes, aims at improvements in jurisdictional, substantive, and procedural principles for straddling fish stocks and highly migratory fish stocks. These negotiations are underway at this writing (they were concluded in 1995). Under FAO auspices, states have adopted an agreement to improve compliance with high-seas management measures that is designed to resolve the difficulties caused by vessel reflagging in order to avoid compliance with otherwise applicable conservation and management measures. Reference is also made to some specific international agreements aimed at straddling stocks and to unilateral actions relating to these stocks.

It is useful initially to identify the deficiencies of the existing high-seas legal regime that help account for the continuing controversies about high-seas fisheries. The basic problem is that the regime provided for in the 1982 Convention on the Law of the Sea, or in customary international law prevailing for high seas fishing, is not an effective one. An effective regime is one that would adequately provide for the following:

- the acquisition, collection, analysis, and dissemination of information necessary for making conservation and allocation decisions about the stocks harvested in high-seas fishing operations;
- a regulatory system that (a) would maintain acceptable (sustainable) levels of abundance of target and incidentally caught species and (b) would deal with overcapitalization and allocation of benefits; and
- the authority and the practical means to enforce regulatory measures.

It should go without saying that such a regime should be implemented rather than exist on paper only.

Past Attempts to Modify or Eliminate Freedom of Fishing on the High Seas

A key to understanding the major trends in high-seas fisheries management is to consider the last half-century of actions by states in relationship to fisheries. Over this entire period of time, we have witnessed, and continue to witness, a series of attempts to escape from the principle of freedom of fishing on high seas. Therefore, the most noticeable trend today, consisting of the various attempts to restrict freedom of fishing on the high seas, is only a continuation of one of the main characteristics of international decision making since 1945. In these remarks, I briefly document this point, although surely it

is beyond serious cavil, and then examine the present array of efforts in that same direction, which are still for the most part underway.

There is one critical difference between the current and previous contexts. In some ways, this problem is at its most critical stage because failure to resolve the conflicting regimes now applicable to the harvest of marine fish could undermine broader understandings about the general regime of law governing events at sea. This assessment is not a product of a fevered academic imagination. It is the expressly stated view of diplomats now struggling to develop solutions to the primary controversies involving straddling stocks (Nandan 1993).

The effort to modify and control high-seas fishing began, as all know, almost simultaneously with the end of World War II when the USA issued, unilaterally, its fisheries and continental shelf proclamations. The former has often been portrayed as nothing more than an effort to encourage international agreement on regulation of fisheries adjacent to national territory, whether of the USA or of any other nation.

That is an incomplete and somewhat distorted view of the object in mind in the fisheries proclamation. This proclamation was clearly an effort to extend unilateral U.S. control over salmon harvesting in the adjacent high seas, a possibility which had agitated the U.S. government since the 1930s when great excitement followed reports that Japanese vessels were entering the fishery off Alaska (Hollick 1981). The control to be established under the proclamation did not depend on the acquiescence of foreign fishing states unless they were *already* participating in the affected fishery, a circumstance that definitely did not apply to salmon. If the proclamation is read carefully, keeping in mind that it had been extremely carefully drafted, it is obvious that it includes the intent of exercising unilateral control over foreign fishing although this is not expressly stated. However, the underlying memorandum distributed by the U.S. Department of State to support this proclamation made it clear that a major objective was to prevent the entry of Japan into the northeast Pacific Ocean salmon fishery (Hollick 1981).

It is helpful to recall that the USA never had to exercise unilateral control over salmon. By the time Japan was again able to fish outside Japanese waters after WW II, the USA had gotten their agreement, if it can be called that considering the disparity between the two countries' positions in 1950–51 to fish west of longitude 175°W despite their freedom to fish east of that line under prevailing international law. Scheiber (1989) provides an insightful account of the evolution of the 1952 Convention for the High Seas Fisheries of the North Pacific Ocean. The USA felt so strongly about the abstention principle that it attached a declaration to its ratification of the 1958 Geneva Convention on Fishing and Conservation of the Living Resources of the High Seas to the effect that the convention did not affect the principle.

Shortly after the Truman Proclamation in 1945, the west coast states of South America extended their jurisdiction to at least 200 nautical miles in language that appeared to be more consistent with extension of the territorial sea than for fisheries or oil and gas alone. Although these claims were not generally observed by other nations, they did lead to more moderate claims to exclusive fishing zones and thus to recognition of the special interest of the coastal state to fisheries beyond a narrow limit of national jurisdiction. The later widespread agreement on the exclusive economic zone (EEZ) resulted in the adoption of the 200-mile (370-km) limit espoused in the Chile, Ecuador, and Peru claims, but not their expansive view of the scope of national jurisdiction within that limit.

The first United Nations (UN) Conference on the Law of the Sea (UNCLOS) in 1958 was the next attempt to reduce freedom of fishing in disregard of efforts to establish conservation measures. At that time, the pressure against freedom of fishing had two consequences. The first, and most important, was that it forced the negotiating parties to separate the concept of jurisdiction over fisheries from the concept of territorial sovereignty. Although no precise formula to that effect was endorsed at the 1958 UNCLOS conference, the only serious proposals for agreement on a fixed width for the territorial sea provided for an additional area of limited fishery jurisdiction, subject to a grandfather clause for a period of 10 years (McDougal and Burke 1985).

This approach was continued in the second United Nations Conference on the Law of the Sea in 1960 where it was even clearer that the principal problem was to agree on a limit on the territorial sea and a separate limit for fisheries (McDougal and Burke 1985). Although agreement narrowly eluded the conference, the stage was set for later successful negotiations as well as for the proliferation of unilateral claims to fishery jurisdiction beyond the territorial sea.

The second consequence of the pressure on freedom to fish in the 1958 conference was the successful conclusion of the 1958 Convention on Fishing and the Conservation of the Living Resources of the High Seas (17 UST 138, TIAS 5969, 559, UNTS 285),[1] which is currently in force for 36 states, including the USA, Canada, Iceland, and France. This agreement has never been implemented, but it does have continued significance for contemporary straddling-stock controversies, as discussed later. One possible implication of this treaty was that there was no need for extended fisheries jurisdiction or the establishment of exclusive fishery management zones since a coastal state could achieve its conservation objectives either by international agreement or unilateral action. However, the ensuing years show that this interpretation did not carry much weight with coastal states, virtually all of whom extended their jurisdiction for fisheries reasons and purposes.

After the Second UNCLOS in 1960, the data show that, through 1975, national claims to fisheries jurisdiction took mainly the form of extending the territorial sea to 12 miles or, in a few instances, establishing fisheries zones beyond the territorial sea. The latter were normally rejected before 1975. Some states extended the territorial sea beyond 12 miles (Burke 1994), sometimes much beyond. By and large, the encroachment on freedom of fishing was relatively modest during this period, but it does show the continuing tendency to separate fishing from territorial jurisdiction. During this period, almost all fisheries management, such as it was, was high-seas management. Virtually all the international fishery agencies, with minor exceptions, and all the bilateral and multilateral agreements on fisheries were aimed at high-seas fisheries, although territorial seas were included in the coverage of some. The reason for this focus was that significant harvesting occurred beyond the territorial sea where regulation could only be by international agreement. There was no basis for national jurisdiction outside the territorial sea, and there was no need for international fishery agencies to deal with territorial seas in most instances except for bilateral agreements. Carroz (1984) summarized as follows: "Over the years, more than

---

[1]United Nations, multilateral treaties deposited with the Secretary-General, status as of Dec. 1992, p. 755.

twenty regional fishery commissions were established to cover nearly all the world's seas and oceans. This network developed when the application of conservation measures and regulation of the conduct of fishing operations on the high seas could only be achieved through international agreements."

Since 1975, the most obvious trend in law of the sea is the great extension of national jurisdiction, and it is now universally accepted that the area out to 200 miles is subject to coastal state jurisdiction. The freedom to fish without effective restraint, but subject to the obligation to conserve, was gradually pushed farther and farther from the coast.

The preoccupation with high-seas fisheries has not yet abated. As was recognized by some states during the UNCLOS negotiations, but glossed over for the sake of more general agreement on other issues, the extension of national jurisdiction was not a complete solution to the high-seas management problem (Burke 1989). Left over as a result were the new issues arising from the fact that some fish stocks, partially within national jurisdiction, were still available for harvesting outside the new EEZs. The result is the embarrassingly difficult problem of fish stocks that are subject to contradictory jurisdictional principles, the effect of which has been to obstruct effective management.

What have attracted so much attention lately are the controversies generated by this lacuna in jurisdictional arrangement and the gap left by the creation of the EEZs. By lacuna I mean the absence of management owing to the concept of freedom of fishing applicable to purely high-seas stocks although a duty to conserve does apply to such stocks. No single nation can regulate such stocks nor can any group of nations unless agreement is reached, and the general obligation to conserve is difficult or impossible to enforce without agreement. Thus, no agreement, no management—usually. The gap is represented by the 200-mile EEZ limit, which only bounds part of a fish stock, some of it being subject to freedom of fishing while the inside stock is subject to the sovereign rights of the coastal state. The lacuna is the absence of management over a portion of a stock, which is also detrimental to management of the portion subject to a single decision-making process. Here again, no agreement, no management. And sometimes in this instance, where there is agreement, there is still no effective management (such as under the North Atlantic Fisheries Organization [NAFO]).

The Concept of Straddling Stocks and Highly Migratory Stocks

A few background facts about this pattern of fishing are useful to know. These stocks are called straddling

stocks for obvious reasons, but the precise coverage of this term is not so simple. Some straddling stocks occur predominantly within the EEZ, but straddle out to some degree. Others occur predominantly outside the EEZ but straddle in to some degree. This relationship is not necessarily stable, but apparently it tends to one or another orientation over time. In each instance, the specific stock characteristics determine whether fishing in one area does or does not affect the same stock in another area. Discussion and illustrations of high-seas fish and different types of straddling stocks are provided by the U.N. Food and Agriculture Organization (FAO 1993) and in a Russian Federation document (U.N. 1993a).

Moreover, some straddling stocks move more than others and do so over great distances outside national jurisdiction, such that some perceive them as having little real relationship to coastal states within whose jurisdiction they also appear. These stocks are labeled highly migratory species—a group of species identified and listed in the UNCLOS treaty, albeit with numerous omissions and errors. A distinction among these different categories of straddling stocks is urged by some states, especially the USA, arguing that legally, biologically, and politically, highly migratory species differ from other straddling stocks.[2] From a management perspective, it is difficult to understand that this distinction makes a difference except in the number of coastal states whose measures must be coordinated for effective management. In either instance, effective management requires agreement of coastal and high-seas fishing states. In both instances, coastal states assert interests in high-seas fishing on straddling and highly migratory stocks, interests that are recognized in article 116 of the UNCLOS treaty.

The aggregate landings of straddling stocks are not trivial. FAO statistics are considered to give a "crude order of magnitude," with a range of 11.4–13.7 million metric tons (mt) between 1988 and 1991, excluding "highly migratory species" (Burke 1989, FAO 1993).[3] For example, if all tuna in the latter category were included, the total world landings would be increased by about 4 million mt, amounting in total to as high as 17 million mt, which is greater than 20% of the total world marine landings. No one should think these estimates are very accurate, just as the total world landings figure is not very accurate, but the best that imperfect information will disclose.

---

[2]U.S. objective for sessions of the United Nations Conferences on Straddling Fish Stocks and Highly Migratory Fish Stocks, 5-6 Feb. 1994. Manuscript prepared by Office of Marine Conservation, Bureau for Oceans and International Environmental and Scientific Affairs, U.S. Department of State.
[3]FAO Fisheries Circular, note 11 above, at 5.

The species that composes the largest single part of the reported world fish catch, and the largest in the U.S. annual harvest, is the Alaska walleye pollock (*Theragra chalcogramma*), which is a straddling stock in the central Bering Sea as well as in the Okhotsk Sea. In the former instance, the high seas enclave (the "donut" hole) is surrounded by the EEZs of Russia and the USA. In the Okhotsk Sea, the "peanut" hole is an enclave wholly within the Russian EEZ.

This same general problem arises in several places around the globe, notably off eastern Canada, Argentina, Chile, Iceland, Norway, and New Zealand, and there is a potential for considerably more instances as the search for high-seas stocks intensifies (Meltzer 1994). If tuna are included as straddling stocks, a very large number of additional states are involved. As noted previously, an alleged difference between tuna and other straddling stocks was an important U.S. criticism of the Negotiating Text resulting from the second substantive session of the U.N. Fish Stock Conference.

The reason for controversy in this context is that the combined species and zonal approaches that prevailed in the 1982 treaty do not provide (or have not been perceived to provide) adequate or clear guidance for decisions about straddling stocks. Under that treaty, coastal states have sovereign rights over fish within 200 miles of their coastlines, but beyond that all states continue to have the right to exercise the principle of freedom of fishing on the high seas. Thus, sovereign rights apply to a portion of a stock and freedom of access to another portion.

A truism of fishery management is that to be effective, regulations must apply to an entire stock and to the entire range of its movement. This assumes that appropriate regulations are actually adopted and enforced—a hazardous assumption.

Continuing Efforts to Resolve High-Seas Fishing Management Problems

The contemporary actions regarding this problem range from the broadest forums, such as the General Assembly and large diplomatic conferences such as the United Nations Conference on Environment and Development (UNCED) (Burke 1993) and the follow-up U.N. conference on fish stocks, to specific multilateral efforts and unilateral actions by particular states. I believe the involvement of the U.N. General Assembly in addressing high-seas fisheries issues is probably idiosyncratic and not indicative of a trend toward such intervention. In 1989 and 1991, the U.N. General Assembly adopted resolutions directed at the use of large-scale driftnets on the high seas. Resolution 46/215, adopted in 1991, rec-

ommended that this use be terminated. For an argument that these resolutions have fragile scientific support, citing available scientific data, see Burke et al. (1994). Irrespective of the merits of this particular intervention, it is unlikely that the U.N. General Assembly delegations as a whole have the background and expertise to cope with issues of fisheries conservation and management and it is inappropriate that they attempt to do so. In the following sections, I discuss the 1993–94 U.N. Conference on Straddling and Highly Migratory Fish Stocks, the FAO Agreement to Promote Compliance with International Conservation and Management Measures by Fishing Vessels on the High Seas, the 1994 Convention on Conservation and Management of the Pollock Resources of the Central Bering Sea, and recent unilateral actions by Argentina, Canada, Chile, and the USA.

The 1993–94 United Nations Conference on Straddling Fish Stocks and Highly Migratory Fish Stocks

Addressing high-seas fisheries management involves three sets of international law principles: procedural, jurisdictional, and substantive. The current negotiating text (revised) of the U.N. fish stock conference preserves the main procedural principle that flag states continue to have exclusive jurisdiction over their fishing vessels on the high seas, except as agreed otherwise, but seeks to make more explicit and sharpen flag state obligations over flag fishing vessels. In addition, the related new international pact—the FAO Compliance Agreement—addresses new obligations in the registry and re-registry of fishing vessels related to compliance with high-seas fishery conservation measures, as noted in following text. Flag states are particularly important for enforcement and compliance by fishing vessels on the high seas, which are also provided by traditional international law.

The main thrust of the negotiations in the U.N. fish stock conference concerns the jurisdictional and substantive principles that determine who makes conservation decisions and what their substantive content must be.

Jurisdictional principles.—Thus, in the specific context of the U.N. fish stock conference, obviously the foremost jurisdictional question is how to get around the constraint of freedom of fishing and to address who should be authorized to prescribe and to apply law for high-seas straddling and highly migratory stocks (i.e., stocks that are found within and beyond national jurisdiction). The key issue is the relative influence or competence of coastal and flag states in determining conservation and allocation of stocks. The lack of a single entity with competence to make decisions has been one obstacle, perhaps not the most important, to effective conservation.

In traditional international law, the high-seas principle of freedom of fishing insulates any fishing vessel from regulation or enforcement by other than the flag state. This is certainly a formidable constraint for a conference whose terms of reference specifically include the requirement that its outcome must be consistent with the UNCLOS treaty, which provides for continued freedom of fishing on the high seas. Part 3a of U.N. General Assembly Resolution 47/192 which convened the conference, reaffirmed that

> the work and results of the conference should be fully consistent with the provisions of the U.N. Convention on the Law of the Sea, in particular the right and obligations of coastal states and states fishing on the high seas, and that states should give full effect to the high-seas fisheries provisions of the convention with regard to fisheries populations whose ranges lie both within and beyond exclusive economic zones (straddling fish stocks) and highly migratory fish stocks. . . .

In general, at the stage of negotiations in 1994, the U.N. fish stock conference had made no large change in the existing distribution of authority to prescribe conservation measures. However, the Revised Negotiating Text (RNT) for the fish stock conference (U.N. 1994) has provisions that mildly, but definitely, erode the sole competence of the flag state. These strive to place greater weight on coastal state measures on the same stock and to strengthen regional organizations relative to the flag state.

The first of these provisions is para. 1 ("Objective"), which declares that conservation of straddling and highly migratory stocks requires that measures on the high seas and within national jurisdiction must be compatible. In achieving compatibility, the states concerned are to "respect any measures and arrangements adopted by relevant coastal states in accordance with the Convention [UNCLOS] in areas under national jurisdiction" and, inter alia, "shall . . . ensure" that high-seas measures are no less stringent than those for areas of national jurisdiction. Note that this provision applies to both categories of stocks, straddling and highly migratory.

If the states involved are unable to agree on compatible measures for the high-seas area, they shall resort to dispute settlement. But pending the completion of that process, the high-seas fishing states shall observe measures equivalent in effect to those applying within the EEZ.

In combination, these arrangements give priority to coastal state measures *beyond* national jurisdiction. The interesting development is the provision on dispute settlement which, to the extent all states are parties to UNCLOS, is as provided there unless agreed otherwise. According to Part XV of UNCLOS, disputes about high-seas fisheries are subject to compulsory dispute settlement. This arrangement means that disputed coastal state measures are in effect applicable to high-seas fishing unless overturned by a review in a dispute settlement proceedings.

Two other sets of arrangements temper flag state dominance: the provisions on mechanisms for international cooperation and on flag state responsibilities for compliance and enforcement (Sections IV and V). If these arrangements are implemented successfully, they either lessen the need to rely on flag state authority to regulate its own vessels or improve the likelihood of successful flag state regulation.

In the high-seas context, it is not surprising that a new agreement would stress the use of international mechanisms as means of cooperating for high-seas conservation. Such cooperation is the major obligation of states under the UNCLOS provisions on high-seas fishing. In the RNT, the central position of the regional fishery organization (RFO) or arrangement is evident in the provision—but not cast in mandatory form—that only participating or cooperating states should have access to the managed fishery (para. 15). Complementing this are the mandatory provisions aimed at inducing new entrants into a fishery to become members of an existing RFO (para. 21).

Paragraph 21 states that these new members of a RFO "shall be entitled to accrue benefits in exchange for the obligations that they undertake." However, these benefits need not necessarily take the form of direct participation in the fishery. An allocation of a right to participate is dependent on a decision by the RFO that takes into account a variety of factors, including the following:

- status of the stock;
- existing levels of fishing effort;
- the interests of existing participants;
- the needs of coastal communities that depend mainly on fishing for the stock involved;
- the fishing patterns and practices of the new participants and their prior contributions to conservation, including collection and provision of accurate data and the conduct of scientific research; and
- the special requirements of developing states from the region or subregion, particularly where they are culturally or economically dependent (or both) on marine resources.

The thrust of the provisions on flag states (para. 24–26) is to require the exercise of effective control over their flag fishing vessels, including an authorization for specific fishing operations through a license, permit, or some other form. Flag states are to issue an authorization only if they can exercise their responsibilities with respect to such vessels under UNCLOS and the new document (para. 25). As noted below, flag states are required to participate in a system designed to achieve effective compliance and enforcement, rather than act as an obstacle to it.

The other jurisdictional principles in the RNT concern compliance and enforcement. The main emphasis is on clarifying the duties of flag states in ensuring that their flag vessels comply with applicable high-seas measures. The RNT spells out the domestic measures that must be taken by flag states to implement their duties to support effective management (para. 24) and specifies the roles of the RFO and flag states in arrangements for compliance and enforcement (para. 31). In general, the exclusive jurisdiction of flag states on the high seas is maintained, but within that principle the RNT spells out clearly how that jurisdiction must be exercised in order to have effective high-seas fishery management.

The required flag state measures run the gamut of procedures for ensuring that fishing vessels observe conservation and management measures. The flag state is specifically obligated to exercise effective control over its fishing vessels (paras. 24 and 25). This is achieved by prohibiting fishing practices that undermine high-seas conservation and management measures. This prohibition is implemented by requiring authorizations for high-seas activities through licenses or permits or other forms. The flag state is required to undertake monitoring, control, and surveillance of the operations of its fishing vessels and related activities, including monitoring in accordance with regional schemes. The flag state must require regular reporting by its vessels of position, catch, and effort information. It must implement national and regionally agreed observer and inspection schemes.

The flag state must effectively enforce applicable measures wherever violations occur, including physical inspections of vessels (para. 25). If violations are established and sufficient evidence is available, the flag state must institute proceedings and possibly detain the vessel. Penalties shall be sufficient to secure compliance and act as a deterrent. In the event of a serious breach of applicable measures, the flag state must ensure that the vessel is prohibited from fishing until all outstanding criminal and civil judgments are satisfied. In addition, the flag state compliance measures shall include cancellation or suspension of authorizations to serve as vessel masters or fishing masters.

Although the flag state retains ultimate authority, a member of an RFO to which it belongs may by agreement stop, board, inspect, arrest, and detain a vessel for violation of applicable measures (para. 31), and may provisionally cancel a vessel's authorization to fish in the region concerned until the flag state takes appropriate enforcement actions (para. 32).

The RNT envisages the use of port states as a means of promoting effective conservation and management measures (Part VI). In this scheme, a port state may assist flag states, which may be remote from the area of operation of their flag vessels, to enforce flag state laws by inspecting documents and catch onboard vessels voluntarily in port. Such action might detect unlawful catches or fishing without an authorization or permit, which would be reported to a flag state and the offending vessel detained for such reasonable period as is necessary for the flag state to take over.

The RNT Part VII addresses the problem of fishing states who do not choose to participate in an RFO. Such states are obliged to cooperate in conservation of high-seas stocks even if they choose not to join the relevant RFO. The RNT seeks to encourage participation by directing that nonparticipating flag states not authorize their flag vessels to operate in the regulated fishery and that they not fish contrary to the measures established by the RFO.

Part VIII deals with dispute settlement, an essential element of an effective management system. Paragraph 45 provides that if disputing states are parties to UNCLOS, provisions of Part XV are applicable unless the parties agree otherwise. If UNCLOS does not apply, the dispute process called for by the RNT would still be compulsory and binding.

Substantive principles: conservation and allocation measures.—While adequate jurisdictional bases for management action are necessary, it is at least as important to ask how the RNT proposes to deal with overfishing and overcapitalization, recognizing that we are talking here about high-seas fishing, not all fishing. The main contribution of the RNT in this respect is to provide some specific content for the general provisions and principles that are found in UNCLOS and to introduce the notion of a precautionary approach to fishery management.

Section A of Part III of the RNT, entitled "General Principles," addresses "The nature of conservation and management measures" where language from article 119 of UNCLOS is repeated and enlarged upon. Specific conservation measures are mentioned in para. 3(b), including total allowable catches (TACs) and quotas, limits on fishing effort, size limits, gear and operational restrictions, and area and seasonal closures. This listing makes clear, inter alia, that management by TACs is only one of the approaches that can be employed, a proposition not that clearly stated in UNCLOS.

Paragraph 3(c) is a verbatim extract from article 119(b) and is notable for its reference to species, rather than "fish stocks." To this extent the outcome of the U.N. conference concerns marine mammals and is not limited solely to fish. Also worth notice is that this formulation concerns conservation of the species affected, as opposed to protection against any and all mortality.

Section B of Part III addresses "precautionary approaches to fisheries management," which "shall be applied widely." A set of appended guidelines for applying the precautionary approach declares that in the conservation context, "maximum sustainable yield [MSY] should be viewed as a minimum international standard." However, this is followed immediately by the statement: "Conservation-related reference points should insure that fisheries mortality does not exceed, and stock biomass is maintained above, the level needed to produce the maximum sustainable yield." This sentence is significant also because it strengthens and reaffirms the UNCLOS formulation of the conservation obligation to restore and maintain the "maximum sustainable yield as qualified by relevant environmental and economic factors. . . ." Moreover, the qualifying words are significant because they make it evident that the conservation obligation is not solely a problem of biological determination of stock abundance and "safe" yields. These words are frequently omitted in discussion of Articles 61 and 119 of CLOS.

In essence these guidelines also support the coastal state position that high-seas measures must also aim at maintaining a stock biomass that is favorable to maximizing the economic benefit of fishing. Thus, such a level is greater than that associated with MSY. Specifically, this appears to support Canada's position on the appropriate level of fishing mortality in the high-seas fishery off Canada's east coast.

These provisions on the precautionary approach do not go as far as those in the five-power draft convention.[4] Annex II, entitled "Selective Precautionary Measures on the High Seas," provides that a coastal state may assume management authority for an initial interim period "at the outset of the development of a fishery directed at a newly discovered stock." This provision is to be followed immediately by consultation with interested states that have fished in the region on interim catch and effort levels and appropriate interim management and conservation measures. On termination of the interim period, the duration of which is unspecified, the fishery is to be subject to the authority of the relevant RFO or, if none exists, "shall be subject to consultation" between coastal and other states interested in participating in the fishery. The Annex also prescribes some specific precautionary measures.

A similar approach is also provided for existing fisheries on straddling and highly migratory stocks when a coastal state determines that an emergency exists. The coastal state may prescribe emergency measures "for a reasonable period." During this period, the coastal and other interested states are to consult about the measures to be applied after a reasonable period.

The provisions of the RNT aimed at the problems of overcapitalization are pretty skimpy compared with those for conservation. Paragraph 3(c) in Section C (General Principles) provides that states shall "take measures to deal with overharvest and overcapacity and to ensure a level of fishing effort commensurate with the sustainable utilization of fisheries resources." However, to the extent that the basic reference points of the precautionary approach are also applicable to the problem of overcapacity, there is reason to be hopeful. Paragraph 4(f) indicates that this is indeed the case for new or exploratory fisheries, so presumably the approach applies to all fisheries.

The important point is that it is not enough to make provision for overall catch quotas. The level of fishing effort allowed is significant for the level of benefits available from the fishery, especially for coastal states fishing a stock that regularly mixes across the EEZ, which makes stock density important for the catch per unit of effort in the EEZ. The reverse also applies for high-seas fisheries.

These considerations also bear on the issue of new participants or new parties. This is not simply a question of avoiding overfishing. The addition of new entrants into the fishery is most important for its impact on the value of the harvest to existing participants. The effect of adding more fishing vessels is to reduce the share of all vessels and to increase the cost of taking that diminished share. Paragraph 21(a) specifies that in allocating participating rights to newcomers, account is to be taken of existing levels of fishing effort in the fishery. Unless this is done, there would be little hope for an effective regime.

The FAO Agreement to Promote Compliance with International Conservation and Management Measures by Fishing Vessels on the High Seas

The objective of this agreement, which was concluded within the framework of FAO under Article XIV of the FAO Constitution (no official citation for it yet), was to prevent the use of transferring registry of fishing vessels to avoid compliance with international conservation and management measures. This is done by specifying the obligations and responsibilities of flag states for their registered vessels in implementing the obligation to cooperate for conservation of fisheries on the high seas. The flag state is the key actor because it alone can take action on the high seas affecting its flag vessels that seek to avoid compliance with international conservation and management measures.

---

[4]Draft Convention on the Conservation and Management of Straddling Fish Stocks on the High Seas and Highly Migratory Fish Stocks on the High Seas, submitted by Argentina, Canada, China, Iceland and New Zealand, U.N. Doc. A/CONF.164/L.11/Rev. 1, 28 July 1993.

In addition to obtaining registry with a particular flag state, fishing vessels must also be authorized to undertake fishing on the high seas. No party is to allow its flag fishing vessels to fish on the high seas unless it has authorized them to do so. The fishing vessel must fish in accordance with the conditions of its authorization. For a vessel previously registered under another flag, the new flag state cannot issue an authorization to fish unless it is satisfied that the suspension of any previous authorization by another party has expired and that no authorization to fish has been withdrawn within the past 3 years. These conditions also apply to previous registrations with a non-party provided that sufficient information is available on the circumstances of the suspension or withdrawal. These conditions are not applicable where ownership of a fishing vessel changes and the new owner provides sufficient evidence that the old owner has no further interest (legal, beneficial, financial) in, or control of, the fishing vessel.

An authorization to fish on the high seas ceases when the fishing vessel is no longer entitled to fly the flag of the authorizing state. Accordingly, loss of registry means loss of authorization to fish on the high seas. Presumably, authorization to fish can be withdrawn or suspended or terminated for undermining the effectiveness of international conservation and management measures.

When a fishing vessel seeks registry, a party must ensure that the fishing vessel provides it with such information as it needs to fulfill its responsibilities under the agreement, including its areas of fishing operation, and its catches and landings. Therefore, a fishing vessel transferring its registry must supply the new flag state with any information relevant to whether it has previously undermined international conservation and management measures. A fishing vessel's previous operating history must be disclosed if it bears on its record of compliance with such measures.

Assuming the flag state issues an authorization to fish on the high seas, it is obligated to take steps to ensure that the authorized fishing vessel does not engage in any activities that undermine the effectiveness of international conservation and management measures. These steps should ensure that the undermining activities cease.

Before issuing an authorization to fish on the high seas, a flag state must be satisfied that it is capable of effectively exercising its responsibilities under this agreement. The flag state is obligated to maintain a record of all fishing vessels registered under its flag. More importantly, the flag state is obliged to ensure that its fishing vessels provide it with information about the area of their fishing operations and their catches and landings. The flag state is obliged to report promptly to FAO "all relevant information" about fishing activities by its fishing vessels that undermine the effectiveness of international conservation and management measures. The FAO compliance provides that, where a non-party fishing vessel is believed to have undermined effectiveness, a party shall draw this to the attention of the flag state and FAO and provide full supporting evidence to the flag state and a summary to FAO.[5]

Convention on Pollock Resources of the Central Bering Sea

After several years of negotiation, the Convention on the Conservation and Management of the Pollock Resources of the Central Bering Sea was concluded in February 1994 between the USA, the Russian Federation, Japan, Republic of Korea, the Peoples Republic of China, and Poland (no official citation for this agreement available at time of writing). The agreement established an international regime for this area, comprising an Annual Conference of the Parties to decide upon the allowable harvest level in the Central Bering Sea based upon an assessment of the total Aleutian basin pollock biomass by a Scientific and Technical Committee.

The Annual Conference also establishes the individual national quotas for parties fishing in that area. No fishing is allowed unless the Aleutian basin pollock biomass level is determined to exceed 1.67 million mt. This determination is to be made by consensus, failing which either the USA and Russia jointly, or in the absence of adequate scientific information, the USA alone effectively decides. The annual harvest level is graduated in accordance with the increase in the biomass level above the minimum.

Each party is authorized to enforce the regime in the Convention area, although only the flag state may conduct the trial and penalty phase. In the accompanying Record of Discussion, the USA and Russia state their intention to suspend fishing within their respective zones if the central basin biomass is below 1.67 million mt. They also said they should set appropriate harvest levels in their zones, taking into account the annual harvest levels in the Convention area.

What can one say about this in relation to the Straddling Stock Conference? An obvious observation is that this appears to be more like a regional fisheries arrangement than a formal institutional mechanism with a set of entities assigned various functions. At the least, this signifies that a formal institutional structure is not regarded by some states in one context as necessary to cope with a specific problem such as the harvest of pollock in the central Bering Sea. An obvious question is, "What factors are responsible for this outcome? Why didn't the participants create a formal institution?"

---

[5]FAO Compliance Agreement, Art. VI(8).

One answer might be that the states concerned with fishing in the central Bering Sea have been working together for several years to understand the nature of the stocks there—distribution, abundance, catch, natural and fishing mortality, age structure, spawning behavior, etc.—and that they are confident of their collective capacity to deal with the problem. Another answer is that there are two major coastal states involved, the USA and the Russian Federation, who have significant negotiating power and were able to use that strength to negotiate an arrangement that does not involve a significant institutional element. The treaty effectively delegates the most significant decision to these states, the determination of the pollock biomass in the central Bering Sea. A more complicated institutional setup might also complicate this decision.

Another component worth notice is that the parties to the agreement are assigned the task of creating an international regime for the high-seas portion of the central Bering Sea. The treaty does not confer any authority on the Annual Conference to make provision for pollock stocks within the Russian and U.S. zones of jurisdiction. In other words, the agreement contemplates measures applicable to the high seas alone, as opposed to measures that would be applicable to pollock stocks throughout the Bering Sea, including the U.S. and Russian EEZs.

It appears, at least, that this arrangement differs somewhat from the position the USA has taken in its assessment of the Negotiating Text for the U.N. Straddling Fish Stock Conference. In a paper released in February 1994, the U.S. Department of State[6] stated:

> Conservation and management measures for straddling fish stocks and highly migratory fish stocks, to be effective, should address the entire fishery stock as a biological unit throughout its entire geographical area or range of distribution, not just when the fish are on the high seas. We believe that management measures taken on the high seas have a strong interrelationship to the same or other measures taken in EEZs. Furthermore, some fundamental management measures, such as TACs, should apply to the stock as a whole. The current text, by contrast, does not promote this relationship strongly enough. Instead, it focuses almost exclusively on the need to improve conservation and management of these stocks only on the high seas.

The pollock agreement has an appended Record of Discussion that does not commit the coastal states to extend or adopt any of the high-seas measures to the areas of the EEZ, but it does contain statements that indicate the close relationship between high seas and coastal

measures. Thus, Part B records several "shared views" pertinent to what needs to be done on the high seas and in the areas of national jurisdiction. All the states record their view that the Bering Sea is a large marine ecosystem and that because the pollock occur within the EEZ as well as in the enclave "the effectiveness of management measures adopted for the Aleutian Basis pollock within the zones and the Convention Area respectively cannot be ensured unless both of such measures are based upon the best scientific information available and fully compatible with each other."

The USA and the Russian Federation also record their views that the

> measures to be taken in their respective 200-n.mi. zone should include, inter alia, the following:
>
> i. suspension of directed fishing for the Aleutian Basin pollock, when the Aleutian Basin pollock biomass established in accordance with the relevant provisions of the Convention is less than 1.67 million metric tons, and
>
> ii. limiting of the allowable harvest for the Aleutian Basin pollock to an appropriate level, taking into account any allowable harvest level to be set in accordance with the relevant provisions of the Convention.

In the final paragraph of the Record of Discussion, all the parties declare that

> in adopting conservation and management measures for the Aleutian Basin pollock within the Convention Area, all Parties to the Convention should take into account the compatibility of such measures with measures adopted within the zones and should, as well take into account the effect of measures taken in the Convention Area on ecologically related species throughout their range based upon the best scientific information available.

These statements are not provisions that impose binding obligations on those making them. Thus, we have very important statements about the nature and impact of conservation and management measures excluded from the actual agreement between the parties. The USA and Russia "should" prohibit fishing and "should" limit their allowable harvest level, but these are not obligations that other parties might dispute or use as the basis for challenge. It would be interesting to know why this arrangement was concluded in this form, in light of the negotiations simultaneously underway on general principles for straddling stocks.

Another interesting feature of the Bering Sea pollock agreement is the provision for producing fishery science and other information for the Annual Conference. The agreement establishes a Scientific and Technical Committee, which is the only subgroup created other than the Annual Conference of the Parties, and it is to compile, exchange, and analyze fisheries information; receive such information from the parties; and make recommenda-

---

[6]U.S. Department of State, Office of Marine Conservation, Bureau for Oceans and International Environmental and Scientific Affairs, U.S. objectives for conference sessions in 1994 (Feb. 1994), p. 9 (paper in file of author).

tions to the Annual Conference for the conservation of pollock and for the annual harvest level in the Convention Area. The Committee is to work in accordance with a Plan of Work established by the Annual Conference.

It is interesting to note that the contracting Parties did not make reference to, for any scientific or other purpose, the North Pacific Marine Science Organization (known as PICES), which was established also in 1992. It is to be recalled in this connection that the creation of PICES was foreshadowed by amendments made in 1978 to the Convention on the High Seas Fisheries of the North Pacific Ocean, which anticipated the creation of just such a scientific body. The purpose of PICES is to coordinate marine scientific research in the North Pacific, and its members include most of the members of the Bering Sea Pollock Convention, omitting only Poland and Korea.

Unilateral Actions

Recent unilateral state actions relating to straddling and highly migratory stocks include Canadian legislation authorizing enforcement of NAFO regulations against specifically named flag of convenience (FOC) states whose vessels have been operating in the NAFO área (Canada 1994). This is an area beyond the 200-mile Canadian fishing zone. It is well known that fishing entities from member states of the European Union have sought to escape NAFO regulations by flagging out. There is no doubt of the Canadian government's intention to seek to deter these and other vessels from operating in the NAFO region contrary to NAFO regulations.

Thus far, this effort has been successful since all challenged vessels of the FOC states have departed from the NAFO area upon being told that they would otherwise be arrested. The legislation does not spell out the jurisdictional bases for this action, which is bound to be controversial. The European Union has already responded by a letter expressing the view that the unilateral action calls into question the principles of management and exploitation found in the 1982 Convention on the Law of the Sea and that it deeply regrets this action.[7]

In another sector of the Atlantic Ocean, Argentine legislation provides that its national conservation legislation applies beyond 200 nautical miles to migratory species and to species "which form part of the food chain of species of the EEZ of Argentina" (Argentina 1991). Within the past year, Argentina has pursued and sunk two Taiwanese vessels fishing in the region. In each instance, the crews were removed before sinking the fishing vessels.

Chile has recently promulgated legislation providing for the establishment of an area beyond 200 miles within which Chile may exercise an undetermined authority for fishing and other purposes (Chile 1991). It is not entirely clear what Chile proposes to do in this so-called "mer presencial," but it is believed likely that actions will be taken unilaterally if agreement is not reached on resolving conservation problems in the area.

The U.S. unilateral action concerns highly migratory stocks, not straddling stocks. The International Dolphin Conservation Act (IDCA) of 1992 prohibits any person or any vessel from setting nets on dolphins in the eastern tropical Pacific after 28 Feb. 1994 (Pub. L. 102–523, 106 Stat 3425, 16 U.S.C.A. §§ 1411–118 (West Suppl. 1993)) (Pedrozo 1993). The prohibition applies to U.S. captains aboard foreign flag vessels and subjects an offender both to civil and criminal penalties (and a vessel to forfeiture) for this substantive offense and also for failure to permit an authorized official to conduct an inspection or search in enforcement of the legislation.

There seems clearly to be a tendency toward unilateral actions to resolve high-seas fisheries management problems, and the question now is whether the U.N. negotiations will be able to satisfy coastal states' and high-seas states' concerns sufficiently to inspire general concurrence in the resulting document, whatever its form might be. The current Revised Negotiating Text is a step or two forward, but whether it is enough is unclear. A key issue is whether a compulsory dispute mechanism can be devised that will be adequate for the purpose. It seems to me that this will require recognition of the interim authority of the coastal state to apply temporary conservation measures, or otherwise the system is not likely to work.

References

Argentina. 1991. Article 5, ACT 23.968. Boletin Oficial, 5 December.

Burke, W. T. 1993. UNCED and the oceans. Marine Policy 17:519-533.

Burke, W. T. 1989. Fishing in the Bering Sea donut: straddling stocks and the new international law of fisheries. Ecology Law Quarterly 16:285-310.

Burke, W. T. 1994. The new international law of fisheries. UNCLOS 1982 and beyond. Clarendon Press, Oxford, UK.

Burke, W.T., M. Freeberg, and E. Miles. 1994. United National resolutions on driftnet fishing: an unsustainable precedent for high seas and coastal fisheries management. Ocean Development and International Law 25:127-185.

Canada. 1994. Section 5.2, Coastal Fisheries Protection Act of Canada as amended.

Carroz, J. 1984. Institutional aspects of fishery management under the new regime of the oceans. San Diego Law Review 21:513-540.

Chile. 1991. Ley 19.080. Modificaciones a la Ley 18.892,

---

[7]Letter of May 20, 1994 from European Commission to Canadian Minister of Foreign Affairs Andre Ouellet and Minister of Fisheries and Oceans Brian Tobin (letter in file of author).

General de Pesca y Acuicutura. Diario Oficial de la Repub-
lica de Chile, 6 September.

FAO (Food and Agriculture Organization). 1993. World review
of high seas and highly migratory fish species and straddling stocks. FAO Fisheries Circular 858, FRI/C858, Rome.

Hollick, A. 1981. U.S. foreign policy and the Law of the Sea.
Chapter 2. Princeton University Press, Princeton, New Jersey.

McDougal, M., and W. Burke. 1985. The public order of the
oceans. New Haven Press, New Haven, Connecticut.

Meltzer, E. 1994. Global overview of straddling and highly
migratory fish stocks. Ocean Development and International
Law 25:255-344.

Nandan, S. 1993. Opening remarks to U.N. conference on straddling fish stock and highly migratory fish stocks, 2nd session. United National Doc. A/CONF.164/11, p. 4.

Pedrozo, R. 1993. The International Dolphin Conservation Act
of 1992: unreasonable extension of U.S. jurisdiction with
eastern tropical Pacific Ocean fishery. Tulane Environmental Law Journal 7:77-130.

Scheiber, H. 1989. Origins of the abstention doctrine in ocean
law: Japanese–U.S. relations and the Pacific fisheries, 1937–
1958. Ecology Law Quarterly 16:23-99.

UN (United Nations). 1993a. Definition of straddling stocks of
marine life and list of their main species. United National
Document A/CONF.164/L.18, 20 July.

UN (United Nations). 1993b. Draft convention on the conservation and management of straddling fish stocks on the high
seas and highly migratory fish stocks on the high seas. Submitted by Argentina, Canada, China, Iceland, and New
Zealand. United Nations Document A/CONF.164/L.11/Rev.
1, 28 July.

UN (United Nations). 1994. Revised negotiating text. United
Nations Document A/CONF. 164/13/Rev. 1, 30 March.

The European Common Fisheries Policy and Its Evolution

A. LAUREC AND D. ARMSTRONG[1]

Abstract.—The Common Fisheries Policy (CFP) covers all aspects, from resource conservation to marketing, of fisheries management in the Member States of the present European Union. The CFP covers a wide variety of situations, which this paper first describes in terms of resources, fleets, jobs, industry and the administration organization, and markets. The CFP is also the result of a specific history, the benchmarks of which are recalled. Special attention is paid to the description of the decision-making processes, to the part played by the various European institutions, and to the analysis of the difference between prerogatives that remain those of the Member States and decisions that must be taken at the European level.

The main frame of the CFP was established in 1983. Total allowable catches (TACs) divided into national quotas by the application of fixed allocation keys were adopted as the simplest way to allocate fishing rights between the Member States, which could and can define in their own way access rules to the various quotas. A mid-term review led to a revision in 1992. This revision aims to complement output management (TAC and quota) with input management (fishing efforts and capacities), achieving a better integration among the various elements, defining a multiannual framework beyond the annual decision-making process, and securing more efficient control and monitoring. The effective implementation of the reform based on principles decided in 1992 has, however, not yet been achieved. We analyze what still has to be decided to complete this reform, and indicate how the Member States might prepare for possible further evolutions.

While the European Common Fisheries Policy (CFP) is imperfect, some criticisms it receives result from misunderstandings about its purposes. The CFP cannot be compared with a policy developed by one country in isolation. In this paper, we describe what the CFP is, and what it is not and cannot be.

The first two sections—Fishery Resources in European Waters, and The European Fishing Industry—are an overview of the fisheries within the European Union (EU).[2] The third section, Basic Principles of the CFP, takes stock of the rules governing the CFP defined in the 1970s and implemented since 1983; the CFP is also subject to regular changes. The fourth section, The Second Decade of the CFP, outlines the main features of its evolution since 1992, which is still in progress.

Fishery Resources in European Waters

Large vessels in distant waters operate from most Member States of the EU. They operate in tropical tuna (*Thunnus* spp.) fisheries, in bottom fisheries off Africa, and in the north Atlantic. However, most of the catches

come from fishing grounds surrounding Europe. A very brief review of the most important resources in the various zones follows. For the Atlantic fisheries, more detailed information than that contained in the following pages can be found in Salz (1991). Such a review unfortunately does not exist for the Baltic or the Mediterranean seas. Nevertheless, information can be obtained in Salz (1993) or from the European Commission, Directorate Générale XIV.

Pelagic Species

Of the small pelagic species, Atlantic herring (*Clupea harengus harengus*) and sprat (*Sprattus sprattus*) are predominant in the Baltic and in the North seas, whereas anchovy (*Engraulis encrasicolus*) and sardine (*Sardinops pilchardus*) are important stocks from the Bay of Biscay to the Iberian Peninsula, and within the Mediterranean. Mackerel (*Scomber scombrus*) and horse mackerel (*Trachurus trachurus*) are found on most of the European continental shelf. Blue whiting (*Micromesistius poutassou*) occurs on the continental slopes in most areas. In the Baltic and around the British Isles, apart from Atlantic salmon (*Salmo salar*), and sea bass (*Dicentrarchus* spp.) in the western English Channel, medium-sized pelagic species do not support significant fisheries. In more southerly areas, sea bass and sparids may be important, especially in the Mediterranean Sea. Fisheries for tuna, primarily albacore (*Thunnus alalunga*) and bluefin (*Thunnus thynnus*) and tuna-like species, mainly swordfish (*Xiphias gladius*), are geographically limited

---

[1]The views expressed in this paper are those of the authors and do not necessarily represent those of the European Union.
[2]The European Union comprises the following Member States: Belgium, Denmark, Germany (post-reunification), Greece, Finland, France, Ireland, Italy, Netherlands, Portugal, Spain, Sweden, and the United Kingdom.

(in European waters) to the Mediterranean and to the oceanic waters of the Atlantic south of Ireland.

Roundfish Stocks

The main roundfish species are cod (*Gadus morhua*), haddock (*Melanogrammus aeglefinus*), pollock (saithe, *Pollachius virens*), whiting (*Merlangius merlangus*), and hake (*Merluccius merluccius*). In the Baltic Sea, only cod is important. The Celtic Sea, to the south of Ireland, is the southern limit for cod, saithe, and haddock, the abundance of the two latter species being low in this area. Hake, which is not a major species in the North Sea or west of Scotland, is the most important roundfish stock from the Celtic Sea to the Mediterranean. Red mullet (*Mullus* spp.) is important in southern fisheries, especially in the Mediterranean. Finally, Norway pout (*Trisopterus esmarkii*) is an important species in the northern part of the North Sea for industrial (fish meal) fisheries.

Benthic Resources

Of the flatfish in the Atlantic and adjacent seas (Irish Sea, English Channel), sole (*Solea solea*) and plaice (*Pleuronectes platessa*) are important in many fisheries from the Bay of Biscay to the north, particularly in the North Sea. Sole is also fished in the Mediterranean. Anglerfish (or monkfish, members of the family Lophiidae), and megrim (*Lepidorhombus* spp.) are important in most European Atlantic benthic fisheries. Anglerfish is also important in the Mediterranean. Sandeels (mainly *Ammodytes* spp., *Gymnammodytes* spp.) are of paramount importance in the North Sea for industrial (fish meal) fisheries.

Various species of shrimp and prawn, apart from *Nephrops norvegicus*, which has a very broad distribution, are taken by several local fisheries. Crab and lobster are important in the Atlantic (south of Ireland) and English Channel fisheries. Lobster are also important north and west of the British Isles. Fisheries for bivalves are limited to mussels (*Mytilus* spp.) and scallops (*Pecten maximus, Chlamys* spp.) in northern waters but are much more diversified in the south. Around the Iberian Peninsula and in the Adriatic Sea, bivalves support very important fisheries. Apart from sporadic events, fisheries for cephalopods are negligible in northern waters, but they are important in more southerly areas, including the Mediterranean.

Status of the Stocks

The status of the various stocks is described in great detail within various reports of the International Council for the Exploration of the Sea (ICES) and the International Commission for the Conservation of Atlantic Tu-

nas (ICCAT). The Mediterranean stocks have not been subject to systematic assessments. However, the reports from the General Fisheries Council for the Mediterranean offer a very useful source of information, as well as various recent reviews (Anonymous 1991a; Lleonart 1993; Caddy 1996).

Large roundfish species (cod, hake, haddock, saithe) are the most severely overfished stocks. Flatfish, *Nephrops*, and most stocks of small pelagic species have been considered until recently to be within safe biological limits. However, scientific diagnoses have become more pessimistic for various stocks of flatfish and small pelagic species (Anonymous 1996a), including sardines off the Iberian Peninsula, and various stocks of mackerel and herring.

Biogeographic Variation and Its Consequences

Even in waters under the jurisdiction of a single EU Member State, there may be much biogeographic heterogeneity. Although it is widely accepted that the Mediterranean exhibits the greatest ecological diversity, similar diversity also exists in other areas. Atlantic waters south of the Iberian Peninsula exhibit many biological similarities to the western Mediterranean and Adriatic seas, while the more eastern part of the Mediterranean has quite different species compositions. Diversity also increases from the Baltic Sea to the North Sea and, hence, to the west of Scotland, from north to south in Atlantic waters, and from west to east in the Mediterranean.

The breadth of the continental shelf is also important. It is wide in the North Sea and adjacent seas, and from west of Scotland to the Bay of Biscay, but it is quite narrow around the Iberian Peninsula and in most of the Mediterranean. Apart from fleets operating outside European waters and Spanish vessels operating in Atlantic waters north of the Bay of Biscay, offshore fisheries (wherein vessels make trips from a few days to a few weeks) are not significant in southern Europe whereas they are very important from the Bay of Biscay to northern Europe, including the North Sea. Coastal fisheries predominate on fishing grounds in southern Europe.

The consequences of this situation are numerous and important. In large-scale fisheries, catches are mainly taken by relatively few ships (from a few hundred to a few tens), and landings are made at a limited number of sites. The opposite is the case for coastal fisheries, which are conducted by many more vessels landing at numerous locations, making the monitoring of fisheries much more difficult. Alternatively, the narrowness of the continental shelf can give a local character to certain fisheries, simplifying their management in comparison to fisheries requiring international organization and regulation.

However, collecting fishery statistics is more difficult for small-scale fisheries.

The European Fishing Industry

Fleets

It was only at the beginning of the 1990s, after a difficult debate on the issue of overcapacity, that it was possible to establish a central computerized fleet register regularly updated by the Member States. Some standardization problems are still evident for the measurement of tonnages and horsepower. For instance, tonnage figures correspond to a mixture of gross tonnes and gross registered tonnes. Information on vessel equipment is also limited. Nevertheless, this fleet register is a major tool for the CFP.

Most of fleet-related elements in Table 1 come from this register. This table illustrates the greater importance of small-scale fisheries in southern Europe, especially in the Mediterranean. The level of motorization expressed in terms of kilowatts (kW) per crew member, which is an indicator of the levels of capitalization, also illustrates discrepancies among Member States. This would also be apparent when considering equipment for navigation or fish detection.

Vessels are, on average, more specialized in northern than in southern Europe. Within the North Sea, there are specific vessel designs for trawlers targeting flatfish (beam trawl), roundfish (otter trawl), and small pelagic species (mid-water trawl and purse seine). In the Atlantic fisheries north of the Iberian Peninsula, multipurpose vessels employ various gears according to their target.

In Spain or Portugal, or in the Mediterranean, a trawler can simultaneously catch species ranging from anglerfish to various pelagics, such as mackerel, horse mackerel, and blue whiting, while purse seiners concentrate on small pelagic species such as anchovies and sardines.

Vessel age composition and profitability are also very different among the various fleets. Some fleets, such as the North Sea beam trawlers, have been so profitable that, without any subsidy, they could be renewed rapidly. On the other hand, some fleets are quite old and in need of modernization. Still other groups of vessels, sometimes quite new and heavily subsidized when they were built, have been facing severe losses in recent years.

These differences make it more difficult to discuss the issue of overcapacity. Some Member States consider that they must promote the modernization of their fleets, through subsidies if required, giving their fishers access to safer and more comfortable fishing vessels that are technologically competitive with vessels from other Member States. Other Member States do not subsidize shipbuilding and nevertheless have to face a continuing overcapacity situation.

Employment and Crews

Table 1 gives the estimated number of fishers in the various Member States. These figures are to be taken with caution due inter alia to the existence of part-time fishers, especially in southern Europe. These figures indicate the varying social importance of fisheries. Remote areas such as islands, where alternative sources of income would be very difficult to create, are of special sensitivity. It must be recognized, however, that the

TABLE 1.—Key statistics for fisheries of the Member States of the European Union.

| Member State | Landings[a] | | Fishing fleets[b] | | | Jobs at sea[c] (number) | Key ratios | | | |
	Weight (metric tons)	Value (000 ECU)	Number of vessels	Total tonnage (T_{jb})	Total horse-power (kW)		Average mt per vessel	kW/ man[d]	KECU/ man[e]	ECU/ kilogram
Belgium	21,819	57,500	157	23,014	65,889	720	146.6	91.5	79.9	2.64
Denmark	1,853,436	380,200	4,993	98,772	412,723	6,567	19.8	62.8	57.9	0.21
Germany[f]	119,626	97,600	2,452	76,890	167,692	4,142	31.4	40.5	23.6	0.82
Greece	171,055	325,400	20,354	120,325	662,768	40,164	5.9	16.5	8.1	1.90
Spain	1,162,963	1,786,300	19,103	613,521	1,849,993	79,369	32.1	23.3	22.5	1.54
France	335,005	675,800	6,650	181,760	997,548	23,000	27.3	43.4	29.4	2.02
Ireland	291,215	136,600	1,421	55,235	190,501	4,919	38.9	38.7	27.8	0.47
Italy	353,197	1,020,100	16,434	259,981	1,513,871	45,000	15.8	33.6	22.7	2.89
Netherlands	544,562	316,000	508	152,928	436,197	2,834	301.0	153.9	111.5	0.58
Portugal	240,717	271,200	12,317	131,123	416,010	34,454	10.6	12.1	7.9	1.13
United Kingdom	667,584	1,041,900	9,983	239,783	1,104,406	23,000	24.0	48.0	45.3	1.56
Finland	103,421	25,900	2,959	174,608	174,608	2,750	59.0	63.5	9.4	0.25
Sweden	379,014	96,400	4,349	59,642	328,686	3,000	13.7	109.6	32.1	0.25
EU	6,243,617	6,230,900	101,680	2,187,582	8,320,892	269,919	21.5	30.8	23.1	1.00

[a]1995 figures for all (estimated).
[b]Figures as of 1 July 1995; for Italy, Sweden and Finland, provisional figures.
[c]Two part-time fishers are considered as equivalent to one full-time job ; figures as 1993 except Greece (1990) and Ireland (1991).
[d]Total horsepower divided by the number of equivalent full-time jobs.
[e]Value of landings divided by the number of equivalent full-time jobs.
[f]Post-reunification.

present level of knowledge is especially poor with regard to social issues in fisheries within the EU.

Differences among fishers in Member States are considerable, for instance in terms of educational level, age structure, and other variables. In some Member States, fishers cannot retire before the age of 65, but in others the age limit is 55 or even 50 where early retirement schemes exist. Labor costs are also quite different among Member States. These differences make it difficult to obtain a full assessment at the European level of the socioeconomic consequences of management decisions. They may sometimes also lead to conflicting interests, such as when some Member States promote the development of new fishing technologies to alleviate high labor costs, while for other Member States such innovations would disrupt profitable fisheries associated with a high number of jobs at sea.

Relationships Between Public Authorities and "Fishers"

The structure and the relative "weight" of professional organizations are heterogeneous within Europe. For example, in northern Europe, fishers who are members of producer organizations, which benefit from special recognition within the CFP, are responsible for a large part of the landings, but this is by no means the case in the Mediterranean.

The political relationships between the fishers and the authorities vary among Member States, and in each Member State it varies over time. In most Member States, the political impact of fisheries issues goes far beyond their economic importance. Some governments are more sensitive and responsive than others to concerns of the fisheries sector. The ability of fisher groups, and other sectors, to promote publicly and politically their point of view is also highly variable. The definition of a fishery policy is of interest not only to the fishers and primary producers, but also to processors, traders, related shipbuilding industries, and consumers. Furthermore, the taxpayers who provide the financial resources for public action and citizens who are sensitive to ecological questions or who sympathize with the economic difficulties of fishers are also instrumental. Clearly, the relative influence of the processing industries and of the fishers differs radically from one Member State to another. The same is true for the balance of influence between fishers and environmentalists.

With regard to public financial interventions in the fisheries sector, some Member States have favored subsidies. Others, especially in northern Europe, are eager to limit interference in the free market economy. Taxation regimes also vary considerably.

The way in which monitoring and control (enforcement) are organized is again variable. In some cases, the same services and even the same people within an administration deal with monitoring and control and also provide social assistance to fishers and to their families. In other Member States, some of the control tasks are dealt with by another administration (e.g., the navy) rather than a ministry in charge of fisheries. In still other Member States, control may be devolved entirely to a special administration. Legal systems also differ among Member States, resulting in different possibilities for applying administrative sanctions.

Markets

The EU market for fish and shellfish is very important, with total value in excess of 13 billion ECUs (first-sale values equivalent). About one half of this total is accounted for by imported products. Food consumption patterns differ from one part of Europe to another, and essentially from north to south, as suggested by the last column in Table 1. Seafood plays a much more important role in southern Europe, where consumers are prepared to pay a higher price. The demand also covers a broader range of products. There exists in southern Europe specific markets for small finfish and shellfish, often considered delicacies, which are not widely marketed in northern Europe. The diversity of landings from the Mediterranean is due to a combination of biological and market factors.

There is a strong synergy with the preponderance of coastal fishing in the south where, in small-scale fisheries, shellfish and small fish assume important roles. Landings that include a large variety of species constitute a safeguard against the variations in abundance of individual species. However, the existence of attractive markets for the smallest fish does not facilitate the protection of juveniles.

Basic Principles of the CFP

Need for a CFP

Despite regional differences in European fisheries, a common fishery policy is essential. The shared stocks that migrate from one exclusive economic zone (EEZ) to another constitute the main part of the catches (apart from small-scale fisheries). The fleet of a Member State may have had traditional access to other Member States' waters, both before and after 1976, when EEZs were extended. The overexploitation of crucial stocks required then, and still requires, a form of action exceeding any national framework. In addition, distant-water fisheries operating outside the waters of the Member States of the EU (Community waters) create similar requirements for the participant Member States, and the definition of common positions increases the possibility of achieving satisfactory fishery agreements with third countries. Except for some strictly local markets, the relative importance of which has decreased over the years, price-setting in a harbor depends on events largely beyond the region to which it belongs. In several cases, a fishery developed by a fleet of one Member State supplies the market of a second state, and sometimes operates in the EEZ of a third. For example, important fishing grounds exist in the northern European waters, but the most attractive markets are located in the south. Finally, it is inappropriate that a Member State would ignore the risk associated with fleet overcapacity in another state when the two states' fleets are in direct competition.

Even if it is impossible to build a totally uniform policy, it is necessary for the reasons just cited to lay down a common fisheries policy covering at least the management of resources located in the EEZ of the Member States, negotiations with third countries on the management of high-seas fisheries and straddling stocks, market regulations, and the policy regarding fleets.

Historical Dates

The current CFP, defined in 1970s, was implemented in its present form in 1983 for a period of 20 years. A detailed analysis of the first 10 years of the CFP can be found in Holden (1994). A mid-term review was envisaged, based on a report to be drawn up by the European Commission (hereafter the Commission) in 1991. Spain and Portugal joined the European Community (EC) in 1986, and this led to the definition of a regime that should have persisted until 2002, except for adjustments that were to be decided on the basis of a report by the Commission in 1992. However, Spain and Portugal were eager to obtain before the year 2003 their full integration within the CFP. The principle of such an integration was agreed upon in 1994. The corresponding effort management rules were defined in 1995 (Anonymous 1995a, b; see ensuing subsection entitled The Mid-Term Review).

The Scope of the CFP

The CFP covers markets, structures (including fleets), access to "external" resources located in international waters or in the waters of third countries, and the management of internal resources fished in the EC waters.

The market organization was first put in place in the 1970s. Unlike the Common Agricultural Policy in its initial version, the CFP does not aim to guarantee price levels, but simply to remedy excessive fluctuations in prices. Its mechanisms were updated in 1994. The structural policy was originally designed to assist the modernization of the fishing fleets and the processing and marketing sectors. Reducing overcapacity of the fishing fleet was not originally a priority. Biological advisors had repeatedly suggested reducing exploitation rates, but other arguments were successfully pushed forwards (e.g., existence of non-overfished stocks, needs for fleet modernization, scientific uncertainties, socioeconomic and political factors). The reduction of overcapacity has only gradually become a major target.

This paper focuses on the so-called conservation policy, which covers the management of stocks occurring within Community waters. The first principle concerns free access to Community waters, such that any ship flying the flag of a Member State may fish in the waters under the sovereignty or jurisdiction of its own or any other Member State. This first principle, however, was conditioned by exemptions keeping the coastal zone for the regional, and therefore national fleets, and by special provisions controlling access to an area surrounding the Shetland Islands (the Shetland Box). Moreover, at the time of the accession in 1986 of Spain and Portugal, access to waters was subject to additional limitations (the North Sea and the so-called Irish Box surrounding Ireland were not accessible to Spanish and Portuguese vessels).

Technical measures relate to rules regulating the use of the various fishing gear, aiming mainly at improved fishing patterns. They include mesh sizes, minimum landing sizes, closed areas (boxes), and other measures. Exploitation rates have been mainly managed by annual total allowable catches (TACs), shared among Member States via national quotas through fixed allocation keys. Those keys quantify the principle of relative stability—a political keystone of the CFP. The decision to rely on management of output (catches) through TACs and not of input (fishing effort) is sometimes astonishing to external observers. It must be understood first as a political choice making it possible to arrive at relative stability on as simple—and therefore politically readable—a basis as possible, and leaving to each Member State a broad margin of freedom to decide in its own way how to use its fishing opportunities. In addition, antecedents connected with the international fishing commissions in place before the new Law of the Sea, and also in scientific practice, made it possible to provide advice on annual TACs, which would not have been the case in 1983, for example, for a management regime based on control of fishing effort.

Direct effort regulation did not form part of the arsenal set up in 1983, except for minor exceptions (e.g., in the Shetland Box). However, at the time of the accession of Spain and Portugal, rules were enacted to combine quotas and effort limitations by area, gear, and target species. Those limitations mainly affected mutual access in their respective EEZs for Spain and Portugal and the Spanish fleet operating in the EEZs of France, the United Kingdom, and Ireland. Access to the groundfish fisheries was subject to an effort limitation regime: only vessels from a basic list were allowed to fish, and the number of vessels that could fish simultaneously was limited.

Finally, it must be stressed that the Mediterranean was not initially covered by the said conservation policy. The first package of technical measures for the Mediterranean was not adopted until 1994.

Decision Mechanisms

Within the EU (previously the European Community), the most important decision-making body is the Council, which holds meetings of the ministers in charge of fisheries within the various Member States. Decisions require a "qualified" (large) majority, each Member State having a certain number of votes (from 10 each for the United Kingdom, Germany, Italy, and France to 2 for Luxembourg). Within the Council, Member States assume the presidency in rotation for periods of 6 months.

Council decisions are based on proposals from the Commission. The Commission is led by a college of commissioners, designated by Member States' governments. The Commission is divided into a number of specialized general directorates of which Direction Générale XIV is in charge of fisheries. The Commission is in charge of taking initiatives. This corresponds essentially to various reports, or "communications," and proposals for Council decisions.

Before introducing proposals, the Commission consults the Scientific Technical and Economic Committee, which groups experts from the Member States designated by the Commission. The ICES also plays a very important part. Its Advisory Committee for Fisheries Management is systematically consulted for all relevant biological issues.

In most cases, the European Parliament must be consulted, but the final decision is made by the Council. The Council discussions may lead to departures from the original proposal from the Commission. A so-called presidency compromise may then be necessary. Unless it obtains unanimity from the Member States, this compromise must also be accepted by the Commission.

This complex decision-making process can be simplified. In some cases, it is not necessary to consult the European Parliament (e.g., for annual TACs and quotas). The Council can also agree to rely on "Commission regulations," in which case the Commission can submit a proposal to a management committee, where decisions can be made more easily. However, even for what would appear as minor issues, the Council has often been eager to maintain making decisions at the highest political level.

Most of the decisions covered by the conservation policy require a Council decision. Agreement with third countries must also be submitted to the Council. The evolution of the fishing capacities for the various Member States is guided by a multiannual guidance program established for 5-year periods. The decision-making process on fishing capacities was originally that of a "Commission regulation"; it now requires a Council decision.

Subsidiarity

According to the subsidiarity principle, a decision should be made at the lowest possible level. This implies that what can be decided within a Member State should not be brought up within the previously described European mechanisms.

Important responsibilities are not covered at the European level and correspond to the competence of the Member States. Well before the subsidiarity principle became one of the most common topics of European debates, the CFP had left essential prerogatives to the Member States.

The most important question for economists is the definition of fishing rights, including possible individual fishing rights, and transfer rules governing these rights. Here again, situations are so different among the Member States that homogeneous rules could not be adopted. The TACs are established and allocated among Member States according to CFP rules. Further allocations of quotas or other fishing rights between individual fishers or groups of fishers (e.g., producer organizations) are addressed at a national level. The decisions made by the Council must solve debates between Member States, not between individual fishers. This question touches a major prerogative of Member States—the possible leasing of a public resource—and any discussion of such a subject causes echoes that reverberate far beyond the fisheries sector. The problem was tackled in different ways by the various Member States. The distribution of national quotas may or may not be done by producer organizations, and may or may not go as far as individual fishing rights. In some Member States, no such procedures have been initiated.

Practically all conceivable systems have been or are being used, including that of individual transferable quotas (e.g., flatfish fisheries in the Netherlands). Complete systems have been devised in which quotas are successively smaller than what fleets could freely catch. In a number of fisheries, Member States underutilize their quotas for various reasons. In such a case, no precise allocation procedures have been built. On the other hand, when constraints are too severe, fishers tend not to comply with any rule.

Another issue that deserves specific comment is that of enforcement. The credibility of any fisheries policy relies on its effectiveness. The various groups of fishers always fear that the CFP is applied with unequal rigor from one Member State to another. Thus, there are important arguments for implementing direct European control. However, the concept of subsidiarity implies the avoidance of any expansion of interventions by the Commission that do not appear strictly necessary. One touches here, as for the delimitation of the rights of private property, on major prerogatives of Member States with respect to policing and justice. This is why the role assigned to the Commission within the CFP was not that of direct control, but that of supervision of controls. One sometimes speaks about a control of controls, which must guarantee each Member State, with overall transparency, the equity of the efforts made by everyone.

The Second Decade of the CFP

The Mid-Term Review

As requested, a report was prepared by the EC in 1991 (Anonymous 1991b). The "91 Report," as it was called, was discussed and unanimously approved by the Council in 1992. In the EU, just as in many other areas throughout the world, fisheries could be summarized by the phrase "too many fishers, using too efficient fishing vessels, chasing too few and too small fish."

Although TACs were mostly in line with the "scientific" recommendations within the Commission's proposal, the Council often agreed on less conservative figures. In addition, actual catches were sometimes much larger than the agreed quotas. Not surprisingly, it appeared that a fishery policy relying on catch limitations had not solved the overfishing problem, which had worsened from 1983. Up to 1991, this had not resulted in a major stock collapse nor in an abrupt socioeconomic crisis, but such crises could be feared. In 1993 and 1994, especially in France, such a crisis occurred, owing to a combination of structural factors (overcapacity resulting through reduced stock levels in lowered catch rates) and unpredicted external events (e.g., changes in currency rates).

Without denying the merits of what was done and attempted, it was concluded that the CFP needed major improvements. Among others it appeared necessary to
- complement output (TAC) management by an increased use of input (effort) management,
- establish a better integration between the "conservation policy" (resource management) and the structural (fleet) policy,
- set the annual decision-making process within a longer-term framework, and
- implement much more efficient monitoring of the fisheries.

While the so-called 91 Report was being discussed, a difficult debate took place about the overcapacity question. On the basis of comparisons between current exploitation rates and biological reference point, mainly F_{MAX}, an independent group of experts had concluded in a previous report (made in 1990 at the request of the EC, and known as the "Gulland Report") that fishing capacities within the EC should be reduced by 40%.

A multiannual guidance program had to be established for the period 1992–96. Difficult discussions took place, on the basis of a proposal from the Commission that attempted to firmly address the overcapacity question, even if it did not appear reasonable to reduce the European fleet by 40%. Member States could only accept the principle of a moderate reduction of fishing capacities. They were also eager to avoid any "unilateral disarmament," and wanted a community decision ensuring balanced restrictions for all Member States. The final reductions agreed upon (20% for roundfish trawl fisheries, 15% for towed gear fisheries for flatfish and other benthic species, stabilization for pelagic species, and fixed gears) could not bring back exploitation rates in the vicinity of reference points, such as F_{MAX}, F_{OPT}, or $F_{0.1}$, but the debate established the necessity for active management of fishing capacity and effort.

Another report prepared in 1992 (Anonymous 1992a) reviewed the consequences of the special regime resulting from the accession of Spain and Portugal. The report concluded that the corresponding special effort management (sub)regime should be integrated within an overall effort management regime applied to all fleets.

First Steps Following the Conclusions of the Mid-Term Reviews

By the end of 1992, a new basic regulation was adopted (Reg. 3760/92; Anonymous 1992b). It established a framework for achieving the first three improvements to the CFP (described previously under "The Mid-term Review")—effort management, integration between conservation and structural policies, and a multiannual framework.

In 1993, a new monitoring and control regulation was adopted (Reg. 2847/93; Anonymous 1993a). It confirms the role to be played by the Commission monitoring services (mainly supervision of the national administrations) but creates better conditions for it (e.g., the possibility of unannounced controls). It extends the competence of the European inspectorate, previously limited to the monitoring of the conservation policy (TAC and quotas, technical measures), in order to cover all aspects of the CFP, including the structural policy through monitoring of fishing capacity. Furthermore, national administrations should establish computerized databases, storing the information issued from various sources such as logbooks and landing declarations. Member States must define validation and cross-checking procedures for the various data, the Commission services having full access to the databases and the possibility of evaluating the validation procedure efficiency. Where derogations from this system for the smallest vessels are applied, sampling schemes should be established by national administrations to estimate globally the catches and effort of such fleets. The way has been paved for effort management, since the required information will be collected and stored. Concurrent approval was given for pilot projects to evaluate the potential use and cost-benefit relationship of satellite monitoring of fishing vessel activity.

The new basic regulation (Reg. 3760/92), and the new monitoring regulation (Reg. 2841/93) were necessary first steps for achieving the improvements suggested by the 91 Report. Further steps have been achieved recently.

- Various regulations were adopted in 1993 and 1994, which relate to licenses and fishing permits, in order to establish legal bases for effort management and to put an end to the "open entry" situation at the European level.
- Regulations were adopted in 1994 and 1995 (Anonymous 1995a, b) to manage the "western fisheries" from west of Scotland to the Strait of Gibraltar and also the Irish Sea and the English Channel. This overall effort management regime applies to all Member States and eliminates, in a homogeneous way, specific previous effort regulations resulting from the accession of Spain and Portugal, even if Spanish vessels still have no access to some areas (e.g., Irish Sea). The keystone of the new effort regime corresponds to the definition of effort quotas (expressed in kW days) per Member State and per fishery, fisheries being defined by the combination of fishing areas, gears, and target species. It does not, however, replace TACs and catch quotas, which remain the basic management tools.

On the other hand, the Council still has to decide how to establish a multiannual framework for stock management. A communication from the European Commission (Anonymous 1993b), analyzing how the new elements included in the new basic regulation (Reg. 3760/92) could be used in practice, was submitted to the Council. In accordance with the content of this communication, two complementary proposals were made.

The first proposal (Anonymous 1993c) deals with medium-term objectives, mainly expressed in terms of spawning biomass thresholds and exploitation rate targets, established on the basis of stock assessments as conducted mainly within the ICES framework. It also suggests associated strategies for achieving the medium-term objectives. This proposal results from repeated scientific advice stressing the need to reduce exploitation rates on a number of overfished stocks, and to rebuild or protect minimum spawning stock biomasses. It has deliberately avoided quantitative references to long-term objectives such as maximizing yield profits, rents, or jobs. Scientists would sometimes be eager to obtain a single objective from managers so that strategies could be defined in order to maximize this objective. However, this would imply choices, between producers and consumers for instance. Such choices would vary between the various EU Member States. Thus, it would be impossible to obtain an immediate consensus. In fact, even within a Member State, priorities are likely to change over time. Before a long-term objective is reached, it could well become obsolete. It is not now possible to find a consensus for the definition of long-term objectives (J. Horwood and D. Griffith, MAFF, Directorate for Fisheries Research, Fisheries Laboratory, Lowestoft, United Kingdom, and Department of the Marine, Fisheries Research Centre, Abbotstown, Dublin, Ireland, respectively, unpubl. rep.; Anonymous 1993b). On the other hand, the first priority is to ensure a step in the right direction (e.g., reducing significantly the exploitation rate of stocks of large roundfish). As progress is made toward first-step targets, further analyses will make it possible to define long-term objectives, or at least to establish a second step.

The second proposal (Anonymous 1994a), about which the Council reached a decision in 1996, introduced flexibility for the management of annual quotas. A Member State could benefit in the following year from the underutilization of some annual quotas. Conversely, if the State overshoots an annual quota, a deduction should apply in the following year, using a penalty coefficient that increases with the magnitude of the overshoot. The mechanism includes some safeguards, especially to prevent an accumulation over the years of the discrepancies between quotas and catches. This system will be more than a minor adjustment since it will facilitate management within Member States. Previously, for all fishers concerned when a quota has been allocated among groups within a Member State, it was necessary to stop fishing when the national quota had been caught because some group(s) had exceeded their share(s) even though others had not. The more flexible approach will make it possible to avoid such drawbacks. It will facilitate quota allocation within Member States, as well as co-management involving fishers' organizations, as is being presently attempted in several Member States. In addition, a more flexible approach for annual quota management could complement the more ambitious mid-term objectives.

What Has Yet to Be Achieved

Completion of regulations.—The first item, completion of regulations, corresponds to decisions to be taken by the Council on the basis of existing proposals from the Commission, as previously mentioned. If the Council reached a decision in 1996 introducing flexibility in annual quotas(Anonymous 1996b), this had not yet been achieved for medium-term objectives. Once decisions have been made, they will have to be implemented. This is especially true in terms of enforcement. The existing regulation (2841/93) offers a number of possibilities, which up to now have been only partially exploited. Some elements were to become compulsory only in 1996 (computerized databases) or 1999 (logbooks and landing declarations in the Mediterranean). A number of specific complementary regulations also have to be decided.

Moreover, in some Member States, progress still has to be made to establish more reliable enforcement of the various components of the CFP. The EC will help, especially in terms of financial assistance for purchasing the proper equipment, but in terms of personnel or administrative organizations, as well as for ensuring sanctions that act as deterrents, the responsibility lies with the Member States. The Commission will, however, report annually on the results achieved by Member States. This will offer the basis for a public debate, which should stimulate the less efficient Member States.

Revisiting effort management and overcapacity.—If the first impetus for a change in the CFP can be associated with the mid-term review, the adaptation process is continuous. For instance, technical measures have to be adapted to improve the protection of juvenile fishes and to reduce bycatches of nontargeted species, including non-commercial ones (Anonymous 1995c).

Since the decision-making process is more complicated within the CFP than it is for the management of other fisheries, even more persistence is necessary. This is especially true for the overcapacity issue. Regardless of the management tools applied, no sustainable solution will be found before this question is resolved. When the next multiannual guidance program is discussed, it will be possible to make use of important progress, such as a much larger acceptance of the overfishing diagnosis and the collection of more precise data for effort management. However, precise, quantitative decisions have still to be made. Between annual TACs and decisions on the evolution of the fishing fleet, a multiple-year time lag exists: TACs are calculated mostly for a single species in a specific area for a single year, while a vessel is operational for several years, if not decades, and operates in various areas for various species. To some extent, regulating fishing effort corresponds to an intermediate management tool. Effort can be managed in fisheries that

group various species. It is easier to obtain scientific advice regarding the evolution of fishing mortalities over several years, which in turn can be related to fishing efforts, than to establish the level of TACs several years in advance. A well-balanced policy will be achieved when agreement is reached on harmonized, multiannual frameworks covering capacity reduction and effort management schemes that are in line with the fishing mortality targets (see medium-term objectives as explained in the previous discussion of the first proposal submitted by the Commission in 1993).

Toward a better debate among scientists, fishers, and other partners.—Flanking measures can provide funding, for instance, for pre-retirement schemes or development of alternatives to fishing. This will make it easier to achieve the difficult adjustments still facing the European fisheries, but painless evolution appears impossible. This is why it is more necessary than ever to promote dialogue among the various partners involved in fisheries, including fishers, administrators, and scientists. Effective management cannot be imposed on fishers if they are not convinced that constraints are necessary and equitable.

More sophisticated research may be less important for successful management than communication with non-scientists about basic concepts, such as yield per recruit or spawning biomass. Until now, within the CFP, scientific results have been largely "underexploited" because they have not been accepted by nonscientists. A number of basic scientific conclusions are not yet accepted by numerous fishers, or at least fishers' representatives. The need to reduce exploitation rates, and thus fishing capacities, is still denied by too many people. A year-to-year increase in a stock abundance because of some improvement in recruitment is commonly interpreted as proof that the scientific "pessimism" was unfounded.

Securing better acceptance of the basic scientific conclusions is a priority, but it will not be easy. Fishers eager to avoid constraints underestimate their influence on stock abundances. Conversely, if they feel overwhelmed by competition with other fishers, especially those operating under a different flag, they will tend to exaggerate the overfishing problems. If scientists are too close to fishers, it may become more difficult for them to make unbiased assessments. If they are not close enough, they will not be trusted. Moreover, public opinion tends to favor black and white answers. A subtle scientific diagnosis in which a population, whether it corresponds to a target species or to bycatches of marine mammals, is depleted by fishing but not put in danger of collapse will be disappointing. The recognition of scientific uncertainties, although absolutely necessary, will make it even more difficult to establish a dialogue between scientists and nonscientists.

Improving dialogues is absolutely necessary. It implies direct discussions involving fishers, scientists, and managers. Whenever possible, such debates should take place at the smallest possible geographical scale. For instance, within the CFP, discussions can be established among fishers, administrators, and scientists at the scale of the Irish Sea.

The establishment of appropriate fora must also account for the growing worries of the public concerning the environmental impact of fisheries. The CFP must take the corresponding effects into consideration. Fishers are not the "owners" of the sea. On the other hand, it is sometimes too easy for nonfishers to promote restrictions on fishing activities. It will never be possible, because of conflicting interests, to build solutions that will please all groups, but compromises will be easier to reach if the proper discussions take place, taking into account the various points of view. Solutions will be facilitated if each group is aware of the expectations and fears of the other partners, and if scientists can quantify the likely consequences of the various decisions. It will also make it possible for scientists to compare their analyses with practical experience.

Paving the Way for the Long-Term Future

Limited entry schemes and their consequences.—It might seem premature to anticipate the evolution of European fisheries. Nevertheless, the evolution of the CFP corresponds to a classical process observed in other fisheries. Overfishing is well known as the symptom of overcapacity, which is due to the open-access regime. The first management attempts tend to be limited to remedies for the symptoms, corresponding with output limitations and the definition of TACs. The second step usually focuses on the reduction of fishing capacities and on effort management. The third level, which may not be the final one, establishes limited entry schemes. Within the European Union, the evolution is made more complicated by the existing differences among the Member States and by the necessity to combine Community decisions, made at a European level, and decisions that fall under the Member States' responsibility. The CFP, established in 1983, focused on TACs and quotas and took a low profile on the overcapacity question, leaving Member States responsible for deciding when it was appropriate to apply effort regulations or limited entry schemes. More emphasis has now been put on capacity reductions and effort management while a symbolic end to the open-entry principle has been enacted by license and fishing permit regulations. As detailed previously, much has yet to be done to ensure an efficient combination of TACs and quotas, effort management, and capacity regulation. Going further and fully defining rules for the allocation of fish-

ing rights probably will not be achieved at the EU level. This does not imply that the Commission ignores the likely evolution toward more refined systems that allocate fishing possibilities. It simply corresponds to the subsidiarity principle. Nevertheless, at Community levels it will be necessary to do the following:

- ensure that the common framework does not hamper attempts by Member States to efficiently manage their fishing possibilities;
- ensure that the rules established within each Member State do not contradict the basic CFP and EU principles—for instance, distorting competition among fishers from the various Member States;
- take advantage of the experience gained by any Member State, which should be made available to the other Member States; and
- promote the development of proper socioeconomic research.

Such research should include the classical analyses of the bases to be chosen for defining fishing rights (catches, effort, territorial rights, etc.) as well as possible keys to establish the initial allocation and transfer rules to be applied later. It should also compare systems that allocate individual rights with those that limit the allocation procedure to a sharing of the total possibilities among groups of fishers. The research should analyze the role that public authorities might play in regulating the market of fishing rights, protecting employment in specific areas, or ensuring that part of the rent can be recovered for public budgets.

Securing efficient research.—Research is by no means the panacea. As previously noted, much needs to be done to make use of the most basic research results available. Nevertheless, no management is possible without a proper understanding of the likely consequences of the decision to be made. Research must be continued, enlarged, and improved. The situation in this respect within the EU was reviewed by a report from the Commission (Anonymous 1993d). Priorities were recently adopted for the 4 years from 1995 to 1998, corresponding within the European terminology to the fourth framework program (see Anonymous 1994b and the corresponding work program). The reader can refer to the corresponding documents for a more complete review.

The priorities retained for the fourth framework correspond to the following:

- the effects of environmental factors on fish and fisheries,
- the ecological impact of fisheries and aquaculture,
- the biology of species of interest for optimization of aquaculture,
- socioeconomics, and
- methodology.

Beyond these priorities, research on the processing of

fish products will be promoted in scientific programs not necessarily limited to seafood. (This last point, as well as the third point in the previously mentioned priorities will receive no further comment in the present paper.) The first two priorities correspond to the necessity to anticipate future dialogues covering fisheries management and environmental issues, and to establish a solid scientific ground for those dialogues. The priority given to socioeconomics corresponds to the urgent need for precise answers to specific questions (illustrated in the preceding section, "Limited entry schemes and their consequences") and to the present weakness of this domain within the EU. The last topic acknowledges that much of what is presented as fishery research is not innovative in nature; rather, it is a routine use of existing methods.

While innovative research must be promoted, it is absolutely necessary to ensure collection of basic data covering biological and socioeconomic issues, as well as data related to fleet structure and activities. In this respect, the situation is far from satisfactory within the EU. In some areas, such as the Mediterranean, the improvements have been very slow, and much basic information is still unavailable. In other areas (e.g., the North Sea) where the situation had been more satisfactory, the situation is in danger of worsening because of the unreliability of some official statistics and because of a decrease in some Member States' data collection budgets. This development could well be the most severe danger facing fishery research. If it has to be more innovative, it cannot exist without the proper time-series of basic statistics, the collection of which is unavoidably costly, especially when research vessels are required. The famous sentence from J. Gulland, according to whom "the right to fish implies the duty to provide data," remains valid, which is why this matter has been given a high priority within Directorate Générale XIV, including in terms of budgets. This is also why discussions about improved administrative statistics, which cannot be separated from the improved monitoring and control, must take into account the needs of reliable, disaggregated, comprehensive statistics for research.

Conclusion

The CFP is evolving rapidly on the basis of the experience gained since 1983. The limits of certain approaches have been recognized. No one is naive enough to believe that once the adaptations suggested by the mid-term review have been adopted all problems will disappear. But it is clear that within such a heterogeneous domain as the European fisheries, there cannot be a simple panacea. Moreover, imposing solutions imagined by any group of experts in a top-down way will never result in efficient management.

There is, however, almost a consensus about the directions to choose for improving the CFP, and it is clear that significant progress is being made. The key question is whether this evolution moves fast enough, taking into account that, simultaneously, some problems could become more and more difficult to solve, such as increases in overcapacity due to technological improvements. The faster the progress, the less difficult will be the adjustments faced by fishers and associated sectors. The progress rate will depend on the ability to better assess the whole range of consequences of various decisions, and on social or political acceptance of the corresponding conclusions.

References

Anonymous. 1991a. 19th Report of the Scientific and Technical Committee for Fisheries, 27th May 1991. Commission of the European Communities, Directorate General for Fisheries, Brussels.

Anonymous. 1991b. Report 1991 from the Commission to the Council and the European Parliament on the common fisheries policy. Office for Official Publications of the European Communities, SEC (91) 2288 final, Luxembourg.

Anonymous. 1992a. Report 1992 by the Commission to the Council and Parliament on the application of the act of accession of Spain and Portugal in the fisheries sector. Office for Official Publications of the European Communities, SEC (92) 2340 final, Luxembourg.

Anonymous. 1992b. Council Regulation (EEC) No. 3760/92 of 20 December 1992 establishing a Community system for fisheries and aquaculture. Page 1 in Office for Official Publications of the European Communities, Official Journal L 389, 31/12/92, Luxembourg.

Anonymous. 1993a. Council Regulation (EEC) No. 2847/93 of 12 October 1993 establishing a control system applicable to the common fisheries policy. Page 1 in Office for Official Publications of the European Communities, Official Journal L 261, 20.10.1993, Luxembourg

Anonymous. 1993b. The new components of the common fisheries policy. Office for Official Publications of the European Communities, COM(93) 664 final, Luxembourg.

Anonymous. 1993c. Proposal for a Council Regulation (EC) fixing management objectives and strategies for certain fisheries or groups of fisheries for the period 1994 to 1997. Office for Official Publications of the European Communities, COM(93) 663 final, Luxembourg.

Anonymous. 1993d. European fisheries research. Current position and prospects. Commission of the European Communities, Office for Official Publications, COM(93) 95 final, Luxembourg.

Anonymous. 1994a. Proposal for Coungil Regulation (EC) introducing additional conditions for year-to-year management of TACs and quotas. Office for Official Publications of the European Communities, COM(94) 583 final, Luxembourg.

Anonymous. 1994b. Council Decision of 23 November 1994 adopting a specific programme of research, technological development and demonstration in the field of agriculture and fisheries (including agro-industry, food technologies, forestry, aquaculture and rural development) (1994-1998).

Page 73 *in* Office for Official Publications of the European Communities, Official Journal L 334, 22.12.1994, Luxembourg.

Anonymous. 1995a. Council Regulation (EEC) No. 685/95 of 27 March 1995 on the management of the fishing effort relating to certain Community fishing areas and resources. Page 5 *in* Office for Official Publications of the European Communities, Official Journal L 71, 31.03.1995, Luxembourg,

Anonymous. 1995b. (3) Council Regulation (EEC) No. 2027/95 of June 1995 establishing a system for the management of fishing effort relating to certain Community fishing areas and resources. Office for Official Publications of the European Communities, Official Journal L 199, 24.08.1995, Luxembourg.

Anonymous. 1995c. Implementation of technical measures in the common fisheries policy. Commission of the European Communities, Office for Official Publications, COM (95) 669 final, Luxembourg.

Anonymous. 1996a. Report of the ICES Advisory Committee. International Council for the Exploration of the Sea Cooperative Research Report 214, Copenhagen.

Anonymous. 1996b. Council Regulation (EC) No. 847/96 of 6 May 1996 introducing additional conditions for year-to-year management of TACs and quotas. Page 3 *in* Commission of the European Communities, Office for Official Publications, Official Journal L 115, 9.5.1996, Luxembourg.

Caddy, J. F., editor. 1996. Resource and environmental issues relevant to Mediterranean fisheries management. General Fisheries Council for the Mediterranean 66, Food and Agriculture Organization of the United Nations, Rome.

Holden, M. J. 1994. The Common Fisheries Policy (origin, evaluation and future). Fishing News Book, Oxford, United Kingdom.

Lleonard, J., editor. 1993. Northwestern Mediterranean fisheries. Scientia Marina 57(2-3):105-271.

Salz, P. 1991. The European Atlantic fisheries; structure, economic performance and policy. Agriculture Economics Research Institute (LEI-DLO), The Hague.

Salz, P. 1993. Regional, socio-economic studies in the fisheries sector. Summary report. Agriculture Economics Research Institute (LEI-DLO), The Hague.

Fisheries Management in Canada: The Case of Atlantic Groundfish

L.S. PARSONS AND J.S. BECKETT

Abstract.—Canada's marine fisheries have undergone major changes in recent decades. In just 25 years, these fisheries went from underdevelopment to overcapacity. Regulatory interventions have mushroomed. Despite the benefits that flowed from extension of fisheries jurisdiction, Canada's marine fisheries continue to be plagued by instability because of a combination of factors, including (1) natural resource variability, (2) the common-property nature of fisheries resources leading to overcapacity and overfishing, (3) market fluctuations, (4) heavy dependence on fisheries as the employer of last resort in isolated coastal communities, and (5) conflicting objectives for fisheries management. Atlantic cod (*Gadus morhua*) stocks have collapsed in recent years, necessitating the imposition of moratoria on fishing. We must err on the side of caution to promote stock rebuilding. There is an urgent need to bring harvesting and processing capacity into balance with sustainable resource levels.

There has been a multinational fishery in the northwest Atlantic for nearly 500 years. The present status of fish stocks is, however, in marked contrast to the abundant resources exploited by early visitors from the east. The declines in Canadian Atlantic groundfish fisheries as shown by total allowable catches (TACs) and catches from 1988 to 1994 (Figure 1) should not, however, mask the fact that most Canadian fisheries do not show these drastic declines—for example, salmon on the Pacific coast (Figure 2) or lobster (*Homarus americanus*) (Figure 3) on the Atlantic coast. In the case of lobster, there was a rapid increase in the 1980s to levels higher than those seen in the previous 100 years.

Management measures were first introduced in Canada as early as the 1700s for Atlantic salmon (*Salmo salar*). Salmon can be seen readily as they ascend rivers, which attracts attention to possible problems. Indeed, by the end of the 1700s, there were concerns about the need for fishways, about the impact of effluents, and—even at that early time—about driftnets in the province of New Brunswick. The early development of fisheries management in Canada was facilitated by the introduction of the Fisheries Act in 1868, which set out the federal government's jurisdiction over fisheries. Federal jurisdiction is not all-encompassing, however, because the provinces license fish plants and stimulate boatbuilding, both of which impact the level of fishing effort. Quotas were introduced as early as 1920 in some freshwater fisheries and individual transferable quotas (ITQs) were even implemented before World War I in one freshwater fishery.

After World War II, there was euphoria in Canada about fisheries; reports extolled the great potential of the fisheries and advocated expansion of trawling and seining. As Canada entered a period of expansion, so did the rest of the world. Distant-water fleets from many western European countries and the USSR came to the western Atlantic, and there was an enormous build-up of effort in the 1960s and 1970s. Catches increased rapidly through the 1960s, followed by a decline in the 1970s, until the extension of jurisdiction to 200 mi (320 km) (Figure 4). The Canadian groundfish catch also declined through the 1970s but increased considerably as stocks recovered up

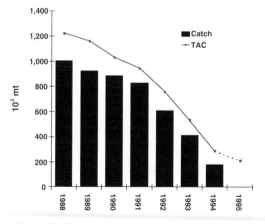

FIGURE 1.—Recent trends in Atlantic groundfish catches.

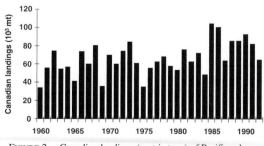

FIGURE 2.—Canadian landings (metric tons) of Pacific salmon.

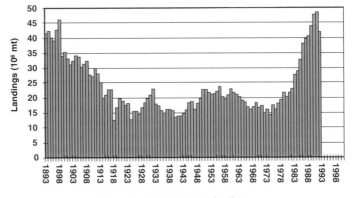

Figure 3.—Atlantic lobster landings.

to the early 1980s (Figure 4). After that, a different set of difficulties began. The operations of all enterprises, particularly large vertically integrated companies, were influenced greatly in the early 1980s by changes in market conditions, currency exchange rates, and interest rates. These factors resulted in major restructuring of the industry. Indeed, the Atlantic groundfish fishery went through a series of boom-and-bust periods from the 1960s to the 1990s—some due to market downturns, some due to resource downturns, and others due to both (for a detailed analysis of these trends and their underlying factors, see Parsons 1993).

Management Strategies and Problems

Canada has experimented with virtually all available management techniques: annual quotas, seasonal quotas, allocation by gear sector, restricted fishing power of vessels, limitations on fishing gear type, limitations on fishing gear amount, limitations on the gear specifica-

tions, requirements for sorting grids, closing spawning areas, closing nursery areas, management based on constant fishing mortality, ITQs, stock enhancement, restriction on vessel size, and strict vessel replacement rules (Parsons 1993).

Modern management has focused as much on controlling the behavior of fishers as on the method of capture and amount of fish caught. Other important lessons have been that fishing practices are not constant; rather, there are steady increases in fishing efficiency, and attempts to control fishing practices can have unintended results. The development of fleets that are specialized or licensed for single fisheries creates problems when the stock declines. Allowing choices of target fisheries in multispecies fisheries can also engender problems if the system is too flexible and effort concentrates on high-value species. In addition, problems are generated by discarding, dumping, and underreporting that have a dramatic impact on the accuracy of stock estimates and confound attempts to achieve target exploitation rates (Parsons 1993).

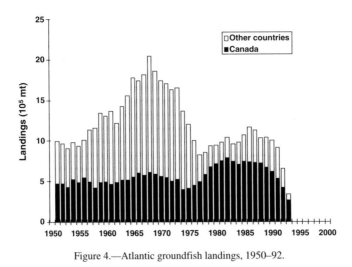

Figure 4.—Atlantic groundfish landings, 1950–92.

The Canadian experience also demonstrates the difficulties of achieving the objective of conservation while maintaining employment in the fisheries. In many inshore fisheries, the seasons can be very short; thus, the problem becomes how to balance continued use of small boats and processing plants that handle the catches with their replacement by fewer, larger vessels that can fish year-round and service fewer plants. Another complication for conservation is industry's search for increasing efficiency. This results in the conundrum of trying to control harvesting effort while trying to design better nets, better fishing boats, better fish-finding equipment, better positioning equipment, and so on. Standard controls on effort can be undermined by increased fishing efficiency. Even when each competing fishery sector has its own share, changes in efficiency for one sector will disrupt the balance by changing the distribution of the catch among different fleets. The distribution pattern can be distorted further if priority of access changes for one fishery component relative to others, or if market factors change the behavior of the fishery.

A further problem arises from the dependency of many communities on the resource. In Canada, there are many small isolated rural communities, especially in Newfoundland, where the dependency on the fishery is extreme. This dependency is particularly vulnerable to changes in resource availability, whether as a function of changes in resource abundance or of other factors, such as the environmental lobby that caused the loss of markets for seal skins in the early 1980s. Impacts have been dramatic because there were large numbers of people dependent on the seal fishery. The northern cod (*Gadus morhua*) fishery, which produced a catch of 800,000 metric tons (mt) in 1968, was closed down in July 1992. For 1993, all other cod quotas were reduced drastically, and some were reduced further during the season. In 1994, most of the remaining cod fisheries were closed. The government response has been an expenditure of CAN$1.7 billion over 5 years for the Atlantic Groundfish Adjustment Strategy for those affected by the moratoria. Measures being taken include income support and, more importantly, programs to train fishers for other activities, to reduce effort when the fisheries do reopen, and to develop a more resilient and rational industry. Whether these efforts will be successful remains to be seen. Recent survey assessment results indicate that the northern cod stock may not recover for another decade.

Assessing the Stocks

The patterns of change in the abundance of the Atlantic cod stocks are reasonably well documented, but the causes are not. The growth rate of individual cod de-

clined through the 1980s, recruitment has been poor since the early 1980s, spawning stocks are low, and apparently predators have increased, food species have decreased, and natural mortality has increased. The latter factor is subject to scientific debate, but, given that the northern cod population continued to decline after not being harvested for 2 years (Figure 5) and that many lightly fished stocks are also declining (Figure 6), there appears to be an environmentally driven component to the decline in the cod stocks. The decline in the individual growth rates for six cod stocks is shown in Figure 7. The cod were much smaller in the early 1990s than they were at the beginning of the 1980s. They appeared to be thinner in the early 1990s (Figure 8). This means that a given catch by weight implies an increase in the number of fish harvested and an increase in fishing mortality, if catch is held constant. The increasing number of grey and harp seals has led to concern about the impact of such predators. Grey seals consume 138,000 mt of prey of which 17,000 mt is Atlantic cod, while harp seals consume 88,000 mt of Atlantic cod, mostly 1- and 2-year-old fish, and hence could impede stock recovery (Mohn and Bowen 1994; Canada Department of Fisheries and Oceans [DFO] 1995).

Adverse environmental conditions off Newfoundland and Labrador during the late 1980s and early 1990s may be partly responsible for the decline of the northern cod stocks. Mean annual air temperatures from St. John's, Newfoundland, were relatively low from 1880 to the early 1900s, rose sharply in the late 1920s, peaked in the 1950s, and remained high through the 1960s (Drinkwater and Mountain, in press). Since the 1960s, temperatures have declined gradually and also fluctuated at 10-year intervals; minima occurred in the early 1970s, the mid-1980s, and the early 1990s.

Cold temperatures occur in years when northwest winds push Arctic air masses farther south. The strong northwest winds result from a deepening of the Icelandic Low, which in turn produces an increase in the North Atlantic Oscillation index (Dickson et al. 1988). Conversely, warm

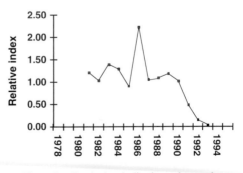

Figure 5.—Continuing decline in northern cod.

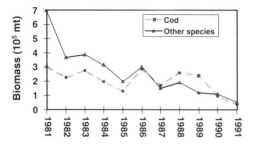

Figure 6.—Decline in lightly fished stocks (survey biomass index in Division 2J).

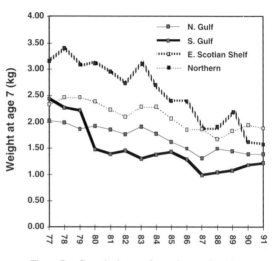

Figure 7.—Growth changes for various cod stocks.

years occur when northwest winds and the Icelandic Low are weak and the North Atlantic Oscillation index is low. Ocean temperatures, monitored since the late 1940s at Station 27 (near St. John's, Newfoundland), were near normal through the early 1960s and have been declining ever since, with the lowest bottom temperatures in the series being observed during the early 1990s (Drinkwater and Mountain, in press). The adverse environmental conditions probably affected recruitment, since recruitment levels of Atlantic cod off west Greenland, Labrador, and Newfoundland have generally been high when ocean temperatures are warm and decrease when temperatures are low (Petrie and Anderson 1983). In the late 1980s and early 1990s, when temperatures were extremely low in the northern regions, recruitment from Labrador to the Grand Bank was poor.

Some people have argued that the collapse of Atlantic cod stocks has been solely the result of excessive fishing pressure. Clearly, the high stock levels off Labrador and eastern Newfoundland during the early 1960s were coincident with environmental factors that were highly favorable. The low stock levels in the 1990s occurred at a time when the environmental conditions in the area were extremely harsh. While fishing was clearly a major factor, it was not the only factor. A more likely explanation is that the combination of high fishing mortality and the emergence of harsh environmental conditions contributed to the collapse of some stocks (e.g., northern cod).

It is interesting to examine fishing mortality over time as it illustrates the problems fisheries management faces. Exploitation rates increased through the 1960s as foreign fleets came to the northwest Atlantic, and they continued to increase steadily in the 1970s until extension of jurisdiction. At that time, the sense was that the fisheries would be controlled and exploitation rates would be maintained at reasonable levels. Fishing mortality did drop, but not as low as had been hoped. Then it crept slowly up (Figure 9).

The initial drop was as expected, given the major reductions in TACs associated with the extension of coastal

state fisheries jurisdiction. The basis for TACs changed from seeking the catch that would cause no further stock decline to setting removals at the level commensurate with the fishing mortality at the $F_{0.1}$[1] level. The reduction in effort necessary to bring fishing mortality down to this level was achieved by cutting the fishing opportunities of distant-water fleets that previously had access to fishing grounds that were, as of January 1, 1977, inside the Canadian 200-mile fisheries management zone. In theory, subsequent TACs were set commensurate with the $F_{0.1}$ level of fishing mortality, but in practice actual fishing mortality was often well above the $F_{0.1}$ level. The trend increased with time despite annual recalculation, for many years, of the expected catch at the $F_{0.1}$ level.

There were a number of possible reasons for the almost universal—for groundfish stocks—pattern for fishing mortalities to exceed the expected level. Different reasons were applicable depending on the stock in question. Thus, for example, underestimation of the real catch, because of unreported discards, or misreporting of landings, would not only have produced an underestimate of fishing mortality but an overestimate of population size for the following fishing year. Furthermore, uncertainties in the scientific database, the need to make assumptions about recruitment levels and productivity (growth), and changes in fishing efficiency of the fleet compromised

---

[1]$F_{0.1}$ is the level of fishing mortality at which any further increase in fishing effort would yield only a 10% increase in the catch per unit of effort that would have been realized if the same effort had been applied in a very lightly exploited fishery. This represents the value of fishing mortality beyond which any increase in fishing effort (mortality) would not be worthwhile.

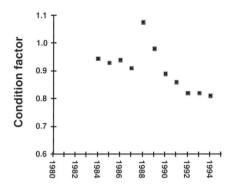

Figure 8.—Changes in condition factor (length/weight^3) for cod of the northern Gulf of St. Lawrence.

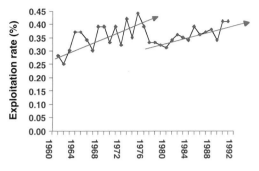

Figure 9.—Exploitation rate of major groundfish stocks.

the accuracy of the forecasts, which were exacerbated by the tendency for fisheries management decisions to be made in favor of the fishers, not the fish. This is similar to problems experienced in Europe (Parsons 1993).

It was not until the mid-1980s that it became clear that fishing mortalities in previous years had not only been higher than expected when the TACs were set, they were also higher than had been estimated in the assessment immediately subsequent to the fishing year. The scientists were not able to quantitatively isolate the source of the divergences, nor could they forecast how much influence the factors might have in the future. Hence, no "corrections" could be devised to adjust for what proved to be continuing problems. Management decisions in Atlantic Canada exacerbated the divergence from the $F_{0.1}$ level harvest by reason of the "50% rule" (only half the adjustment necessary to bring fishing mortality back to $F_{0.1}$ was made in one year, subject to provisions about fishing mortality level associated with maximum sustainable yield) and by reason of multiyear plans in which the TAC was maintained unchanged for several years, usually 3.

The increase in effective fishing power, despite efforts to prevent it, distorted the stock assessments, which assumed that a unit of fishing effort was constant from year to year. The more effective the effort, the more rapidly TACs are reached. TACs, together with the need to offset the capital costs of improving efficiency, increase fishers' motivation to maximize the value of their catch within the controls in place by high-grading[2] and misreporting catch species or volume, particularly as TACs are reduced. The uncertainties in the amount of real fishing effort were compounded by other management measures, such as seasonal quotas and ITQs, that changed the pat-

tern of effort over the total fishing season. As an example, measures were imposed by government or industry itself to distribute the catch more evenly over time to avoid glutting the market, hence reducing prices. These measures meant that there was less effort concentrated in the period of maximum catch rate, which would bias downward the calculations of overall catch rate as a measure of abundance.

The net result is that fishing capacity greatly exceeds the available resource, and as fishers have sought to maximize the value of landings, particularly as stocks decline, all the factors noted previously have combined. Consequently, fishing mortality increased steadily.

New Initiatives

Many of the difficulties experienced in managing groundfish fisheries in Atlantic Canada underscore the need to make industry an integral part of fisheries management. Canada has undertaken a number of initiatives to address this need, including using "index fishers" to collect biological and catch-per-unit-effort data. In 1994, on the Atlantic coast, the DFO initiated a sentinel fisheries program for stocks under moratoria in which limited numbers of fishers use a limited amount of commercial gear to fish according to a scientific survey design. In addition, fishers participate in research vessel surveys. A major initiative is the establishment of the Fisheries Resource Conservation Council (FRCC) on the Atlantic coast. The council consists of people knowledgeable of the fisheries, university researchers, representatives of provincial governments, and DFO scientific and management staff. In the past, DFO scientists provided catch projections to managers, who then made recommendations to the Minister of Fisheries and Oceans on catch levels after consultation with the industry. The FRCC receives scientific conclusions from the annual stock assessments and consults with industry on implications of the stock assessments. Subsequently, there is a further round of consultations as to what conservation measures

---

[2]The practice of discarding less valuable fish so fishers can get the most money possible for their catch.

should be taken. The independent FRCC then makes public recommendations on TACs and other conservation measures to the Minister of Fisheries and Oceans. This process results in a much greater transparency, as advice on conservation is completely open to public scrutiny. The council also makes recommendations on scientific research, assessment priorities, and methodologies.

The creation of the FRCC has given more visibility to conservation requirements for groundfish on the Canadian Atlantic coast. The FRCC members, by their participation in various public events, have also contributed toward developing a better understanding of science and of conservation issues. The FRCC has broadened the forum for discussing conservation and reaffirmed the importance of erring on the side of caution when making management decisions.

The biggest task facing Canadian fisheries management, however, is how to reduce excess capacity in both the harvesting and processing sectors of the Atlantic groundfish fishery, irrespective of the current resource crisis. Efforts to reduce excessive capacity have been foiled in the past. As noted earlier, the control of fishing effort proved very difficult. Attempts to limit the number of licenses were compromised by the reactivation of inactive licenses when the economic conditions became attractive because of good fishing or reduced employment in other sectors. This has applied mainly, but not exclusively, to small-boat fisheries, particularly where multiple licenses were available based on historical involvement. Attempts to cancel inactive licenses produced extremely negative reactions, as did attempts to freeze fishing power on the basis of vessel size, hold capacity, or engine power. Naval architects have proved to be very successful in designing vessels that meet the rules for vessel replacement reductions while achieving increased fishing power. Modern technology has led to increased fishing power—bigger and stronger nets, and improved fish-finding gear, positioning equipment, and gear handling. Even attempts to limit the amount of gear have been frustrated by human behavior. The limit (number of traps, gillnets, etc.) is usually chosen to inconvenience only the most aggressive fishing units and is well above the level used by most licensees. Following the introduction of the measure, all participants considered that they must use the upper limit.

During the 1980s, the concept of individual quotas (enterprise allocations; IQs) as a management tool for Canada's marine fisheries was widely debated and tested in several major fisheries on the Atlantic coast. These experiments met with varying degrees of success. The success of enterprise allocations in the Atlantic offshore groundfish fishery indicates that a system of individual

quotas can have considerable benefits, chiefly the dampening of the incentive to race for the fish to maximize an enterprise's share of the TAC. Experience in several fisheries has confirmed that IQs provide flexibility as to when, how, and whether an enterprise will harvest its allocation during a given year. The Atlantic offshore groundfish fishery has provided some evidence that IQs foster fleet rationalization. However, some disadvantages exist for IQs. Chief among these is the problem of high-grading, misreporting, and underreporting of catches and the consequent difficulty of ensuring compliance with IQ management regimes. As a consequence, IQs must be carefully tailored to the different characteristics of particular fisheries. In appropriate fisheries and with appropriately designed compliance mechanisms, IQs constitute a useful addition to the wide array of fisheries management tools.

In October 1996, a new Fisheries Act was introduced into the House of Commons. This new bill will substantially modernize the legal basis for fisheries, conservation, and habitat management in Canada. The bill will allow industry a direct voice in fisheries management through partnering agreements. The proposed legislation emphasizes self-regulation and self-reliance, and sets a climate for long-term stability in the industry. It is hoped that this bill will help eliminate the "gold rush" mentality that has plagued Canada's fisheries for too long.

Conclusions

Twenty-five years ago, many Canadian Atlantic fisheries already had more harvesting and processing capacity than was needed, particularly given that foreign fleets had free access to fisheries within the 12-mile (19-km) territorial limit. The problem is worse today, yet regulatory interventions have mushroomed, including TACs, allocation of access among fleet sectors, limited entry licensing, and IQs. Major benefits flowed from the 1977 extension of fisheries jurisdiction to 200 miles. However, Canada's marine fisheries continue to be plagued by instability because of various problems and constraints. These include the following:

- natural resource variability, often environmentally determined;
- the common-property nature of fisheries resources and resultant overcapacity;
- market fluctuations;
- heavy dependence on fisheries in isolated coastal communities;
- recurrent conflict among competing users; and
- conflicting objectives for fisheries management.

Despite abundant resources, various combinations of these factors have contributed to recurrent boom-and-bust patterns in many marine fisheries. Extended jurisdiction and various post-extension initiatives have not solved the problems of the fisheries sector. There is an urgent need to bring harvesting and processing capacity into balance with sustainable resource levels. It is unclear whether recent efforts to reduce capacity through government-funded withdrawal of some vessels from the fleet will substantially reduce overall capacity. As the first signs of stock recovery began to appear in 1996, pressures to reopen certain fisheries before the stocks had recovered to sustainable levels began to intensify. This situation will be compounded by the termination of compensation payments to groundfish fishers in 1998. There is considerable risk that the benefits of the substantial investments in conservation (moratoria for several years) will be dissipated by premature opening of certain fisheries. Periodic fisheries crises and demands for government assistance can be expected to continue unless viable alternative economic opportunities can be developed in coastal regions.

References

DFO (Canada Department of Fisheries and Oceans). 1995. Report on the status of harp seals in the northwest Atlantic. DFO Atlantic Fisheries Stock Status Report 95/7.

Dickson, R. R., J. Meicke, S.-A. Malmberg, and A. J. Lee. 1988. The great salinity anomaly in the northern North Atlantic, 1968–82. Progress in Oceanography 20:103-151.

Drinkwater, K. F., and D. B. Mountain. In press. Climate and oceanography. *In* J. Boremen, B. Nakashima, J. A. Wilson, and R. L. Kendall, editors. Northwest Atlantic groundfish: perspectives on a fishery collapse. American Fisheries Society, Bethesda, Maryland.

Mohn, R., and W. D. Bowen. 1994. A model of grey seal predation on 4VsW cod and its effects on the dynamics and potential yield of cod. Canada Department of Fisheries and Oceans, Atlantic Fisheries Research Document 94/64, Ottawa.

Parsons, L. S. 1993. Management of marine fisheries in Canada. Canadian Bulletin of Fisheries and Aquatic Sciences 225.

Petrie, B., and C. Anderson. 1983. Circulation on the Newfoundland continental shelf. Atmosphere–Ocean 21:207-226.

Management Procedures:
A Better Way to Manage Fisheries?
The South African Experience

D. S. BUTTERWORTH, K. L. COCHRANE, AND J. A. A. DE OLIVEIRA

Abstract.—Whether the costs of the conventional fishery management process—an annual assessment that leads, for example, to a total allowable catch (TAC) based on a biological reference point—are justified by the benefits is questionable. Perhaps fisheries should instead be regulated by means of "management procedures": pre-agreed sets of possibly quite simple rules for translating data from the fishery into a TAC each year. Selection by managers between candidate management procedures should be based upon inspection of the trade-offs among anticipated levels of medium-term reward (catch/profit), risk of stock "collapse," and interannual catch variability, where these are calculated by simulation. The concept is illustrated by reference to two of South Africa's major fisheries that have been regulated on this basis since the late 1980s. For example, an ambitious procedure for the mixed-species South African pelagic fishery was put in place at the beginning of 1994; this procedure makes allowance for operational interactions by taking into account the inevitability of a juvenile pilchard (sardine, *Sardinops sagax*) bycatch whose magnitude will be related to the size of the anchovy (*Engraulis capensis*) TAC awarded. The approach has been well received by the fishing industry and is a cornerstone of the draft marine resource policy put forward by the majority party in the new South African government. The greatest problems of the approach are considered to be in defining risk in comparable manner for different fisheries, and in interpreting the results of simulations that test candidate procedures for robustness of performance to uncertainties about the model structure assumed to describe the system's dynamics. The future role for assessments is seen not as a basis for management advice but to assist in defining the simulations used to compute anticipated procedure performance as procedures typically are refined on a 3- to 5-year time scale.

The Present Norm—Annual Fishery Assessments

In many of the world's fisheries, the typical process followed currently for providing scientific recommendations for management is roughly as follows. First, scientists will assemble on an annual basis to argue out—usually at some length—their current best assessment of the status of the resource concerned. Then, for example, a total allowable catch (TAC) will be recommended arising out of that assessment, usually based upon some biological reference point in whose choice scientists have typically (though not entirely properly) played a greater role than managers.

A considerable portion of the resources of government fishery research institutes is expended on this exercise. But in times of funding cutbacks, this practice is starting to come under scrutiny. For example, at the International Council for the Exploration of the Sea (ICES), which coordinates an enormous set of assessment exercises in the eastern north Atlantic for report to the European Union, there are the beginnings of questions from some countries whether they can afford to have their scientists invest so much of their time in this process. For example, N. Daan (Netherlands Institute for Fishery Investigations, CP IJmuiden, the Netherlands, pers. comm.) comments that while requests for advice from ICES are increasing, available manpower is effectively reduced as a consequence of the privatization process of many research institutes. He adds that although improving the logistics and efficiency of assessment meetings would bring some relief, there is also a clear need to prioritize assessment requirements.

Do the costs of such scientific efforts justify the benefits? How frequently is the often cumbersome process of an annual update of management measures really necessary? Though not immediately aware of any quantitative study in this regard, we suspect that the benefits obtained from such annual reviews are usually not large. For developed fisheries, the impact of but one further data point on parameter estimates is likely to be small. This is not a general argument for TACs set at a fixed level a number of years ahead because annual adjustments are clearly needed in fisheries for which annual recruitment constitutes a sizable fraction of the exploitable biomass. But does the method used for assessing that recruitment need to be changed so much from year to year that it could not be pre-agreed upon for a multiyear period?

We suggest that substantive changes in the scientific understanding of developed fisheries occur on a time scale much closer to 5 years than 12 months. A particular problem of the present norm of an annual cycle in the assessment process is that fisheries scientists focus too much time on the short term instead of on more important problems that can be addressed properly only over a longer period. Subsequent arguments in this paper seek to show that a valuable spin-off of the alternative "management procedure" approach that we advocate is the automatic achievement of this refocusing of research on the longer term.

South Africa's Most Valuable Fisheries

The application of a management procedure approach to provide scientific recommendations for the regulation of some of South Africa's major fisheries was an initiative with roots in research along these lines that commenced in the International Whaling Commission (IWC) in the mid-1980s (IWC 1989). But before we go into more detail about these procedures, a few brief comments about the fisheries themselves are needed for perspective.

South African fisheries are of medium size on the world scale, with the largest annual catch of any one species seldom exceeding a few hundred thousand metric tons (mt) (Table 1). The west coast rock lobster fishery (*Jasus lalandii*) (Pollock 1986) provides a valuable export but has been going through increasingly difficult times recently (Figure 1A). The pelagic fishery (Butterworth 1983; Butterworth and Bergh 1993) followed the pattern of its Californian counterpart—albeit about a decade later—with a collapse of the pilchard (sardine, *Sardinops sagax*) resource in the mid-1960s, after which the less valuable anchovy (*Engraulis capensis*) has become this fishery's mainstay (Figure 1B). The hake (*Merluccius capensis, M. paradoxus*) resource (Punt 1994; Payne and Punt 1995) was overexploited as a result of rising catches by foreign fleets during the 1960s and early 1970s, but it has shown a steady though slow recovery under cautiously regulated local harvests since the implementation of a 200-nautical mile (320-km) exclusive fishing zone in 1977 (Figure 1C).

Since 1990, the hake fishery has been managed under $f_{0.2}$[1] harvesting strategy TACs calculated by applying a non-equilibrium production model to catch-per-unit-effort (CPUE) data and biomass indices from research surveys (Payne and Punt 1995). The particular model used provides a reasonable fit to the CPUE data (Figure 2).

This may sound like a standard fisheries annual assessment–management process, but it differs from that in two important ways. First, the model fit and $f_{0.2}$ TAC

calculation is an automatic process repeated annually—Hilborn and Luedke (1987) would call it "clockwork." The model fit does not pretend to necessarily correspond to the "best" possible assessment of the status of the resource at any one time. Second, this process was chosen over other possibilities for regulation based on extensive simulation studies of the anticipated performance of the fishery in the medium term (Punt 1991, 1993). These characteristics shift the process into the realm of a "management procedure."

Management Procedures

What exactly is a management procedure? The underlying philosophy is that all parties (scientists, industry, managers) should agree upon clearly defined rules before the management game is played.

These rules specify exactly how the TAC (or the level of some other regulatory mechanism, such as fishing effort) is to be computed each year and what data are to be collected and used for this purpose. The rules need not necessarily relate to a fairly complex assessment process, as was described briefly for hake, but can be quite simple. For example, under the present procedure for the South African anchovy and pilchard fisheries (Anonymous, Sea Fisheries Research Institute WG/JAN94/PEL/3, South Africa, unpubl. rep.; Figure 3), the annual TAC for pilchard-directed fishing is simply 10% of the biomass estimate of fish of age-1 and older, which is obtained from the most recent November hydroacoustic survey, but subject to the constraints of a maximum reduction of 25% from the previous year and a minimum of 25,000 mt. Nevertheless, an important difference exists between the anchovy–pilchard and the hake examples: the parameters of the "catch-control laws" of the anchovy–pilchard procedure remain fixed, at least until the procedure as a whole is revised after a number of years; in contrast, the hake procedure incorporates a more developed form of feedback control, with the control-law parameter automatically adjusted each year as further data become available.

What is the basis for such rules? Not the so-called biological reference points (such as the $F_{0.1}$ or F_{MED} values evaluated from age-related data) of the traditional fisheries assessment and management process but, rather, consideration of outputs of direct interest to industry and managers. These relate to the anticipated performance of the fishery and resource in the medium term (e.g., 10 to 20 years[2]), which is evaluated by Monte Carlo simulation.

---

[1]An f strategy is a constant effort strategy, with the effort level calculated from a surplus production model. For the $f_{0.2}$ strategy, this effort level is that for which the slope of the equilibrium yield vs. effort plot is 20% of the slope of this curve at the origin.

---

[2]Periods longer than this are unlikely to be of much concern to industry or politicians and become important in regard to the resource only for very long-lived species (e.g., whales).

TABLE 1.—Some statistics for South Africa's most valuable fisheries. Catches are in metric tons (mt).

Type	Species	Largest annual catch (year)	1993 catch
Inshore (west coast)	Rock lobster (*Jasus* sp.)	18,900 (1951)	2,200
Pelagic	Anchovy (*Engraulis* sp.)	596,000 (1987)	236,000
	Pilchard (*Sardinops* sp.)	410,000 (1962)	51,000
Demersal	Hake (*Merluccius* sp.)	295,000 (1972)	141,000

"Performance" encompasses three essentially conflicting objectives, which have pertinence in most fisheries: maximize rewards in terms of catches or profits; minimize the risk of something nasty happening, such as the collapse of the resource; and maximize stability. We emphasize the desirability of showing the trade-offs between measures related to these different objectives when a choice between different candidate procedures is discussed with industry and managers. By choosing measures that are operationally meaningful to the industry, we feel that much greater understanding is achieved. This leads to more sensible choices than would an appeal to some utility function approach that attempts to combine these qualitatively different features into a single measure.

Thus, the management procedure chosen for South African hake is based on a non-equilibrium production model rather than virtual population analysis (VPA) because simulations indicated that the latter would lead to much greater interannual catch fluctuations with no real corresponding gains in terms of average catch levels or risk (Punt 1991, 1993).

Attention has been drawn to a disadvantage of this overall approach (P. Sullivan, International Pacific Halibut Commission, Seattle, Washington, pers. comm.): given the importance that models underlying any TAC-setting process remain appropriate, the approach can sacrifice the valuable "reality check" of the annual exposure of an assessment to critical review. Of course, this begs the question, given limited human resources in practice, of whether such comprehensive "checks" can be entertained for every stock—priorities have to be set based upon some combination of resource value and the level of uncertainty associated with the assessment. Nevertheless, this would seem to be outweighed by the many advantages, both scientific and practical:

- With the possible exception of some fisheries for very short-lived species, risk cannot be meaningfully evaluated for a catch limit for a single year, but only for the repeated application of some TAC-setting process over a number of years.
- Simple pre-agreed formulae allow for quick decisions: this is important in, for example, the South African anchovy fishery, which is based predominantly on the recruits of the year. Environmental

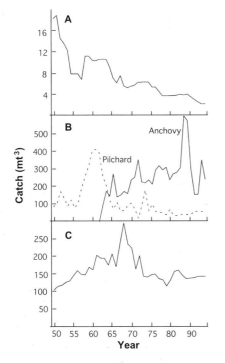

FIGURE 1.—Annual catch trends for South Africa's most valuable fisheries, 1950–93: (A) west coast rock lobster, (B) pilchard and anchovy (the major components of the pelagic fishery), and (C) hake.

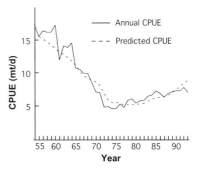

FIGURE 2.—The CPUE for the South African west coast hake fishery, 1955–93, together with the trend predicted by the fit of a non-equilibrium production model to both these data and biomass indices from research surveys.

and weather conditions usually make these available to the fishery for only 2 to 3 months after the annual recruitment survey so that any TAC increase awarded in the light of the survey results needs to be announced speedily.

- The demarcation between scientific and policy responsibilities becomes clearer: scientists calculate the anticipated performances for different procedures, managers select one of these procedures based on the trade-offs reported, and the TAC is then calculated automatically over the next few years. This has the particular advantage of less scope for politicking (i.e., industry introducing questionable arguments to attempt to increase the TAC to meet their short-term requirements, and scientists guilty of the identical practice in trying to frustrate them; see, for example, Hilborn and Luedke [1987]).
- A firm basis for TAC evaluation, coupled with scientific predictions of the likely outcomes, facilitates industry planning under clearer, longer-term perspectives and security.
- A management procedure can be viewed as a testable hypothesis: one can retrospectively evaluate how well alternative procedures would have performed, unlike the situation for year-to-year decisions.
- Fishers are likely to be happy to agree to an approach that specifies how TACs will increase if things get better, and vice versa, because their necessary optimism leads them to think that the former is the much more likely to occur.

Despite such optimism, what if things *do* get worse: will the lower TACs indicated by the procedure be acceptable? Our approach to this potential problem has been to have procedures include a limit to the percentage by which the TAC may be reduced in any one year. The tighter this limit, the more conservatively the overall procedure needs to be tuned, as shown by the corresponding lower anticipated average annual catches for the South African anchovy fishery (Table 2). The local pelagic fishing industry was prepared to agree to raise the maximum percentage reduction allowed for the anchovy TAC from 25% to 40% in anticipation of an associated 10% increase in the average annual catch.

Given such rules, some allowance also needs to be made for freak situations where keeping to the rules could damage the resource. For example, for the South African anchovy, "exceptional circumstances" can be invoked by managers under the pre-agreed criterion of a hydroacoustic survey indicating a spawning biomass below 500,000 mt. Though the anchovy procedure is designed to manifest a low probability of the spawning biomass falling below 20% of its mean pre-exploitation

A

Following the November biomass survey

$$TAC_{anch}^{init} = 0.7 \times 300 \left(0.7 + 0.3 \frac{\hat{B}_{1+}}{\bar{B}_{1+}} \right)$$

Following the May/June recruit survey

$$TAC_{anch}^{rev} = 300 \left(0.7 \frac{\hat{R}}{\bar{R}} + 0.3 \frac{\hat{B}_{1+}}{\bar{B}_{1+}} \right)$$

Subject to:

I. $200 \leq TAC_{anch}^{init/rev} \leq 600$

II. Annual drop in $TAC_{anch}^{init/rev} \leq 40\%$ (relative to previous TAC_{anch}^{rev})

III. $0 \leq TAC_{anch}^{rev} - TAC_{anch}^{init} \leq 150$

IV. Exceptional Circumstances: $\hat{B}_{1+} < 500$

B

Following the November biomass survey

$$TAC_{dir} = 0.1 \, \hat{B}_{1+}$$

$$TAC_{byc}^{init} = 7.5 + 0.06 \, TAC_{anch}^{init}$$

Following the May/June recruit survey

$$TAC_{byc}^{rev} = 7.5 + x \, TAC_{anch}^{rev}$$

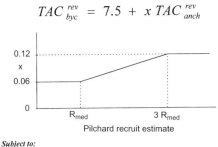

Subject to:

I. $TAC_{dir} \geq 25$

II. Annual drop in $TAC_{dir} \leq 25\%$

III. Exceptional Circumstances: $\hat{B}_{1+} < 150$

FIGURE 3.—A diagrammatic representation of the joint anchovy-pilchard management procedure adopted at the beginning of 1994 for the South African pelagic fishery. The component that pertains to the anchovy TAC (1,000s mt) is set in (A). $\hat{B}_1$ and $\bar{B}_1$ refer to estimates of current and average past 1+ biomasses from hydroacoustic surveys, with a similar convention for $\hat{R}$ and $\bar{R}$ from the recruitment surveys. The pilchard component is set out in (B), where "dir" refers to the directed pilchard fishery and "byc" to pilchard bycatch in fisheries directed at other species (primarily anchovy). The pilchard bycatch TAC depends both on the anchovy TAC and on the current pilchard recruitment estimate (from a hydroacoustic survey) in relationship to the median of past estimates (R_{MED}).

TABLE 2.—The present management procedure for the South African anchovy (Anonymous, Sea Fisheries Research Institute WG/JAN94/PEL/3, Cape Town, South Africa, unpubl. rep.) was selected on the basis of the trade-off between predicted average annual catch and interannual catch variability across a number of variants of a candidate procedure. The performance statistics in question were computed for the same level of risk (a 30% probability of spawning biomass falling below 20% of the mean pre-exploitation level within a 20-year period) for each variant, by adjusting a control parameter of the procedure. The variant selected incorporates the following constraints: maximum annual TAC, 600,000 mt; minimum annual TAC, 200,000 mt; maximum downward adjustment of TAC from one year to the next, 40%; maximum increase of initial TAC following recruitment survey results, 150,000 mt. The table shows the performance statistics for this and the other variants considered, on the basis of which the final selection was made.

Variant	Average annual catch (mt)	Average interannual catch variability (%)
Variant selected	315,000	25
Maximum TAC = 450,000 mt	314,000	23
Minimum TAC = 150,000 mt	328,000	25
Maximum downward adjustment = 50%	321,000	25
= 25%	285,000	22
Maximum increase after recruit survey = 200,000 mt	326,000	28

level, allowance must also be made for stringent remedial measures to be taken if this should nevertheless occur. The criterion chosen takes account of survey sampling error, trading off probabilities of not taking further action when it is actually necessary with the disruptive possibility of taking action that is not needed. We look for criteria that should amount to invoking "exceptional circumstances" no more frequently than once every 10 to 15 years on average, and we are currently evaluating options for meta-rules to be applied in such circumstances. We have found that a major difficulty in these circumstances is that industry will argue to define an acceptable risk level to be whatever will maintain their present level of catches.

Facilitating User Participation

When developing or updating management procedures, we consider that facilitating a sensible choice by industry and managers between alternative candidates is of particular importance. This is especially difficult for short-lived pelagic species such as anchovy, for which the high level of recruitment fluctuation renders comparisons of deterministic projections valueless. Our experience is that the summary statistics of the probability

distributions necessary to describe such circumstances are not immediately meaningful to industry and managers. Therefore, we have concentrated on first getting them to play computer simulation games, which show individual realizations of catch and biomass time-series under alternative procedures, to give them a better feel for the trade-offs involved.

This overall approach has generally been well received by our industry, particularly in the case of a recent update of the procedure for anchovy (Table 2). However, we are perhaps fortunate, at least in this context, in dealing with fisheries dominated by a few large industries with long-period planning horizons. Whether this approach would work as well with a large number of small operators, for whom cash-flow considerations dictate a shorter-term focus, is debatable.

The Problem Area: "Risk"

The aspect of management procedures that we find the most problematic is the definition of "risk" although the same problem does also arise in the conventional assessment–management approach. The statistic we have been using for risk for the South African pelagic fishery is the probability of spawning biomass falling below 20% of the mean pre-exploitation level within a 20-year period. We note that Hilborn (1997) strongly criticizes the use of criteria of this form, suggesting instead that analyses should specifically incorporate a stock–recruit relationship and uncertainties about its form, with a quantitative basis for the latter provided by the information available on similar species. If these uncertainties encompass the possibility of depensatory effects, then the probability of stock collapse can be computed explicitly instead of using measures of "risk" such as that adopted here as a surrogate for this probability. Our concern is only that such an approach should not be seen as a defensible way to reduce the three fundamental attributes of procedure performance (i.e., average catch, risk, and short-term variability of catches) to two by absorbing risk into average catch. The reason is evident from consideration of the following two scenarios: very high catches for a short period, followed by a rapid "collapse" and subsequent stable but low catches; and stable catches at an intermediate level throughout the overall period of interest. These two scenarios would have nearly identical average catches and short-term catch variability, but industry and managers would almost certainly have a strong preference for the second. Essentially, what is lost in such an attempt to reduce the number of attributes to two is a measure of long-term catch variability, for which "risk" is a surrogate. Hilborn (1997) implicitly acknowledges this, by suggesting four attributes: "average yield, short- and long-term variability in catch, and the probability of stock collapse."

Recently, we have as scientists been pulling the wool a little over the South African pelagic industry's eyes by pre-specifying a risk level ourselves and giving them options from which to choose that involve only the average catch versus interannual TAC variability trade-off. For anchovy, the risk level specified has been 30%. However, Figure 4 shows that there is actually a wide scope for alternative choices, and these could be considered explicitly. It also indicates, incidentally, the much greater rewards available from an approach that varies TACs in response to resource survey results in comparison with a constant catch policy, which would be the only possibility were surveys to cease.

There are a number of difficulties related to risk:
- An acceptable level of risk, if defined in this manner, cannot be argued to be invariant across different fisheries. The larger the extent of recruitment variability, the lower the level to which the population might drop in the absence of harvesting as a consequence of these natural fluctuations; therefore, presumably, the population would be more resilient to depletion to a certain level. This means that as appraisals of recruitment variability change, so too should acceptable levels of risk as we have defined it.
- While performance computations can readily incorporate the consequences of survey imprecision, there is no straightforward approach to account for "quality" aspects, which are difficult to quantify. For example, the primary reason we have set a much lower acceptable risk level for our pilchard procedure, compared with the 30% for anchovy, is concern that target identification problems in analyzing hydroacoustic survey results are much greater for the subdominant pilchard resource.
- At the beginning of 1994, a joint management procedure (Anonymous, Sea Fisheries Research Institute WG/JAN94/PEL/3, South Africa, unpubl. rep.; Figure 3) was put in place for our pilchard and anchovy resources. This proved necessary because we needed to take account of the unavoidable bycatch of juvenile pilchard in the anchovy fishery, where the magnitude of this bycatch depends in part on the size of the anchovy TAC. However, we have the difficulty of not being sure whether we are dealing with recruitments for both species that fluctuate independently about fixed average levels or a possible recovery of pilchard at the expense of anchovy. This is compounded by concerns about the reliability of estimates of juvenile pilchard numbers from hydroacoustic recruitment surveys, which seem not to correlate well with subsequent estimates from VPA.

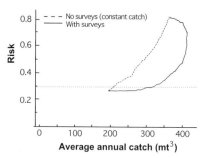

FIGURE 4.—Plots of risk (see text for definition) against reward (average annual catch) for simulations of alternative management procedures for the South African anchovy fishery (Butterworth and Bergh 1993, Butterworth et al. 1993). The dashed line reflects results in the absence of any surveys, which would necessitate a constant catch strategy; the achieved average catch is actually less than the constant catch set because the increasing probability of stock collapse as the level of the intended catch is raised results in a greater frequency of failure to achieve that catch. The solid line shows results for different values of a control parameter for the procedure operative at present; the actual value of the control parameter set corresponds to a 30% risk level (see "Variant selected," Table 2, for more details).

Risk relates to uncertainty, which has two components. The first is measurement error, and the consequent imprecision of estimates of model parameters, which can be taken into account fairly straightforwardly in computations of management procedure performance. The second component involves the much more difficult question of exactly how best to deal with model structure uncertainty. Particularly important concerns relate to the form of a stock–recruitment relationship and questions about the number of stocks present and the demarcation of their boundaries.

It is easy to state the "in principle" approach to address this problem: the performance statistics for a candidate management procedure need to be robust across the range of plausible hypotheses for the system's dynamics. A procedure may well include a feedback mechanism (as does that for the South African hake fishery; see Management Procedures), which adjusts appropriately even though the system's dynamics ultimately behave somewhat differently from what the model underlying the procedure might assume. However, this still leaves the question of how to interpret the results of the simulations corresponding with such alternative hypotheses, particularly as performance must degrade to an increasing extent for more extreme hypotheses—how much degradation in performance can be tolerated if the procedure under examination is to be considered still adequately robust? The IWC Scientific Committee seems to be moving towards a "worst-case scenario" basis in a risk context,[3] but we do not see this as a viable approach. Procedures can always be made to "fail" on this basis simply by including a sufficiently extreme scenario; this is readily accomplished if, for example, no limits are placed on the number of stocks that may be postulated to be present in the fishery. What is or is not plausible? Perhaps Sir Isaac Newton's rules of reasoning in his "Principia" (Cajori 1934) give us some guidelines. His Rule IV states:

> In experimental philosophy we are to look upon propositions inferred by general induction from phenomena as accurately or very nearly true, notwithstanding any contrary hypotheses that may be imagined, till such time as other phenomena occur, by which they may either be made more accurate, or liable to exceptions.

> This rule we must follow, that the argument of induction may not be evaded by hypotheses.

This suggests that we should be guided by data in preference to pure speculation.

If a man considers crossing a road, the hypothesis that he will be knocked down by a bus and killed is perfectly plausible. This is effectively his worst-case scenario, but the fact that such an eventuality cannot be excluded does not mean that he, therefore, decides not to cross the road. The reason he does decide to cross is that he assesses the probability of this undesirable event to be small. The moral in the context of management procedures is that the only defensible approach for interpreting the results from a range of models for system dynamics is a probabilistic one. But how then are relative weights (probabilities) to be assigned to the different models? The management procedure approach does not solve all problems; some it merely translates into another form.

The Future

The application of management procedures in South African fisheries thus far has essentially been de facto at the scientific level. It is encouraging that the draft Marine Resource Policy of the majority party in the new government seeks to change this to a de jure situation. This document incorporates the following principles:

i. Stocks are to be harvested to optimize societal benefits without placing them at undue risk.

ii. Management plans, including harvesting strategies, are to be developed for each fishery.

iii. Plans are to be gazetted with procedures to allow amendment by agreement.

iv. Plans are to aim to stabilize the TAC, while recognizing the inherent variability of fish stocks.

It, thus, also gives high weight to the objective of lessening interannual TAC variability—presumably with stability of employment in mind—though it recognizes the inherent variability of fish stocks nevertheless. The formulation of a management procedure approach for managing the other major South African marine resource, the west coast rock lobster, is now in hand. Indeed, one can argue strongly for the management procedure approach instead of year-by-year, assessment-based scientific recommendations from the South African experience. The argument is based on the relative ease with which recommendations for TACs for anchovy and hake have been agreed upon by scientists over recent years, in contrast to ongoing disagreements for west coast rock lobster (though this difference is admittedly not the only factor that distinguishes these cases).

Does any role remain for assessments in this new approach? Certainly, but not for the direct provision of management advice; rather, the results of assessments can be used to bound the range of possibilities considered in the simulation evaluations of procedure performance. The aim of research becomes the reduction of the overall range of plausible hypotheses, allowing revised procedure choices that will achieve greater rewards for the same perceived risk. We see this updating of procedures as having a typical time scale of 3 to 5 years (i.e., a requisite time period for substantive improvements in the scientific understanding of developed fisheries), and we are attempting to dovetail local, longer-term research projects with such a schedule.

For example, an update of the procedure for South African hake fishery is now in progress. In recent years, the model-predicted and observed CPUE trends in this fishery have increasingly diverged (Figure 2). Although the procedure used has broadly self-corrected for this (as it was designed to do) by not increasing TACs as rapidly as projected a few years ago, the divergence suggests that the new data now available will indicate better performance from a procedure different from that selected some years back.

On the wider front, we see a major goal as being able to treat risk in a manner that allows for consistency of evaluations across different fisheries (i.e., something akin to the role that has been played by the $F_{0.1}$ reference point). The justification for an effective choice for risk level, as for $F_{0.1}$, will likely be empirical (it works; i.e., it does not collapse fisheries too often), but comparability is a necessary prerequisite to allow such an evaluation.

Thus, do we really yet know how to deal properly with uncertainty in fisheries management decisions? We doubt

---

[3]As indicated, for example, by their recommendation from among the variants of the Revised Management Procedure (Kirkwood 1992, 1997; IWC 1994a) for implementation for southern hemisphere minke whales, following simulation trials for this specific resource (IWC 1994b).

this, and instead are awaiting first a practical basis for a consistent approach to the matter of model structure uncertainty—a key issue which is addressed further in this volume (Hilborn 1997).

Acknowledgments

We are appreciative of the comments of R. Hilborn, P. Sullivan, and an anonymous referee on an earlier version of this manuscript.

References

Butterworth, D. S. 1983. Assessment and management of pelagic stocks in the southern Benguela region. Pages 329-405 in G. D. Sharp and J. Csirke, editors. Proceedings of the expert consultation to examine changes in the abundance and species composition of neritic fish resources. FAO Fisheries Report/FAO Informe de Pesca 291, Vol. 2, Rome.

Butterworth, D. S., and M. O. Bergh. 1993. The development of a management procedure for the South African anchovy resource. Pages 83-99 in S. J. Smith, J. J. Hunt, and D. Rivard, editors. Risk evaluation and biological reference points for fisheries management. Canadian Special Publication of Fisheries and Aquatic Science 120.

Butterworth, D. S., J. A. A. De Oliveira, and K. L. Cochrane. 1993. Current initiatives in refining the management procedure for the South African anchovy resource. Pages 439-473 in G. Kruse, D. M. Eggers, R. J. Marasco, C. Pautzke and T. J. Quinn II, editors. Proceedings of the international symposium on management strategies for exploited fish populations. Alaska Sea Grant College Program Report 93-02, University of Alaska, Fairbanks.

Cajori, F. 1934. Sir Isaac Newton's mathematical principles of natural philosophy and his system of the world. (Translated into English by A. Motte in 1729. Translations revised and supplied with a historical and explanatory appendix by F. Cajori.) University of California Press, Berkeley.

Hilborn, R. 1996. Uncertainty, risk and the precautionary principle. Pages 100-106 in E. K. Pikitch, D. D. Huppert, and M. P. Sissenwine, editors. Global trends: fisheries management. American Fisheries Society Symposium 20, Bethesda, Maryland.

Hilborn, R., and W. Luedke. 1987. Rationalising the irrational: a case study in user group participation in Pacific salmon management. Canadian Journal of Fisheries and Aquatic Science 44:1796-1805.

IWC (International Whaling Commission). 1989. Report of the comprehensive assessment workshop on management, Reykjavik, 23-25 March 1987. Report of the International Whaling Commission (Special Issue 11):21-28.

IWC (International Whaling Commission). 1994a. The revised management procedure (RMP) for baleen whales. Report of the International Whaling Commission 44:145-152.

IWC (International Whaling Commission). 1994b. Report of the sub-committee on management procedures. Report of the International Whaling Commission 44:74-92.

Kirkwood, G. P. 1992. Background to the development of revised management procedures. Report of the International Whaling Commission 42:236-243.

Kirkwood, G. P. 1997. The revised management procedure of the International Whaling Commission. Pages 91-99 in E. K. Pikitch, D. D. Huppert, and M. P. Sissenwine, editors. Global trends: fisheries management. American Fisheries Society Symposium 20, Bethesda, Maryland.

Payne, A. I. L., and A. E. Punt. 1995. Biology and fisheries of the South African Cape hakes (*M. capensis* and *M. paradoxus*). Pages 15-47 in J. Alheit and T. J. Pitcher, editors. Hake fisheries, products and markets. Chapman and Hall, London.

Pollock, D. E. 1986. Review of the fishery for the biology of the Cape rock lobster *Jasus lalandii* with notes on larval recruitment. Canadian Journal of Fisheries and Aquatic Science 43:2107-2117.

Punt, A. E. 1991. Management procedures for the Cape hake and baleen whale resources. Benguela Ecology Programme, Foundation for Research Development, Report 23, Pretoria, South Africa.

Punt, A. E. 1993. The comparative performance of production-model and ad hoc tuned VPA based feedback-control management procedures for the stock of Cape hake off the west coast of South Africa. Pages 283-299 in S. J. Smith, J. J. Hunt, and D. Rivard, editors. Risk evaluation and biological reference points for fisheries management. Canadian Special Publication of Fisheries and Aquatic Science 120.

Punt, A. E. 1994. Assessments of the stocks of Cape hakes *Merluccius* spp. off South Africa. South African Journal of Marine Science 14:159-186.

The Revised Management Procedure of the International Whaling Commission

G. P. KIRKWOOD

Abstract.—Prior to its 1982 moratorium on commercial whaling, the International Whaling Commission (IWC) based catch limits on stock assessments developed by its Scientific Committee. These analyses were similar in nature to standard fishery assessments of the day. For each stock, all available data were used to obtain best estimates of current and historical stock sizes and productivity. Catch limits were then set with the aim of keeping the stock at or above the level at which the maximum sustainable yield could be taken, or moving it toward that level. One major reason for deciding to impose the moratorium was the difficulty the Scientific Committee experienced in reaching consensus on the status of stocks, given the prevailing uncertainties in the data and their interpretation. Over the past 7 years, the IWC Scientific Committee has developed a revised management procedure designed to resolve these difficulties. The development process, described in this paper, involved a thorough reexamination of management objectives, a realistic view of the uncertainties inherent in current and future data, and very thorough testing via Monte Carlo simulation of the robustness of proposed procedures in the face of these uncertainties. Effectively, allowances for risk and uncertainties have been directly incorporated into the management procedure.

Before the 1980s, most scientific advice to fishery management was presented in terms of "best" estimates. Uncertainties often were glossed over in the interest of reaching consensus on international fishery commissions, and they typically were acknowledged only when the advice transpired to be clearly wrong. Even enlightened institutions apparently have failed to heed messages of uncertainty and risk, as judged by the management measures they actually adopted. (Throughout this paper, the term "risk" is used to denote the probability of something bad happening. It is not used in its technical sense in a decision theoretic framework as meaning expected loss.)

This failure to take proper account of uncertainty occurred because fishery management is an inherently political process, particularly in international fishery commissions. Biological advice on the current and likely future state of fish stocks is essential information needed for management, but it is by no means the only information needed. Thus, the final decisions on catch quotas may appear to fly in the face of scientific advice. Sometimes this is an unavoidable feature of the conflicting objectives or conflicting pressures on managers. I also believe that it occurs in part because the scientific advice is given in such a way that managers find it very difficult to interpret.

A simple example illustrates the issue. Advice that a 95% confidence interval for a recommended total allowable catch (TAC) is 8,000–12,000 metric tons (mt) may be better than a single estimate of 10,000 mt, but not a lot. How is the manager to choose an appropriate number within that interval? Different pressures may incline managers to opt for a TAC near the middle of the range, or toward either the lower or upper end of the range.

Lobbyists will almost certainly argue for either extreme. The situation becomes much worse when more than one possible range is presented (corresponding, for example, to the results from two assessment techniques or two interpretations of data). These were exactly the types of difficulties faced by the International Whaling Commission (IWC) and its Scientific Committee in the late 1970s and early 1980s prior to the imposition of the moratorium on commercial whaling.

The recent emphasis on assessing risks of breaching biological thresholds in fisheries (Smith et al. 1993), rather than just assessing statistical uncertainty, is certainly a big step in the right direction, but it is not the end of the road. What is really needed is clear scientific advice, couched in understandable terms, of the risks of failing to meet the management objectives. It is the relating of biological (or economic) risks to management objectives, which usually are not couched in terms of such risks, that is crucial to bridging the communication gap.

If agreed-upon, quantifiable management objectives can be developed (which may be asking a lot, since many managers appear reluctant to articulate their objectives), there are at least two ways of proceeding. The obvious one is routinely to associate appropriate risk statements with each element of scientific advice, leaving the managers to deal with that information as they will. Even if desirable, this is not always straightforward. The approach taken by the IWC, apparently quite paradoxically, was to seek to return to the days of consensus advice. This was to be achieved by seeking a management procedure, applicable across all stocks and species of baleen whales (*Balaenoptera* spp.), that could be demonstrated to produce catch limits for which the risks of not meeting the objectives were acceptably low. Application of such a management procedure would then produce a single recommendation for catch limits.

In describing the IWC's revised management procedure (RMP) in this paper, I concentrate primarily on the approach the IWC used to develop and test the RMP, rather than on the procedure itself, and I omit most of the technical details. It is the approach, rather than the procedure itself, that could with merit be considered for use in other fishery management situations. For similar reasons, most discussion is devoted to development of management procedures for a single biological stock.

The International Whaling Commission

The IWC (the Commission) is an intergovernmental organization whose member countries are signatory to the 1946 International Convention for the Regulation of Whaling. The Commission meets annually, and major decisions, such as the setting of annual catch limits, are made by a three-quarters majority. The IWC is advised by a standing Scientific Committee consisting of scientists nominated by member countries and invited experts.

The first attempt to specify a formal set of rules for calculating catch limits for commercial whaling was made in 1974, when the so-called New Management Procedure (NMP) was adopted. Details of this procedure are unimportant here, and only two features will be mentioned. The first is that stocks estimated to have been reduced to less than 54% of their unexploited level were classified as Protection Stocks, and no commercial catches were allowed. The 54% cutoff point was chosen on the basis that it was 10% below the conventionally assumed stock level at which the maximum sustainable yield (MSY) could be taken. It was a device designed to ensure that stocks did not fall much below their "optimum" level, and if they did to rectify the situation as soon as possible. In no sense was the protection level intended to be associated with the stocks being in danger of extinction, though frequently it has been interpreted that way. The second feature of interest is that even as far back as the mid-1970s, the need to make some allowance for uncertainty had been recognized in that the maximum allowable catch limit was only 90% of the estimated MSY, thus explicitly incorporating a 10% allowance for uncertainty, albeit a fixed one.

The NMP apparently worked reasonably well for a few years, but problems began to be noted in the late 1970s. Most problems arose because of difficulties in reliably estimating the MSY and the current and initial stock levels. Even when acceptable estimates were made for a particular stock, the changes in the estimates as they were updated annually often led to widely fluctuat-ing catch limits. By the early 1980s, the IWC Scientific Committee found it almost impossible to reach consensus on recommendations for stocks subject to commercial whaling, other than for Protection Stocks. Partly as a result of these difficulties, in 1982 the IWC agreed to implement a pause in commercial whaling (the so-called moratorium), which would take effect from the mid-1980s. This provision was to be kept under review based on the best scientific advice. Subsequently, the Commission would undertake a comprehensive assessment of the effects of this decision on whale stocks and consider its modification.

In 1986, the Scientific Committee agreed that a key element of a comprehensive assessment should be the development of an RMP. The development process occupied the years 1986–93. A procedure for setting catch limits for single baleen whale stocks was adopted by the IWC at its 1992 meeting, and a comprehensive management procedure for possible multiple stocks in a region was completed in 1993.

Management Objectives

As I have already asserted, an essential element in the search for an improved management procedure is a specification of the objectives that must be met. This specification must be sufficiently precise to determine objectively how well the management procedure meets the objectives.

In principle, the primary source for IWC objectives is the convention governing its operations. However, the wording in that document is very general. A number of joint meetings of scientists and commissioners finally led the IWC to accept the following statement of its objectives (IWC 1988:36):

i. stability of catch limits, which would be desirable for the orderly development of the whaling industry;

ii. acceptable risk that a stock not be depleted (at a certain level of probability) below some chosen level (e.g., some fraction of its carrying capacity), so that the risk of extinction of the stock is not seriously increased by exploitation;

iii. making possible the highest continuing yield from the stock.

These objectives are partly incompatible in that they cannot be fully satisfied simultaneously. Trade-offs between these objectives are inevitable, especially between the first two and the third. Generally, the higher the level of continuing yield, the higher will be the risk of depletion below acceptable levels, and vice versa. Similarly, the greater the catch limit stability required, the lower will be the continuing yield. Further progress required that weightings be given to the three objectives.

After further discussion, the IWC agreed that an acceptable management procedure must first satisfy the conservation objective (ii). Subject to that, the management procedure was free to maximize catches under (iii) while performing satisfactorily in (i). One gap in the specification of the conservation objective was filled by the requirement that the RMP should not be apparently less conservative than the old NMP. A stock assessed to be below 54% of its initial level (i.e., below the old protection level) should have a zero catch limit. Acceptable risk was then to be judged in terms of the possible effects of inadvertently setting non-zero catch limits when the stock was actually below the protection level but was assessed to be above it. If the risk were deemed sufficiently low, then objective (ii) would be satisfied.

Development of a Management Procedure for a Single Stock

For an RMP to be judged satisfactory, it must meet the IWC's management objectives, and it must do so regardless of existing and continuing uncertainties in the basic data, stock identity, and whale population dynamics. A management procedure was sought that was robust in the presence of these uncertainties. Whether a potential management procedure is robust can only be decided by examining its performance across a wide range of plausible situations. Development of a revised management procedure proceeded on two fronts: (1) identifying and refining potential management procedures, and (2) specifying means for testing these procedures for robustness and ability to meet objectives. No attempt was made to develop an optimum management procedure in the sense of one that maximized some utility function derived from the management objectives. Given the preceding objectives, it is unlikely that such a utility function could be derived; in any case, the relevance of such an approach is unclear in view of the robustness requirements. Rather, the aim was to develop a procedure that exhibited satisfactory, robust performance in meeting the management objectives.

For this paper, a management procedure is taken to be a set of rules for calculating annual catch limits from available stock information. Traditionally, for both fisheries assessment in general and whale assessment in particular, setting rules has involved fitting models of the stock dynamics to available historical catch and absolute or relative abundance data. Catch limits were then calculated based on the resulting estimates of model parameters and their sizes in relationship to appropriate biological reference points. These models may be of varying complexities, but typically the more that is (thought to be) known, the more complex has been the model fitted. In terms of the NMP, however, this approach did not

seem particularly successful for the reasons already outlined.

Five potential management procedures were formulated and investigated by members of the IWC Scientific Committee. Three procedures (developed by Butterworth and Punt, by Cooke, and by de la Mare) involved the fitting of simple stock production models. The other two procedures (developed by Magnusson and Stefansson and by Sakuramoto and Tanaka) at least initially eschewed fitting of any population model; rather, they set catch limits primarily on the basis of recent trends in relative abundance. Later versions of the Magnusson and Stefansson procedure did, however, incorporate a population model. Space does not permit a proper description of the five procedures. Instead, interested readers are referred to IWC (1992a:93–103), in which the five procedures are described as they stood when one was selected.

Simulation Trials of Management Procedures

Since experimental application of potential procedures for managing actual whale stocks was out of the question, the IWC applied a computer simulation of whale stock management using techniques similar to those introduced by Hilborn (1979). An initially unexploited whale population was set up on the computer and subjected to a series of historical catches prior to the onset of management. The dynamics of the simulated stock were governed by models similar to those used regularly by the IWC Scientific Committee. Abundance data were simulated in such a way that they had the same nature and properties believed to occur in observed data of those types. Computer programs then applied potential management procedures to this simulated stock. This approach has several advantages: it allows many tests to be done relatively quickly, the state of the simulated stock is known exactly at any time so that how well the management procedure performs can be accurately determined, and extinction of a simulated population is of no consequence.

All five potential management procedures were subjected to a lengthy series of computer-based trials. Each trial examined management of a simulated whale stock over a 100-year period. This was repeated 100 times (400 for some trials) for each trial scenario. Summary statistics monitoring the procedure's performance in relationship to the three management objectives were collected for each trial. These statistics are described in a later section.

The prime concern in the screening trials was to subject the potential management procedures to a set of severe performance tests. However, it was also intended to assist in developing and improving the procedures. When a procedure was found to perform worse than expected on one or more trials, it was often possible to modify it to rectify the problem. The modified version then had to be subjected again to the full set of trials to ensure that improvements for one trial did not lead to unacceptable performance on others. This feedback, and the development of common approaches, was a key feature of the development process.

It was also recognized that some trials may be so severe that all potential procedures may fail. This was not to be taken as a signal to abandon further development. Rather, if it became clear, for example, that the management procedures would only perform satisfactorily if true stock boundaries were well known, then this would clearly signal that research to delineate stock boundaries was an essential prerequisite to successful management.

Assumptions Common to All Trials

Stock dynamics.—In early trials, the dynamics of the simulated stock or stocks followed a simple population model similar to that used in past assessments of baleen whale stocks. In the later development stages, a fully age-structured model was used, and in robustness trials a variety of alternative forms of population models were used. In all trials, however, the form of the underlying dynamics model actually used and its parameter values were hidden from the management procedure.

Data availability.—The extent and types of data available for different whale stocks vary widely. Seeking maximum applicability, the Scientific Committee concentrated on procedures that made minimal data demands and did not require prior estimates of biological parameters that led to past difficulties in applying the NMP. Consequently, the following data were assumed to be available to the procedures:
- the annual catches (including, in most trials, all historical catch data for a simulated stock exploited prior to application of the management procedure);
- estimates of abundance, which were available at the beginning of the first year of management and then at regular intervals thereafter (typically every 5 years), regardless of whether catches were taken.

The most common estimates of whale abundance result from sightings surveys. The variability and possible biases for these estimates examined in the trials were based on past experience of such data. In early trials, catch-per-unit-effort (CPUE) data were also assumed to be available; however, they were subsequently not used in view of interpretation difficulties (IWC 1989). Management procedures were allowed to use all, part, or none of these data as the developer desired. The levels of variability and bias in relationship to the true abundances were not known by the management procedure.

Specification of Individual Trials

Specification of trials was an iterative process, as was development of the management procedures, and the set of trials has been revised and extended over time. Eventually, two categories of trials were identified: base case and robustness. Base case trials consisted of a short series of relatively mild trials that examined the ability of procedures to manage unexploited, moderately depleted, and heavily depleted whale stocks with different productivity levels in cases where the stock dynamics followed conventional models. The robustness trials were longer series of much more stringent trials designed to examine the effects of a wide variety of failures in assumptions.

Base case trials.—The base case trials examined all combinations of the following scenarios:
- At the onset of management, the stock was either unexploited (a "development" case), reduced to 30% of its unexploited abundance (a "rehabilitation" case), or reduced to 60% of its unexploited abundance (a "sustainable" case).
- The MSY rate (the MSY, as a percentage of the MSY stock level, and thus a measure of potential productivity) for the population was either 1%, 4%, or 7%. Only the 1% MSY rate was used in the sustainable case.
- The estimates of abundance were unbiased and available in the first year of management and every fifth year thereafter.

Robustness trials for single stocks.—Robustness trials examined a very wide range of plausible departures from assumptions commonly made in past assessments. Following the initial specification, each trial was repeated for a selected subset of the base cases mentioned above. The robustness trials examined the following (for a full list, see IWC 1993a:224a):
- Incorrect assumptions about the dynamics of the true stock. This formed the largest category. Cases examined included widely differing MSY levels and density-dependent responses, differing ages at maturity, differing time lags in density-dependent responses, trends and cycles in the carrying capacity of the population, and cyclic changes in the MSY rate.
- Initial abundance at different levels than examined in the base case trials.
- Upward and downward bias in the abundance estimates, and trends in that bias, as well as differing collection frequencies and precision levels. Another trial examined a case in which the decision on whether to undertake an abundance survey (an expensive undertaking) was dictated by the likely catch quota that would arise from the anticipated results of the survey.

- Uncertain or inaccurate catch histories prior to exploitation, and long periods of protection before management starts.
- Irregular episodic events (e.g., occasional occurrence of epidemics).
- Deterioration of the environment with declining trends in both carrying capacity and MSY rate.
- Randomly chosen parameters.

These trials tested the effects of individual failures of assumptions. In reality, more than one failure may occur simultaneously. It is possible that the net effect of simultaneous failure of a number of assumptions may be much greater than their individual effects. This possibility was covered by conducting further trials that examined interactions among those factors that were most important on their own.

Statistics for Evaluating Ability to Meet Management Objectives

For each trial, statistics were collected to evaluate the performance of management procedures in meeting the three management objectives. The primary statistics and management objectives to which they referred were as follows:

Objective (i): the average year-to-year variability in catch limits.

Objective (ii): percentiles of the lowest population size during the 100 years of management.

Objective (iii): percentiles of the total catch over 100 years, and of a measure of continuing catch, which in most cases was the average catch over the final 10 years.

Similar statistics were also collected for the final population size after 100 years. This was effectively used as a proxy for a target population size.

Interpretation of these statistics is straightforward. To meet objective (i), the average year-to-year variability in catch limits must be sufficiently low. For objective (ii), both the median and lower percentiles of the lowest population size are important and facilitate assessment of the likelihood of depletion to unacceptably low stock levels. For objective (iii), consideration of both median and low percentiles allows judgments to be made not only about a procedure's average performance, but also the spread of total catches it can produce. The continuing catch statistic was viewed as an indicator of the long-term sustainable yield allowed by a procedure, though in some trials with an initially heavily depleted stock with low productivity, even after 100 years the population had not reached an equilibrium level.

Two further statistics were used to allow advice to be given on the probability of whaling being inadvertently allowed when stock levels were significantly below the protection level of 54%. These were the realized protection level (RPL) and a measure of relative recovery (RR). For a single simulation, the RPL was defined as the lowest population level at which a non-zero catch limit was set. Over a trial, percentiles of the distribution of RPL values were calculated. Note that with the incorporation of an internal protection level, in the presence of perfect information the RPL would never be less than 54%. With increasingly imperfect information, there is a greater chance that the stock will be assessed to be above the protection level when it is not, and thus that a non-zero catch limit will be set inadvertently

The RR statistic complemented the RPL statistic by measuring the extent to which inadvertent setting of non-zero catch limits delayed stocks from recovering from levels below the protection level. Specifically, for each simulation, the first year, T, at which the stock recovered to just above the protection level under a regime of zero catches was determined. The RR statistic was then the stock level achieved under management (and therefore possibly with some non-zero catches set) in year T, measured as a proportion of the stock level achieved under zero catches in year T. Perfect performance would be reflected by an RR statistic of 1.0. Again, percentiles over sets of simulations were collected. The complementary nature of the two statistics is illustrated by the fact that a management procedure may inadvertently set non-zero catch limits at stock levels considerably below the protection level, but they may be so small that any delay in stock recovery is also very minor.

For each of the statistics collected, and for each of the simulation trials, comparisons were made of the performance of the five potential management procedures, and judgments were made as to whether each procedure performed satisfactorily in terms of meeting the relevant management objective. However, there were many trials, each producing a number of performance statistics that addressed different aspects of the three management objectives. While attempts were made, using the techniques of multi-criteria decision making, to develop a single objective ranking of procedures in terms of performance across all trials, these attempts proved unsuccessful. Inevitably, therefore, there was some degree of subjectivity in the final selection of the "best" procedure; however, consensus was achieved in this selection.

The Revised Management Procedure for a Single Stock

By 1991, the development and testing process for management procedures applicable to single known stocks of baleen whales was considered complete. At its 1991 meeting, the IWC Scientific Committee reviewed the performance of the five procedures (IWC 1992a).

While judging that all five had performed well enough in meeting the management objectives so that any one could, in principle, be suitable for adoption, the Committee agreed that the procedure developed by J. Cooke showed the best performance across the trials. At its 1991 meeting (IWC 1992b:47–48), the IWC adopted the procedure by resolution, subject to satisfactory advice on the probability of inadvertently allowing whaling when the stock was significantly below the protection level. This advice was transmitted and accepted at the IWC's 1992 meeting. The management procedure for single stocks, subsequently renamed the "catch limit algorithm," formed the core of a more elaborate procedure applicable to management of possibly multiple stocks in a management area, which is briefly described in the next section.

The Catch Limit Algorithm

Each time the catch limit algorithm is to be applied to a given stock of baleen whales, the following input data are required:
- one or more estimates of absolute abundance derived from sightings surveys and their variance–covariance matrix; and
- all previous known catches for that stock by year.

In brief, the catch limit algorithm involves the following steps. First, a simple stock production model is fitted to these data. The model has two estimable parameters: (1) the current degree of depletion of the population (the ratio of current population size to unexploited population size), and (2) a productivity parameter determining the MSY rate. An additional parameter is also estimated, representing the bias in estimates of absolute abundance. These three parameters are assigned fixed, independent, prior probability distributions, and estimation then proceeds in a Bayes-like manner to produce a posterior distribution of the nominal catch limit, which is a simple function of the estimated parameters provided that the estimated depletion level is above 54%; otherwise it is zero. When the posterior distribution of the nominal catch limit is determined, the likelihood of the abundance data is downweighted by a fixed scaling factor. The final resulting catch limit is a selected percentile (41.02%) of the posterior distribution, the percentile having been selected to achieve a nominated median final population size on a specified simulation trial.

Readers wishing to see more details of the catch limit algorithm are referred to its technical description (first given in IWC 1992b:148–149), but it should be noted that an understanding of the algorithm's technical details is not essential to what follows. There are, however, some important points that require further clarification.

The first point is that the stock production model used in the algorithm, while superficially very similar to those used previously in whale stock assessments, does not claim to give an accurate representation of real baleen whale population dynamics, either in its functional form or in its parameter values. Rather, it is a model that, when used as an integral part of the catch limit algorithm, has been demonstrated to allow robust calculation of catch limits. Should it transpire that an alternative model more accurately reflects the true population dynamics of baleen whales, it does not follow that this new model should automatically be substituted for the existing one in the catch limit algorithm. On the contrary, that would only be appropriate if it could be demonstrated that the algorithm with the "improved" model performs at least as well on the full set of simulation trials as the existing catch limit algorithm did.

In the same vein, the prior probability distributions for the three parameters to be estimated, despite their appearances, are not intended actually to reflect prior beliefs about the true values of these parameters, as is usually the case with Bayesian estimation methods. Instead, the prior distributions and their ranges were adjusted to provide optimum performance on the simulation trials. As indicated in annotations to the specification of the revised management procedure (IWC 1993a:152), should perceptions change on likely distributions and ranges of biological parameters, the appropriate way of accounting for them is to amend the corresponding simulation trials and adjust the "tuning" of the algorithm in light of performance in the trials, rather than directly to adjust the prior distributions themselves to reflect the changed perceptions.

A final important property of the algorithm results from the fact that a key input to the estimation process is the variance–covariance matrix of the absolute abundance estimates (a measure of their precision). The less precise the absolute abundance estimates, the wider is the spread of the posterior distribution of catch limits, thus the lower is the final catch limit (it being a fixed percentile of that distribution). Indeed, it is possible with very imprecise abundance data for the final catch limit to be set at zero almost solely because of this imprecision. Automatic reduction of catch limits as the precision of the primary input data decreases is a key feature of this procedure. Similarly, the scaling factor that effectively downweights the information from the abundance data prevents the algorithm from getting too "excited" when it is suddenly fed a larger-than-usual abundance estimate that could very well have occurred by chance. Whale dynamics are sufficiently slow that it is far harder to recover from an inadvertently high catch limit than from one that is too low. In both cases, there is a strong incentive, in terms of getting higher catch limits, to conduct surveys that are as thorough as possible.

Management of Multiple Stocks in a Region

The catch-limit algorithm and the trials used to test its performance make the strong assumption that the population of whales being managed is a single biological stock with known boundaries, such that estimates of abundance and catch data apply to that stock alone. There are, in fact, few whale stocks for which those assumptions are justified. A more common situation is one in which management areas are defined, with fixed geographical boundaries, and catch limits apply to any catches within that management area. The selection of management areas usually takes account of existing knowledge about stock identity in and around the area, but it is quite likely that whales from more than one biological stock will from time to time be found in a single management area.

Early in the development process, it was recognized that uncertain stock identity was likely to pose substantial problems in seeking a procedure that was robust in the presence of this type of uncertainty. Initially, generic trials were devised to investigate the seriousness of these problems. The first of these trials, mimicking a possible coastal whaling scenario, quickly demonstrated that problems could be very serious indeed. Coastal whaling operations are typically carried out from vessels that have a rather restricted operating range from their home port. This restricted operating range may be much smaller than the distribution area of whales being taken by those operations. In such circumstances, it is common for abundance surveys to be carried out over the larger distribution area of the whales.

Provided that the whales in the survey area belong to a single stock that distributes itself randomly throughout the area each year, and provided that the only catches taken from that stock are taken by the coastal whaling operation, direct application of the single-stock catch-limit algorithm is unlikely to cause any difficulties. But if, in the extreme case, there are two stocks of the same species, one of which is found only in the area of coastal whaling and the other in the remainder of the survey area, then it is obvious that extreme overexploitation of the stock near the home port could arise from application of the single-stock algorithm.

Simulation trials to examine this latter scenario duly demonstrated that the single-stock algorithm did not provide anything like robust management. However, they also suggested a possible solution, albeit a rather conservative one: When catching takes place in only a part of a species' distribution area, and when stock identity is unknown, treat the operation area as defining the boundary of a single stock and apply the single-stock algorithm to

that restricted area only. This guards against the worst case while being conservative in the best case when there is a single stock in the larger area.

The solution also demonstrates the key role of further research to resolve the uncertainties. The conservative approach requires that, when the stock identity is unknown, one should in a sense assume the worst. However, should research demonstrate that the true situation was not nearly so bad (and maybe even that there really was only one stock in the larger area), the catch limit algorithm could be applied to a larger area with a corresponding increase in catch limits.

Subsequent, more complex trials of multistock management were based loosely on previous commercial pelagic whaling operations for minke whales in the southern hemisphere (IWC 1993b:185–188), and mixed coastal and pelagic minke whaling in the north Atlantic Ocean (IWC 1993b:189–195). An approach that appeared to avoid inadvertent overexploitation of biological stocks revolved around identification of three geographical areas:

1. regions, which typically corresponded to major ocean basins, containing stocks that did not mix with stocks in other regions;
2. medium areas, of approximately similar geographical scales to existing IWC management areas, which corresponded to known or suspected ranges of distinct biological stocks within a region; and
3. small areas, which were small enough to contain whales from only one biological stock, or were such that if whales from more than one biological stock were present, catching operations would not be able to harvest them in proportions substantially different from their proportions in the small area.

The catch-limit algorithm is applied to each of the small areas within each medium area, and separately to each medium area. In most cases, the operative catch limits would be those calculated for the small areas. Two other possibilities were envisaged. The first, called catch cascading, involved basing small-area catch limits within a medium area on the catch limit for the medium area treated as a whole, with that medium area catch limit being distributed (cascaded) among the constituent small areas in proportion to the estimated relative abundance in them. The second, called catch capping, involved calculating catch limits for each small area within a medium area and for the whole medium area. Catching in a small area would cease in a season as soon as the catch limit for that small area was reached, or when the sum of catches to date across all small areas in the medium area exceeded the medium-area catch limit.

The key to success in this strategy for handling uncertain stock identity naturally lies first in identifying appropriate small areas, but it also depends critically on a previously mentioned property of the catch-limit algorithm: catch limits decrease with decreasing precision of abundance estimates. The nature of sighting survey abundance estimates, which tend to be conducted on the scale of medium-sized areas, is such that the estimates of abundance for only part of the medium area are less precise than for the whole medium area. The extent of this loss of precision increases as the size of the small area decreases. It follows that catch cascading normally produces larger catch limits than setting small-area limits only, and in turn, catch capping produces smaller catch limits again. Which of these three approaches is to be adopted in any one case would depend on the results of appropriate simulation trials.

Up to this stage, attention had been concentrated on developing a generic management procedure applicable to all baleen whale stocks, and testing of single-stock procedures via simulation trials was also conducted using generic scenarios. The original aim was to develop a single management procedure that could be applied across the board, rather than a different procedure for each stock or region. It was clear, however, that this would no longer be completely possible in cases of uncertain stock identity.

Identification of appropriate regions, medium areas, and small areas in any application of multistock management depends on the available knowledge of the stocks and regions to be managed. More importantly, testing the robustness of the multistock management rules depends critically on devising trials that mirror the extent (or lack) of knowledge of stock identity, migration, and mixing in the areas under consideration and trials that incorporate the full range of plausible alternative hypotheses about these factors. The final nail in the coffin of strictly generic trials is the requirement in the multistock management rules for a choice to be made as to whether catch capping or catch cascading should be invoked. This again requires consideration on a case-by-case basis.

Accordingly, the idea of implementation simulation trials was introduced. Before the multi-stock management procedure could be applied to any region and stock, comprehensive trials had to be carried out specific to that stock and region. Such trials were completed for minke whale stocks in the southern hemisphere and the north Atlantic Ocean (IWC 1993a:153–196). Trials for north Pacific minke whales are under way but have not yet been completed.

A full description of both single-stock and multi-stock rules is given in the specification for the calculation of catch limits in a revised management procedure for baleen whales (IWC 1994:145–152). This specification also describes rules for gradually phasing out catches in cases

where more than 8 years have elapsed since the last abundance estimate was obtained and for setting separate catch limits by sex to take account of unequal numbers of male and female whales in catches.

Relevance to Management of Other Fisheries

Despite its good properties, there is no suggestion that the IWC's revised management procedure should be applied, as is, to the management of other marine species. This is not just because the dynamics of whale stocks and their responses to exploitation are so different from those of most fish stocks (though they are). Far more fundamental is the fact that the RMP was designed to meet the IWC's particular management objectives.

The three objectives described earlier could serve as an admirable first draft of objectives for any commercial fishery. However, where the IWC's approach differs is in the explicit requirement that the conservation objective have absolute primacy and in the interpretation of when that objective has been violated (depletion to 54% or less of unexploited abundance). Satisfying the conservation objective alone is trivially easy: all one has to do is set catches low enough. Even the most committed environmentalist will not seriously try to mount an argument that a catch of a few whales out of a population of hundreds of thousands is likely to cause conservation problems. What is difficult is trying to produce reasonable catches while still satisfying the conservation constraint.

The relative balance between the three objectives and certainly the interpretation of the conservation objective are likely to be considerably different from the IWC's for most domestic and international fishery management regimes. Mission statements and management plans frequently contain pious statements along the lines of "conservation of fish stocks has the highest priority," but it is very rare that this statement can ever be justified on the basis of actual decisions taken, except possibly after a stock collapse, which should have been avoided in the first place if the assertion were true. Changes to the relative priority of the three objectives and their interpretation are likely to demand a somewhat different form of management procedure from that of the IWC, though doubtless some elements of that procedure could form a part of it.

In my view, it is the process adopted in developing the RMP, and especially in testing its robustness, that is of greatest relevance to fishery management of other resources. In recent years, increasingly faster computers have become available, and computationally intensive Bayesian approaches to stock assessment have been developed in which uncertainties in stock assessment are explicitly allowed for through incorporation of prior probability distributions for uncertain parameters. Such an approach, however, still stands or falls on the extent to which the full range of uncertainties has been incorporated into the assessment. Computers may be fast, but we are a very long way from being able to allow for all the factors examined in the robustness trials described here.

The alternative approach, in which the robustness of a rather simpler management procedure is tested in the presence of as full a range of plausible uncertainties as possible, has much to commend it. It has practical advantages, too: every stock assessment scientist will be familiar with the cry, "But you haven't allowed for such and such!" You may know that that particular factor will not have the slightest discernible effect, but convincing its proponent that you are right can be another kettle of fish. The IWC's approach is ideal for handling such situations: if persuasion fails, conduct a simple trial. Either you will gain irrefutable proof of the irrelevance of the factor in question, or you will learn not to be quite so arrogant next time. For all its advantages, however, it cannot be denied that the IWC approach is computationally very intensive, and to be followed thoroughly, it does take considerable time. Obviously, the time and resources required will vary with the complexity of the fishery and the fish stocks, but experience elsewhere suggests that this approach is practicable for government fishery agencies with responsibility for important commercial fisheries.

The revised management procedure of the IWC has been demonstrated to perform admirably in managing stocks of computer whales. However, it has not been applied to management of real whale stocks, and I believe it most unlikely that it ever will be, at least for commercial whaling. If I am correct, we will never know whether it would work in practice. The first tests of the approach will have to come from real fishery applications. The most advanced of these is in South Africa, and readers are referred to the case study presented by Butterworth (1997).

Acknowledgments

Perspicacious readers may have noticed that my name appears nowhere as a developer of one of the five potential management procedures, nor is it in the list of refer-

ences. My association with the development of the RMP was initially as chairman of the IWC Scientific Committee when the process first started, and then as chair of the various subcommittees and working groups of the IWC Scientific Committee, which guided the development process. None of this would be possible without the great dedication and innovation of the developers of the potential procedures: D. Butterworth, J. Cooke, W. de la Mare, K. Magnusson, A. Punt, K. Sakuramoto, G. Stefansson, and S. Tanaka. The computing skills of C. Allison of the IWC Secretariat were essential to the development process, and the other members of the Scientific Committee who participated in the discussions also played key roles.

References

Butterworth, D. S. 1997. Management procedures: A better way to manage fisheries? The South African experience. Pages 83-90 in E. K. Pikitch, D. D. Huppert, and M. P. Sissenwine, editors. Global trends: fisheries management. American Fisheries Society symposium 20, Bethesda, Maryland.

Hilborn, R. 1979. Comparison of fisheries control systems that utilize catch and effort data. Journal of the Fisheries Research Board of Canada 33:1-5.

International Whaling Commission (IWC). 1988. Report of the scientific committee. Report of the International Whaling Commission 38:32-61.

International Whaling Commission (IWC). 1989. Report of the comprehensive assessment workshop on catch per unit effort (CPUE). Report of the International Whaling Commission (Special Issue 11):15-20.

International Whaling Commission (IWC). 1992a. Report of the scientific committee. Report of the International Whaling Commission 42:51-267.

International Whaling Commission (IWC). 1992b. Chairman's report of the forty-third annual meeting. Report of the International Whaling Commission 42:11-50, 148-149.

International Whaling Commission (IWC). 1993a. Report of the scientific committee. Report of the International Whaling Commission 43:55-218, 224a.

International Whaling Commission (IWC). 1993b. Report of the fifth comprehensive assessment workshop on revised management procedures. Report of the International Whaling Commission 43:185-195, 229-240.

International Whaling Commission (IWC). 1994. Report of the scientific committee. Report of the International Whaling Commission 44:41-201.

Smith, S. J., J. J. Hunt, and D. Rivard, editors. 1993. Risk evaluation and biological reference points for fisheries management. Canadian Special Publication of Fisheries and Aquatic Sciences 120.

Uncertainty, Risk, and the Precautionary Principle

RAY HILBORN

Abstract.—Management of commercial and recreational fisheries is pervaded by uncertainty in stock abundance, product price, costs of fishing, political constraints, and budgets. Unfortunately, the assessment of many fisheries depends on biological characteristics that cannot be readily determined from the data available on the fish stock; this is particularly true when dealing with concerns of fish stock response and the potential for collapse at low abundance or high fishing pressure. Scientists often present decision makers with a measure of "risk," which may be the probability that the abundance would drop below a threshold or the fishing mortality rate would increase above a threshold. Often, specific consequences of crossing these thresholds are not given, nor are the probabilities of alternative consequences provided. I argue it is far better to include alternative hypotheses about how the stock will behave when these thresholds are crossed and, using all available data, assign probabilities to these alternative hypotheses. If scientists do not do so, there is no scientific guidance on the consequences of passing these thresholds for decision makers. The "precautionary principle" has gained popularity as a method for providing guidance in uncertain situations. During a developing fishery, the precautionary principle would suggest a slow development of fishing capacity, which minimizes the chance of stock depletion and economic overcapacity. However, when a stock is fully developed or overdeveloped, precautionary reductions in catch would reduce biological risk but increase the risk of economic collapse. Thus, the precautionary principle provides little guidance when stocks are fully exploited.

Perhaps the most ubiquitous theme in world fisheries is uncertainty—it pervades almost all aspects of fishing and fisheries management. Commercial fishers operate with considerable uncertainty about the price they will receive, government regulations such as seasons and quotas, and the costs of fishing, including fuel, interest rates, and license values. If you were to ask commercial fishers what they would like most in their fishery, I suspect most would say stability. Recreational anglers also face changing regulations and fluctuating abundance of fish. Government managers are also beset by uncertainties: changing demands and political power of competing user groups, estimates of the stock size and appropriate quotas for recreational and commercial fisheries, fluctuating governmental resources and budgets to perform their work, and interactions with other issues such as marine mammals and international relationships.

In this paper, I concentrate on how fisheries agencies deal with uncertainty in the stock assessment process, and I consider some of the recent approaches to what is now often called "risk assessment." My comments naturally reflect my own experience, and I can only speak about the management agencies and systems I know. I also consider the appropriate management response to uncertainty and discuss the concept of the "precautionary principle" in relation to uncertainty.

Uncertainty in Stock Assessments

There are four major sources of uncertainty in fisheries stock assessments: (1) measurements, (2) model parameters, (3) model structure, and (4) physical and biological processes affecting the stocks.

Almost all stock assessments depend a great deal on data, derived either from survey results; from biological samples of length, age, maturity, and so on; or from measures of total catch. The reliability of the assessments is greatly influenced by the reliability of the data. Many data sources have internal measures of variance associated with the sampling scheme—in surveys, the internal variability of the survey will determine the confidence limits on the survey result. Beyond this, however, there is uncertainty about the overall reliability of the survey, particularly whether it is an absolute measure of abundance or should be treated as a relative index. Obviously, surveys are much more informative if they provide absolute measures of abundance, but in some organizations it is accepted practice to treat research surveys as relative indices of abundance. To use survey results as absolute indices of abundance makes a very strong assumption about the lack of bias in the survey methodology.

Almost all stock assessments have some form of model at their core, and these models have parameters that are estimated from the data. Most agencies now compute and report uncertainty in model parameters although the methods for computing uncertainty, and the extent and format for reporting it, differ greatly between agencies and localities. The most common methods for computing uncertainty include bootstrapping (Restrepo et al. 1992), maximum likelihood (Polacheck et al. 1993; Punt and Butterworth 1993), and Bayesian statistical analysis (Givens et al. 1993; Hilborn et al. 1994). All these methods have a common goal of helping stock assessment scientists and managers to understand the range of possible "states of nature"—what types of stock dynamics are consistent with the measurements available.

Uncertainty in model structure is much less frequently dealt with in fisheries stock assessments. Almost all assessments I know of are based on a single model, and any uncertainty that is reported to managers is uncertainty in the parameters of the model. For instance, many commonly used assessment procedures are based on age-structured models that have parameters for natural mortality rates, growth rates, age-specific selectivities, and stock recruitment functions. It is common practice to assume that all these parameters are time-invariant—however, the more adventurous assessment groups are now allowing for temporal changes in some parameters.

Examples of allowing for changing parameters or alternative model structures include the assessment of yellowfin tuna (*Thunnus albacares*) by the Inter-American Tropical Tuna Commission (IATTC), which is divided into two time periods that correspond to perceived changes in oceanic conditions (IATTC 1991), and the consideration of a wide range of alternative models regarding stock structure in the robustness trials used by the Scientific Committee of the International Whaling Commission (IWC) to develop the revised management procedures (Kirkwood 1997).

Explicit alternative recruitment hypotheses were used in the analysis of the status of the southern bluefin tuna (*Thunnus maccoyii*), where three possible spawner–recruit relationships were presented to the 1994 Scientific Meeting of the Commission for the Conservation of Southern Bluefin Tuna (CCSBT): (1) a traditional Beverton–Holt curve, (2) a depensatory Beverton–Holt curve and (3) a recruitment function that did not decrease at low spawning stock sizes (CCSBT 1994). The second and third recruitment options can be thought of as alternative models, and since the spawning biomass of the stock is thought to have decreased well below any levels seen in the historical record, the policy implications of depensatory versus non-declining recruitments are quite different. Under depensatory recruitment, stock collapse is quite likely; under non-declining recruitment, stock collapse is unlikely. Thus, when one admits the possibility of alternative models of recruitment, the range of uncertainty expands greatly.

Random or systematic variation in biological or physical processes is the fourth type of uncertainty. Many assessment models allow for some form of stochastic recruitment (process error). However, interdecadal climate shifts, as described by Hare and Francis (1995) pose even greater threats to our models. If such shifts are common enough that they need to be incorporated into our stock assessments, we really must consider completely different models of fish production. While such climatic changes may often be represented by different parameter combinations of the same models, this is only a cosmetic covering of the fact that completely different production hypotheses need to be considered in our stock assessments. This has been the subject of a small number of published papers (Parma and Deriso 1990); also, while a number of laboratories have working groups on climatic change (e.g., the IWC Scientific Committee included robustness trials with environmental shifts, and the IATTC considered the change in oceanographic regimes), allowing for major environmental change remains unusual in stock assessments.

Statistical Decision Theory

When uncertainty in stock assessments is presented to decision makers, at least three different approaches are used. The first approach relies on statistical decision theory and computes the outcomes under different possible "states of nature" (Hilborn and Walters 1992). States of nature are most often different parameters of a common model, but they may include different models. The most straightforward method uses decision tables (see Table 1) that list both different possible states of nature and different management actions. Each cell of the table contains the expected consequences of taking any particular action if a given state of nature is true. If you assign probabilities to the states of nature, then the expected consequences of each possible action can easily be computed.

When the number of states of nature considered is quite high (for instance, when considering uncertainty in several parameters), presenting discrete alternative states of nature is often difficult. Instead, the expected consequences may be presented by integrating across the uncertainty. An example of this is found in the 1994 assessment of hoki (*Macruronus novaezelandiae*) in New Zealand (Table 2; Punt et al. 1994).

Assessment groups almost universally incorporate sensitivity analysis in the basic assumptions of their main assessment. For instance, expected consequences of management actions might be presented for changes in assumed natural mortality rate. Sensitivity analysis is a valuable tool for scientists with which to explore how robust their results are in comparison with their assumptions. However, assuming that the results were sensitive to a parameter, how would a manager use the results of a sensitivity analysis unless the scientists assigned probabilities to each case?

Indicators of Risk

A third method to incorporate uncertainty in stock assessment advice is to calculate and pass on to managers an indicator sometimes labeled as "risk." In New Zealand assessments, two forms of risk are calculated: risk to the fishery, which is the probability that the quota will be at least 80% of the vulnerable biomass, and risk to the stock,

TABLE 1.—Sample decision table showing the ratio between stock size in the year 2001 and the virgin stock size after applying different quotas annually between 1992 and 2001. Source: Hilborn et al. 1994.

Virgin stock size	Probability this was the true virgin stock size	Future quota (10^3 metric tons)		
		100	150	200
500	0.0000	0.22	0.22	0.22
550	0.0000	0.26	0.22	0.22
600	0.0002	0.33	0.22	0.22
650	0.0020	0.40	0.22	0.22
700	0.0089	0.46	0.23	0.22
750	0.0246	0.51	0.26	0.22
800	0.0493	0.54	0.32	0.23
850	0.0775	0.58	0.37	0.23
900	0.1017	0.60	0.42	0.24
950	0.1159	0.63	0.45	0.26
1,000	0.1184	0.65	0.49	0.31
1,050	0.1106	0.67	0.51	0.35
1,100	0.0963	0.68	0.54	0.39
1,150	0.0791	0.70	0.56	0.42
1,200	0.0619	0.71	0.58	0.45
1,250	0.0466	0.72	0.60	0.47
1,300	0.0339	0.74	0.62	0.50
1,350	0.0240	0.75	0.63	0.52
1,400	0.0166	0.76	0.65	0.54
1,450	0.0113	0.77	0.66	0.55
1,500	0.0075	0.77	0.67	0.57
1,550	0.0050	0.78	0.69	0.59
1,600	0.0032	0.79	0.70	0.60
1,650	0.0021	0.80	0.71	0.61
1,700	0.0013	0.80	0.71	0.63
1,750	0.0008	0.81	0.72	0.64
1,800	0.0005	0.81	0.73	0.65
1,850	0.0003	0.82	0.74	0.66
1,900	0.0002	0.82	0.75	0.67
1,950	0.0001	0.83	0.75	0.68
	Expected value	0.65	0.49	0.34

TABLE 2.—A decision analysis output that does not explicitly show alternative states of nature. Source: Punt et al. (1994), Table 9.

Performance indicator	Annual quota (1,000 metric tons)			
	200	250	300	350
Probability stock drops below 20% of virgin within 5 years (RISK)	0.0	0.001	0.004	0.013
Probability quota is set to be 80% or greater of vulnerable biomass	0.0	0.0	0.002	0.014
Expected ratio of biomass in 1998 to biomass for MSY	6.48	6.05	5.61	5.17

which is the probability that the vulnerable biomass will be less than 20% of the virgin biomass. The probability that the stock will drop below some specified level is commonly used and discussed in stock assessment groups (Mace 1994), and it is designed to reflect concern about possible consequences of low stock size, such as stock collapse due to recruitment failure, species replacement, or depensatory predation processes. This measure of risk is one way of capturing model uncertainty since stock assessment scientists are much less confident that model predictions will be correct in the range of low spawning stock sizes than in the range of high spawning stock sizes.

The National Marine Fisheries Service has adopted the term "probability profile" to describe an analysis that plots the probability of exceeding some threshold on the Y axis and alternative actions on the X axis (Rosenberg and Restrepo 1994). This is a more detailed presentation than computing a single probability.

Francis (1993) has proposed that the 20% risk measure be incorporated formally into management planning by not accepting any fishery policy that would push the stock to less than 20% of virgin biomass more than 10%

of the time. The use of the 20% virgin measure of risk is fraught with difficulty because (1) it is totally arbitrary, (2) virgin biomass (and therefore 20% of virgin biomass) is often very difficult to estimate, and (3) many stocks appear to be less than 20% of virgin biomass and provide considerable sustainable yield at those levels.

The three preceding points are closely related. Why not choose 10% or 30% of virgin biomass? Such assessments should be based on an analysis of data on probability of recruitment declines, species replacements, and so forth, rather than on an arbitrary number. Sissenwine and Shepherd (1987), Clark (1991), and Mace and Sissenwine (1993) have used eggs-per-recruit data to estimate appropriate levels of concern about recruitment overfishing for a number of stocks. McAllister et al. (1995) used data from a number of stocks in order to estimate the expected sensitivity of recruits to spawning stock for the New Zealand hoki stock. In New Zealand, three of the four most important commercial species illustrate the problem with a 20% rule. Snapper (*Chrysophrys auratus*) and rock lobster (*Jasus edwardsii*) are both assessed as well below 20% of virgin biomass, yet the abundance of both fish stocks is growing, and the optimum biomass for MSY is either less than 20% of virgin biomass or the difference in yield at 20% of virgin and at MSY is trivial. The New Zealand hoki stock is currently at 60–70% of virgin biomass and growing, yet the biomass at MSY is assessed to be at 18% of virgin biomass (because of relatively low vulnerability to gear). For all three of these stocks, presenting the probability the stock will fall to less than 20% of virgin biomass and calling it "risk" seems nonsensical.

Any measure based on virgin biomass poses severe problems since many stock assessments make no reliable estimate of the virgin biomass. In these cases, one can simply use thresholds based on the largest observed population size.

Explicit Estimation of Risk of Stock Collapse

There is a far more rational approach to incorporating

uncertainty in stock assessments. If, for instance, the only concern about low spawning stock sizes is the possibility of reduced recruitment, then including uncertainty about the spawner–recruit relationship would be sufficient. The chance of reduced recruitment at low spawning stock sizes will be reflected in the expected consequences of alternative policies. The obvious consequences one must calculate include average yield, short- and long-term variability in catch, and the probability of stock collapse. There is absolutely no need to calculate an additional indicator called "risk." Extensive analysis of spawner recruit data (Myers et al. 1994) is an excellent starting point in evaluating the relative probability of alternative spawner–recruit relationships.

If the stock assessment group believes there is a risk of stock collapse at low spawning stock sizes for reasons other than recruitment overfishing (e.g., Thompson 1993), this concern should be put directly into the analysis. I would suggest that stock assessment groups perform a literature review of the available data on species similar to the one of interest and calculate the proportion of stocks that collapsed when driven below 10% of virgin, when driven below 20% of virgin, etc., and incorporate these probabilities directly into the stock assessment model. Mace and Sissenwine (1993) have attempted to incorporate this type of information, but a great deal has yet to be learned about how to quantify such possibilities; I merely suggest what we should aim for rather than describing a well-defined method.

Stock assessment groups should also include the probability the stock will collapse at 50% of virgin biomass or higher. This is to say that we must be careful when discussing stock collapse. If we define stock collapse as a circumstance in which the stock continues to decline after fishing pressure is stopped, we would undoubtedly find that fishing will rarely be stopped until the stock is quite low, and we would be tempted to say that stocks greater than 50% never collapse. Yet, if we believe that the collapse is due to external ecosystem changes, then the collapse may be entirely unrelated to stock abundance. Sorting out the relative importance of stock abundance and external change in environmental conditions poses difficult challenges but deserves a high priority.

The principal advantage of explicitly including the probability of stock collapse is that it makes good scientific sense. We must use the data available to make the probability estimates, which are something managers can use. If we simply say that a certain harvest policy has a risk of 25%, where risk was defined as the probability the stock would drop lower than 20% of virgin biomass, the representative of a conservation group might consider this unacceptable while a commercial fisher might consider it just fine. The conservationist could argue that it is too dangerous to push stocks that low; the fisher

would cite all the fisheries that have been sustainably managed at less than 20% of virgin biomass. Indeed, it is quite likely that many people would assume that a 25% risk is a 25% chance of stock collapse. If the scientists were to provide their best estimates of the probability of stock collapse, it would remove one level of the political debate and alternative interest groups would directly discuss the key issue of acceptable levels of probability that a stock will collapse. Since commercial fishers are almost always interested in long-term sustainable yields from the resource (when evaluating long-term management plans), the differences in perspectives may disappear. The best hope for reaching such a consensus is to agree on how historical data on other stocks could be used to determine the probability of stock collapse.

Many assessments pose problems to the scientific staff because more than one hypothesis is consistent with the data, and these competing hypotheses have different policy implications. Examples include the debate over whether to believe catch per unit effort (CPUE) or surveys in the northern cod (*Gadus morhua*) fishery in the mid-1980s (Harris 1990), whether southern bluefin tuna recruitment will decline dramatically as spawning stocks decline (CCSBT 1994), and whether the commercial fishery in the Bering Sea affected the population of Steller sea lions (*Eumetopias jubatus*) because of a reduction of forage fish (Pascual and Adkison 1994). I suspect that stock assessment scientists can cite competing hypotheses in almost every assessment they perform.

When the competing hypotheses are all compatible with the existing data from the stock, scientists have two choices: they can either simply pass the alternatives along to managers without assigning probabilities to alternative hypotheses, or they can use their experience with exploited populations, the scientific literature, and all their personal knowledge to arrive at a best estimate of the probabilities of alternative hypotheses. The first option would seem most attractive because it does not necessitate possibly subjective assessments of what data to use in assigning degrees of belief to competing hypotheses. However, if scientists do not assign probabilities to competing hypotheses, the decision makers will do so, either explicitly or, much more likely, implicitly. Further, they will use their experience to determine their belief in the alternative hypotheses, or, has been as suggested from experiences in the South African fisheries (D. Butterworth, Univ. Cape Town, Rondebosch, South Africa, pers. comm.), they will choose the total allowable catch they wanted for economic or political reasons if it is compatible with any of the hypotheses put forward. Therefore, the question is whether scientists or decision makers are better equipped to evaluate the collective experience of fisheries science in evaluating the major uncertainties in fisheries stock assessments. This assumes of course that the decision makers actually use the scientific advice. I am sure most stock assessment scientists can name many decision-making bodies who use their advice and many others who ignore it.

Many of these issues can be resolved from published scientific literature. We have data on the frequency of stock declines at low spawning stock sizes, on the reliability of different measures of stock abundance, and on the frequency of density-dependent growth. The danger is that the scientists think too narrowly about what data are useful in their stock assessment. If the spawning stock size of a particular stock has gone below any level seen before, you must look to similar experience with other stocks—not simply pass two alternatives to managers and say "we have never been there, therefore we don't know how this stock will respond."

This is also the solution to the problem of sensitivity analysis posed earlier. While it is useful to know that uncertainty in a particular parameter may have little effect on predicted outcomes (results not sensitive), how should a manager treat a sensitivity analysis that shows the outcomes *are* sensitive to the parameter selected? The answer is, of course, that the scientific staff should assign probability distributions to the parameter based both on the data available from the stock and again the collective experience of fisheries science with other stocks.

There will be cases where different scientists will evaluate the data quite differently and come to different conclusions about the probabilities of different hypotheses. There is no easy solution to this problem. My experience suggests that in many cases the scientists will be able to agree on the appropriate probabilities by agreeing on a procedure to determine the probabilities prior to actually doing the computations of probabilities.

Readers familiar with Bayesian statistics will recognize that all this discussion is simply saying is that in many cases the results of the assessment will depend on the prior probability distributions assigned to some of the parameters. Determining prior probabilities is the single most difficult aspect of Bayesian statistics (and decision making in general), and at some point we may have to admit priors based on experience rather than quantitative analysis. I have been involved in two recent stock assessment meetings where each scientist present simply wrote down her/his own assessment of the probability of alternative hypotheses, all of which were included in the report of the meeting.

Strategies for Dealing with Uncertainty

Once we have identified all the uncertainties and done our best to assign probabilities to them, what options do managers have in dealing with uncertainty? From many perspectives the most desirable option is more research. We would like to collect more data and resolve the uncertainty. This is almost never a short-term payoff—the benefits of added research are usually several years away. In my experience, accumulation of fisheries-independent measures of stock abundance will resolve many uncertainties over time. However, it is probably more common that even with more research and several additional years, major uncertainty will persist and will probably always be with us.

Butterworth et al. (1997) describe searching for management procedures that are robust to the uncertainties. There are two important parts to this approach: the first is recognizing that a management policy should be a specified feedback procedure, not an annual quota-setting process. There is simply no way to evaluate the consequences of yearly quotas unless the procedure for setting quotas is specified. The second key ingredient in the approach of Butterworth et al. is the search for robustness—search long and hard to find something that performs acceptably under all the different possible states of nature. One can always define alternative hypotheses that will cause a management procedure to fail, and it has been argued (D. Butterworth, Univ. Cape Town, Rondebosch, South Africa, pers. comm.) that members of the IWC's Scientific Committee with links to international conservationist/protectionist organizations have indeed searched long and hard. If we wish to judge the robustness of a proposed management procedure, we must assign probabilities to each alternative hypothesis; no procedure can be robust to all alternatives.

Sainsbury et al. (1997) describe an application of adaptive management in spatially replicated experimental policies for a specified number of years. An adaptive management plan is simply an elaborate management procedure with explicitly recognized experimental components and unlike the usually non-experimental procedures described by Butterworth. A key difference in the problem described by Sainsbury is the potential for spatial replication. This is possible in some fisheries and should be used when available. Generally, the more you can implement different policies in different areas, the more quickly you will resolve major uncertainties and the better the long-term expected yield from the fishery will be (Walters 1986).

The Precautionary Principle

One possible approach to uncertainty is the precautionary principle. A pollution-related statement of the precautionary principle is that "potentially damaging pollution emissions should be reduced even 'when there is no scientific evidence to prove a causal link between emissions and effects'" (Peterman and M'Gonigle 1992). The obvious fisheries variation on this is that one should reduce catches unless there is good evidence that the current level of catch is sustainable. This would be the opposite of what Ludwig et al. (1993) consider the norm— not reducing catches unless there is overwhelming evidence that they are not sustainable. The precautionary principle says that you should act cautiously in the face of uncertainty.

How does one actually apply such a principle in real fisheries? First, one must recognize that any change in harvests has expected consequences both to the fish population and to the fishery—the economic and social community of people who harvest the fish. Any society must weigh the consequences of a reduction or increase in catch to both the fish stock and the fishery. I believe that in most of such analyses, recommendations based on the precautionary principle will be very different depending upon the state of the fishery and the discount rate of the fishers. In a developing fishery, the precautionary principle will call for caution in expanding catches or, more particularly, in expanding fishing capacity. Increase capacity slowly unless there is strong evidence that much larger catches are sustainable. The most serious long-term threat to both fish stock and fishery is overcapacity. Such biological caution may have economic costs in forgone yield but will not lead to serious dislocation of existing users.

However, when stocks are fully exploited (or overexploited), precautionary reduction in catches poses serious problems. We must carefully weigh the risks of stock collapse against the risk of large-scale economic or social change.

Thus, while the precautionary principle might provide reasonable advice in developing fisheries, it provides less guidance when stocks are intensively exploited. The precautionary principle has received criticism from other authors (Broadus 1992) in other fields for similar reasons. We must rely on good analysis of historical data and statistical decision theory to guide our analysis of the benefits of reduced catches.

References

Broadus, J. M. 1992. Creature feature too: Principium precautionarium. Oceanus, Spring 1992:6-7.

Butterworth, D. S., K. L. Cochrane, and J. A. A. De Oliveira 1997. Management procedures: a better way to manage fisheries? The South African experience. Pages 83-90 in E. K. Pikitch, D. D. Huppert, and M. P. Sissenwine, editors. Global trends: fisheries management. American Fisheries Society Symposium 20, Bethesda, Maryland.

Clark, W. G. 1991. Groundfish exploitation rates based on life history parameters. Canadian Journal of Fisheries and Aquatic Sciences 48:734-750.

CCSBT (Commission for the Conservation of Southern Bluefin Tuna). 1994 Annual Report. Canberra, Australia.

Francis, R. I. C. C. 1993. Monte Carlo evaluation of risks for

biological reference points used in New Zealand fishery assessments. Pages 221-230 in S. J. Smith, J. J. Hunt and D. Rivard, editors. Risk evaluation and biological reference points for fisheries management. Canadian Special Publication in Fisheries and Aquatic Sciences 120.

Givens, G. H, A. E. Raftery and J. E. Zeh. 1993. Benefits of a Bayesian approach for synthesizing multiple sources of evidence and uncertainty linked by a deterministic model. Report of the International Whaling Commission 43:495-500

Hare, S. R., and R. C. Francis. 1995. Climate change and salmon production in the Northeast Pacific Ocean. Pages 357-372 in R. J. Beamish, editor. Climate change and northern fish populations. Canadian Special Publications in Fisheries and Aquatic Sciences 121.

Harris, L. 1990. Independent review of the state of the northern cod stock. Canada Department of Fisheries and Oceans, Ottawa.

Hilborn, R., E. K. Pikitch, and M. K. McAllister. 1994. A Bayesian estimation and decision analysis for an age-structured model using biomass survey data. Fisheries Research 19:17-30.

Hilborn, R., and C. J. Walters. 1992. Quantitative fisheries stock assessment: choice, dynamics and uncertainty. Chapman and Hall, New York.

IATTC (Inter-American Tropical Tuna Commission). 1991. Annual Report. La Jolla, California.

Kirkwood, G. P. 1997. The revised management procedure of the International Whaling Commission. Pages 91-99 in E. K. Pikitch, D. L. Huppert, M. P. Sissenwine, editors. Global trends: fisheries management. American Fisheries Society 20, Bethesda, Maryland.

Ludwig, D., R. Hilborn, and C. Walters. 1993. Uncertainty, resource exploitation, and conservation: lessons from history. Science 260:17/36.

Mace, P. M. 1994. Relationships between common biological reference points used as thresholds and targets of fisheries management strategies. Canadian Journal of Fisheries and Aquatic Sciences 51:110-122.

Myers, R. A., P. M. Mace, N. Barrowman, and V. R. Restrepo. 1994. In search of thresholds for recruitment overfishing. ICES Journal of Marine Science 51:191-205.

Parma, A. M., and R. B. Deriso. 1990. Experimental harvesting of cyclic stocks in the face of alternative recruitment hypotheses. Canadian Journal of Fisheries and Aquatic Sciences 47:595-610.

Pascual, M. A., and M. D. Adkison. 1994. The decline of northern sea lion populations: demography, harvest or environment. Ecological Applications 94:393-403.

Peterman, R. M., and M. M'Gonigle. 1992. Statistical power analysis and the precautionary principle. Marine Pollution Bulletin 24:231-234.

Polacheck, T., R. Hilborn, and A. E. Punt. 1993. Fitting surplus production models: comparing methods and measuring uncertainty. Canadian Journal of Fisheries and Aquatic Sciences 50:2597-2607.

Punt, A. E., and D. S. Butterworth. 1993. Variance estimates for fisheries assessments: their importance and how to best evaluate them. Pages 145-162 in S. J. Smith, J. J. Hunt, and D. Rivard, editors. Risk evaluation and biological reference points for fissures management. Canadian Special Publication in Fisheries and Aquatic Sciences 120.

Punt, A. E., M. K. McAllister, E. K. Pikitch, and R. Hilborn.

1994. Stock assessment and decision analysis for hoki *(Macruronus novaezelandiae)* for 1994. New Zealand Fisheries Assessment Research Document, Ministry of Agriculture and Fisheries, Greta Point, Wellington.

Restrepo, V. R., J. M. Hoenig, J. E. Powers, J. W. Baird, and S. C. Turner. 1992. A simple simulation approach to risk and cost analysis, with applications to swordfish and cod stocks. Fisheries Bulletin 90:736-748.

Rosenburg, A. A., and V. Restrepo. 1994. Uncertainty and risk evaluation in stock assessment advice for U.S. marine fisheries. Canadian Journal of Fisheries and Aquatic Sciences 51:2715-2720

Sainsbury, K., R. A. Campbell, R. Lindholm, and A. W. Whitelaw. 1997. Experimental management of an Australian multispecies fishery: examining the possibility of trawl-induced habitat modification. Pages 107-112 *in* E. K. Pikitch, D. D. Huppert, M. P. Sissenwine, editors. Global trends: fisheries management. American Fisheries Society Symposium 20, Bethesda, Maryland.

Sissenwine, M. P, and J. G. Shepherd. 1987. An alternative perspective on recruitment overfishing and biological reference points. Canadian Journal of Fisheries and Aquatic Sciences 44:913-918.

Thompson, G. G. 1993. A proposal for a threshold stock size and maximum fishing mortality rate. Pages 303-320 *in* S. J. Smith, J. J. Hunt and D. Rivard, editors. Risk evaluation and biological reference points for fisheries management. Canadian Special Publication in Fisheries and Aquatic Sciences 120.

Walters, C. J. 1986. Adaptive management of renewable resources. Macmillan, New York.

Experimental Management of an Australian Multispecies Fishery: Examining the Possibility of Trawl-Induced Habitat Modification

K. J. Sainsbury, R. A. Campbell, R. Lindholm, and A. W. Whitelaw

Abstract.—The North West Shelf of Australia supports a diverse tropical fish fauna. Changes in species composition were observed following the introduction of fishing. Several different ecological hypotheses to explain the changed species composition were consistent with the available data. These hypotheses included combinations of interspecific interactions, intraspecific interactions, and trawl-induced modification of benthic habitat. Some hypotheses indicated that a considerable improvement in catch value was possible. It was shown that an experimental or actively adaptive management approach with spatial and temporal manipulation of the trawl fishery effort was scientifically and economically viable for resolving key management uncertainties. Experimental periods of less than approximately 5 years were not expected to provide sufficient hypothesis discrimination to allow significantly improved management decisions, and experimental periods longer than around 15 years cost more in research and forgone catches than the resulting hypothesis discrimination is worth.

Three contrasting management zones were established on the North West Shelf; one area was closed to trawling in 1985, a second was closed to trawling in 1987, and the third remained open to trawling. Research surveys were used to monitor fish abundance and the benthic habitat. The North West Shelf management experiment provided close to the expected level of hypothesis discrimination. The results increased the probability placed on hypotheses involving habitat modification mechanisms. Consequently, the possibility of improved catch value is judged more likely than was the case before the experiment. However, the results also indicate that habitat recovery dynamics are slower than previously thought, so that resources recovery will be slow. Furthermore, direct observations of trawl–habitat interactions showed a high rate of damage to the habitat on encounter with the trawl gear. Consequently, a high-yield fishery is expected to be slow to attain and difficult to maintain if existing trawl fishing methods are used.

This paper provides an overview of an experimental or actively adaptive (Walters 1986; Hilborn and Walters 1992) approach to fishery management in a situation of very high uncertainty in resource dynamics and fishery economics. The research described has extended for over 10 years and is continuing. This paper outlines the background to the management issues being faced, summarizes the approach taken to analysis and provision for advice to fishery managers, describes the management actions taken and some subsequent developments in the fishery, and summarizes some main observations and results.

Background and Management Issues

The North West Shelf region of Australia supports a diverse Indo-West Pacific fish fauna comprising several hundred species (Sainsbury et al. 1985). A brief period of trawl fishing occurred between 1959 and 1963; intensive pair-trawling began in 1971 (Sainsbury 1991). Attempts to operate domestic trawlers in the fishery in the early 1980s were not economically successful, but a small domestic Australian trap fishery began in 1984.

Research surveys conducted by various nations between 1962 and 1983 (Sainsbury 1987, 1988, 1991) showed that the abundance of high-valued fish (from the genera *Lethrinus* and *Lutjanus* in particular) declined with

the development of trawl fishing, and that the abundance of some lower-valued fish (*Nemipterus* and *Saurida* in particular) increased. They also showed that the catch rate of epibenthic organisms, such as sponges, greatly decreased between 1963 and 1979. Photographic surveys of the seabed conducted in association with trawl surveys during the early 1980s showed a significantly higher probability of occurrence of *Lethrinus* and *Lutjanus* in areas where large (>25 cm) benthic organisms were present than in areas with no large epibenthos. Conversely, *Nemipterus* and *Saurida* showed a significantly higher probability of occurrence in areas without large epibenthos (Sainsbury 1991).

In 1979, the Australian 200-mile fishing zone was declared with the broad aim of obtaining the greatest return from the resource, if possible from development of domestic fisheries. The historical fish community structure, dominated by *Lethrinus* and *Lutjanus*, would be expected to have given improved value of catches and increased opportunity for development of domestic fisheries compared with the community structure as it existed in the early 1980s. This gave rise to the following related research and management issues:

- Can the change in community composition be reversed (linking to the question of what caused the change)?

- If it appears that the change could be reversed, is it worth attempting given the uncertainties in outcome, time frame, and value of the catches?
- If an attempt is to be made, exactly what management measures should be used?

The Analytical Approach

Four alternative mechanisms were developed that could explain the observed changes in abundance of the four key genera (*Lethrinus*, *Lutjanus*, *Nemipterus*, and *Saurida*) being considered. The methods of analysis, described in detail in Sainsbury (1988, 1991), were as follows:

1. An intraspecific mechanism, under which the observed changes are regarded as independent single-species responses.
2. An interspecific mechanism in which there is a negative influence of *Lethrinus* and *Lutjanus* on the population growth rate of *Saurida* and *Nemipterus*, so that *Saurida* and *Nemipterus* experience a competitive release as the abundance of *Lethrinus* and *Lutjanus* is reduced by fishing.
3. An interspecific mechanism in which there is an negative influence of *Saurida* and *Nemipterus* on the growth rate of *Lethrinus* and *Lutjanus*, so that *Lethrinus* and *Lutjanus* are inhibited as the abundance of *Saurida* and *Nemipterus* increases for other reasons.
4. Habitat determination of the carrying capacity of each genus separately, so that trawl-induced modification of the abundance of the habitat types alters the carrying capacity of the different genera.

The available data were insufficient to allow unique parameterization of the models, and this parameter uncertainty was encompassed by choosing different parameter sets with about equally high likelihood and treating model–parameter set combinations as separate models. While all of these models are ecologically reasonable and consistent with the available data, they have different management implications. Models 1 and 2 imply a relatively low productivity of *Lethrinus* and *Lutjanus*; the historical reduction in abundance seen is interpreted as being due to overfishing, and yields would have to be kept low to prevent this from happening again even after the resource had been rebuilt. However, models 3 and 4 imply a relatively high productivity of *Lethrinus* and *Lutjanus* under some circumstances; selective harvesting of *Lethrinus* and *Lutjanus* under model 3 and harvesting without removal of large benthic organisms under model 4 would result in high sustainable catches.

The development of scientific fishery management advice for the North West Shelf used the analytic framework for evaluating adaptive management regimes

(Walters 1986; Hilborn and Walters 1992; and described in Sainsbury 1988, 1991). In summary, the analysis considers a short-term management regime (W) that is applied during a "learning period" of duration (t). During this learning period, certain revenues are obtained, monitoring costs are incurred, and observations are made. At the end of the learning period, the observations are used to guide a long-term policy choice. At that time, a risk-neutral manager selects one from the optimal strategies relating to each of the models considered and applies that strategy for all future time. The economic and observational consequences of failure to be able to implement either the learning period regime or the selected long-term strategy were also included because there was considerable uncertainty about the ability to implement the chosen management regimes on the North West Shelf (especially those requiring expansion of the domestic fisheries). The choice of long-term strategy depends in part on the probability placed on each model at the end of the learning period, when the decision is made, and the power of the observations to discriminate among the alternative models. A Bayesian updating method was used to calculate changes in the probabilities placed on the alternative models. Repeated simulation of the revenues, observations, and decisions was used to determine the expected present value, which was conditional on each alternative model being true. These simulations were combined by the probabilities initially placed on each model to give an overall expected present value from the learning period and experimental regime being evaluated. The key issue is which combination of W and t gives the best overall expected revenue. A good experimental regime (1) would result in a high probability being assigned to each alternative model, when that model was true, so that the appropriate long-term policy is selected, and (2) would do this at low net cost (e.g., cheap observations and little or no catch forgone during the learning period).

These evaluations indicated that there was relatively low expected present value from continuation of the existing licensed foreign trawl fishery. Even with the relative probabilities on each model as they were at that time, a higher expected present value would be obtained from an immediate switch to the long-term policy of a domestic trap fishery. Moreover, continued trawl fishing while more observations were made to support a later decision resulted in a lower expected present value the longer the experiment was continued. Essentially, the observations of continued trawling were relatively uninformative. There was a cost to obtaining the observations, and they did not greatly improve the final decision. However, it was also shown that some experimental management regimes, involving the cessation of trawling in some areas and the introduction of trap fishing in some of the areas closed to trawling, could provide a higher expected present return from the resource. For learning periods less than roughly 5 years, the discrimination between hypotheses obtained was insufficient to greatly improve decision making. For learning periods greater than approximately 15 years, a high level of discrimination is obtained, but the cost of obtaining this discrimination is greater than the value of improved management. However, for experimental periods between 5 and 15 years, the experimental regime gave the greatest expected present value of any approach examined.

Management Actions and Some Subsequent Developments

Three contrasting management zones were established on the North West Shelf, each involving the continental shelf adjacent to about 80 nautical miles of coastline (Figure 1). One area was closed to foreign trawling in 1985, the second was closed to foreign trawling in 1987, and the third remained open to trawling. Trap fishing was permitted throughout. Annual research surveys were planned to monitor fish abundance and the benthic habitat for the first 5 years, after which the situation would be reviewed and the appropriate survey interval reexamined.

The resulting contrasts in trawl fishing effort were not exactly as planned. The trawl effort was reduced in the first area and maintained in the remaining areas as intended, but the closure of the second area also resulted initially in a reduction in fishing effort in the remaining open areas. This was because the foreign fleet initially responded to the closure by moving to alternative fishing regions. Trawl effort increased again in later years. However, the main surprise was the development of domestic trawl effort in 1989 and especially 1990. This possibility had not been considered seriously because earlier attempts at domestic trawling had proved to be uneconomic. Apparently, a number of factors were involved in development of the domestic trawl fishery:
- the improved resource condition following the reduction in foreign trawling,
- a down-turn in inshore shrimp catches and the consequent need for alternative resources,
- improved domestic markets for tropical species (as a result of changes in consumer awareness and the supply of trap-caught fish since the early 1980s).

The main effect of this development on the management experiment was a rapid increase in trawling effort in the central closed area in 1990. This was possible because the foreign trawl controls were established under federal management jurisdiction, but management control was transferred to state jurisdiction in 1988 and the state regulations allowed domestic trawling east of 116°45'E. This was not appreciated to be a problem at

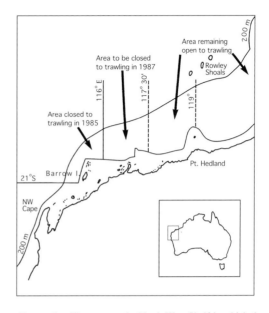

FIGURE 1.—The area on the North West Shelf in which the Commonwealth Scientific and Industrial Organization research was conducted. The zoning used during the experimental management between 1986 and 1991 is also shown.

the time. However, as the trawl fishery developed and targeted the central closed area, it was agreed that this development should not be impeded. The change of management jurisdiction also carried with it a change in the research agency primarily responsible for management advice, from federal to state agencies. This and other changes in the federal research agency resulted in a cessation of annual trawl surveys after 1992. Despite all these changes, the experimental approach is being continued. The design will be different, with the central zone left open to trawling and the easternmost zone closed to trawling from 1993 (that is adaptive management!), and resource surveys planned for roughly 3-year intervals.

Observations and Results from the First 5 Years of Experiment

The observations from the eastern and western zones are most simply interpreted because they are essentially trawled or untrawled for the whole 5 years between 1985 and 1990. The catch rate of all fish and the abundance of large (>25 cm) and small (<25 cm) benthos in the untrawled and trawled areas (Figures 2 and 3, respectively) show that, in the area closed to trawling, the density of fish increased, the abundance of small benthos increased, and the abundance of large benthos stayed about the same or increased slightly. In the area open to trawling, the abundance of fish decreased, and the abundance of both large and small benthos decreased.

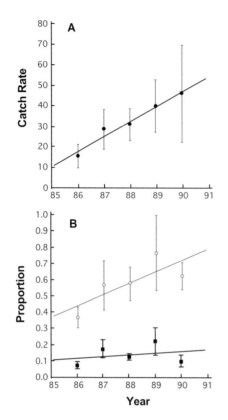

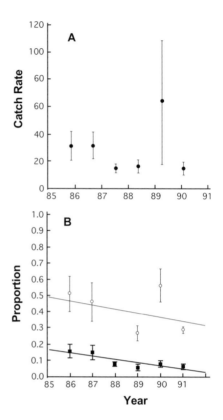

FIGURE 2.—(A) Total catch rate of *Lethrinus* plus *Lutjanus* (kg/30 min trawl) in the zone closed to trawling in October 1985 based on the annual research data. (B) Proportion of seabed with large (closed square) and small (open circle) benthos in the zone closed to trawling in October 1985 based on the annual research data. Standard errors and line of best fit are also shown for (A) and (B).

FIGURE 3.—(A) Total catch rate of *Lethrinus* plus *Lutjanus* (kg/30 min) in the zone left open to trawling based on the annual research data. (B) Proportion of seabed with large (closed square) and small (open circle) benthos in the zone left open to trawling based on the annual research data. Standard errors and line of best fit are also shown for (A) and (B).

The relative probability that was placed on each of the alternative hypotheses before and after the experiment was calculated by applying the same Bayesian updating method used in the initial analysis (Sainsbury 1991) to the data from all three experimental zones. Where P is a Bayesian probability, L is a likelihood, O is a set of observations, and m represents the alternative models:

- initially all four hypotheses are considered to have equal probability so $P_{INITIAL}(m_i) = 0.25$,
- the probability on each hypothesis at the start of the experiment in 1985 was calculated by Bayesian updating from research survey observations available at that time (observations O_{1985}) from

$$P_{1985}(m_i) = P_{INITIAL}(m_i) L(O_{1985}| m_i) / \Sigma^j P_{INITIAL}(m_j)$$
$$L(O_{1985}| m_j)$$

where $L(O_{1995}| m_j)$ is the likelihood of the observations to 1985 fitted to model m_j.

- The probability on each hypothesis at the end of the initial 5 years of the experiment in 1990 was calculated by Bayesian updating from research survey observations made in all areas during that period (observations $O_{1985-90}$) from

$$P_{1990}(m_i) = P_{1985}(m_i) L(O_{1985-90}| m_i) / \Sigma_j P_{1985}(m_j)$$
$$L(O_{1985-90}| m_j).$$

The estimates of probability for each hypothesis in 1985 and 1990 were calculated (Table 1). The additional information from the experiment considerably reduced the probability placed on one of the interspecific mechanisms, which was previously given a high probability. In that mechanism, *Lethrinus* and *Lutjanus* negatively influence the population growth rate of *Nemipterus* and *Saurida*, resulting in competitive release of the latter pair with fishing. The probability placed on the interspecific (i.e., single species) and the other intraspecific mechanisms increased slightly, but the probabilities placed on both of these mechanisms remain low. The probability placed on the habitat limitation model increased substantially, with the probability now placed on the habitat modification model being about double the probability of the closest other contender. There are clearly further analyses that could and should be performed. However, by the methodology originally used to design the experiment, the results after 5 years indicate a considerable increase in probability placed on the habitat modification model. This indicates a substantially increased possibility that a high-valued *Lethrinus* and *Lutjanus* fishery could be established on the North West Shelf if the habitat could also be protected.

The observed change in abundance of large (>25 cm) and small (<25 cm) benthic organisms was found to be inconsistent with the assumed growth and settlement rates that were assumed in the initial analysis (Sainsbury 1988, 1991). A constant settlement rate was assumed and estimated; it was also assumed that epibenthic organisms could grow to 25 cm in roughly 6–10 years. The results obtained from the experiment indicate clearly that for the estimated settlement rate the time to grow to 25 cm is at least 15 years, and that settlement rates could also be lower than estimated. Further analysis of settlement rates is planned, but by either of the mechanisms examined it is clear that habitat dynamics are slower than anticipated and, thus, that recovery times are longer than anticipated.

Some additional research on the interaction between benthic organisms and trawl gear was conducted on the North West Shelf during the experimental period. Video cameras were mounted on the trawl net to observe interaction between epibenthic organisms and the trawl groundline (for this trawl, the groundline consisted of a wire cable threaded with 15-cm-diameter punchings from car tires). The result of each encounter with a benthic organism was recorded, with the possibilities being (1) no removal (i.e., the trawl ran over the epibenthic organism without apparently dislodging it from the seabed),

(2) removal with the organism then rolling under the trawl net, (3) removal with the organism entering the trawl net, and (4) an unknown result. The results for benthic organisms larger than about 15 cm are given in Table 2. For a large number of observed encounters, the result is unknown because of poor visibility and other visual obstructions to observing the organism's fate. However, of those encounters where the fate was observable, 89% resulted in the removal of the organism from its substrate. Bounds for this removal probability were 43% if all encounters with unknown results did not result in removal and 95% if all encounters with unknown results did result in removal. The very low occurrence of removed organisms being retained by the trawl means that most removals would not be apparent from trawl catches of benthic organisms. It is not known what happens to sponges that are broken from the substrate or whether they regrow from small portions that might remain in the substrate. However, the absence of observations of sponges showing the changed orientation that would be expected if reattachment was common suggests that loose sponges disintegrate rather than reattach.

Discussion

The analysis carried out for the North West Shelf proved very useful in guiding management actions despite the initial high levels of uncertainty. Furthermore, the use of an experimental approach, even if not implemented exactly as intended, provided some useful results: it increased resolution of the key features affecting resource dynamics and increased credibility of the habitat model in particular. It increased attention to management actions that take habitat modification into account, including trawl designs and zoning for non-trawl areas. It provided improved information on the dynamics of both the fish resource and the benthic habitat for use in present and future fishery assessments. It aided in the empirical resolution of the economic viability of domestic fisheries in the North West Shelf region (although other factors also played a large part in this, as they prob-

TABLE 1.—The probability placed on each alternative hypothesis prior to 1985 and after 1990, the first 5 years of the North West Shelf experiment.[a]

Hypothesis	Prior to experiment	After 5 years of experiment
Interspecific control	0.01	0.02
Intraspecific control	0.52	0.33 (L,L>N,S)
Intraspecific control	0.01	0.03 (N,S>L,L)
Habitat and interspecific	0.46	0.62 control

[a]The probabilities were calculated using a Bayesian updating method described in Sainsbury (1991). The hypotheses are described in general terms in the text, whereas Sainsbury (1991) provides the mathematical details.

TABLE 2.—Observed results of encounter between sponges and a demersal trawl groundline.[a]

Result on encounter	Number	Percent
No removal	19	5
Removal and rolled under net	154	39
Removal and caught in net	15	4
Unknown	205	52
Total observations	393	100

[a]A wire cable onto which rubber punchings were threaded. The observations were made from a video camera mounted on the trawl during seven 30-min demersal tows. The result of encounter was scored only for individual sponges that were larger than the diameter of the groundline itself (~15 cm).

ably always will). It encouraged further research on habitat effects in this fishery and the development of a follow-up experiment. It provided empirical resolution of the initial major uncertainty about the economic viability of an Australian domestic fishery on the North West Shelf.

The experimental approach is continuing, despite a complete change in management jurisdiction and an almost complete change in research agencies. This indicates that—at least sometimes—long-term experiments can overcome the difficulty of organizational "turn-over time," which tends to act against any coherent long-term experimental strategy.

Results to date indicate that the relative composition of the multispecies fish community on the North West Shelf (and possibly other tropical areas of northern Australia) is, to an important extent, habitat dependent, and that historical changes in species composition in this region are in part a result of trawl-induced modification of the epibenthic habitat. Furthermore, continued alteration of the demersal habitat due to trawling will most likely continue to alter species composition. However, in the case of the North West Shelf, an implication of this situation is that high-value yields would be possible if the habitat modification could be avoided or greatly reduced. The slow dynamics of habitat recovery, combined with the apparently high probability of the larger elements of the habitat being removed on encounter with a trawl, mean that protection measures will have to be very effective to provide and maintain the community structure that will support this high-valued yield.

Interestingly, this concern about trawl-induced damage to the broader community of marine organisms is almost as old as the trawl itself. An early reference to this problem appeared in Bellamy (quoted by Street 1961:34-35):

> Fishing, taken generally, interferes in the slightest way with the habits of the creatures in question; but the employment of a trawl, during a long series of years, must assuredly act with the greatest prejudice towards them. Dragged along with force over considerable areas of marine bottom, it tears away, promiscuously, hosts of inferior beings there resident, besides bringing destruction on the multitudes of smaller fishes, the whole of

which, be it observed, are the appointed diet of those edible species sought as human food. It also disturbs and drags forth the masses of deposited ova of various species. An interference with the economical arrangement of creation, of such magnitude and of such duration, will hereafter bring its fruits in a perceptible diminution of these articles of consumption for which we have so great necessity. The trawl is fast bringing ruin on numbers of poorer orders requiring the most considerable attention.

Today, sustainable management of our marine resources has become a high priority. While the concerns about sustainability may not be new, the need to deal with them effectively has perhaps never been greater because of increasing fishing power of fleets and the multi-use demands placed on many marine resources.

References

Hilborn, R., and C. J. Walters. 1992. Quantitative fisheries stock assessment. Chapman and Hall, New York.

Sainsbury, K. J. 1987. Assessment and management of the demersal fishery on the continental shelf of northwestern Australia. Pages 465-503 in J. J. Polovina and S. Ralston, editors. Tropical snappers and groupers: biology and fisheries management. Westview Press, Boulder, Colorado.

Sainsbury, K. J. 1988. The ecological basis of multispecies fisheries, and management of a demersal fishery in tropical Australia. Pages 349-382 in J. A. Gulland, editor. Fish population dynamics, 2nd ed. John Wiley and Sons, New York.

Sainsbury, K. J. 1991. Application of an experimental approach to management of a tropical multispecies fishery with highly uncertain dynamics. ICES Marine Science Symposium 193:301-320.

Sainsbury, K. J., P. J. Kailola, and G. G. Leyland. 1985. Continental shelf fishes of northern and northwestern Australia. Clouston and Hall, Canberra, Australia.

Street, P. 1961. Vanishing animals. Preserving nature's rarities. Faber and Faber, London.

Walters, C. J. 1986. Adaptive management of renewable resources. Macmillan Press, New York.

Walters, C. J., and R. Hilborn. 1976. Adaptive control of fishing systems. Journal of the Fisheries Research Board of Canada 33:145-159.

Walters, C. J., and R. Hilborn. 1978. Ecological optimization and adaptive management. Annual Review of Ecology and Systematics 9:157-188.

MULTIPLE SPECIES AND ECOSYSTEM CONSIDERATIONS

Global Assessment of Fisheries Bycatch and Discards: A Summary Overview[1]

DAYTON L. ALVERSON

Abstract.—Global estimates of discarded fish range from 17.9 to 39.5 million metric tons per year. Limited data suggest that survival of most discarded species is low, that discarding causes declines of some nontarget species, that bycatch often contributes strongly to overfishing, and that discarding partly accounts for shifts in species dominance and occupation of certain ecological niches. The value of discarded fish plus the costs of monitoring and preventing discards amount to billions of dollars annually. Success in reducing bycatches and discards has varied with the species managed and the cooperation given by industry. Effort reduction, incentive programs, and individual transferable quotas are emerging as potentially viable control techniques, but they will have to be adapted to particular fisheries and regions, and their efficacy may be limited by a paucity of observer programs. Policies for bycatch and discard reduction must take sociocultural attitudes into account, both nationally and internationally, but they should be based on sound conservation principles.

Over the past 15 years, bycatch has become a topic of discussion in a variety of scientific, technical, and political forums. It has emerged as the "fishery management issue of the 1990s" (Tillman 1992). It is, therefore, somewhat ironic that the term "bycatch" is so frequently undefined and misused by managers, politicians, advocacy groups, and frequently even fishery biologists. Its contemporary application among most of these groups is generally synonymous with the capture and discard of marine life by fishing fleets resulting in waste, unreported fishing mortality, and threat to the survival of populations of birds, marine mammals, and other sea life.

The definition of bycatch used in a vast majority of technical/scientific papers, which are often the factual bases for formulating conclusions on the extent of biological loss and unreported catches, is all species captured other than target species. Thus, bycatch discards may constitute a small-to-significant fraction of the identified bycatch, depending on the nature of the fisheries and local customs. For example, Andrew and Pepperell (1992) and the authors of this paper estimate the upper range of bycatch taken in the world shrimp fisheries to be in excess of 16 million metric tons (mt).

Yet throughout much of Asia and many of the world's artisanal fisheries, a variable share (perhaps 5% to 80%) of the shrimp bycatch may be species retained for food or industrial purposes. Although discards in tropical shrimp fisheries are generally high, they are considerably less than the total reported bycatch, so it is important not to associate estimates of total bycatch with world discards of marine fishes. Further, bycatch discard rates

and numbers may misrepresent the impacts because for a number of species some fraction of the discard survives. Without good estimates of the biomass discarded, the fraction of which survive, unobserved mortality and other fishery-related losses, and the landed catch of a particular species, it will be impossible to assess overall impacts of fishing. We need to know the portion of the natural turnover that is killed.

To achieve a global estimate of discards, we have at times been forced to use total reported bycatch estimates and "back out" rather subjective estimates of retained species from these data. Our best estimates may seem rather staggering, but it is possible they may be an underestimate in that world recreational fishery discards are not included, the database for most areas of the world is incomplete, and discard weights are not included for marine mammals, seabirds, and turtles, and for many areas, invertebrates.

The discard ratios and rates are likely to appear obtuse to many readers, but it is not our intent to condemn those who harvest the oceans' living resources. However, it has been the our goal to clarify the character of world bycatch problems and, where possible, provide information that may be helpful to managers examining potential solutions. The level of losses for many fisheries may not be particularly excessive in light of other industries, based on the use of natural resources (Natural Resource Consultants [NRC] 1990). Rather, a substantial opportunity apparently is available for improving use of marine living resources and protection of overfished, threatened, or endangered species.

Data Sources and Limitations

We examined several hundred articles concerned with bycatch and discards in world fisheries. Over 800 pa-

---

[1]Those interested in evaluating data sources, limitations, methodology, and fuller explanations of the details presented herein should refer to Alverson et al. (1994), the comprehensive document upon which this summary overview is based.

pers containing quantitative and qualitative information were used to characterize the nature and scope of regional and global bycatch discard problems. Mortalities associated with discarding practices were also reviewed. The total volume of records analyzed by number, weight, and gear types is provided in Table 1.

The potential for errors in calculating estimated bycatch discards is enormous in light of reporting procedures in which some equate bycatch with (1) total discards, (2) secondary target species and discards, and (3) selected species within the bycatch complex. Different operational definitions and failure to define what sector of the bycatch is involved may have in some instances complicated our analysis and made calculations less precise.

Although the major objective of our study has been to estimate regional and global levels of bycatch discards, we recognize a considerable portion of the report deals with northern temperate fisheries. This was not the intent of the authors, but acquiring data from many developing areas of the world proved difficult. Our estimates are based on numerous research and observer records made throughout the world. Nevertheless, there is a paucity of data from many regions, and many observations involve data taken over short time spans by a small fraction of the fleets or even single sampling efforts made by research vessels. Further, the quantitative data used span a number of years and may not accurately portray the present situation.

These and other data problems noted make it frivolous to attempt to establish hard statistical parameters around regional and gear-type estimates. In reality, the estimates constitute "snapshots" based on collages of observations having various degrees of reliability taken over different seasons and years. The ultimate understanding of the true scope, distribution, and magnitude of the bycatch discard problem will require extensive documentation and acquisition of additional data from many regions of the world.

Thus, we urge our global and regional discard estimates be used as a provisional "best guess" of the potential magnitude of the discard problem resulting from fishing in the world's oceans. Further, we would hope this "best guess" will stimulate fishery researchers to collect and report adequate data leading to a more precise estimate.

Estimates of Regional and Global Bycatch Levels

A provisional estimate of global discards in commercial fisheries is 27.0 million mt, with a range of 17.9 to 39.5 million mt. The region defined by the Food and Agriculture Organization of the United Nations (FAO Sta-

TABLE 1.—Total number and number of records in weight-based and numbers-based formats for each gear type in the Natural Resource Consultants[a] (NRC) bycatch database.[b]

Gear type	Total number of records	Number of weight-based records	Number of numbers-based records
Trawl	966	571	75
Net	232	2	107
Line	150	58	33
Purse seine	82	6	5
Troll	16	0	1
Danish seine	24	21	22
Pot	83	56	41
Other	81	12	0
Not stated	89	1	1

[a]Natural Resources Consultants, 4055 21st W., Seattle, WA 98199, Tel: 206-285-3480; FAX: 206-283-8263, email: NRCSeattle@aol.com.
[b]All ensuing table sources for NRC refer to NRC's bycatch database.

tistical Area 67) as having the highest discard estimate is the northwest Pacific (Figure 1, Table 2). Shrimp trawl fisheries, particularly for tropical species, were found to generate more discards (volume and number) than any other fishery type and account for just over one-third of the global total (Table 3). On a weight-per-weight basis, 14 of the highest 20 discard ratios were associated with shrimp trawls. The fisheries associated with the 20 highest numbers-based ratios represented a more eclectic mix of shrimp trawl, pot, fish trawl, and longline fishery gear types. At the opposite end of the scale, fish trawl, seine, and high-seas driftnet fisheries accounted for the majority of the gear types in our list of the 10 lowest discard ratios (Tables 4 and 5). Discards by major species groups (weight) are provided in Table 6.

Although data are tremendously variable, four major gear groups stand out. Shrimp trawls are alone at the top of the list, while relatively low levels are recorded for pelagic trawls, purse seines targeting on menhaden (*Brevoortia* spp.), sardines, and anchoveta (*Cetengraulis mysticetus*), and some of the high-seas driftnet fisheries. Between these two extremes lie two other groups. The first of these comprises bottom trawls, unspecified trawls, longline gear, and the majority of the pot fisheries. The final group fits between the very low ratios of the pelagic trawl group and the moderate ratios of the aforementioned bottom trawl, pot, and line assemblage. Fisheries in this last group include the Japanese high-seas driftnet fisheries, Danish seines, and purse seines for capelin (*Mallotus villosus*). Discard rates by numbers result in a reordering of the highest and lowest discard rates per kilogram of target species (Tables 7 and 8). Specific discard rates by weight and number for gear types are shown in Table 9 and 10.

Case studies are provided for bycatch discard problems in the northeast Pacific, as well as the northeast and northwest Atlantic. Bycatch and bycatch issues have

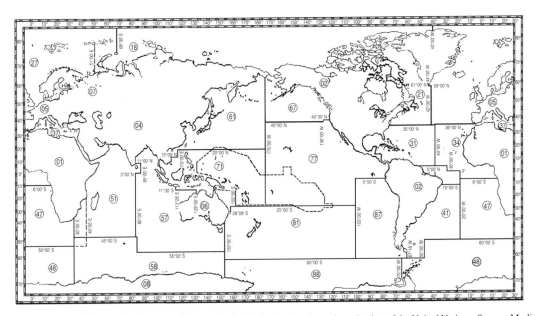

FIGURE 1.—Statistical areas based on annual reports of the Food and Agriculture Organization of the United Nations. Source: Modified from FAO (1992b).

been intensively studied in these locations relative to other areas. In the northeast Pacific, a suite of fisheries produces a bycatch total exceeding 1 billion individuals annually. Impacts appear low on most species (Table 11) although they may be significantly higher for Pacific halibut (*Stenolepis hippoglossoides*) and king (*Paralithoides* spp.) and Tanner crab (*Chionoecetes* spp.) (Table 12).

Discard problems in the Northwest Atlantic are classified into four groups: (1) marketable species too small or otherwise prohibited from landings; (2) species for which no current market exists, but which are caught along with commercial or recreational species; (3) species-specific fleet sectors discarding another fisheries target species; and (4) non-fishery bycatch species, including marine mammals, turtles, and birds. Regulatory approaches and management actions to address these problems are also discussed. On the basis of 1988–92 data, the estimated catch (landings and discards) of the 1987 year class of southern New England yellowtail flounder (*Limanda ferruginea*) is presented for landings and discards separately (Figure 2).

TABLE 2.—Bycatch discard weight by major world region. Source: NRC, Seattle, Washington.

Area	Discard weight (mt)
Northwest Pacific	9,131,752
Northeast Atlantic	3,671,346
West central Pacific	2,776,726
Southeast Pacific	2,601,640
West central Atlantic	1,600,897
West Indian Ocean	1,471,274
Northeast Pacific	924,783
Southwest Atlantic	802,884
East Indian Ocean	802,189
East central Pacific	767,444
Northwest Atlantic	685,949
East central Atlantic	594,232
Mediterranean and Black seas	564,613
Southwest Pacific	293,394
Southeast Atlantic	277,730
Atlantic Antarctic	35,119
Indian Ocean Antarctic	10,018
Pacific Antarctic	109
Total	27,012,099

TABLE 3.—Estimated bycatch and discards from world shrimp fisheries derived from reported bycatch levels and estimated amount of bycatch retained. Source: NRC, Seattle, Washington.

Area	Estimated bycatch (mt)	Estimated discard (mt)
Northwest Atlantic	81,665	80,031
Northeast Atlantic	210,297	206,091
West central Atlantic	1,310,653	1,271,334
East central Atlantic	123,636	61,818
Mediterranean and Black seas	257,859	250,124
Southwest Atlantic	253,446	245,842
Southeast Atlantic	39,143	19,571
Western Indian Ocean	1,871,075	748,430
Eastern Indian Ocean	482,879	289,727
Northwest Pacific	4,284,408	4,155,903
Northeast Pacific	28,269	27,421
West central Pacific	1,450,352	1,377,835
East central Pacific	590,955	561,416
Southwest Pacific	19,446	18,863
Southeast Pacific	203,677	197,567
Total	11,207,760	9,511,973

TABLE 4.—Top 20 fisheries with the highest recorded bycatch-to-discard ratios by weight (discard weight per landed target catch weight). Source: NRC, Seattle, Washington.

Fishery description	Kilograms discarded per kilograms landed
Trinidadian shrimp trawl	14.71
Indonesian shrimp trawl	12.01
Australian northern prawn trawl	11.10
Sri Lankan shrimp trawl	10.96
U.S. Gulf of Mexico shrimp trawl	10.30
Sea of Cortes shrimp trawl	9.70
Brazilian shrimp trawl	9.30
West Indian shrimp trawl	8.52
U.S. Southeast shrimp trawl	8.00
Northwest Atlantic fish trawl	5.28
Persian Gulf shrimp trawl	4.17
Southwest Atlantic shrimp trawl	4.10
East Indian shrimp trawl	3.79
Bering Sea sablefish pot	3.51
Malaysian shrimp trawl	3.03
Senegalese shrimp trawl	2.72
Bering Sea rock sole trawl	2.61
British Columbia cod trawl	2.21
Gulf of Alaska flatfish trawl	2.08
Northeast Atlantic dab trawl	2.01

TABLE 5.—The ten lowest observed weight-based discard ratios in fisheries other than shrimp (discard weight per landed target catch weight). Source: NRC, Seattle, Washington.

Fishery description	Kilogram discarded per kilogram landed
Northwest Atlantic hake trawl	0.011
West central Atlantic menhaden seine	0.029
Bering Sea cod pot	0.041
Northeast Pacific whiting trawl	0.043
Northwest Atlantic cod trawl	0.058
Bering Sea pelagic pollock trawl	0.062
Northwest Atlantic redfish trawl	0.063
Northeast Atlantic groundfish trawl	0.083
Gulf of Alaska midwater pollock trawl	0.086
Northwest Atlantic plaice trawl	0.118

The study covering the northeast Atlantic focuses on discarding in the mixed-species trawl fisheries for North Sea gadoids. The impact of discards on mortality rates for haddock (*Gadus morhua*) and whiting (*Merluccius bilinearis*) and the effect of reductions in fishing effort draw particular scrutiny. Most of the discard problems noted pertaining to northwest Atlantic trawl fisheries are also noted in the fisheries of the northeast Atlantic. Local variations associated with misreporting and environmental effects are, however, discussed in some detail (Alverson et al. 1994). Also provided is a review of regulatory and gear management measures commonly applied

in the region. We point out that many supposed technical solutions can generate unsuspected side effects that may impair their effectiveness. Further, these solutions remind the reader that voluntary bycatch reduction measures are unlikely to be successful if they are not in the short-term economic interest of the affected fisher. Reasons for discards reported for the northeast coast of the USA and for the West Coast ground fisheries are provided (Figures 3 and 4).

In the Northeast Pacific, the added fishing mortality resulting from discards does not appear significant for most gadoids, flounders, and other species, but the impacts of trawl, trap, and line fisheries on halibut are relatively large (~0.08) and, in terms of the allowed fishing mortality, are around 0.28 (total fishery-induced mortality is ~0.3, of which bycatch accounts for 0.08). Further, the potential impacts of the king and Tanner crab pot fishery on king crab populations may be significant. Reeves (1993) suggests Bering Sea red king crab discards

TABLE 6.—Global marine bycatch discards on the basis of the FAO International Standard Statistical Classification of Aquatic Animals and Plants (ISSCAAP) species groups.

ISSCAAP	Mean discard weight (mt)	Landed catch weight (mt)	Ratio of discarded weight to landed weight	Ratio of discarded weight to total weight
Shrimp, prawn	9,511,973	1,827,569	5.2	0.84
Redfishes, bass, conger	3,631,057	5,739,743	0.63	0.39
Herring, sardine, anchovy	2,789,201	23,792,608	0.12	0.1
Crab	2,777,848	1,117,061	2.49	0.71
Jack, mullet, saurie	2,607,748	9,349,055	0.28	0.22
Cod, hake, haddock	2,539,068	12,808,658	0.2	0.17
Miscellaneous marine fishes	992,356	9,923,560	0.1	0.09
Flounder, halibut, sole	946,436	1,257,858	0.75	0.43
Tuna, bonito, billfishes	739,580	4,177,653	0.18	0.15
Squid, cuttlefishes, octopus	191,801	2,073,523	0.09	0.08
Lobster, spiney-rock lobster	113,216	205,851	0.55	0.35
Mackerel, snook, cutlassfishes	102,377	3,722,818	0.03	0.03
Salmon, trout, smelt	38,323	766,462	0.05	0.05
Shad	22,755	227,549	0.1	0.09
Eel	8,359	9,975	0.84	0.46
Total	27,012,098	76,999,943	0.35	0.26

TABLE 7.—Top 20 fisheries with the highest recorded bycatch-to-discard ratio by number (discard number per landed target catch number). Source: NRC, Seattle, Washington.

Fishery description	Number discarded per number landed
West Central Atlantic shrimp trawl	12.13
Bering Sea king crab pot	9.71
California halibut net	4.83
Northeast Atlantic whiting trawl	2.83
Bering Sea tanner crab pot	2.34
Northeast Atlantic haddock trawl	1.94
Arabian Gulf finfish trawl	1.75
Northeast Atlantic nephrops trawl	1.70
East central Pacific spiny lobster pot	1.68
East central Pacific swordfish longline	1.58
East Indian Ocean finfish trawl	1.27
Northeast Atlantic hake trawl	1.18
East Indian Ocean tuna longline	1.13
Northeast Atlantic cod Danish seine	0.79
East central Pacific slipper lobster pot	0.37
Northeast Atlantic plaice trawl	0.42
Caribbean tuna longline	0.40
Japanese high-seas squid driftnet	0.39
Northeast Atlantic sole trawl	0.33
Northeast Atlantic herring seine	0.20

TABLE 8.—The 10 lowest observed numbers-based discard ratios in fisheries other than shrimp (discard number per landed target catch number). Source: NRC, Seattle, Washington.

Fishery description	Number discarded per number landed
Bering Sea midwater trawl pollock	0.005
Northeast Atlantic tuna driftnet	0.009
Gulf of Alaska midwater trawl pollock	0.018
Northwest Pacific squid driftnet (Korean)	0.037
Northwest Pacific squid driftnet (Taiwanese)	0.068
Subtropical Convergence Zone tuna driftnet	0.080
Tasman Sea tuna driftnet	0.123
Bering Sea king crab pot	0.132
Eastern tropical Pacific tuna purse seine	0.180
Bering Sea cod pot	0.180

amounted to approximately 16 million animals in 1990, more than five times the number landed. Many of these discards are sub-legal (juvenile) individuals. The economic and biological implications of these discards, depending on discard mortality, may be a serious problem for red king crab stock dynamics and management.

Impacts

It must also be recognized that some portion of discards survives and thus is not lost from the ecosystem. In terms of finfishes, we see little evidence that discard survival constitutes a significant portion of the discard for many commercial and recreational finfish species. Nevertheless, survival of flounder, dab, invertebrates, and fishes not affected by rapid change in depth shows some promise for improved survival under constrained operational practices and appropriate handling (E. Pikitch, Univ. Washington School of Fisheries, Seattle, pers. comm.).

The consequences of bycatch discards, varying between regions, include significant biological waste, biological overfishing of target and bycatch species, economic losses imposed on target fisheries, modification of biological community structures in ecosystems, and impacts on severely depleted, threatened, or endangered species. These impact categories are similar to those outlined by Fowle and Upton (1992).

Alverson et al. (1994) provide scientific evidence supporting assertions that significant biological losses and ecological shifts in the biotic communities occur as a result of bycatch discarding. Reports of bycatch discards

TABLE 9.—The top weight-based, discard-to-landed-target catch ratios by gear type. Source: NRC, Seattle, Washington.

Fishery description	Kilogram bycatch per kilogram landed
Non-Pelagic fish trawl	
Bering Sea rock sole	2.61
British Columbia pacific cod	2.21
Gulf of Alaska flatfish	2.08
Northeast Atlantic dab	2.01
Northeast Atlantic flatfish	1.60
Pelagic fish trawl	
Bering Sea pollock (1988)	0.01
Northeast Atlantic cod	0.00
Bering Sea pollock (1989)	0.00
Gulf of Alaska pollock (1989)	0.00
Bering Sea pollock (1987)	0.00
Shrimp trawl	
Trinidad	14.71
Indonesia	12.01
Australia	11.10
Sri Lanka	10.96
U.S. Gulf of Mexico	10.30
Longline	
Eastern central Pacific swordfish (1990)	1.13
Bering Sea Greenland turbot	1.03
Eastern central Pacific swordfish (1991)	1.00
Bering Sea sablefish	0.50
Gulf of Alaska cod	0.26
Purse seine	
Northwest Atlantic capelin (1983)	0.81
Northwest Atlantic capelin (1981)	0.37
East central Atlantic sardine	0.03
U.S. Gulf of Mexico menhaden	0.03
West central Pacific tuna	0.00
Danish seine	
Northeast Atlantic haddock	0.50
Notheast Atlantic whiting	0.45
Northeast Atlantic cod	0.36
Pot/trap	
Bering Sea sablefish	3.51
Bering Sea king crab	3.39
Beirng Sea tanner crab	1.78
Northwest Atlantic capelin	0.80
East Central Pacific spiny lobster	0.36

TABLE 10.—The top numbers-based discard-to-landed target catch ratios by gear type. Source: NRC, Seattle, Washington.

Fishery description	Bycatch number per target number
Trawl	
Caribbean shrimp	12.13
Dutch shrimp trawl	7.30
Northeast Atlantic whiting	2.83
Northeast Atlantic haddock	1.94
Northeast Atlantic *Nephrops*	1.70
High-seas driftnet	
Japanese north Pacific squid	0.37
Japanese Tasman Sea tuna	0.12
Japanese Subtropical Convergence Zone tuna	0.08
Taiwan north Pacific squid	0.07
Korean north Pacific squid	0.06
Longline	
Eastern central Pacific swordfish	1.58
Eastern Indian Ocean tuna	1.13
Danish seine	
Northeast Atlantic cod	0.79
Northeast Atlantic haddock	0.70
Northeast Atlantic whiting	0.64
Pot/trap	
Bering Sea king crab	8.78
Bering Sea Tanner crab	2.35
Eastern central Pacific spiny lobster	1.68
Eastern central Pacific slipper lobster	0.67
American lobster	0.22

suggest that major problems occur throughout the Atlantic. However, to some extent this may reflect the relative intensity of fisheries and the high level of reporting of discards. Other fisheries of the world having high bycatch discard rates include (1) most tropical and subtropical shrimp fisheries, (2) trawl and seine bottom fisheries, as well as northern shrimp and *Nephrops* fisheries in the north and south Atlantic; (3) tuna seine fisheries in the eastern tropical Pacific (ETP) that set on logs;[2] (4) North Pacific king and Tanner crab pot fisheries, as well as the yellowfin sole (*Limanda aspera*) and rock sole

TABLE 11.—Estimated fishing mortality rates for key species in the Bering Sea in 1992 resulting from discarding major commercial target species (assumes 100% mortality of discards). Source: National Marine Fisheries Service 1992; also, NRC, Seattle, Washington.

Species	Mortality due to discards	Mortality due to landings
Pollock	0.016	0.171
Pacific cod	0.013	0.190
Atka mackerel	0.008	0.053
Rockfish	0.004	0.008
Yellowfin sole	0.012	0.046
Sablefish	0.001	0.053
Greenland turbot/arrowtooth flounder	0.012	0.014
Rock sole	0.015	0.027
Flounder	0.020	0.024
Pacific Ocean perch	0.035	0.035

TABLE 12.—Discard ratio for Bering Sea crab fisheries, 1991 and 1992. Source: Beers (1992), Reeves (1993), Tracy (1993).

Year	Species/fishery	Number discarded per number retained
1990	Adak golden king crab	5.126
1990	Dutch Harbor golden king crab	9.948
1991	Dutch Harbor golden king crab	7.23
1991	Bering Sea snow crab	0.106
1991	Bering Sea Tanner crab	0.895
1991	Bristol Bay red king crab	1.583
1991	St. Matthew blue king crab	2.036
1992	Bristol Bay red king crab	5.889

(*Lepidopsetta bilineatus*) trawl fisheries; and (5) a variety of pot fisheries for invertebrates throughout the world.

For many of the above-noted fisheries, discards may equal or exceed in number or weight the quantities of what is landed and marketed. In total, millions of tons and billions of dollars of loss probably occur as a result of accidental capture and discard of unmarketable and marketable species. Much of the world's discarded fish appears to be small juveniles of commercially important species which, if left to mature, would most likely produce significantly higher weight yields compared with the discarded weight. Finally, fishing discards affect a broad range of aquatic species other than finfish and shellfish.

We need to recognize that the ratio of discarded catch to the retained catch frequently may have little to do with observed and documented biological or ecological impacts. Such impacts must be evaluated on a case-by-case basis in terms of the discard mortality imposed on target and non-target species populations. Low bycatch discard rates may still generate serious impacts, particularly if the fisheries of concern involve considerable and geographically dispersed fishing effort. For example, the observed bycatch discard rate for turtles in the Gulf of Mexico and southeast U.S. shrimp fisheries is very low and the actual encounter of turtles in the nets is infrequent. However, the take in number of animals may be several tens of thousands, resulting in a discard mortality on turtles exceeding all other sources (Tillman 1992; National Research Council 1992).

Conversely, large observed bycatch of a species by number or weight may not constitute serious biological problems. For example, the large discard of pollock (*Theragra chalcogramma*) in the Bering Sea involves a very small fraction of the pollock population (on average <1% of the exploitable biomass), and managers require the bycatch take to be tallied as a part of the authorized harvest. However, a much smaller catch of halibut

---

[2]Tuna purse seine fisheries setting on dolphins have a very low bycatch-to-discard ratio while school fish sets have an intermediate bycatch-to-discard ratio.

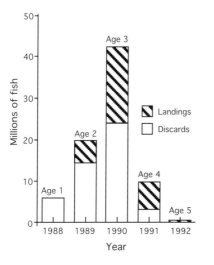

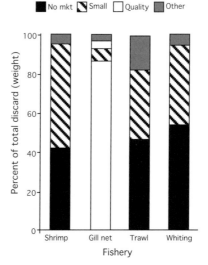

FIGURE 2.—Estimated catch (landings and discards) of the 1987 year class of southern New England yellowtail flounder, 1988–92. Total catch of the year class was 77 million fish (age 1–5), with 46.5 million (60%) discarded. Source: Alverson et al. (1994).

FIGURE 3.—Reasons for discards in four Gulf of Maine groundfish fisheries from sea sampling trips conducted in 1991. Reasons are as follows: No mkt = species for which no market existed; small = fish smaller than minimum legal size or minimum market size; quality = fish of poor market quality; other = various other reasons for discards. Source: Murawski (1993).

in the same region by trawls, line, and pot gear has constituted a serious loss of fish that could be taken in the halibut line fishery. For both species, a significant economic loss occurs to the involved fisheries.

Policy and Issues

The codification of international bycatch discard policies, which varies regionally, has in many instances been in response to conservation and environmental groups concerned with impacts on marine mammals, turtles, and

Solutions

A variety of techniques have been attempted by managers, engineers, and scientists to reduce bycatch discard levels, including traditional net selectivity approaches, the development of fishing gear taking advantage of differential species behavior, and time and area fishing restrictions. These methodologies have worked with varying degrees of success depending on the species being managed and the willingness of industry to work together for positive solutions. The successful reduction of dolphin mortalities (Figure 5) in the ETP tuna seine fisheries and the reduction of fish discards in the northern shrimp fisheries (Table 10) are classic examples of a program emphasizing industry-proposed technology, education, and effective monitoring of operation results.

In addition to these technological and educational efforts, managers are experimenting with time–area–bathymetric fishing patterns, incentive programs, and altered operational modes, bycatch quotas, and gear limitations. Most scientists, fishers, and managers agree that no universal panacea to the discard problems exists and that fisheries must be examined by regions and species in relationship to behavioral responses to gear and alternative options for solutions. Local industry expertise is frequently key to defining appropriate solutions.

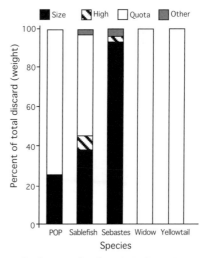

FIGURE 4.—Reasons for discards in five U.S. west coast groundfish fisheries from sea sampling trips conducted in 1991. Reasons are as follows: Size = below minimum acceptable market size; high = high-grading of species catch; quota = trip quota for species already exceeded; other = various other reasons for discards. Source: Pikitch (1991).

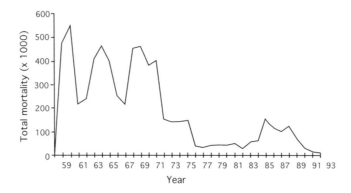

FIGURE 5.—Discard mortalities (number of porpoise) in the Eastern Tropical Pacific tuna purse seine fisheries, 1959–93. Source: M. Hall, Inter-American Tropical Tuna Commission, La Jolla, California, pers. comm.

seabirds. In this respect, the impacts may only indirectly involve discard rates; rather, the type of the bycatch discard, including issues of ethics, may sharply influence regional policies. Further, regional policies on bycatch discards may be more concerned with sociopolitical or socioeconomic consequences of selected discards than with overall bycatch rates of a particular fishery.

In some instances, regulatory policies appear to have been emotionally driven (Burke et al. 1993; Miles 1992) and developed in the absence of available scientific evidence, while in other instances, actions taken to curb discards led to results inconsistent with the managers' expectations. Nevertheless, the magnitude of the problem and the potential range of the consequences have brought bycatch discards to the surface as a legitimate management issue requiring serious national and international attention. To date, most bycatch policy has focused on high-profile fish species, such as salmon (*Oncorhynchus* spp.), halibut, crab, lobster, and also marine mammals, birds, and turtles. However, in the past several years, policy development has included issues involving biological and ecological impacts, biological waste, and economic losses.

We suggest that sociocultural attitudes toward marine resources should be an important consideration in the development of international discard policies. An anonymous reviewer of this paper suggested the following: "In the development of discards policy, public education should be emphasized so that sociocultural attitudes evolve towards an intelligent approach to problems rather than an emotional and irrational approach. The goal should be to harmonize the sociocultural attitudes from different societies and the development of tolerance towards different views of marine resource uses." Unfortunately, to date the policy process has paid too little attention to sociocultural perspectives, which are often influenced by differing national dependencies on marine resources as a protein staple.

We also note the growing importance of non-consumptive uses to fisheries and bycatch policy changes. We urge that evolution of global discard policies be earmarked by minimizing social conflicts; be independent, to the degree possible, of ideological differences; and be based on sound conservation principles.

Bycatch as Part of a Larger Resource Management Problem

During the conduct of our study, we became increasingly interested in the current total biomass of fish and shellfish removed from the world's oceans or killed as a result of fishing activity. At about the beginning of the 1990s, data suggest landings of approximately 98 million mt (FAO 1992a). These landings are frequently gauged against scientific estimates of potential sustainable yield of conventional species from the world's oceans' wild stock of about 100 million mt. As a result, a number of authors have noted current world landings approaching the estimated maximum sustainable yield for world marine fisheries.

In making such reflections, authors frequently fail to recognize that this global catch total includes freshwater and aquacultural production. On the other hand, current reported world commercial fisheries landings omit the following:

1. discard mortalities involved in commercial fisheries,
2. recreational fish catches and bycatches taken from marine waters,
3. fish killed as a result of contact with fishing gear (i.e., mortalities resulting from fish passing through net webbing or resulting from hooking of fish which subsequently escape),
4. ghost fishing mortalities,
5. underreporting,
6. substantial subsistence catches and discards, and

7. landed catches of commercial fishes that are not purchased (rejects).

On the basis of the 1990 landing report for marine fishes (FAO 1992b), with approximately 83 million mt and discard levels of about 27 million mt, we arrive at a world marine catch of 110 million mt without considering the six other noted categories involving fishery-related mortalities in marine waters. Even though a significant fraction of discards may not involve marketable species, it seems very likely the aggregate fishery deaths from fishing may be significantly over the sustainable yield estimates of 100 million . Of course, the 100 million mt questioned by many scientists may be unrealistic, but in what direction? The point raised, however, is that the total marine losses resulting from fishing are much greater than suggested by landing records.

General Discussion

In undertaking our study and discussing bycatch discard problems with various elements of the fishing industry throughout the world, it quickly became apparent that most commercial and recreational fishers see bycatch as a problem confronting other gear types and user groups, but not themselves. Each user group tends to place itself on "high moral ground" and see other groups as the "culprits." Further, most had strong personal feelings on the "dirty" character of various fishing gears although few had ever seen or reviewed data on bycatch discard rates, either for their own or for other fisheries.

Authors, too, had somewhat tainted and preconceived views regarding the magnitude of bycatch discard rates that we were likely to find for various fisheries. We were somewhat surprised to find that the low bycatch discard rates by gear type frequently involve high-seas driftnets, both by weight-to-weight and number-to-number ratios. We noted the highest bycatch discard rates per number of squid taken, observed for the Japanese squid fishery, is 0.37, lower than for any other reported high gear-type fishery noted in the world except for American lobster (*Homarus americanus*) pots (0.22). Further, number-based bycatch rates for the remaining driftnet fisheries are lower than rates for all other gear-type fisheries throughout the world. High-seas driftnet fisheries were commonly listed among the lowest observed bycatch rates by number.

Of course, we were not surprised by the high discard rates for shrimp and some trawl fisheries, but did not expect the very low discard rates for mid-water trawls and the very high discard rates of sub-legals in many invertebrate pot fisheries. With few exceptions, high and low rates occurred for each gear type, depending on area and times.

The fact that actual observations are often at odds with public perception is not surprising and is a reminder that the perception of a gear's impact and whether it is clean or dirty has a strong qualitative overtone, in many instances, in terms of the character of the bycatch discarded. High-seas squid driftnet fishing was in reality being condemned because of the take of birds, marine mammals, and turtles (some of which were considered endangered), plus the association with illegal salmon fishing in the areas to the north of the squid grounds. Further, documentation of the fishery was not transparent to other interested nations. Some of the phrases and words used to describe high-seas gillnetting by the press have little basis in fact for many high-seas driftnet fisheries. Nevertheless, they served to rally national and international political support to condemn the gear and have its use prohibited in ocean space beyond national jurisdictions. At the same time, they created a perception that driftnets and gillnets are destructive fishing gears wherever they are deployed. As a consequence, driftnets and gillnets are now being condemned in areas under national jurisdiction, regardless of the character of their bycatch.

The data suggest that generic characterization of gear types as clean or dirty may easily run into counterintuitive results. For example, in the Bering Sea groundfish fisheries, the average bycatch discard rate for trawlers (0.15 kg per kilogram) is considerably lower than the average for longline fisheries (0.22 kg per kilogram), yet just to the south in the Gulf of Alaska, the average bycatch discard rate for line fisheries is about the same as for trawls (0.21 kg per kilogram for both gear types). Further, inter-gear observations may change from year to year.

Assessments of bycatch discard impacts at the population level must take into account numbers and weight discarded and the survival of the discards, as well as compare the discard mortality in numbers or weight with the subject population. In this regard, the terms "dirty" or "clean," based on observed rates, are rather meaningless, except as they may relate to the issues of biological waste. Bycatch discarding represents too complex an issue to classify it neatly as "good and bad" or "clean and dirty," based on ratios of discards to retained catch or on numbers, weights, or other absolute indices. Unfortunately, when combined with the "spin" placed on reported numbers or weights of discards by advocacy groups, the press, and politicians, such classifications often serve to condemn a particular fishery or gear and frequently may result in generic condemnation of such gear or fishery without regard to biological–environmental, economic, and cultural impacts. Further, this process is too often blemished by inaccuracies and misrepresentation of facts. Taken out of context, a discussion of millions or billions of fish or thousands or millions of metric tons of catch serves as a powerful motivator of public opinion, which in turn has a considerable impact on the evolution of fishery policies.

Emerging ideas include effort reduction, incentive programs, and individual bycatch quotas that move the responsibility for bycatch reduction to the individual vessel level. We suggest that major gains against the global bycatch problem are likely to occur as such shifts towards individual responsibility take place. Progress may be impeded, however, because observer programs, an uncommon characteristic of today's fisheries, are necessary to audit adequately the progress toward bycatch goals.

In greatly overcapitalized fisheries and those in which gross overfishing is obvious, the importance of controlling overall effort as a means of reducing bycatch discards is apparent. However, the authors note that quick, "easy-fix" solutions are unlikely and a dedicated national and international effort will be necessary to secure important conservation and economic goals associated with bycatch. Bycatch discard reduction efforts should involve a clear and focused understanding of actual impacts and desired results. Reduction in bycatch for species suffering from overfishing or otherwise threatened or endangered should rank high among international goals.

Conclusion

In conclusion, we paraphrase the concluding observations of Iudicello and Leape (1994):

There needs to be a shift in the approach and accountability (in fisheries management): (1) Fishers need to prove that their current levels of bycatch in the short term are unavoidable, and (2) management agencies need to use tools currently available and accept discarding as a problem to which government assistance should be directed. If conservation groups, governments, and fishing groups commit to finding solutions, and there is a force of law behind policies directed towards reduction of discarding, a comprehensive program including (1) reduction in fishing on overfished stocks, (2) a combination of incentives and disincentives, and (3) new technologies as well as other alternatives could lead to significant reductions in the level of global discards.

Finally, there is a growing global recognition that the world's fishing effort already exceeds what is necessary to harvest sustainable yields of marine fishes. The single action that will provide the greatest improvement to the bycatch discard problem (other than halting world population growth) will be the reduction of these effort levels. Without such control, other solutions to the bycatch discard problem will be less effective and real success in our efforts to better manage the ocean's resources much more difficult.

References

Alverson, D. L., M. J. Freeberg, J. G. Pope, and S. A. Murawski. 1994. A global assessment of fisheries bycatch and discards. Food and Agriculture Organization of the United Nations, Fisheries Technical Paper 339, Rome.

Andrew, N. L., and J. G. Pepperell. 1992. The by-catch of shrimp trawl fisheries. Pages 527-565 in M. Barnes, A. D. Ansell, and R. N. Gibson, editors. Oceanography and marine biology annual review. Vol. 30. UCL Press, United Kingdom.

Beers, D. E. 1992. Annual biological summary of the Westward Region shellfish observer database, 1991. Alaska Department of Fish and Game Regional Information Report 4K92-33, Kodiak, Alaska.

Burke, W. T., M. J. Freeberg, and E. L. Miles. 1993. The United Nations resolutions on driftnet fishing: an unsustainable precedent for high seas and coastal fisheries. Ocean Development and International Law 25:122-136.

FAO (Food and Agriculture Organization of the United Nations). 1992a. Marine fisheries and the law of the seas: a decade of change. FID/C853, Rome.

FAO (Food and Agriculture Organization of the United Nations). 1992b. FAO Yearbook. Fishery statistics: catches and landings, 1990. Volume 70, Rome.

Fowle, S., and H. Upton. 1992. A national organization's perspective on the shrimp bycatch issue. Abstract. In Proceedings of the international conference on shrimp bycatch, Lake Buena Vista, Florida, 24-27 May 1992. NOAA, National Marine Fisheries Service, Tallahassee, Florida.

Harrington, D. 1992. A Sea Grant perspective on reducing bycatch. Pages 65-82 in Proceedings of the international conference on shrimp bycatch, Lake Buena Vista, Florida, 24-27 May 1992. NOAA, National Marine Fisheries Service, Tallahassee, Florida.

Iudicello, S., and G. Leape. 1994. Conserving America's fisheries. Progress on Magnuson Act. National Coalition for Marine Conservation, Inc., Savannah, Georgia.

Miles, E. L. 1992. The need to identify and clarify national goals and management objectives concerning bycatch. Pages 169-179 in R. W. Schoning, R. W. Jacobson, D. L. Alverson, T. G. Gentle, and J. Auyong, editors. Proceedings of the national industry bycatch workshop, Newport, Oregon, 4-6 February 1992. Natural Resources Consultants, Inc., Seattle, Washington.

Murawski, S. A. 1993. Factors influencing by-catch and discard rates: analyses from multispecies/multifishery sea sampling. North Atlantic Fisheries Organization SCR Document 93/115.

National Marine Fisheries Service. 1992. Targets and discards of the Bering Sea and Gulf of Alaska. NMFS Alaska Fisheries Science Center, Seattle, Washington.

National Research Council. 1992. Dolphins and the tuna industry. National Academy Press, Washington, D.C.

Natural Resources Consultants. 1990. The nature and scope of fishery dependent mortalities in the commercial fisheries of the Northeast Pacific. Natural Resources Consultants, Seattle, Washington.

Pikitch, E. K. 1991. Technological interactions in the U.S. west coast groundfish trawl fishery and their implications for management. Pages 253-263 in Proceedings of a symposium held in the Hague, 2-4 October 1989. ICES Vol. 193, Copenhagen, Denmark.

Reeves, J. E. 1993. Use of lower minimum size limits to reduce discards in the Bristol Bay red king crab *(Paralithodes camtschatica)* fishery. National Marine Fisheries Service, Alaska Fisheries Science Center Technical Memorandum 20, Seattle, Washington.

Tillman, M. F. 1992. Bycatch: the issue of the 90's. Pages 13-18 *in* Proceedings of the international conference on shrimp bycatch, Lake Buena Vista, Florida, 24-27 May 1992. NOAA, National Marine Fisheries Service, Tallahassee, Florida.

Tracy, D. 1993. State of Alaska biological summary of the 1992 Bristol Bay red king crab fishery mandatory shellfish observer program database. Report to the Alaska Board of Fisheries. Alaska Department of Fish and Game, Dutch Harbor, Alaska.

Multispecies Assessment Issues for the North Sea

NIELS DAAN

Abstract.—Progress in multispecies virtual population analysis in the North Sea is reviewed with particular emphasis on tests of the underlying assumption of constant suitabilities (i.e., the probabilities that different prey types will be eaten by particular predators at any point in time are determined by the relative abundances of all potential prey types, weighted by a constant suitability factor for each predator–prey combination). The results of a second year of intensive stomach sampling indicate that year effects in estimated suitabilities are significant but the explanatory power is relatively small. In practice, results of long-term predictions for different management options based on either data set were broadly similar. Therefore, no basis exists for developing a more complex model that incorporates prey switching. Multispecies assessment is also reviewed against its actual impact on fisheries management in the North Sea. So far, the effect has been marginal because the solution to short-term problems related to individual stocks has overruled the formulation of any long-term management objectives for the commercial fishery resources in the area. Methods to evaluate ecosystem effects of fishing have recently attracted much research because of management requests. Although multispecies assessment offers possibilities for extending the scope for advice in this respect, the model has clear limitations. Other approaches appear to be required.

Since the development in the late 1970s of an algorithm for the simultaneous solution of virtual population analyses for more than one stock interacting through predation (Helgason and Gislason 1979; Pope 1979), multispecies assessment has been a major research focus within the International Council for the Exploration of the Sea (ICES), the organization responsible for advice on fisheries management in the northeastern Atlantic Ocean. Extensive reviews of the application of multispecies virtual population analysis (MSVPA) in the North Sea (Pope 1991) and the Baltic Sea (Sparholt 1991) were presented at the Symposium on Multispecies Models Relevant to Management of Living Resources in The Hague in 1989. The primary objective of MSVPA (Sparre 1991) is to quantify feeding interactions among species in relationship to the interaction between fish stocks and fisheries.

Until recently, MSVPA in the North Sea has been based largely on 1 year (1981) of intensive stomach content sampling. One of the crucial assumptions underlying MSVPA—which allows extrapolation of predation rates to other years—is that suitability of each species age-group as prey for each predator age-group is constant. Suitability is defined as the probability that a particular prey type would be eaten by a particular predator when all prey types are present in equal numbers. This factor can be thought of as integrating different aspects of vulnerability, such as prey size preference of the predator and degree of geographical overlap between predator and prey type (Andersen and Ursin 1977). Consequentially, different prey types are assumed to be eaten in proportion to their relative abundance, weighted by a specific suitability factor for each predator and prey combination. By assuming that suitability is constant, the model ignores the possibility of prey switching in response to

changes in relative abundance of prey species or to annual variations in spatial overlap of predators and prey, and the inherent simplification has been subject to much debate within the ICES Multispecies Assessment Working Group. Although additional data sets collected for cod (*Gadus morhua*) and whiting (*Merlangius merlangus*) in some quarters of 1985, 1986, and 1987 suggested relative stability of the suitability matrix for these species (Rice et al. 1991), it was generally agreed that a more complete test was needed (Sissenwine and Daan 1991). Therefore, a full-scale ICES-coordinated stomach sampling project was repeated in 1991 (Anonymous 1992). All species preying on fish and caught during the surveys were included in the samples in order to extend the database for future applications, and in total over 100,000 stomachs were collected. However, at time of writing, only data for the main five predator species had been analyzed.

In this paper, I present an overview of the preliminary results obtained during a recent meeting of the Multispecies Assessment Working Group (Anonymous 1994), which was aimed at testing the constant suitability hypothesis on the basis of a comparison of the 1981 and 1991 data sets. Although I give a personal account of the major findings, I emphasize that the intellectual ownership rests entirely with all members of the group and that some additional meetings are required before the data will be thoroughly scrutinized. I also discuss the prospects of multispecies assessment in relationship to fisheries management in the North Sea.

Is Suitability Constant?

Two sets of independent stomach content data provide various checks on and comparisons of the results obtained by applying the MSVPA model as well as alternative models, but testing the hypothesis of constant suitability is not straightforward. The problem is that suitability is a theoretical concept. The parameters cannot be estimated directly from stomach content data but only by iteration within the model because information on the relative sizes of the different prey stocks is required to estimate suitabilities, and prey stock estimates depend in turn on the suitabilities. Moreover, estimates of suitability are affected not only by the stomach content data but also by the catch data and the terminal fishing mortalities. The latter are derived from tuning against catch per unit of effort (commercial or survey data or both) in a single-species mode. All these input data have largely unknown sampling variances. Therefore, it is not directly obvious how a test for significant differences should be performed.

The problem can be tackled from different angles, and the questions that may be addressed include the following:

- What is the proper test against simpler or more complex models?
- Are changes in suitability (or derivatives thereof) greater than would be expected from sampling error?
- Do suitabilities vary in a systematic way?
- Are the consequences of changes in suitability for management important enough to warrant alternative models to be investigated?

What Is the Proper Test?

Before deciding on proper tests, possible alternative hypotheses and their implications should be considered in relationship to the MSVPA objective of quantifying predation mortalities and fishing mortalities simultaneously. Identifying model complexity involves a certain amount of subjectivity. A simple assumption may imply complex interactions, whereas a more complicated assumption may lead to a more tractable, and therefore simpler, model. However, taking the view that model complexity is directly related to the number of variables, I discuss a number of possible models of increasing complexity.

Model A.—The traditional single-species assumption has been that natural mortality M of species age i is constant:

$$Z_i = F_i + M_i \qquad (1)$$

where Z is total mortality and F is fishing mortality. The available data do not allow a test of this hypothesis be-cause M cannot be estimated on an annual basis. The multispecies concept is based on a split of natural mortality in two parts:

$$Z_i = F_i + P_i + M1_i \qquad (2)$$

where P represents the predation mortality caused by fish predators and M1 represents other natural mortality. Since predation mortalities by fish predators vary with stock size (e.g., Daan 1975), assuming that M is constant implies that changes in P are automatically compensated by equal and opposite changes in M1. This result deviates from the generally accepted assumption that sources of mortality are additive (Beverton and Holt 1957).

Model B.—A more specific assumption would be that the partial predation mortality p caused by each predator of age j on each prey of age i is constant, so that

$$P_i = \int_j p_{ij} \, N_j \qquad (3)$$

where N_j represents the number of predators age j. The assumption of constant partial predation mortality can be viewed as one possible formulation of constant "suitability" whereby a predator would behave entirely as a unit of effort (e) in a fishery:

$$F_i = \int_j q_{ij} \, e \qquad (4)$$

where the catchability q would match suitability in equation (3). However, the implications are that per capita consumption by a predator would increase directly in proportion to increases in abundance of individual prey species. Thus, although the implications are quite tractable, the model does not take into account that a fish cannot eat more than a certain amount, unless additional parameters are included that take account of nonlinearity (MacCall 1976).

Model C.—Therefore, a biologically more realistic model would result from formulating the concept of ration R to estimate partial predation mortality p:

$$p_{ij} = \frac{f_{ij} \, R_i}{n_i \, w_i} \qquad (5)$$

where f$_{ij}$ = fraction of the food of predator j that consists of prey i,
   n = number of prey, and
   w = average weight.

Because food composition and rations refer to weights, division by biomass (n w) is required to obtain the partial predation mortality. Thus, a straightforward assumption would be that food composition by predator age is constant and independent of relative prey biomasses, and that rations are constant. The implication is that predation rates are negatively correlated with prey abundance. This would lead to an extremely unstable system because increased fishing mortalities on prey stocks would be paralleled with increased predation mortalities.

Stomach samples provide information on the amount of food in the stomach, representing some measure of food intake. However, digestion rates depend on ambient temperature, prey type, and meal size and, therefore, the assumption of constant ration cannot be readily tested. In contrast, food composition in a particular year can be directly quantified on the basis of stomach samples and used as input to the MSVPA model to predict the food composition in other years. Therefore, the assumption of constant food composition can be tested against the MSVPA model by comparing the differences between observed and predicted food compositions in two sampling years. Although food compositions in different years are significantly correlated, previous work has shown that correlation coefficients between the MSVPA predictions and individual data sets were better than between the observed values directly (ICES 1989).

Model D.—The MSVPA assumes that suitability s of a particular prey i as food for a particular predator j is constant, but that the food composition f in any one year depends on the relative abundance n of the different prey stocks k:

$$f_{ij} = \frac{s_{ij} n_i}{\sum\limits_{k} s_{kj} n_k} \qquad (6)$$

These assumptions of constant suitability and constant ration allow predation mortality to vary depending on total available food for each predator and numbers of predators. Thus, MSVPA appears to be one of the least complex models to incorporate feeding interactions. Therefore, proper tests might only be made against more complex models, which allow suitability to vary systematically between years in connection with annual measures of predator–prey overlap or with prey abundance by allowing for prey switching.

Model E.—The present MSVPA model uses singularly estimated suitabilities for each predator–prey age combination. Since sampling variance translates directly into individual estimates, there are advantages in smoothing suitabilities (or predation mortalities) in any of the above models. Smoothing is defined here as fitting parameter estimates as a continuous function of predator or prey age (or both) based on a least-square approximation. Such models should also be considered in the present context because removing part of the sampling variance possibly gives a better indication of whether suitability is constant.

The possibilities for examining questions related to species interaction are largely constrained to interpreting the data through the medium of MSVPA since this is the only appropriate technology available. Except for model D underlying MSVPA and model C connected directly to input data, no alternative simpler or more complex formulations are presently available in a modeled form that would allow a proper comparison. However, smoothing of estimated suitabilities can be performed outside the model and the statistics might be revealing.

Are Changes in Suitability Greater than Expected from Sampling Error?

A rigid statistical test was not possible because variances of the different sets of input data to MSVPA are not precisely known. Nevertheless, there are qualitative ways to compare results obtained from different data sets.

The model integrates the observed food compositions in 1981 and 1991 into singular estimates of suitability. Thus, if suitabilities were estimated with high precision and remained constant, the model should reproduce exactly the observed food composition in both years. The analysis clearly indicated marked differences for most prey species. Therefore, these two conditions are apparently not fulfilled. Nevertheless, if marked changes in prey biomasses had taken place (e.g., the 60% reduction in abundance of cod and the fourfold increase in abundance of herring [*Clupea harengus*] in 1991 compared with 1981), the observed changes in food composition were well mimicked by the model predictions, as is exemplified by the data for cod as a predator (Figure 1). The model overestimates the contribution of sandeel (*Ammodytes* spp.) in 1991 and underestimates whiting, sprat (*Sprattus sprattus*), and Norway pout (*Trisopterus esmarkii*). However, the changes in cod and herring are reflected in both the observations and the model predictions. Overall, the correspondence appears to be promising.

A plot of the observed minus predicted proportions for all prey age–predator age combinations (Figure 2A) shows that the differences between the two ages are centered about 0 and range from -0.3 to 0.3, with a few outliers beyond this range. There was no obvious trend between years, indicating that the model fits the stomach content data from all years equally well.

A plot of the observed minus predicted proportions against the number of stomachs involved in each estimate indicates that outliers are associated with relatively small sample sizes (Figure 2B). Differences declined rapidly with increasing sample size until this figure reached about 400, suggesting little gain in precision beyond this value. The observed trend suggests that variance is largely due to sampling error.

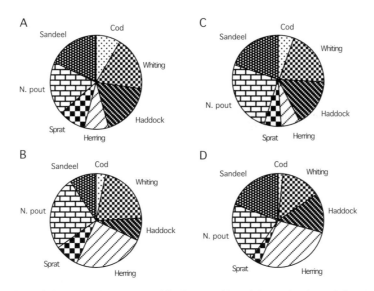

FIGURE 1.—Comparison of the observed and expected food composition of the total cod population: (A) stomach content observations in 1981, (B) stomach content observations in 1991, (C) model predictions for 1981 from the key run, and (D) model predictions for 1991 from the key run. The key run is based on all available stomach content data for the years 1981, 1985, 1986, 1987, and 1991. The haddock is *Melanogrammus aeglefinus.*

Scatterplots of suitabilities estimated from two runs of the model using the 1981 and 1991 data indicated a large variation in suitability estimates based on different data sets (e.g., Figure 3A). However, there was no consistent pattern of change between the runs. Moreover, a plot of the corresponding partial predation mortalities estimated from the two runs shows that relatively few of all possible predator–prey combinations contribute significantly to the estimated predation mortalities (Figure 3B). Although suitabilities may differ considerably, the associated predation mortalities are often negligible. This suggests that variation is particularly related to prey that are seldom eaten, again indicating that sampling variance plays an important role.

Apart from these qualitative comparisons, the hypothesis that changes in suitability arise only from chance was also more formally tested by fitting smoothing functions to the suitabilities estimated by the two runs based on the 1981 and 1991 stomach content data. These smoothing functions represented least-square approximations of a multiplicative model of the 'lognormal size preference function' (Ursin 1973), 'predator species' and 'prey species x predator species x quarter-year' scaling effects, coupled with the Poisson log-link function. Additional terms were included for predator species size and for possible skewness in the size preference function. More than 50% of the variation in suitability estimates could be explained by a single model fitted to the estimates of both years. Fitting separate year effects to the scaling improved the fit by another 5–10%. The de-

grees of freedom available to test the significance of different effects were sufficiently large that even minor effects were statistically significant. Only the skewness effects and year x size suitability factors failed to attain the 5% level of significance.

Do Suitabilities Vary in a Systematic Way?

The estimated suitabilities were subjected to regression analysis against predator and prey biomass, the rationale being that there should be no systematic patterns in the observed changes if all changes were due to sampling error. If prey switching were important, then a systematic relationship should exist between changes in suitability and changes in either predator or prey biomass, or both, depending on the type of switching. The results indicated that very little variance could be explained by fitting overall slopes of change in suitability to change in predator or prey biomass. Models improve by fitting separate slopes for each species and by including quarter-year. However, explanatory power was still only around 10%. A few species slopes were significantly different from zero, particularly for sprat. Suitability of sprat was lower in 1991 than in 1981, coinciding with a reduction in sprat biomass. Although this is consistent with less use of sprat by predators when abundance was lower, it is not conclusive evidence for prey switching because the distribution of sprat may have been more restricted. There was no evidence for strong switching

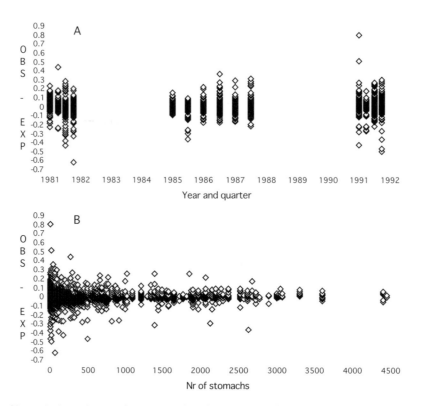

FIGURE 2.—Observed (obs) and expected (exp) proportion of prey in the diet for all quarterly predator age–prey age combinations (A) by year in which the observations were made, and (B) by the number (Nr) of stomachs on which the estimates are based.

towards herring relative to other prey, despite substantial increase in herring biomass from 1981 to 1991.

Do the Variations in Suitability Matter?

The long-term steady state of the model in the forecast mode is a function of the input recruitments, fishing mortalities, and suitabilities. The sensitivity to estimated suitabilities can be evaluated by comparing the results generated from the 2 years of stomach data. Two preliminary tests have been carried out by applying (1) a general 10% reduction in the fishing mortality for all species and (2) a set of altered fishing mortalities corresponding to the estimated effect of an increase in mesh size to 130 mm in the human consumption fisheries for groundfish. The baseline run for each year of stomach data was the steady state associated with unchanged fishing mortalities. Differences in predicted catches by species between the runs (Figure 4A), which were due to the combined effects of different suitabilities, different fishing mortalities, and different average recruitment generated by each data set, were relatively small.

A general reduction in fishing mortality by 10% re-sulted in smaller catches relative to the baseline for all species except saithe (*Pollachius virens*) and cod (Figure 4B). The saithe catches were predicted to increase slightly, and the results for cod varied depending on the data set used. Pronounced discrepancies between the runs (based on the two years of stomach content data) were only observed for haddock (*Melanogrammus aeglefinus*) and, to a lesser extent, for Norway pout. However, only in haddock could the differences be a matter of concern.

The introduction of a 130-mm mesh size can be considered as a much stronger perturbation of the system than the 10% reduction in effort, and the effects varied substantially for some species (Figure 4C). Again, the most pronounced differences were observed for haddock. This sensitivity of haddock is probably related to variations, due to relatively low sample sizes, in its contribution to the stomach content data for saithe, its most important predator.

Despite minor quantitative differences, the long-term steady state and changes therein, in relationship to changing exploitation rates, are apparently not very sensitive to the choice of either year of stomach data, and the overall trends were remarkably consistent in a qualitative sense. This suggests that these variations would hardly

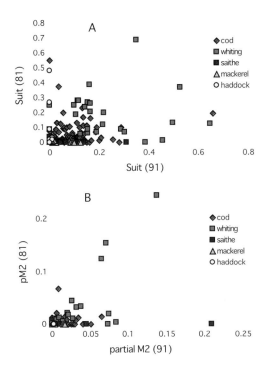

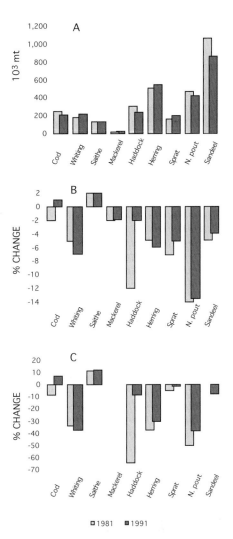

FIGURE 3. A comparison of (A) the suitabilities (suit) of whiting as prey for individual predator age-groups and (B) partial predation mortalities (pM2) of whiting caused by individual predator age-groups as estimated from the 1981 (81) and the 1991 (91) data sets. The predator species are indicated. Mackerel is *Scomber scombrus* and the saithe is *Pollachius virens*.

affect overall management advice, even though estimated suitabilities may vary.

Conclusions

If MSVPA is to be tested with any statistical rigor, it is imperative that the statistical distributions of both stomach content data and survey data be known. Without such knowledge, one cannot accurately establish expected values to test model predictions. A particular problem is the smearing of stomach content data across ages, resulting in interdependence of the estimated suitabilities and partial predation mortalities.

Nevertheless, the conclusion from the analyses is that the two independent stomach content data sets are consistent. Many of the larger changes in estimated suitabilities could be explained by either undersampled predator categories or predators having a limited impact on the prey in question. Although estimated suitabilities do vary, sampling error does evidently play an important role. Application of a smoothing function to estimated suitabilities confirmed that (1) a large part of the variation between the 2 years could be explained by fitting a common model

FIGURE 4. Predicted steady-state catch by species for three management scenarios based on the 1981 and the 1991 data sets: (A) absolute catches when exploitation is continued at recent levels (baseline); (B) predicted percentage change in catch compared with the baseline when all fishing mortalities are reduced by 10%; (C) predicted percentage change in catch compared with the baseline when a 130-mm mesh size is introduced in all human consumption demersal fisheries.

to the 2 years of data and (2) the year effect explained only a relatively small though significant proportion. Evidence for prey switching is absent for most species and inconclusive for the others.

The long-term steady-state results appear not to be critically dependent on the choice of input stomach data from either year. This feature of robustness of the forecasting properties of the model is reassuring with regard to the consistency of advice based on the results of multispecies assessment.

Multispecies Assessment and Fisheries Management

To control total fishing mortality, fisheries management in European waters is largely based on a total allowable catch (TAC) regime, which has been developed under the Common Fisheries Policy (Holden 1994) of the European Union since 1983. Technical measures are also taken to solve particular problems in the exploitation of the different stocks by specific fisheries (mesh size and bycatch regulations, closed areas and seasons, etc.). Last, a long-term program is being developed to limit the size of national fleets because gross overcapacity in many fleets causes great problems in enforcing quota regulations.

The ICES Advisory Committee for Fisheries Management annually provides TAC advice in the form of catch options. These options are still based on single-species assessment because short-term catch predictions are independent of whether a single-species or a multispecies model is applied (Pope 1991). It is only in mid- and particularly long-term projections that differences become significant. However, as yet, no decisions have been taken on longer-term management objectives, resulting in little use of the multispecies assessment model. Applications have virtually been restricted to mesh assessments, which indicates that long-term gains predicted by single-species models are often largely counteracted by increased predation rates. However, there has been a large indirect effect of MSVPA because the estimated average predation mortalities have been incorporated in single-species assessment.

It must be stressed that MSVPA takes only predation on post-recruits into account. Although interactions in the early life stages do exist, these are complex and not easily quantified. Moreover, it is unlikely that pre-recruit life stages can be effectively linked to MSVPA because predation processes operate at a different time scale. Still, the extent to which recruitment of individual stocks is affected by management measures aimed at regulating spawning stock sizes and predator stocks is an unknown factor in long-term predictions.

Recently, more emphasis has been put on conserving the marine environment and particularly on reducing adverse effects of fishing on the ecosystem. This aspect is probably going to play an increasing role in fisheries management and may become even more important than economic considerations. Particular problems (e.g., bycatch of marine mammals) may be addressed directly, but evaluation of overall ecosystem effects from intensive exploitation will demand even more complex models than the present MSVPA. An extended MSVPA model might provide a firm basis for studying the integrated effects of fishing and inter- and intraspecific predation

within the entire fish community—provided that data requirements are fulfilled. Since stomach sampling was extended in 1991 to include other predators of lesser economic importance, and the stomachs were analyzed in great taxonomic detail, there is scope for extending the number of other predators in the model, including sea birds and marine mammals. Discard sampling has also intensified in recent years to enable rough estimates to be made of true catches of noncommercial species. In combination with biomass estimates from surveys, the number of prey species in the model could also be increased.

If the age-structured MSVPA could be replaced by a model that is structured according to both size and age, this would obviously be a great refinement because predation is essentially a size-related process. Another improvement envisaged is the development of a spatially structured MSVPA to account for the dynamics of spatial heterogeneity of fish fauna and fisheries.

A new application of MSVPA is as a tool to study possible effects of fishing on community parameters such as biodiversity. A sensitivity analysis of different parameters within the model may help in selecting suitable parameters and interpreting these when applied to field monitoring studies.

The MSVPA is less a goal in itself than a technology allowing tests of hypotheses and evaluation of consistency in available information. In this sense, one may not expect an entirely new perspective for fisheries management. However, MSVPA serves as a powerful tool in solving important scientific problems.

Acknowledgments

I wish to thank the organizers of the symposium for inviting me to present this overview and for providing financial support. I am also indebted to my colleague-members of the ICES Multispecies Assessment Working Group for allowing me to use the results of the last meeting for this contribution, but even more for the many thought-provoking discussions over many years of cooperation. Referees and editors made many valuable suggestions for improvement of the manuscript.

References

Andersen, K. P., and E. Ursin. 1977. A multispecies extension to the Beverton and Holt theory of fishing, with accounts of phosphorus circulation and primary production. Meddelelser Danmarks Fiskeri- og Havundersøgelser 7:319-435.

Anonymous. 1989. Report of the Multispecies Assessment Working Group, Copenhagen, 7–16 June 1989. ICES (International Council for the Exploration of the Sea), Council Meeting Document 1989/Assess:20, Copenhagen.

Anonymous. 1992. Progress report on the ICES 1991 North Sea stomach sampling project. ICES (International Council for the Exploration of the Sea), Council Meeting, Document 1992/G:12, Copenhagen.

Anonymous. 1994. Report of the Multispecies Assessment Working Group, Copenhagen, 23 November–2 December 1993. ICES (International Council for the Exploration of the Sea), Council Meeting, Document 1989/Assess:9, Copenhagen.

Beverton, R. J. H., and S. J. Holt. 1957. On the dynamics of exploited fish populations. Fishery investigations series II, marine fisheries, Great Britain Ministry of Agriculture, Fisheries and Food 19, London.

Daan, N. 1975. Consumption and production in North Sea cod, *Gadus morhua*: an assessment of the ecological status of the stock. Netherlands Journal of Sea Research 9:24-55.

Helgason, T., and H. Gislason. 1979. VPA analysis with species interactions due to predation. ICES (International Council for the Exploration of the Sea), Council Meeting, Document 1979/G:52, Copenhagen.

Holden, M. 1994. The Common Fisheries Policy. Origin, evaluation and future. A Buckland Foundation Book, Fishing News Books, Oxford.

MacCall, A. D. 1976. Density dependence of catchability coefficient in the California sardine purse seine fishery. CALCOFI (California Cooperative Oceanic Fisheries Investigations) Report 18:136-148.

Pope, J. G. 1979. A modified cohort analysis in which constant natural mortality is replaced by estimates of predation levels. ICES (International Council for the Exploration of the Sea), Council Meeting, Document 1979/H:16, Copenhagen.

Pope, J. G. 1991. The ICES Multispecies Assessment Working Group: evolution, insights, and future problems. Pages 22-33 *in* N. Daan, and M. P. Sissenwine, editors. Multispecies models relevant to management of living resources. ICES Marine Science Symposia 193. International Council for the Exploration of the Sea, Copenhagen.

Rice, J. C., N. Daan, J. G. Pope, and H. Gislason. 1991. The stability of estimates of suitabilities in MSVPA over four years of data from predator stomachs. Pages 34-45 *in* N. Daan, and M. P. Sissenwine, editors. Multispecies models relevant to management of living resources. ICES Marine Science Symposia 193. International Council for the Exploration of the Sea, Copenhagen.

Sissenwine, M. P., and N. Daan. 1991. An overview of multispecies models relevant to management of living resources. Pages 6-11 *in* N. Daan, and M. P. Sissenwine, editors. Multispecies models relevant to management of living resources. ICES Marine Science Symposia 193. International Council for the Exploration of the Sea, Copenhagen.

Sparholt, H. 1991. Multispecies assessment of Baltic fish stocks. Pages 64-79 *in* N. Daan, and M. P. Sissenwine, editors. Multispecies models relevant to management of living resources. ICES Marine Science Symposia 193. International Council for the Exploration of the Sea, Copenhagen.

Sparre, P. 1991. Introduction to multispecies virtual population analysis. Pages 12-21 *in* N. Daan, and M. P. Sissenwine, editors. Multispecies models relevant to management of living resources. ICES Marine Science Symposia 193. International Council for the Exploration of the Sea, Copenhagen.

Ursin, E. 1973. On the prey size preferences of cod and dab. Meddelelser Danmarks Fiskeri -og Havundersøgelser 7:85-98.

The Effects of Future Consumption by Seals on Catches of Cape Hake off South Africa

ANDRE E. PUNT

Abstract.—The Cape hakes *Merluccius capensis* and *M. paradoxus* are estimated to constitute some 10–20% of the diet of the Cape fur seal *Arctocephalus pusillus pusillus*. This seal population was subject to intense harvest mortality prior to the 20th century, but has been increasing since then. Concern has been expressed in some quarters about the impact that hake consumption by seals may have on future hake catch rates and catch levels in the southern African region. The qualitative effects of the consumption of Cape hake by seals is examined using a minimal realistic model, which incorporates hake, seals, "other predatory fish," and the fishery. Over quite widely ranging sets of parameter values and assumptions within models that assume the presence of a single hake species only, there are consistent indications that, as the size of a possible seal cull is increased from zero, the average catch level and catch rate, and hence profitability of the hake fishery, increase slightly while the average annual consumption of hake by seals decreases. However, extensions to more realistic models that include both hake species indicate qualitatively different conclusions, suggesting that a seal cull could lead even to slight negative effects for the fishery.

The Cape hakes, which comprise the two morphologically similar species *Merluccius capensis* and *M. paradoxus*, are caught in shelf and slope waters from close inshore to more than 800 m, from northern Namibia to south of Durban on the south coast of South Africa (Payne 1989; Payne and Punt 1994). The fishery off South Africa commenced around the turn of the century and annual catches have been at least 50,000 metric tons (mt) since 1948 (Payne and Punt 1994). The Cape hakes have constituted 70–80% of the catch by the South African demersal fleet historically although recently this percentage has dropped as improved catch rates, and hence greater ease in reaching the annual total allowable catch (TAC), have allowed some diversification of the available fishing effort towards other species. Further details about the biology of, and fishery for, the Cape hakes can be found in Botha (1980) and Payne (1989).

The harvesting of Cape fur seals *Arctocephalus pusillus pusillus* commenced in the 17th century. By the time legal controls were implemented in 1893, seals had been eliminated from over 20 island colonies (Shaughnessy 1984) and the total population may have been as low as 100,000 animals. Since then, harvests have generally been smaller than sustainable yields and the seal population has consequently increased. The current annual pup production is of the order of 300,000 (Butterworth et al. 1995). There are now 25 breeding colonies on the islands and the mainland, and the seals' distribution extends from Cape Frio to Port Elizabeth (Figure 1).

Butterworth and Harwood (1991) calculated the annual consumption by various predator species of six commercially important marine species (anchovy [*Engraulis capensis*], the Cape hakes [*Merluccius capensis* and *M. paradoxus*], Cape horse mackerel [*Trachurus trachurus capensis*], round herring [*Etrumeus whiteheadi*], pilchard [*Sardinops sagax*], and squid species) for three subdivisions of the southern Africa coast (Namibia, the South African west coast, and the South African south coast). These estimates indicate that seals are probably not the most important natural predator for any commercially exploited fish stock, and that even where seals appear to be an important predator, there are other predator species whose consumption is of a similar magnitude. Nevertheless, the consumption of Cape hake off the South African west coast by seals was estimated to be double the size of the commercial take (the data upon which the analyses of Butterworth and Harwood [1991]) are based have since been updated and it now appears that the current consumption of Cape hake by seals is roughly the same size as the current commercial take). It is, therefore, possible that the increasing seal population may have an impact on future yields of hake. Arguments in favor of possible culls of marine mammal populations because of possible impacts on fishery yields have also been made in Canada, Iceland, and Norway (Butterworth and Harwood 1991; Anonymous 1992).

This study evaluates the impact of the future consumption of Cape hake by seals for the South African west coast only. This is because the catch of Cape hake off the west coast is twice that off the south coast, and because almost three times as many seals breed off the west coast. The framework employed to compare the implications of alternative seal culls on the future trends in the hake fishery was developed during the Benguela Ecology Programme Workshop on Seal–Fishery Biological Interactions (Butterworth and Harwood 1991). Punt and Butterworth (1995) should be consulted for detailed de-

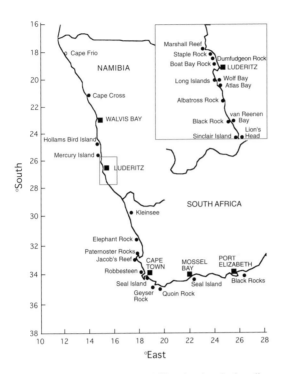

FIGURE 1.—Map of southern Africa showing the breeding colonies of the Cape fur seal.

scriptions of the models considered and for the methods employed to estimate their parameters. Butterworth and Harwood (1991) draw attention to the limitations of the modeling approach being undertaken, which participants in the Benguela Ecology Programme Workshop on Seal–Fishery Biological Interactions considered to be only qualitatively rather than quantitatively reliable.

Methods

The Evaluation Framework

The consequences of different levels of future consumption of hake by seals are investigated in the context of the change in the level of hake TACs and catch rates. The existing hake management procedure is used to calculate future TACs, which are assumed to be taken exactly. This procedure is the combination of the Butterworth–Andrew observation error, Schaefer form production model estimator, and the $f_{0.2}$ harvesting strategy (see Punt [1994] for details, and Payne and Punt [1995] for the rationale for the use of this procedure for this fishery). The actual process of evaluating the consequences of alternative levels of seal cull involves carrying out the following three steps.

1. A number of operating models of the biological system and the fishery are constructed. An operating

model is a mathematical statistical model of the fishery and the component species of the system. Each alternative model reflects, inter alia, an alternative (yet plausible) representation of the system. In this case, each operating model reflects a different level of future cull of seals, a different level of predation or cannibalism for hake, or different values for some of the parameters of the population dynamics model that are poorly known. The operating model is used to generate artificial data sets (such as annual catch rates) required to implement the management procedure and to determine the effects of a series of management decisions on the system over time.

2. A number of simulations are carried out. Each simulation involves projecting the system represented by one of the artificial data sets generated by the operating model forward for 20 years. The level of man-induced removals each year is determined by applying management procedures for each component species. The different data sets reflect the uncertainties (modeled by statistical distributions) in model parameters and future observations for each alternative overall scenario (i.e., combination of an operating model and a management procedure) considered. Most of the results of this paper pertain to the case in which recruitment is taken to be deterministic and future data are not subject to observation error, so only a single simulation is needed to evaluate the consequences of alternative cull options.

3. The results of the simulations are summarized by means of a small number of performance indices. These indices are chosen so that it is straightforward for decision makers to assess whether different levels of consumption of hake by seals are likely to have a substantial impact on the future prospects for the hake fishery.

The key feature of this minimal realistic model approach is that the operating models selected consider only those species likely to have important interactions with the species of interest (Cape hakes). Thus, the only species considered are the Cape hakes, the Cape fur seal, and other predatory fish. It is this restriction to key species only that makes it feasible to construct a model for which sufficient information is available to admit reasonable specification of all the parameter values.

The Operating Models

Each operating model incorporates four components: the seal population, the hake population, other predatory fish, and the fishery (Table 1). These four components were selected by Butterworth and Harwood (1991) because they account for most of the mortality on hake.

TABLE 1.—Qualitative overview of the operating models.

Specification	Cape fur seal	Cape hake	Other predatory fish
Preys on	Cape hake Other predatory fish	Cape hake Other predatory fish	Cape hake Other predatory fish
Removals through management	Depends on level of cull	Determined by applying the hake management procedure	Pre-specified time trajectory of fishing mortalities
Annual per capita food consumption (independent of year and abundance of prey)	Punt et al. (1995)	Punt and Leslie (1995)	Not required (Punt and Butterworth 1995)
Fraction of diet consisting of hake (changes according to Holling Type II relationship)	15% in 1991	*M. capensis* 29% and *M. paradoxus* 17% in 1991	10% in the absence of exploitation

The entire seal population from northern Namibia to the south coast of South Africa is modeled using a deterministic age- and sex-structured population dynamics model. Natural mortality of pups is assumed to be both density-dependent and sex-specific. The parameters of the seal model are determined from the results of biological studies and by fitting to aerial counts of pup abundance (Butterworth et al. 1995). The total consumption of food off the west coast of South Africa by seals is taken to be 27% of that off the whole of southern Africa. The range considered for the fraction of the seal diet in 1991 consisting of Cape hake is 10% to 20%, with 15% being taken as a base-case value (Punt et al. 1995).

The culling options consider cow and harem bull culls, and range from no further removal of seals to culls double that necessary to keep the total food consumption by seals the same in 2012 as it was in 1993. A cow cull is emphasized because Anonymous (1990) noted that it is the most effective means of reducing the total seal population. Bull culls are considered because these animals make a disproportionate contribution to the total consumption owing to their large mass (Butterworth et al. 1995).

The hake population is modeled either as a single species or as two species, and allowance is made for predation of hake by seals, hake, and other predatory fish. Large *M. capensis* and juvenile *M. paradoxus* are located in roughly the same depth range, and the former feed extensively on the latter (Botha 1980; Payne et al. 1987; Punt et al. 1995). The relationship between hake density and the fraction of the diet of the various predators (seals, hake, and other predatory fish) consisting of hake is taken to be of the Holling Type II form (Holling 1965), again following the recommendation of Butterworth and Harwood (1991). The values of the parameters of the hake model are obtained from the literature (e.g., Punt and Leslie 1991) and by fitting the operating model to the historical data for the Cape hakes (Figure 2). The values of the parameters related to the diet of Cape hake and seals were obtained from analyses of stomach con-

tent data collected during research cruises (Punt et al. 1995; Punt and Leslie 1995).

The "other predatory fish" component is described using a model that divides this population into adult (mature and recruited) and juvenile (immature and unrecruited) animals. The juvenile component is subject to predation by hake, seals, and other predatory fish, while the adults are subject to a fishery. The dynamics of predatory fish are currently only poorly understood, so that values for the parameters of the predatory fish model are set using the results of stock assessments for similar species.

Measuring Performance

Four measures are used to quantify the effects of different levels of future consumption of hake by seals: the annual catch of hake, the size of the exploitable component of the hake biomass, the interannual variability of catches, and the net present value. The latter is taken to be the discounted revenue, assuming a constant price-to-cost ratio. It is not intended to be a definitive economic analysis. It is merely a convenient manner for summarizing both catch and catch rate projection information—both of which are important to the industry—in a single number.

Results

A time series of the total number of seals off the South African west coast (27% of the total off southern Africa) under three culling options was estimated (Figure 3). The culling options are as follows: allowing no further removals of seals (no further removals), an annual cow and harem bull cull that forces the total seal consumption in 2012 to be the same as that in 1993 (status quo), and a cull that is 50% larger than that required to force the total seal consumption in 2012 to be the same as that in 1993 (1.5 X status quo). Figure 4 shows time trajecto-

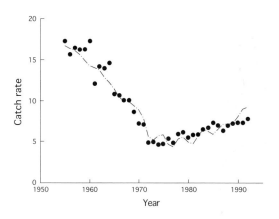

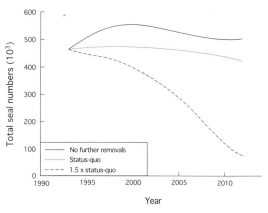

FIGURE 2.—Actual (solid dots) and two-hake-species operating model predicted (dashed line) catch rate time-series for Cape hake off the west coast of South Africa.

FIGURE 3.—Time-series of the total number of seals off the west coast of South Africa under three alternative culling options.

ries of annual consumption of hake by seals (in 1,000s of metric tons [mt]), the hake catch (in 1,000s of mt), the catch rate for hake, and the annual contributions to the net present value (PV, i.e., discounted revenue) for the three culling options for a one-hake-species operating model. These trajectories were then estimated for a two-hake-species operating model (Figure 5). The calculations upon which Figures 4 and 5 are based assume deterministic dynamics and no future observation error.

Discussion

One-Hake-Species Operating Model

If the seal population is not culled over the next 20 years, the hake TAC under an $f_{0.2}$ strategy is expected to drop to roughly half its current size (Figure 4b), and the annual contributions to PV may show a marked reduction over this period (Figure 4d). Much of the latter reduction is due to future catches being discounted relative to current catches. However, it is also due to reduced catches and lower catch rates. If no seal cull takes place, the number of seals off the South African west coast is predicted to continue increasing for a further 7 years, peaking at almost 550,000 animals, before declining slightly (Figure 3). Somewhat counterintuitively, the annual consumption of hake by seals decreases from its current 84,000 mt to 48,000 mt by 2012 under the "no further removals" option (Figure 4a). The reason for this is the decreasing trend in hake abundance (as indicated by the trends in hake catch rate, Figure 4c), which leads to hake constituting a smaller fraction of the diet of the seal population according to the Holling Type II feeding relationship assumed.

A cull that forces the total seal consumption in 2012 to be the same as that in 1993 leads to larger annual catches, higher catch rates, and hence greater profits for the demersal fishing industry (Figures 4b–d) than if the "no further removals" option is exercised. Increasing the level of a future seal cull even further leads to substantial reductions in seal numbers (Figure 3). The reduction in the consumption of hake by seals is less than that in seal numbers. This is because the hake population does not decline to the same extent as before: the Holling Type II feeding relationship assumed leads to an increase in the fraction of hake in the seal diet.

Two-Hake-Species Operating Model

The results for the two-hake-species operating model (Figure 5) are qualitatively different from those for the one-hake-species case. In contrast to the one-hake-species case, the effect of a reduction in the number of seals on average catch levels and PV is negligible while a seal cull leads to a lower hake biomass (and hence catch rate) after 10 years of management. The reason for this behavior is that, even without any seal culls, the biomass of hake is expected to increase markedly over the next 20 years (Figure 5c). By reducing the consumption of hake by seals (through a cull), the biomass of the shallow-water species, *M. capensis*, increases (as seals are assumed to feed on this species only). However, the increase in the biomass of *M. capensis* leads to greater predation mortality on *M. paradoxus*. It is the reduction in the biomass of *M. paradoxus* that offsets any benefits from the increase in the biomass of *M. capensis* caused by the culling of seals.

Sensitivity Analyses

A large number of tests of the sensitivity of these results to the values of the operating model parameters have been conducted (Punt and Butterworth 1995). Qualitatively,

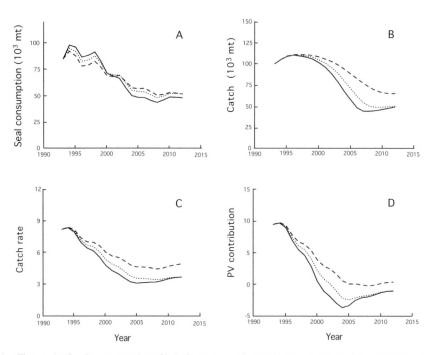

FIGURE 4.—Time-series for the consumption of hake by seals, catch, annual present value (PV) contribution, and hake catch rate for a one-hake-species operating model. Results are shown for three alternative culling options. The results for the "no further removals," "status-quo," and "1.5 X status-quo" strategies are shown by the solid, dotted, and dashed lines, respectively.

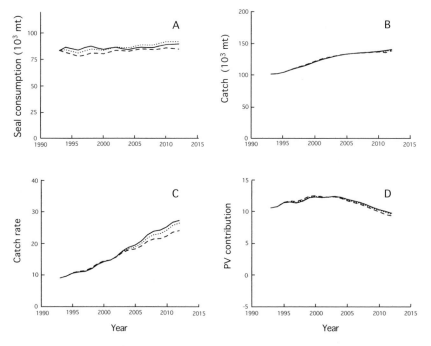

FIGURE 5.—As for Figure 4, except that results are shown for a two-hake-species operating model.

the results of the sensitivity tests are the same as those reported here. (That is, for the one-hake-species operating model, as the size of a future seal cull is increased, the final hake population size, the average annual catch, and the net present value increase, while the average annual consumption of hake by seals decreases. For the two-hake-species operating model, the impact of a seal cull on the prospects for the demersal fishing industry are either negligible or slightly negative.) The quantitative results are sensitive to the values of some of the parameters. For example, for the one-hake-species operating model, increasing the fraction of hake in the seal diet from the base-case value of 15% to 20% leads to culls being more beneficial to the fishing industry, while reducing it to 10% leads to the opposite effect.

The qualitative results of analyses in which parameter uncertainty and future observation error are included, and in which recruitment and natural mortality are taken to be stochastic, are also the same as those in Figures 4 and 5. However, adding noise to the system results in the performance measures having quite wide distributions (Punt and Butterworth 1995).

Final Remarks

The methodology used in this study differs from that used by the Canadian Royal Commission on Seals and Sealing (Malouf 1986) because it takes second-order effects, such as density-dependent natural mortality, into account. Although second-order effects are seldom taken into account when estimating interactions between fisheries (Gulland 1987), their incorporation can make qualitative changes to the decision of whether a seal cull will have a beneficial impact on the yield from a fishery. An example of this is given by Butterworth et al. (1988).

A result similar to that obtained here was obtained by the International Council for the Exploration of the Sea Multispecies Working Group when it examined the impact of a possible increase to the mesh size used by the human-consumption fisheries in the North Sea (Stokes 1992). The group found that, for a single-species model, there were considerable benefits to be had, in terms of catches, by increasing the mesh size. But when biological interactions were accounted for, the catches either declined or stayed the same.

The main reason for the difference between the results for the one- and two-hake-species operating models is the effect of introducing an extra predator–prey interaction. This would suggest that great care needs to be taken when designing minimal realistic models to include all important predator–prey interactions. The model considered here takes into account over 90% of the mortality on hake identified by Wickens et al. (1992).

Another difference between the framework applied here and previous examinations of the impact of seal culls on fishery yields is that the fishery yield is measured by the TACs set using the existing hake management procedure. As noted by Butterworth and Harwood (1991), the TACs set using this procedure will not necessarily be equal to the optimal take from the fishery. This is because the management procedure is based on a simple population dynamics model, the parameters of which have to be estimated from the data collected from the fishery. Thus, the impacts on the fishery from culls are those that will actually be realized rather than merely potential gains. In some situations, the potential gain in yield may be high but the quality of the assessment data too poor to detect this. This is evident to some extent in Figure 5, in which the hake biomass increases markedly over the 20-year management period, but the management procedure for hake is unable to make full use of this increase.

The impact of a seal cull that keeps the total seal consumption at its 1993 size is beneficial to the fishery for the less realistic one-hake-species operating model (Figure 4). However, the size of this effect is not particularly marked because the results of recent (1989 and 1993) aerial surveys suggest that the seal population is approaching its environmental carrying capacity. Furthermore, the projections suggest that the seal population will reach this level during the next 20 years and then oscillate about it (see Figure 3). The conclusion that the seal population is approaching its environmental carrying capacity depends largely on the information from the most recent (1993) survey and may change given even a single additional data point. Continuation of the current aerial survey program would therefore seem essential to confirm this conclusion.

The framework used in this paper requires a considerable quantity of data. For example, in order to parameterize an operating model, stomach content data for hake and seals; relative abundance information for hake and seals; catch information for hake, seals, and other predatory fish; and estimates of biological parameters for all these species are required. Such data are not available for many of the systems for which the impact of future possible seal culls may be required. The data problem is less serious for species, such as hake, that are close to the top of the food chain, but could be considerable for species, such as anchovy, that form the prey species of several predators. For example, in the South African context, anchovy form a large component of the diet not only of hake and seals, but also of many other predators that are less well studied. Applying this framework to anchovy at the present time would, almost certainly, lead to equivocal results because of the large uncertainties associated with predation by species about whose population dynamics there is very little information available at present.

Acknowledgments

Funding for this work was provided by the Sea Fisheries Research Institute, Cape Town, South Africa. A. Badenhorst, R. Leslie, S. Pillar, and A. Payne (Sea Fisheries Research Institute), D. Butterworth and P. Wickens (University of Cape Town) are thanked for their valuable insights into seal population dynamics and feeding behavior, and for their comments on the modeling techniques applied. R. Hilborn, W. Clark, and an anonymous reviewer are thanked for their comments on an earlier version of this paper.

References

Anonymous. 1990. Report of the Subcommittee of the Sea Fisheries Advisory Committee appointed by the Minister of Environment and of Water Affairs, to advise the minister on scientific aspects of sealing. Ministry of National Education and Environment Affairs, South Africa.

Anonymous. 1992. Marine mammal/fisheries interactions: analysis of cull proposals. Report of the Meeting of the Scientific Advisory Committee of the Marine Mammal Action Plan, Liege 27 November–1 December 1992. United Nations Environmental Programme, Greenwich.

Botha, L. 1980. The biology of the Cape hakes *Merluccius capensis* Cast. and *M. paradoxus* Franca in the Cape of Good Hope area. Doctoral dissertation. University of Stellenbosch, Stellenbosch, South Africa.

Butterworth, D. S., D. C. Duffy, P. B. Best, and M. O. Bergh. 1988. On the scientific basis for reducing the South African seal population. South African Journal of Science 84:179-188.

Butterworth, D. S., and J. Harwood, rapporteurs. 1991. Report of the Benguela Ecology Programme workshop on seal–fishery biological interactions. Report of the Benguela Ecology Programme. South Africa.

Butterworth, D. S., A. E. Punt, W. H. Oosthuizen, and P. A. Wickens. 1995. The effects of future consumption by the Cape fur seal on catches and catch rates of the Cape hakes 3. Modeling the dynamics of the Cape fur seal *Arctocephalus pusillus pusillus*. South African Journal of Marine Science. 16:161-183.

Gulland, J. A. 1987. The impact of seals on fisheries. Marine Policy 11(3):196-204.

Holling, C. S. 1965. The functional response of predators to prey density and its role in mimicry and population regulation. Memories Entomological Society of Canada 45.

Malouf, A., editor. 1986. Seals and sealing in Canada. Report of the Royal Commission, 3 vols. Ministry of Supply and Services, Canada, Ottawa.

Payne, A. I. L. 1989. Cape hakes. Pages 136-147 *in* A. I. L. Payne and R. J. M. Crawford, editors. Oceans of life off Southern Africa. Vlaeberg, Cape Town.

Payne, A. I. L., and A. E. Punt. 1995. Biology and fisheries of South African Cape hakes (*Merluccius capensis* and *M. paradoxus*). Pages 15-47 *in* J. Alheit and T. J. Pitcher, editors. Hake: fisheries, ecology and markets. Chapman and Hall, London.

Payne, A. I. L, B. Rose, and R. W. Leslie. 1987. Feeding of hake and a first attempt at determining their trophic role in the South African west coast marine environment. South African Journal of Marine Science 5:471-501.

Punt, A. E. 1994. Assessments of the stocks of Cape hake *Merluccius* spp. off South Africa. South African Journal of Marine Science 14:159-186.

Punt, A. E., and D. S. Butterworth. 1995. The effects of future consumption by the Cape fur seal on catches and catch rates of the Cape hakes 4. Modeling the biological interaction between Cape fur seals *Arctocephalus pusillus pusillus* and the Cape hake *Merluccius capensis* and *M. paradoxus*. South African Journal of Marine Science. 16:255-285.

Punt, A. E., J. H. M. David, and R. W. Leslie. 1995. The effects of future consumption by the Cape fur seal on catches and catch rates of the Cape hakes 2. Feeding and diet of the Cape fur seal *Arctocephalus pusillus pusillus*. South African Journal of Marine Science 16:85-99

Punt, A. E., and R. W. Leslie. 1991. Estimates of some biological parameters for the Cape hake off the South African west coast. South African Journal of Marine Science 10:271-284.

Punt, A. E. and R. W. Leslie. 1995. The effects of future consumption by the Cape fur seal on catches and catch rates of the Cape hakes 1. Feeding and diet of the Cape hakes *Merluccius capensis* and *M. paradoxus*. South African Journal of Marine Science 16:37-55.

Shaughnessy, P. D. 1984. Historical population levels of seals and seabirds off southern Africa, with special reference to Seal Island, False Bay. Investigational report. Sea Fisheries Research Institute, South Africa 127:1-61.

Stokes, T. K. 1992. An overview of the North Sea multispecies modeling work in ICES. South African Journal of Marine Science 12:1051-1060.

Wickens, P. A., D. W. Japp, P. A. Shelton, F. Kriel, P. C. Goosen, B. Rose, C. J. Augustyn, C. A. R. Bross, A. J. Penney, and R. G. Krohn. 1992. Seals and fisheries in South Africa—competition and conflict. South African Journal of Marine Science 12:773-789.

Bycatch Management in Alaska Groundfish Fisheries

STEVEN PENNOYER

Abstract.—The history of the Alaska groundfish fisheries is reviewed with emphasis on the rapid evolution of the domestic fishery and its management based on experience gained through observation and management of the foreign groundfish fisheries since the early 1960s. The stable status of the Alaska groundfish resource is attributed largely to historical and ongoing efforts to collect information on resource status and inseason groundfish harvests, and closure of fisheries when annual quotas are reached. Nonselective harvesting techniques used in the groundfish fisheries result in incidental catches (bycatch) of nontarget species, size categories, or sex. An open-access management of the Alaska groundfish fisheries contributes to bycatch amounts that are greater than what is minimally needed to conduct the groundfish fisheries. Similarly, efforts to control bycatch are hampered by the intense competition for Alaska groundfish resources that result from overcapitalization of the domestic groundfish fleet and increasingly short fishing seasons. The effectiveness of numerous measures implemented to address the bycatch problem is reviewed, and a discussion of future approaches being considered for effective bycatch management is presented.

The Alaska Region of the National Marine Fisheries Service (NMFS) encompasses all federal waters off the state of Alaska. Within this area, NMFS is responsible for managing marine fisheries and a large assemblage of marine mammals. The NMFS also has responsibilities for the habitat upon which these resources depend. The management area encompasses approximately 70% of the United States continental shelf and nearly half of its coastline. Currently, the harvest off Alaska constitutes about 50% of the total U.S. harvest caught in federal waters.

The NMFS manages fisheries under the authority of the Magnuson Fisheries Conservation Management Act of 1976, which came into effect at the time the U.S. extended its jurisdiction out to 200 miles (320 km). Figure 1 shows the exclusive economic zone (EEZ) off Alaska and the three main areas—the Gulf of Alaska (GOA), Aleutian Islands, and the Bering Sea—that NMFS management generally is divided into.

The north Pacific fishery resources are managed by NMFS under a complex system of regulations and relationships with the North Pacific Fisheries Management Council (NPFMC), the state of Alaska, the International Pacific Halibut Commission (IPHC), and various other national and international agreements. This paper does not describe this interactive system in any detail except where it specifically relates to NMFS's ability to manage bycatch in the Alaska groundfish fisheries. The primary purpose of this paper is to present a case history of the bycatch problem in the Alaska groundfish fisheries and to summarize the character of this problem, goals in trying to manage it, and some of the opportunities and difficulties experienced by management agencies in trying to do so.

Overall, the Alaska groundfish fishery is not exceptional with respect to bycatch and discard rates compared with other major fisheries in the world (Alverson et al.

1994). For example, the Bering Sea midwater fishery for pollock (*Theragra chalcogramma*) annually harvests almost 1 million metric tons (mt) of fish, yet has one of the lowest observed bycatch rates—discard of 0.062 kg/ kg landed—in the world. Through 1994, the Bering Sea trawl fishery for rock sole (*Pleuronectes bilineatus*) typically experienced the highest discard rate relative to other Alaska groundfish fisheries. Alverson et al. (1994) estimate a discard rate of 2.61 kg/kg landed for this fishery. However, even this rate is not exceptional high relative to other trawl fisheries in the world. For example, some shrimp trawl fisheries experience bycatch discard rates as high as 15 kg/kg landed. Nevertheless, in the past 5 years, hundreds of millions of dollars have been lost or expended by the Alaska groundfish industry owing to bycatch closures, forgone harvest opportunity, discarded resource that might otherwise have been retained and processed, and administrative costs incurred by management agencies and the industry. While the bycatch rates in most segments of the fishery may not be exceptional, the 2.3-million-mt fishery is so immense that the absolute volume of discards and the forgone opportunity they represent has raised national and industry consciousness and poses a significant concern to other fisheries dependent on some of the bycaught species.

The Groundfish Fisheries

Most of the Alaska groundfish harvest is taken with trawl gear, a nonselective harvesting technique that results in some catch of nontargeted species (Figure 2). Alaska groundfish also are harvested by vessels using more selective gear, such as hook-and-line, pot, and jig.

The annual harvest of Alaska groundfish approaches 2.3 million mt. The associated total annual discard amounts are approximately 450 mt of Pacific salmon

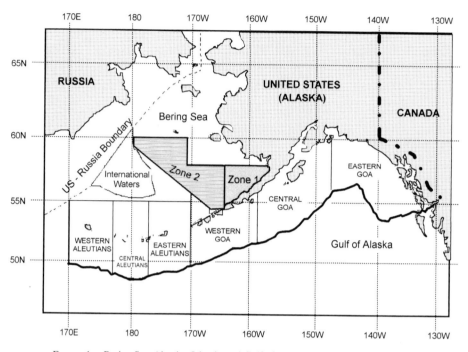

FIGURE 1.—Bering Sea, Aleutian Islands, and Gulf of Alaska (GOA) management areas.

(*Oncorhynchus* spp.), approximately 7,000 mt of halibut (*Hippoglossus stenolepis*), and over 340,000 mt of groundfish. Crab bycatch is monitored in terms of number of animals and totals about 16 to 17 million crab annually. Most of the crab bycatch is composed of Tanner crab (*Chionoecetes* sp.) that weigh less than 0.11 kg and have a carapace width of 80 mm or less (Narita et al. 1994). The bycatch of king crab species is dominated by red king crab (*Paralithodes camtschaticus*) that approach

or exceed size at maturity for females and near legal size for males, roughly 130 mm carapace length (Armstrong et al. 1993; Narita et al. 1994).

Although the magnitude of bycatch amounts in the Alaska groundfish fisheries is high, the ratio of bycatch to retained harvest is relatively low. These low bycatch rates create a problem that defies easy resolution because incremental improvements are increasingly difficult and expensive. Furthermore, the different components of the

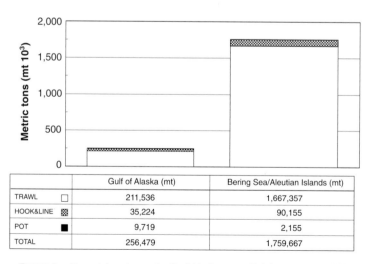

	Gulf of Alaska (mt)	Bering Sea/Aleutian Islands (mt)
TRAWL ☐	211,536	1,667,357
HOOK&LINE ▨	35,224	90,155
POT ■	9,719	2,155
TOTAL	256,479	1,759,667

FIGURE 2.—Harvest (metric tons [mt]) of Alaska groundfish by gear type, 1993.

groundfish fishery and each bycatch species have their own set of problems, concerns, and potential solutions.

The Alaska groundfish resources were harvested primarily by foreign nations until the mid-1980s (Figure 3). The foreign catches declined in the late 1980s and were replaced briefly by joint-venture harvests by domestic fishers delivering to foreign processors. Fully domestic operations escalated in the late 1980s and became the dominant form of operation in 1989 and, by 1990–91, were the only form of operation. This is important because the domestic harvest and NMFS's ability to monitor and manage domestic fisheries are recent developments. So while the harvest has been at current levels for a long time period, the methodologies and procedures needed to regulate the fisheries have changed dramatically. During the years of foreign exploitation, NMFS managed aggregates of fleets fishing for one country or another that could be regulated as units. In contrast, NMFS currently manages a domestic fleet where regulation of an individual vessel's activity must be carried out in a way that will withstand judicial review in U.S. courts.

Overcapitalization

The large increases in processing and harvesting capacity in the domestic sector since 1989 have had a dramatic impact on the amount of bycatch and discard that is occurring. Harvesting and processing capacity has increased to a level that probably exceeds by three to four times the capacity required to harvest the resource on a 12-month basis. Seasons for all species have shrunk dramatically. For example, in 1989 the domestic pollock season lasted for 9 months; in 1994, it lasted only around 3 months in total. Large vessels landing groundfish in the domestic fishery increased from about 45 in 1987 to roughly 120 in 1994. Effort in the pollock fishery increased from 41 catcher processors and 4 motherships in 1989 to 70 catcher processors and 13 motherships in 1994 with fleetwide catch rates of over 10,000 mt/d.

Other examples of this overcapitalization and excess effort abound. Since the mid-1980s, the Alaska halibut fishery has harvested over 20,400 mt of fish with 4,000 vessels during a 1- or 2-d fishing season per year in our major fishing areas. In 1994, over 1,000 vessels participated in the Gulf of Alaska sablefish (*Anoplopoma fimbria*) hook-and-line fishery and harvested approximately 19,000 mt of sablefish in less than 10 d. These[1] and other open-access fisheries have become an extremely competitive race for the fish with every vessel having to do its utmost to harvest its share of the quota before someone else either catches the target species or shuts the fishery down under prohibited species bycatch restrictions.

Resource Status

An important point regarding fishery history is the status of the resource. Groundfish stocks off Alaska are generally in a healthy and stable condition. All Alaska groundfish stocks have fluctuated in abundance over the years, but no widespread trend toward decline is evident (NPFMC 1994a, 1994b). Generally, the fisheries annually have removed some 1.8 million to 2.3 million mt of groundfish from these areas for over 25 years.

Changes obviously have occurred in the ecosystem of the Bering Sea and Gulf of Alaska. Some marine mammal, seabird, and shellfish populations have declined and are continuing to do so (NPFMC 1995a). Some groundfish populations have increased or decreased dramatically over the years (NPFMC 1994a, b). The whole question of

---

[1]In 1995, NMFS implemented an individual fishing quota (IFQ) program that was developed by the North Pacific Fishery Management Council for the sablefish and Pacific halibut hook-and-line gear fisheries off Alaska (Title 50, part 676, Code of Federal Regulations [CFR]).

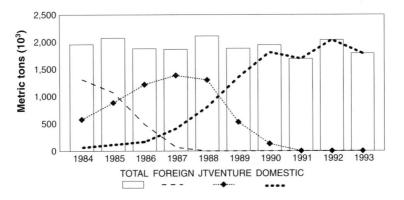

FIGURE 3.—Harvest of Alaska groundfish by foreign, joint-venture, and domestic operations, 1984–93. The 1993 total is based on retained blend catch data. Domestic catch data are only for amounts landed.

managing or even assessing the ecosystem changes off Alaska is one that has and is receiving considerable attention. Management agencies have yet to attribute the bycatch or even the directed harvest to be the cause for any of these fluctuations.

Management System

Another important aspect of the management of the Alaska groundfish resource and the reason for its health is the type of management and monitoring system implemented. An annual process exists for establishing the status of each groundfish species and management of the groundfish fisheries. This process is required by federal regulations under 50 CFR part 602, Guidelines for Fishery Management Plans (602 guidelines). The 602 Guidelines were published by NMFS and require the annual preparation of a stock assessment and fishery evaluation (SAFE) report. The SAFE reports are intended to summarize the best available scientific information concerning the past, present, and possible future condition of the stocks and fisheries under federal management. Typically, the Alaska groundfish SAFE reports are prepared by scientists from the NMFS Alaska Fisheries Science Center, the Alaska State Department of Fish and Game, and other resource management agencies. The stock assessment section of the SAFE report recommends acceptable biological catch (ABC) levels for each stock and stock complex under federal management. The ABC is defined in the 602 guidelines as "a preliminary description of the acceptable harvest (or range of harvests) for a given stock or stock complex." The derivation of an ABC focuses on the status and dynamics of the stock, environmental conditions, and other ecological factors and prevailing technological characteristics of the fishery.

Management of the Alaska groundfish fisheries is directed to maintain total harvest amounts within these ABCs. Annual total allowable catch (TAC) amounts are specified within each species' ABC, and discard amounts of groundfish are charged against the annual TACs. Management policy attempts to account for all harvest. In 1994, the total harvest of Alaska groundfish species (2.05 million mt) accounted for 77% of the total ABC (2.66 million mt).

NMFS conducts extensive stock assessment surveys that provide the basis for annual groundfish ABCs. Ongoing data collection is provided through an industry-funded observer program that, by regulation, implements mandatory observer coverage in this fleet. NMFS monitors catch through mandatory weekly or daily observer reports and extensive industry record-keeping and reporting requirements. The use of groundfish species and Pacific halibut as bycatch from other fisheries is considered when setting and monitoring annual quotas, and

fisheries are closed when annual catch or bycatch quotas are reached.

Bycatch

Types of Bycatch

Three types of bycatch discard occur in the Alaska groundfish fisheries, classified according to the origin of discard. The first type, prohibited species discard, applies to the bycatch of salmon, Pacific halibut, king crab, Tanner crab, and Pacific herring (*Clupea harengus pallasi*). These species have been allocated for the directed harvest in other domestic fisheries, and bycatch of prohibited species in the groundfish fisheries is required to be returned to the sea as soon as possible with a minimum of injury.

With the exception of Pacific halibut and crab, discarded prohibited species are assumed to experience 100% mortality. In general, these mortalities have not created specific conservation concerns for these species because bycatch is taken into account in management of the other fisheries directing harvest on the bycaught species. Nevertheless, non-retainable Pacific halibut bycatch mortality experienced in the Alaska groundfish fishery during recent years represents over one-fourth of the total U.S. and Canadian set-line quota for halibut (IPHC 1992, 1993). The concern for this type of mortality is heightened when stocks are declining even though such declines may not be related directly to fishing mortality. Red king crab stocks in the Bristol Bay area of the Bering Sea have declined to the point that the commercial crab fisheries in the area have been required to be closed since 1994. Although trawl closures have been implemented to reduce the number of female red king crab taken as bycatch in the Bering Sea trawl fisheries, the continued potential for red king crab bycatch outside the trawl closure areas continues to be a controversial issue among management agencies and the crab and groundfish industries.

Regulatory discards are a second type of discard created by the management system. When a directed fishery for a species is closed, regulations specify allowable bycatch amounts of that species, which can be retained onboard vessels when fishing for other species. Vessels must discard excess catch. Retainable bycatch amounts are hopefully set somewhere near the natural level of bycatch that will occur in other fisheries. It is up to the management agency to establish bycatch allowances that prevent unnecessary discards, hold enough fish in reserve to support bycatch needs in other groundfish fisheries, and avoid exceeding TAC amounts. Unfortunately, difficulties exist in meeting these management goals, and discard occurs when catch amounts of species exceed the retainable bycatch amounts specified in regulations. When a species'

TAC has been reached, including both the directed and bycatch portions, the species will be designated prohibited and any bycatch amounts in other fisheries must be discarded for the balance of the fishing season. This does not eliminate the bycatch problem although it may reduce it by taking away the economic incentive that may exist to target on the species.

The third type of discard, economic discards, results from the bycatch of undersized target species, male fish in roe fisheries, and undesirable, low-value groundfish species in the catch.

Discard rates vary considerably in the Bering Sea and Gulf of Alaska groundfish fisheries (Figures 4, 5). What stands out is that some of the major fisheries, such as the pollock midwater trawl fishery, have a very low (5%) discard rate (Figure 6), whereas some other significant fisheries, like the Bering Sea rock sole fishery, have an overall retention of only 31% (Figure 7). Even in hook-and-line fisheries, such as the sablefish fishery in the Gulf of Alaska, the groundfish discard rate is 19% (Figure 8). Nearly all of the discarded amounts in these fisheries reflect a type of economic discard.

Goals of Bycatch Management

Bycatch problems occur when discard mortality (1) results in conservation concerns, (2) is thought to significantly impact resources available to another fishery, or (3) results in unnecessarily high levels of protein and economic waste. The goal of bycatch management for

the Alaskan groundfish fisheries has been to identify and work toward restriction of bycatch discard amounts to levels that would allow the fishers to reasonably harvest available groundfish resources while minimizing bycatch mortality and discard. Obviously, the words "unnecessary," "minimize," and "reasonably harvest" have defied clear definition. After all the data are presented and all the costs of achieving the results are examined, it finally comes down to the political (policy) process to determine the appropriate levels.

Bycatch Management Approach

The NPFMC (the Council) plays a key role in the ongoing development of a bycatch management program for the Alaska groundfish fisheries. The Council provides for analyses of proposed management measures, public review and testimony on these measures, and policy decisions that form recommendations to NMFS. A brief summary follows of the NMFS-approved approach to the bycatch problem in the Alaska groundfish fisheries and the relative success of regulations implemented to address this problem.

Research and monitoring programs implemented to determine the magnitude and character of the bycatch problem are important elements in managing bycatch. The fundamental component of these programs is an industry-funded mandatory observer program. In order to fish, vessels greater than 125 ft (~38 m) in length must have an observer onboard at all times. Vessels of

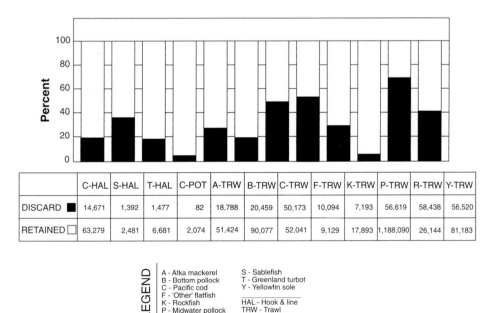

	C-HAL	S-HAL	T-HAL	C-POT	A-TRW	B-TRW	C-TRW	F-TRW	K-TRW	P-TRW	R-TRW	Y-TRW
DISCARD ■	14,671	1,392	1,477	82	18,788	20,459	50,173	10,094	7,193	56,619	58,438	56,520
RETAINED ☐	63,279	2,481	6,681	2,074	51,424	90,077	52,041	9,129	17,893	1,188,090	26,144	81,183

LEGEND
A - Atka mackerel
B - Bottom pollock
C - Pacific cod
F - 'Other' flatfish
K - Rockfish
P - Midwater pollock
R - Rock sole

S - Sablefish
T - Greenland turbot
Y - Yellowfin sole

HAL - Hook & line
TRW - Trawl

FIGURE 4.—Bering Sea and Aleutian Islands area groundfish discard amounts (mt) by gear and target fishery, 1993.

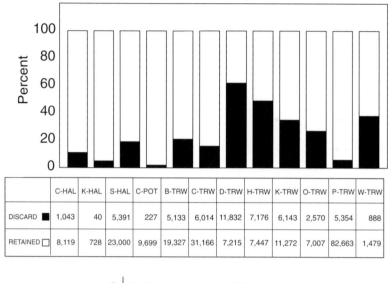

	C-HAL	K-HAL	S-HAL	C-POT	B-TRW	C-TRW	D-TRW	H-TRW	K-TRW	O-TRW	P-TRW	W-TRW
DISCARD ■	1,043	40	5,391	227	5,133	6,014	11,832	7,176	6,143	2,570	5,354	888
RETAINED □	8,119	728	23,000	9,699	19,327	31,166	7,215	7,447	11,272	7,007	82,663	1,479

LEGEND

B - Bottom pollock
C - Pacific cod
D - Deepwater flatfish
H - Shallow water flatfish
K - Rockfish
O - 'Other'

P - Midwater pollock
S - Sablefish
W - Rex sole

HAL - Hook & line
TRW - Trawl

FIGURE 5.—Gulf of Alaska groundfish discard amounts (metric tons) by gear and target fishery, 1993.

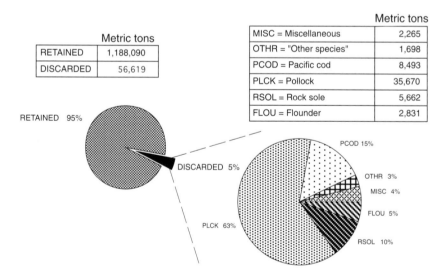

	Metric tons
RETAINED	1,188,090
DISCARDED	56,619

	Metric tons
MISC = Miscellaneous	2,265
OTHR = "Other species"	1,698
PCOD = Pacific cod	8,493
PLCK = Pollock	35,670
RSOL = Rock sole	5,662
FLOU = Flounder	2,831

RETAINED 95%

DISCARDED 5%

PCOD 15%
OTHR 3%
MISC 4%
FLOU 5%
RSOL 10%
PLCK 63%

FIGURE 6.—Groundfish discard amounts in the 1993 Bering Sea and Aleutian Islands area midwater pollock fishery.

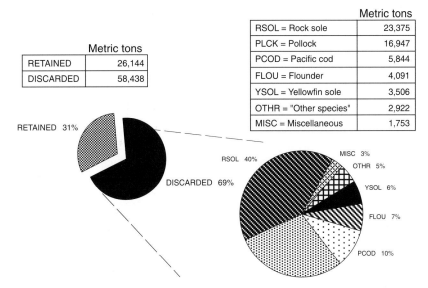

Metric tons	
RETAINED	26,144
DISCARDED	58,438

	Metric tons
RSOL = Rock sole	23,375
PLCK = Pollock	16,947
PCOD = Pacific cod	5,844
FLOU = Flounder	4,091
YSOL = Yellowfin sole	3,506
OTHR = "Other species"	2,922
MISC = Miscellaneous	1,753

FIGURE 7.—Groundfish discard amounts in the 1993 Bering Sea and Aleutian Islands area rock sole trawl fishery.

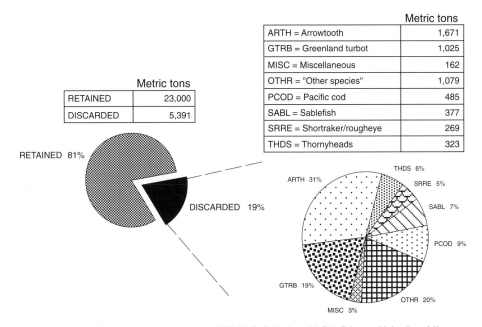

Metric tons	
RETAINED	23,000
DISCARDED	5,391

	Metric tons
ARTH = Arrowtooth	1,671
GTRB = Greenland turbot	1,025
MISC = Miscellaneous	162
OTHR = "Other species"	1,079
PCOD = Pacific cod	485
SABL = Sablefish	377
SRRE = Shortraker/rougheye	269
THDS = Thornyheads	323

FIGURE 8.—Groundfish discard amounts in the 1993 Gulf of Alaska sablefish fishery with hook-and-line gear.

60–124 ft (~18–38 m) length overall must have an observer on board 30% of the days that fishing gear is retrieved and groundfish are retained. Shoreside processors receiving under 1,000 mt of groundfish during a month must have an observer present 30% of the days groundfish are received or processed; those processors that receive greater amounts of groundfish must have an observer present each day of operation. The observer program obviously is costly to maintain (>US$10 million annually). An extensive industry record-keeping and reporting program complements the observer program and requires mandatory logbooks and processor reports. NMFS currently is developing a program that would require real-time information exchange through satellite transmission of observer reports and industry catch statistics.

Considerable research has been conducted on how bycatch operates within the various fisheries and gear types, the mortality associated with discards of prohibited species, the relationship of bycatch in terms of size and abundance to the stock status of the bycatch species, and the effect of bycatch on other fisheries (Armstrong et al. 1993; IPHC 1994; NPFMC 1995b; Queirolo et al. 1995). A detailed description of these studies is beyond the scope of this paper, but suffice it to say a very large part of the NMFS resource management program has been dedicated to the bycatch problem. Additionally, the IPHC, the NPFMC, the Alaska Department of Fish and Game, and universities and private industry groups have expended considerable effort on this problem.

Regulatory Approaches

Currently, the principal regulatory approach implemented to address the bycatch problem consists of fishery closures when specified prohibited species bycatch limits are reached. Halibut, herring, and crab bycatch limits established for the 1994 groundfish fisheries (Table 1) are gear- and area-specific, and many are divided further among fisheries as bycatch allowances that may be seasonally apportioned. When a fishery bycatch allowance is reached, the fishery is closed.

While bycatch limits have not solved the bycatch problem and may have contributed to the race for fish, they have kept the prohibited species bycatch amounts from going higher. Specific concerns regarding bycatch limits include monitoring of bycatch, mortality rate assumptions, and extrapolation of observed bycatch rates against estimates of total groundfish catch weight. Prohibited species bycatch restrictions and groundfish closures also have the potential of leaving significant amounts of groundfish unharvested. Since 1990, costs annually incurred by the groundfish industry due to bycatch closures

have ranged from about US$80 million in 1990 to about US$30 million in 1992 (this NMFS estimate is based on unpublished data on forgone catch during 1991–93 and average 1990–91 first wholesale values of Alaska groundfish.)

Bycatch regulations have also required the following gear restrictions: biodegradable panels in groundfish pots, halibut exclusion devices on groundfish pots, gear configurations and performance standards for pelagic trawl gear to encourage off-bottom harvest of pollock, and groundfish allocations among different types of gear. By and large, these have not been dramatically effective with perhaps the exception of the pelagic trawl gear restrictions and performance standard established for the pollock fishery. When bottom trawling for pollock is closed because a halibut or crab bycatch allowance has been reached, fishing vessels are required to fish midwater with low bycatch rates to meet the performance standard of less than 20 crab onboard a vessel at any time. These regulatory measures still do not solve the problem of how to fish on or near the bottom to take larger pollock and cod while avoiding bycatch of other species.

Season delays or time–area closures attempting to limit bycatch of certain species in specific groundfish fisheries have met with variable success. In some cases, management of time–area closures have limited, but in no case really resolved, the bycatch problem.

A vessel incentive program was implemented in 1991

TABLE 1.—Bycatch limits in 1994 established by NMFS for halibut, herring, and crab.

Groundfish fishery	Bycatch limit
Halibut mortality (mt)	
BSAI[a] trawl fisheries	3,775
BSAI non-trawl fisheries	900
GOA[b] trawl fisheries	2,000
GOA non-trawl fisheries[c]	750
Bering Sea zone 1[d] red king crab (no.)	
BSAI trawl fisheries	200,000
Tanner crab (*Chionoecetes bairdi*) (no.)	
BSAI trawl fisheries in zone 1	1,000,000
BSAI trawl fisheries in zone 2[d]	3,000,000
Pacific herring (mt)	
BSAI trawl fisheries	1,962

[a]Bering Sea and Aleutian Islands area.
[b]Gulf of Alaska.
[c]In 1995, the halibut bycatch mortality limit specified for the GOA hook-and-line gear fisheries was reduced to 300 metric tons because of the implementation of an individual quota program for sablefish and halibut and the concurrent exemption of the hook-and-line sablefish fishery from halibut bycatch restrictions.
[d]Zone 1 and Zone 2 areas of the Bering Sea generally refer to the Bristol Bay and continental slope areas, respectively.

for the Alaska groundfish trawl fisheries; this program specifies allowable bycatch rates for halibut and red king crab and holds individual vessels accountable for their bycatch rates. This approach has developed from the theory that a small number of vessels account for a disproportionately large share of the bycatch. The NMFS observer data show that, in any of the groundfish fisheries, usually only a few vessels experience high bycatch rates that far exceed the rates experienced by most vessels in the fishery. If the existing incentive program worked, rates could theoretically be ratcheted down until it met some desired level of bycatch vs. cost of achievement. Unfortunately, we are a long way from making this type of program work. Prosecuting violators of the incentive program is time-consuming and costly. The U.S. court system requires standards of proof that often exceed agency capability in prosecuting an alleged violation of the incentive program. To date, four cases of violation of the vessel incentive program have been pursued by NMFS: three were brought before an administrative law judge and ruled in favor of NMFS; the fourth was settled out of court.

The most successful bycatch reduction program implemented in recent years is the individual vessel fishing quota (IFQ) program for hook-and-line sablefish and halibut fisheries. In this program, sablefish and halibut may be fished together under individual vessel quota allowances established for nearly 4,000 boats. No longer are halibut caught by sablefish fishers with halibut IFQ discarded. Instead, bycaught halibut must be retained as commercial catch if they are of legal size in the commercial fishery. The same is true for sablefish. Additionally, fishers who participate in other hook-and-line fisheries, such as for rockfish (*Sebastes* spp.) or Pacific cod (*Gadus morhua macrocephalus*), will be required to retain bycaught halibut if they have a quota share of halibut. Fishing seasons will be extended, individuals will be able to either fish in ways to avoid bycatch or take the time to retain and handle it, and waste attributed to lost or excessive gear will be minimized. This program will have gone into effect in 1995 with the hope that it will result in reducing bycatch rates in the Gulf of Alaska and the Bering Sea. At time of writing, preliminary estimates by NMFS indicated that the Pacific halibut discard mortality in the Alaska sablefish IFQ fishery totaled around 136 mt in 1995 compared with about 650 mt in 1994.

Voluntary Industry Initiatives

Numerous voluntary industry initiatives have been undertaken to reduce bycatch. In 1988, U.S. participants in joint-venture fisheries for flatfish in the Bering Sea initiated a program of self-management to help minimize crab bycatch. Similar programs were initiated in the 1990s for the domestic yellowfin sole (*Pleuronectes asper*) and Pacific cod fisheries to help control crab and halibut bycatch rates. In 1994, the Bering Sea trawl industry formed a nonprofit corporation—the Salmon Research Foundation (SRF)—involving both industry and western Alaska interests. The purpose of the SRF was to implement a voluntary fee assessment program to fund inseason data collection and analyses necessary to provide information to the fleet on how best to avoid salmon bycatch in the Bering Sea trawl fisheries. A summary of the SRF's formation, salmon avoidance program, and anticipated research initiatives is contained in the Foundation's April 23, 1994, report submitted to the North Pacific Fishery Management Council, Anchorage, Alaska.

The various industry initiatives are not detailed in this paper, but they reflect industry's attempt to live within the bycatch restrictions and policies adopted by NPFMC while still being able to harvest the groundfish resource. The success of these initiatives has varied and appears to be proportional to the number of voluntary participants. Generally when fisheries are relatively small, industry incentive programs work. As the size of the fleet expands, and as more people enter the fishery who may not subscribe to such voluntary measures, industry incentive programs tend to fall apart.

The Future of Bycatch Management

One of the biggest problems in approaching bycatch management is the lack of adequate incentive programs for individual operators to reduce bycatch rates. Bycatch limits in an overcapitalized fishery result in a race for the fish, which exacerbates bycatch rates and the frequency of groundfish fishery closures before reaching TAC. Bycatch closures have heightened the awareness of the fleet regarding bycatch, have probably led to voluntary industry measures to reduce bycatch rates, have initiated a sizable body of research into the problem including gear modification, and may have limited the level of bycatch. Nonetheless, bycatch limits have not led to a solution to the bycatch problem. From the perspective of the groundfish fishery, bycatch limits appear too restrictive and costly. From the perspective of other fisheries impacted by such incidental takes, bycatch restrictions may limit the problem but do not return bycaught species to the fisheries that target them. With the potential exception of the sablefish and halibut IFQ program, existing bycatch management measures have yet to provide adequate incentive to individual groundfish fishers to take action to reduce bycatch rates in a manner that still allows for the opportunity to harvest groundfish quotas.

Numerous proposals have been submitted to NPFMC to address the bycatch problem in the groundfish fisheries. These proposals include everything from requiring retention and use of catch to giving priorities to gear types with low bycatch rates, to reducing bycatch caps, and to modifying gear configuration, changing seasons, closing areas, and other parameters. These efforts take the form of regulatory proposals in fishery management plans and even federal legislation. Some proposals recommend fishery closures or discard restrictions for specific fisheries, such as the rock sole fishery in the Bering Sea, that have relatively high discard rates. Most of these proposals would involve regulations applied to the fleet as a whole, and most do not deal with the problem of providing a workable economic incentive for the individual operation to modify its behavior. Most industry and management groups recognize that it is technologically not possible to eliminate all bycatch or have full use of catch without eliminating some very important fisheries. The NPFMC strives to evaluate these proposals based on their ability to reduce bycatch and increase catch use to the levels beyond which further changes would increase costs more than they would increase benefits (where costs and benefits are defined from the national perspective). This approach could best be accomplished through a combination of bycatch reduction measures with a program that allows individual fishers to tailor their operations to achieve maximum individual benefit under the rules.

Conclusions

Excessive bycatch and inadequate use are but two symptoms of a major flaw in the way our fisheries currently are managed. The overcapitalized, open-access nature of the Alaska groundfish fisheries results in extreme competition among vessels to maximize individual harvest amounts of groundfish before fleetwide groundfish quotas or prohibited species bycatch limits are reached and fisheries are closed. This race for fish allocates fish among competing fishers and uses. This allocation mechanism tends both to increase harvesting and processing costs and to decrease the value of what is harvested. The individual fisher has little incentive to slow down, care better for his harvest, retain fish with a lower value than others he is catching, or in any way diminish his competitive performance vis a vis others who are racing for the same quotas. In the final analysis, some form of indi-

vidual vessel quota for target species, combined perhaps with individual prohibited species quotas or rate restrictions, will probably be a preferred solution. Until we reach that stage, we will probably not be able to effectively bring this problem under control.

References

Alverson, D. L., M. H. Freeberg, J. G. Pope, S. A. Murawski, and J. G. Pope. 1994. A global assessment of fisheries bycatch and discards. Food and Agriculture Organization of the United Nations, Fisheries Technical Paper 339, Rome, Rome.

Armstrong, D. A., T. C. Wainwright, G. C. Jensen, P. A. Dinnel, and H. B. Andersen. 1993. Taking refuge from bycatch issues: red king crab (*Paralithodes camtschaticus*) and trawl fisheries in the eastern Bering Sea. Canadian Journal of Fisheries and Aquatic Science 50:1993-2000.

IPHC (International Pacific Halibut Commission). 1992. Report of the halibut bycatch work group. IPHC Technical Report 25.

IPHC (International Pacific Halibut Commission). 1993. Annual report 1992. IPHC, Seattle, Washington.

IPHC (International Pacific Halibut Commission). 1994. Report of commission activities—1994. IPHC, Seattle, Washington.

Narita, R., M. Guttormsen, J. Gharrett, G. Tromble, and J. Berger. 1994. Summary of observer sampling of domestic groundfish fisheries in the northeast Pacific Ocean and eastern Bering Sea, 1991. NOAA Technical Memorandum NMFS-AFSC-48, Seattle, Washington.

NPFMC (North Pacific Fishery Management Council). 1994a. Stock assessment and fishery evaluation report for the groundfish resources of the Gulf of Alaska as projected for 1995. November 1994. Anchorage, Alaska.

NPFMC (North Pacific Fishery Management Council). 1994b. Stock assessment and fishery evaluation report for the groundfish resources of the Bering Sea and Aleutian Islands management area as projected for 1995. November 1994. NPFMC, Anchorage, Alaska.

NPFMC (North Pacific Fishery Management Council). 1995a. Ecosystem considerations–1996, November 1995. Anchorage, Alaska.

NPFMC (North Pacific Fishery Management Council). 1995b. Environmental assessment/regulatory impact review/initial regulatory flexibility analysis for proposed alternatives to limit chinook salmon bycatch in the Bering Sea trawl fisheries: Amendment 21b. Anchorage, Alaska.

Queirolo, L. E., L. W. Fritz, P. A. Livingston, M. R. Loefflad, D. A. Colpo, and Y. L. deReynier. 1995. Bycatch, utilization, and discards in the commercial groundfish fisheries of the Gulf of Alaska, eastern Bering Sea, and Aleutian Islands. NOAA Technical Memorandum NMFS-AFSC-58, Seattle, Washington.

TRENDS IN WORLD AQUACULTURE

World Aquaculture Review:
Performance and Perspectives

W. Herbert L. Allsopp

Abstract.—The status of world aquaculture production of food fish during the last 25 years is analyzed from United Nations Food and Agriculture Organization (FAO) data by environmental sector, region, quantity, value, and species. Significant changes and trends are discussed. Problems receiving special attention include biological, economic, and social factors that influence the choice of fish; market requirements; and operating constraints of the systems. Development prospects cited include culturable indigenous species, extensive aquaculture in tropical water retention dams, the aquarium ornamental fish industry, and new biotechnologies. Breeding techniques, gamete cryopreservation, sex control, growth stimulation, and transgenic fishes are briefly described. Opportunities are indicated for university scientists to contribute to world fish supplies by enhancing use of indigenous tropical food fish through overseas partnerships with aquaculturists of the developing world.

Farming of aquatic organisms (fish, mollusks, crustaceans, and plants) has been attempted for several thousand years. The inclusive term aquaculture was adopted about 40 years ago to imply controlled farming or some form of intervention (stocking, feeding, and protection) to enhance production. Such intervention also involves individual or corporate ownership of the organisms being cultivated. Such identifiable aquaculture production was estimated by the Food and Agriculture Organization of the United Nations (FAO) World Conferences in Rome (1966) and Kyoto (1974). In 1974, the World Bank (WB), recognizing a significant increase in aquaculture harvests, requested its Technical Advisory Committee and Consultative Group for International Agriculture Research to review the world's aquaculture potential. Thus was established the Aquaculture Study Group, on which it was my privilege to participate. Available statistics indicated that 6 million metric tons (mt) of finfish and shellfish were then being produced. This was equivalent to 12% of the annual world fish catch for direct human consumption, and it was almost 4% of the world's animal protein supply, excluding milk. Though this was significant, the food importance for Asian countries (Allsopp 1973), especially land-locked communities, was of far greater magnitude than the world average of 4% animal protein production suggests. Given the food crises facing developing countries, aquaculture for food was therefore strongly advocated for efficient multiple water use and in integrated rural agricultural development.

The Aquaculture Study Group (WB and FAO 1974) estimated a potential tenfold increase in production of fish through aquaculture (from 2 to 20 million mt), whereas the estimated increase for capture fisheries was less than twofold (to 90 million mt). Because harvests in capture fisheries depend largely on uncontrollable natural oceanographic variables that ultimately determine the productivity of major fish species, there is a recognized maximum catch limit for the world's oceans. In the case of aquaculture, production harvests are more directly related to controllable inputs and thus are capable of greater increase. Harvests of certain capture fisheries were already indicated to be near their maximum sustainable yields, while opportunistic aquaculture production systems were then at initial stages of promising development.

The FAO (1984) determined for statistical purposes that where aquatic organisms are owned throughout the rearing period before being harvested, they shall constitute aquaculture production (FAO 1984). Thereafter, FAO recorded and published data from such owned resources as aquaculture statistics. Catches of other aquatic organisms that are common or public property were considered production by capture fisheries.

Aquaculture systems aim to achieve rapid, manipulated production of the target species much beyond the natural standing crop, with optimum economy of water space and inputs. The drive for increased food production caused worldwide promotion of aquaculture activities in each geographic region (FAO 1976); aquaculture production has increased in every geographical region, but especially in Asia (Figure 1).

Assessment of Global Performance

Retrospection

The United Nations overview of fisheries also projected trends to the year 2010. The world fishery situation has been exposed to severe constraints for increasing the aggregate world fish production to satisfy world food needs (FAO 1993b). Further growth of capture fisheries is insufficient to maintain needed food supplies sustainably, and few new resources can be brought into exploitation. Major stocks of pelagic species, though

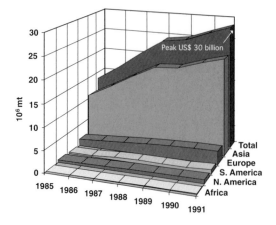

FIGURE 1.—World aquaculture production by major producing continents, 1985–91 (metric tons [mt], billions $US). Source: Food and Agriculture Organization of the United Nations (FAO 1993a).

subject to wide natural fluctuations, appear to have passed their peaks of production. Most of these stocks are used for livestock feeds and not direct human consumption. Major demersal species in all oceans have been overfished, leaving little opportunity for increased total catch in this sector. Inevitably, reduction in fishing effort will be required in this sector so stocks can rebuild to sustainably harvestable levels.

In 1973, FAO estimated that the potential world fish production from all sources ranged from 100 to 150 million mt. Currently, it is recognized that marine capture fisheries are adversely affected if harvests exceed 80 million mt, and inland capture fisheries may be limited to 6 million mt (World Bank et al. 1992). Current analyses disclose serious imbalances in the world's fish harvests. Only 20 countries harvest 81% of the world catch. Twenty-one species groups account for 40% of total catch but only three groups are from freshwater; yet freshwater species make up 40% of the world's fish species. Diligent stock management and environmental conservation measures are needed to maintain water quality in order to sustain current harvests. Consequently, aquaculture has been encouraged to satisfy the increasing worldwide need for fish.

Broad Principles

During its centuries of development, intensive livestock husbandry for food production largely focused on four major herbivorous mammals (cattle, pigs, sheep, and goats), and four omnivorous birds (chickens, turkeys, ducks, and geese). By contrast, aquaculture, starting with variants of common carp (*Cyprinus carpio*) in Asia, has grown to a husbandry of hundreds of fish, mollusk, and crustacean species from widely different climates and

ecosystems and with varied food habits (many predators) and growth rates (Figure 2). The most efficient aquatic organisms for food conversion are mollusks, which are filter-feeders found in plankton-rich marine waters. Most high-value finfish and crustaceans feed high on the food chain. Although poultry are marketable within weeks and some mammals within months, fish take several months or years to rear to marketable size, though tropical aquatic species achieve faster growth.

The 1974 World Bank study (World Bank and FAO 1974) recognized that it was not feasible to concentrate on just a few food species at any single international center, as was being done for tropical livestock or rice. For the intensified efforts, site-specific biotechnology research aspects were necessary in most tropical areas because of the unique environments affecting culture of different food species. In view of significant social factors, particular emphasis was for low-cost food fish.

The freshwater environment is the most readily controlled by human intervention and therefore is broadly considered to offer better opportunity for aquaculture than the marine environment in terms of benefit–cost determinations (Figure 3). Warmer waters are more efficient for productive growth of aquaculture organisms. Therefore, tropical areas have a natural advantage whereas heated systems need to be used in temperate climates. Demographic and social factors have also impelled greater production from inland and coastal brackish-water areas (Figure 4); in marine areas, costs of sea-stable engineered structures are a constraint. Mariculture operations within coves and protected sea areas have been progressively developed where installations are less costly than in open seas.

The 1974 World Bank and FAO (1974) study listed some 300 species that were cultured in significant quantities worldwide. Of these, one-fifth were classified as predatory in food habits; they accounted for 10% of all aquaculture production by weight but approximately 40% of market value. Herbivorous and omnivorous fish, which accounted for 90% by weight of world production in 1974, were cheaper to produce but of lower unit value and therefore offered the better means of massively increasing fish protein production for low-income world populations. Thus, the strategy of "aquaculture for food" advocated by the World Bank urged concentration of production effort on herbivorous and omnivorous species. In Asian countries, such production increase has been progressively gained. In other regions, economic demand and worldwide market forces are also driving production enterprises towards intensive culture of the high-value predatory species.

The FAO study indicated eight broad categories of problems for target food species:

1. Reproduction—mass seed supply, controlled breeding and hybridization

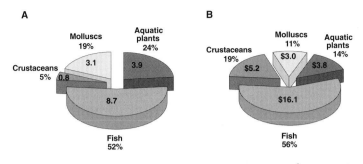

FIGURE 2.—World aquaculture production by species groupings, 1991: (A) tonnage (10^6 mt); (B) value (billions US$). Source: FAO (1993a).

2. Nutrition and feeds—juveniles, larval rearing, growth efficiency
3. Intensifying culture systems—polyculture combinations
4. Aquaculture engineering—maximizing use of water, space, installations, and equipment
5. Aquafarm management—economic and environmental efficiency
6. Fish health maintenance—controlling diseases and parasites to improve yields
7. Environmental impact—avoiding disequilibrium through culture systems
8. Skilled scientific and operational personnel—training researchers and technicians

Very limited scientific data were available to improve growth performance of selected species, particularly indigenous food species. Such culturable species showed great promise because of their food habits, growth rates, and compatible behavior in confinement. There were strong advocates and valid reasons for using indigenous species and against introducing species exotic to a region. Nevertheless, because of rising demand for food fish and immediate investment profitability of proven systems, enterprises have promoted worldwide distribu-

tion of tilapia, carp, salmonids, penaeid shrimp, and oyster species at a relentless pace to satisfy markets.

Aquaculture Production Systems

Productions systems for aquaculture are classified as "extensive," "semi-intensive," and "intensive" (United National Development Programme [UNDP] et al. 1987) though no precise criteria for these terms have been universally accepted. In extensive systems, the cultured fish populations stocked rely mainly on the naturally produced organisms in the aquatic food chain. Natural productivity may be enhanced by adding nutrients in the form of inorganic fertilizers, manures, and organic wastes. With increasing management intervention—including application of nutrients; the manipulation of the fish stocks by numbers, sizes and species' combination; and provision of supplemental feeds for the fish—such systems are described as semi-intensive. With close controls on stocks, high stocking density, and total dependence on artificial feeds for growth of the fish, the culture systems are considered intensive. Most of the traditional Asian

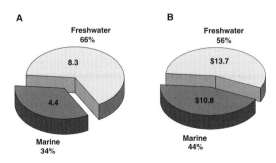

FIGURE 3.—World aquaculture production of inland and marine species, 1991: (A) tonnage (10^6 mt); (B) value (billions US$). Source: FAO (1993a).

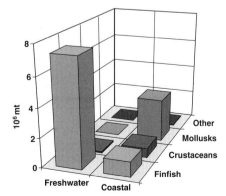

FIGURE 4.—World aquaculture coastal and freshwater production, 1990 (10^6 mt). Source: FAO (1993a).

aquaculture husbandry and rural enterprises have now evolved from extensive to semi-intensive and intensive production systems.

Stock enhancement systems that provide surplus fish for water-retention or irrigation reservoirs have increasingly contributed to extensive aquaculture production in Asia (China, India, Thailand, Malaysia), in the former Soviet Union, in Brazil, and gradually in Africa. By contrast, recent aquaculture in developed countries has concentrated on intensive systems though the historical stock enhancement of lakes and dams for recreational fishing has continued.

Another broad production classification refers to the "monoculture" of individual species in intensive culture systems, which is typically practiced in developed economies for salmonids, catfish, and mollusks, and to the "polyculture" of several compatible species, which has been traditionally practiced in Asia with various Chinese or Indian carp, tilapia, and crustacea. There are also integrated systems that combine aquaculture with plant husbandry (rice and vegetable crops jointly or in rotation) in Asia, or with livestock husbandry (poultry and pigs), which is widely practiced in Asia and Europe. These production systems are capable of high yields per unit enclosed area, and byproducts of the plant or animal husbandry provide nutrients for the cultured fish.

Aquaculture Production Increase

During the past decade, growth of aquaculture production has been rapid, averaging over 10% per year and reaching 12 million mt in 1991. More cultured tonnage has been produced in inland than in coastal waters (Figure 4). Significant increases were achieved with culture of shrimp in tropical areas and salmon in temperate zones. Aquaculture expansion over most of the world is responsible for the increased total food contribution from inland fisheries by developing countries. Intensification of production systems is largely industry-driven and export-oriented for salmon and shrimp. Environmental issues, disease control, and feeds are the main constraints. Rural aquaculture for direct local food consumption has been less successful owing to various managerial, supply logistics, and site-specific problems.

Industrial production systems have focused on intensive culture of relatively few predatory food-fish species (salmon, catfish, etc.) for which rather precise biological requirements are already known. Most food-fish aquaculture in the developing world derives from semi-intensive or extensive systems (which basically rely on the enhanced pond productivity of natural food chain organisms) and is constrained by site-specific ecological conditions. Despite supplemental feeding and enhancement with fertilizers, such extensive systems are still comparable to the early stages of animal husbandry in domesticating wild animals.

Production increases during the 1980s have been broadly due to two main thrusts: China's production and culture of globally high-value species. China's aquaculture in 1990 reached 45% of the world's finfish production (mainly carp), 27% of its shrimp production, and 38% of its mussel production (FAO 1992). Luxury market demands for high-value seafood in developed countries have pushed farmed salmon to 25% of total world salmon production from all sources (capture fisheries plus culture) and farmed shrimp to 24% of all shrimp production (FAO 1993b). Both increases have significantly affected world prices for these species. Mollusk culture systems have increased notably: mussels and clams by 60%, scallops by over 300%.

In developed countries, the industrial food-fish aquaculture systems are constrained by feed prices. Fish feeds contain high proportions of fish meal, which is mainly imported and for which international market prices are difficult to control. For developing countries, aquaculture production emphasis has been on the freshwater herbivorous finfish species (7.4 million mt). Such systems are not much constrained by fish meal supplies and costs.

Regional Assessments

World aquaculture is strongly dominated by Asia, which accounted for about 80% of all such production in 1991 (Figure 1). Europe contributed 10%, North and South America about 2% each, and the balance was spread over Africa, Oceania, and other areas. This regional distribution broadly reflects the world's history of aquaculture: more than 3,000 years in Asia, approximately 200 years in Europe, 100 years in North America, about 70 years in South America, and fewer than 50 years in Africa. Aquaculture remains predominantly a phenomenon of developing countries. Its productivity has continued to grow whereas harvests in capture fisheries peaked in 1989 and have declined since.

World Production by Species Groupings

Half of all aquacultural production (by weight) consists of finfish (Figure 2). Most of the rest is divided between aquatic plants (chiefly red and brown algae) and mollusks (nearly all bivalves). Crustaceans (mainly marine shrimp) contribute less than 1% by weight, but their economic value is far greater than this.

Nearly all finfish production occurs in freshwater although marine species have a somewhat higher dollar value (Figure 5). Various species of carp strongly dominate freshwater production (Figure 6), reflecting the Asian

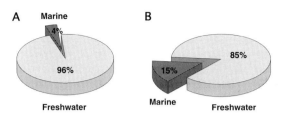

FIGURE 5.—World aquaculture production (%) of inland and marine finfish, 1991: (A) tonnage (10^6 mt); (B) value (billions US$). Source: FAO (1993a).

dominance of aquaculture, and the rest is about evenly divided between the salmonids (some of which now occur in coastal areas) and tilapia. Aquaculture production of Atlantic salmon (*Salmo salar*) now far outstrips the capture of wild fish; hatchery production of Pacific salmon (*Oncorhynchus* spp.), chiefly for stocking, has matched capture fisheries since 1985 (Figure 7). The value of cultured Atlantic salmon now exceeds US$1 billion annually. Salmon and trout have been transplanted to high-latitude or high-altitude culture areas throughout the world, including Mexico, Chile, and New Zealand.

Brown algae dominate the production and value of cultured aquatic plants, but nearly 20% of the production and a third of the value in this class is represented by red algae (Figure 8). Nearly all aquatic plant culture occurs in Asia.

Among mollusks, oyster culture has been about level since 1985, but mussel production increased notably in the late 1980s and clam production has edged up (Figure 9). Scallops represent a relatively minor aquaculture harvest by weight, but they have a higher overall monetary value than mussels.

Crustacean culture is predominantly directed at penaeid shrimp (Figure 10), which now have an annual value close to US$5 billion. Freshwater prawns have contributed nearly 100,000 mt/year.

Ecological Considerations

The earth's waters cover 68% of the planet. Two percent of this water is freshwater and supports more than 40% of the world's fish species and two-thirds of its current aquaculture production (Figure 2). Increasing demand for freshwater has created severe strain on natural water sources for human, domestic, agricultural, livestock, and industrial uses. Additionally, environmental land degradation and various forms of industrial pollution have directly affected adjacent waters with consequent impact on fish life. Such discharges have adversely affected the productivity equilibrium of freshwaters, reduced fish output, and even endangered survival of some

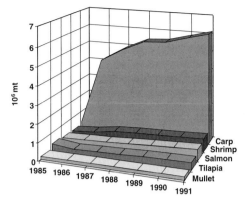

FIGURE 6.—World aquaculture production of major finfish groupings, 1985–91 (10^6 mt). Source: FAO (1993a).

fish species. The 1992 United Nations Conference on Environment and Development (UNCED) in Rio de Janeiro recognized the above aspects and endorsed a series of strategies in which aquaculture systems are directly involved (UNCED 1992).

Target Species and Development Constraints

Culture systems and production outputs are influenced by constraints of space, costs, and local or export market opportunities for each species (World Bank et al. 1992). Consumer preferences have historically determined local market demand for various fish species. However, fish selection and suitability for aquaculture also depend on fish behavior, survival, growth, and feeding in confinement, provided that adequate supplies of the juveniles are available. Aquaculture development in different regions of the world has been affected by several

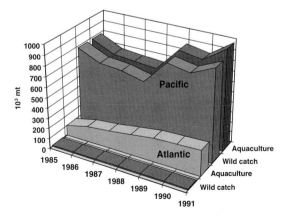

FIGURE 7.—World production of Atlantic and Pacific salmon by culture systems and capture fisheries, 1985–91 (10^3 mt) (FAO 1993a).

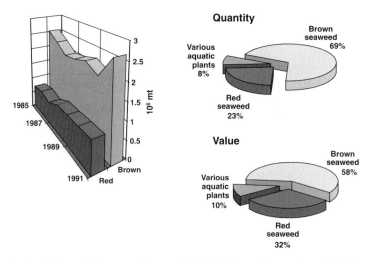

FIGURE 8.—Quantity of world aquatic plant aquaculture production (%) by species groupings, 1991. Source: FAO (1993a).

controlling factors, which may be grouped in biological, economic, and social categories.

Biological factors include breeding and seed supply, feed conversion efficiency in dense culture, disease resistance, and growth performance at ambient water temperatures. Economic factors include costs of seed stock, water (space, supplies, quality, heating), feed, energy, labor, equipment, and duration of production operations (which differs with climate and species). Social factors relate to peoples' cultural traditions regarding husbandry practices, availability of technical skilled labor, local market preferences for species that are more easily cultivated, and year-round operational convenience of

aquafarms. Collectively, these complex factors have determined the progressive worldwide spread of aquaculture following the early historical origins in China and central Europe.

This review does not address constraints of the major sector of health assurance and pathobiology in aquaculture, which requires comprehensive and not cursory coverage. Suffice it to say that bacterial, fungal, viral, neoplastic, and parasitic diseases affect fishes, crustacea, mollusks, and algae. Both infectious and noninfectious diseases have become critically significant—particularly for intensive salmon and shrimp systems—and catastrophic outbreaks have occurred. Specialized toxicologists, parasitologists, and

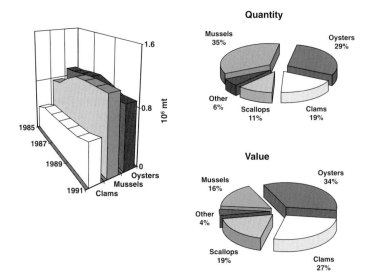

FIGURE 9.—Quantity of world molluskan aquaculture production (%) by species groupings, 1991 (10^6 mt). Source: FAO (1993a).

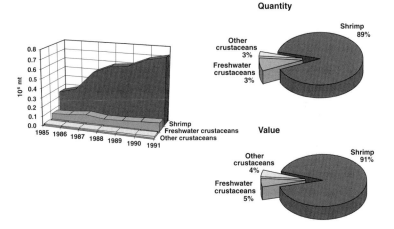

FIGURE 10.—Quantity of world crustacean aquaculture production, 1985–91 (10^6 mt). Source: FAO (1993a).

pathologists now provide veterinary services for salmon, catfish, carp, and shrimp aquaculture.

Operational Constraints

The adoption and spread of various culture systems in each region have been indirectly determined by the cost–efficiency of operations. The most cost-efficient finfish for cultivation are phytoplankton foragers while filter-feeding mollusks are the most efficient producers in nutrient-rich coastal waters. New et al. (1993) showed in their assessment of Asian aquaculture production that 1990 harvests from carnivorous species yielded 0.5 million mt while noncarnivorous finfish production totaled 1.6 million mt. The increasing quantities of shrimp and prawn (high-protein feeders) totaled 0.75 million mt. A significant aspect of the intensive aquaculture industry has been that the most valuable fish species being cultured are predators that require a high protein level in the compounded feeds provided. Efficient fish can convert such feeds at ratios from 2:1 to 5:1. Rations comprise both moist feeds (consisting of a large portion of low-value marine fish) and dry pelleted feeds made with fish meal together with sorghum or other plant ingredients. For grow-out operations, profitability ultimately depends on costs of the protein component of the supplemental feeds.

Consumer preference in affluent societies has been traditionally for predatory species, and this has fueled the export market demand from developing regions for such cultivated species of high unit value. However, in view of the population crisis for food self-sufficiency and the ecological dilemma posed by declining marine fish harvests worldwide, the greatest aquaculture emphasis needed is for maximizing production of noncarnivorous species. Phytoplankton feeders and herbivorous species should

therefore be given much greater attention, particularly in tropical aquaculture research. Accordingly, the candidate species of macrophage and phytoplankton feeders of South America, Africa, and Asia should continue to be targeted for intensive aquaculture enterprises.

Feeds and Nutrition

World fish consumption preferences have influenced international marketing of harvests from capture fishing and impelled commercial culture systems everywhere towards aquaculture of high-value carnivorous species. The profitability of such intensive systems is constrained by feed costs because the protein component relies on fish meal or low-value bycatch fish to satisfy dietary nutrient requirements. Such commercially prepared feeds account for more than 50% of production costs for salmonids and Asian shrimp production. Asian countries provide for 80% of world finfish aquaculture (Figure 1) but depend on artificially prepared commercial feeds for only 10% of that output (New et al. 1993). The remaining 90% comprises noncarnivorous species in polyculture, in which nutrition depends on enhancing productivity of natural food organisms in extensive and semi-intensive aquaculture systems. Notably, the Asian domestic mass consumption preference is traditionally for these species (carp, milkfish [*Chanos chanos*], tilapias) though high-value carnivore species (yellowtails [*Seriola* spp.], eels [*Anguilla* spp.], catfish [*Clarias* spp.], seabream [*Mylio* spp.]) have emerged as significant exports to affluent communities.

Rising costs of supplemental feeds have adversely affected the cost–efficiency of intensive aquaculture systems and encouraged the recent increase in Asia of noncarnivorous finfish with an evident relative decline of the Asian output of carnivorous fish. From 1985 to 1990,

Asian carnivore outputs declined from 52% to 41% of the world catch while Asian polycultures of noncarnivorous fish increased by 3% in the same period (New et al. 1993). However, while this aquaculture increase of noncarnivores was the case in Asia, where there is unsatisfied market demand for cultured food-fish, in other areas markets need mainly carnivorous species that require high-protein feeds. Rising costs and lessening world availability of fish meal have stimulated research studies on vegetable replacements for the fish meal component in the feeds of salmonids and catfish.

Significant progress has been achieved in fish feed technology by using soybean and canola. Experience in salmon aquaculture illustrates these production economies (Prendergast et al. 1994). Feed costs vary from 40% to 60% of farm operational expense. The protein fraction of fish feed cost may account for 64% of the cost, which is being driven higher by declining world supplies of fish meal. Recent research has focused on the use of canola (a variety of Canadian rapeseed [*Brassica* spp.]), which has been shown to have high nutritive value for livestock and poultry feed. Protein from rapeseed (which ranks third in world production of oilseed crops) exceeds the world production of fish meal protein, and is half the cost of Canadian fish meal. Initial results of growth trials demonstrate the successful replacement in trout diets of fish meal by canola protein concentrates. This field experience offers promise for commercial application with salmonids because of the benefits—in contrast to fish meal—of cost savings, reliable supplies, uniform nutrient composition, and storage stability.

Perspectives and Operational Opportunities

In general, marine environments offer greater prospects for expanded culture of finfish and shellfish species than freshwater environments. However, in temperate countries the investment attraction for marketing high-value predatory species has led to concentration of mariculture effort on a few such species monocultured in cages, with low-value fish species provided as their supplemental feed. The potential still remains to be fully developed for expanded tropical culture of filter-feeding mollusks, seaweed, and polyculture combinations. The major contributors of world aquaculture have very different objectives from those of North American producers. There are many other development horizons that may be cited but this review will be confined to finfish and not crustacea and mollusks.

Suitable New Species

Ichthyologists have identified some 1,800 species of freshwater fish in Latin America, of which the major groupings suitable for aquaculture are in the Amazon River system. *Colossoma bidens* is one of the many herbivorous species of proven performance. In Africa, at least 40 among 3,000 freshwater species have been similarly identified. The great lakes of Africa contain the richest lacustrine fish speciation in the world. Of these, more than 50 candidate species, including many tilapia, have been recognized as exceptionally suitable for aquaculture owing to their fast growth, desirable food habits, and favorable comportment in mass confinement. At least 15 indigenous African species are showing great promise for aquaculture (Jhingran and Gopalakrishnan 1994). In Asia, over 30 new species are already the focus of research and field performance studies.

Currently there are hundreds of private farmers who individually breed and culture autochthonous species in scattered, localized fish farming enterprises throughout many Asian countries. These naturalists may not be trained scientific researchers but are keen dedicated observers and efficient purposeful practitioners located in very different but opportunistic locations for culture of particular local species of their choosing. They are most effective collaborators whose different site-specific operations provide variable but controlled parameters for research biologists to examine and standardize in determining the further selection of appropriate indigenous candidate species.

It is noteworthy that Pak Mudjair discovered the African tilapia (*Oreochromis mossambica*), which unexplainedly appeared in coastal pools in Eastern Java around 1939. Its subsequent worldwide promotional use directly resulted from his keen observation and initial trial cultures, which were later validated by scientists' publications describing adabtability in different conditions. There are similar historical precedents for the work of fish culturists in China, India, and Europe that validated work of keenly observant naturalists. Close liaison between fishery research scientists in academic centers and naturalist fish farmers of remote tropical areas with rich abundant ichthyofauna may help to determine the practical aquaculture performance of new candidate species.

Tropical Fish Seed Banks

The opportunity for fishery scientists in developed countries to contribute to the improved efficiency of tropical aquaculture may be most readily facilitated through university research on the mass breeding and larval survival of culturable indigenous fishes. The traditional collection of fry in China and India (after the seasonal floods from monsoon rains have covered the floodplain) gave rise to the deliberately organized fry collection for aquaculture. This is largely now superseded by hatchery production and "seed bank" systems where select broodstock of the desired species are artificially bred to mass-produce "fish seed" or "commercial fry."

Similar collection of "wild" fry from floodplains fisheries has now begun in several countries of Africa to re-stock reservoirs. This is particularly evident in Mali where the floodplains of the Niger river provide enormous quantities of fingerlings. Catfish (*Clarias* and *Heterobranchus* spp.) and tilapia are now gathered and held alive for re-stocking the increasing numbers of community reservoirs. This process is the first activity stage of community awareness through aquaculture intervention that is now supervening in many of the inland areas of Africa. This situation is being driven by the urgent demand for food in Africa and the clear need in integrated rural development programs for the multiple use of the water resources of this vast continent.

Controlling Human Water-Borne Diseases

Tropical aquaculture activities also endeavor to make a virtue out of a necessity. Where watersheds increasingly discharge agricultural fertilizers, eutrophication of warm waters in tropical lakes and large reservoirs can develop rapidly, often causing fish kills. When not controlled, such fertilized waters also promote rapid algal and weed infestations that create suitable habitats for several human disease vectors. These include mosquito larvae (vectors of malaria and filaria), *Simulium* larvae (vectors of onchocerciasis), and snails (vectors of bilharzia). Natural control of these aquatic organisms occurs in equilibrium when they are consumed by various fish species that are abundant in the geographic regions where these human diseases are endemic. Herbivores also consume the vegetation that protects the insect vectors and provides food for snail vectors (Aquaculture Development Coordination Program 1976). The polyculture combinations of herbivore, insectivore, and malacophage fish species (in China and India) have, therefore, been opportunistically designed to produce edible protein when these fish are stocked in large ponds or behind water-retention dams. Such fishes consume these organisms while occupying different compatible ecological niches in the tropical waters where these disease vectors occur naturally. Continuous multiple benefits of cleaner water, disease vector control, and food fish have resulted from such large retention dams by regular strategic stocking with these indigenous fishes.

Large permanent reservoirs or dams for hydroelectric power and numerous small catchment reservoirs have been built throughout many tropical countries. However, unless these are well stocked with fish, they may create some health hazards for riparian communities and spread the particular water-associated diseases mentioned when they become scattered foci of aquatic vectors of human infectious diseases. These dams are often located in semi-arid regions with critical water shortages where settler and nomadic communities are consequently subject to high incidence of bilharzia and malaria. The ancient Chinese sayings that "fish sanitizes water" and "good waters have plenty of healthy fish" refer to early experience with fish, algal blooms, and various diseases when the stocking of fish in natural waters made the fish and water more satisfactory for general human use in riparian communities.

Africa only produced 51,000 mt of aquacultured fish in 1990 while catching 1.9 million mt of food fish from inland waters. Geographical information systems have disclosed over 10,000 small water bodies in Zimbabwe (Coche et al. 1994), and more than 200,000 such retention dams of various sizes throughout Africa were estimated by participants of a 1993 symposium of FAO's Committee for Inland Fisheries of Africa. Kapetsky (1994) estimated that vast areas in some 30 countries have suitable temperature zones (>22°C for >8 months) for aquaculture operations. The enhancement of culture-based fisheries in these widespread small water bodies offers excellent opportunity for sustainable food production increase and is gathering momentum with the stocking of irrigation dams in sub-Saharan Africa.

Accordingly, the beneficial experience of Asia should be fruitfully repeated in Africa and Latin America through the obligatory stocking with suitable "vector-control species" indigenous to the sub-region. As extensive aquaculture systems, these dams give cost-effective yields, particularly since the intensive aquaculture husbandry systems of Asia are not easily transferable for sociocultural reasons. Desirably, standard hatchery systems for the mass production of juveniles of recommended indigenous species should be an essential requirement for stocking tropical water-retention dams. Such "açudes" (artificial lakes), which have been built in northeast Brazil and stocked with fish since the 1930s, now yearly provide over $104 million worth of food fish.

I submit that these extensive systems of culture-based reservoir fisheries can prove to be the most important immediate food contribution of tropical aquaculture, giving large total yields at low unit costs and management inputs in tropical countries that are less aquaculturally advanced. In Thailand, China, and Indonesia (Bhukaswan 1977), the stocking of shallow water-retention dams with suitable combinations of indigenous species has resulted in massive, sustained harvests for nearby riparian populations. This process involves easy management and direct beneficial involvement by communities. In the Ubolrathana reservoir (Thailand), the fish harvests were of greater value than the value of electricity sold to the rural communities. These revenues do not reflect an export value and foreign exchange earning, but they are significant for food sufficiency of inland communities and contribute continually to their social and economic stability and savings of expenditures for fish that were previously imported. These are invisible or little recognized sustained benefits of aquaculture in addition to its public health controls—virtues from necessities!

Aquarium Ornamental Species

Apart from aquaculture of food fish, there has been dynamic growth of the aquaculture of ornamental species as a burgeoning "cash crop." This has helped promote new breeding technologies for food fish. In 1973, the International Development Research Centre scientific study group of Southeast Asian aquaculturists (Allsopp 1973) identified the aquarium fish trade as their most profitable aquaculture enterprise based on operational space, investment, and revenues from assured export markets. The initial practice of seasonal collection of juveniles from wild sources during favorable seasons was progressively being replaced by private breeding centers. Because of foreign exchange earnings, the incentive for efficient mass propagation of highly prized ornamental species attracted privately sponsored research at universities for the breeding, nutrition, and disease controls of such local and exotic species. Universities in Singapore, Thailand, the Philippines, Hong Kong, Indonesia, and Malaysia had special projects on premium-value freshwater and marine species, working in collaboration with public aquaria.

The Asian aquarium fish trade has become one of the major opportunities for developing effective technologies for breeding, feed formulation, mass proliferation, and health maintenance of various tropical species groupings. Though entrepreneurs have concentrated on premium species for the ornamental fish trade, such unpublished research helped solve problems in culture biology and health assurance of juvenile stages of many tropical fish species. University researchers first succeeded with captive breeding of many attractive local species and then progressed to other exotic tropical species in high demand from Africa and Latin America. Moreover, there are attractive professional incentives for such research.

The worldwide value of the aquarium fish trade exceeds $400 million annually. This rapidly enlarging and lucrative international trade in ornamental fish has direct relevance to the food-fish aquaculture industry. It has addressed, in a microcosm, many complex technical aquaculture problems and transferred results from small university laboratory facilities to larger commercial operational centers for profitable replication. Because such experimentation is centered near the equator, the con-

stant temperatures and length of day provide little climatic variation, enabling manipulation of other significant parameters to induce breeding year-round. Many smaller, colorful species for the ornamental fish trade are varieties of larger food species being cultured (e.g., cyprinids, characids, cichlids).

The historical performance of the Singapore aquarium industry has been truly remarkable. Cheong (1993) described how this opportunity has been maximized within space limitations of a land-scarce island (384 km^2). Singapore has 2.7 million people of whom only 13,000 are employed in the total agricultural sector (<1% of the national labor force in 1991). However the aquarium fish industry directly employs 1,200 people (i.e., 0.04% of the national population) but earns 1.8% of the value of national exports in hard currency. Notably, the export value of Singapore's ornamental fish accounts for 20% of the world trade, while aquatic plants are 10% of world exports.

Singapore has established a network of fish breeders, exporters, and brokers specialized in breeding popular target species and a wide array of hybrids. These operations are now relocated at the Ornamental Fish Breeding Centre with development of modern farms in special agro-technology parks. Three hundred and forty varieties of ornamental fish are produced while specialist expertise in selective breeding and health assurance has been progressively developed by concentrating efforts on a few major species groups. Similarly, more than 100 species of ornamental aquarium plants are cultivated while specialized tissue culture of plant species has developed varieties of top value. Advanced biotechnologies of developed countries are being directly applied with aquaculture of ornamental fish species. This augurs well for wider application with tropical food-fish species.

Singapore's ornamental fish industry, government, and the national university collaborate closely to provide technical advice and to support required research on breeding genetics, diets, and health assurance. They have introduced innovative fish health certification and controls that set international standards which have established for Singapore a reputation as "the ornamental fish capital of the world."

New Aquaculture Biotechnologies

Increased aquaculture production has been largely due to the worldwide translocation of species, including carp, tilapia, catfish, and salmonids, that are exotic to the areas where they are now being commercially cultured. These translocations fly in the face of proven historical experiences that introductions of exotic species cause irreversible ecological consequences. We are now recognizing the adverse impacts of many such carp and tilapia introductions in various countries; the tragedy of Nile perch (*Lates nilotica*) in Lake Victoria is well publicized.

Despite the difficulties of incomplete knowledge of the indigenous species, the major opportunity and greatest focus of tropical aquaculture researchers should be on the dozens of promising species in Africa and Latin America. Biotechnology research advances can assist wider use of indigenous species. Some advances were cited by Donaldson et al. (1993), and they will have increasing impact on the aquaculture industry in several direct ways.

Induced breeding for seed production.—Several laboratories are perfecting spawning procedures and genetic selection for various salmon species. Gonadotropic extracts initially used to induce breeding have been replaced by refined synthetic hormones and analogs, with or without dopamine antagonists, to achieve successful spawning and hybrid production. These procedures have achieved operational success with Chinese and Thai carp and Amazon characids. Procedures have focused on broodstock husbandry, genetic selection, induced breeding, gamete storage, fertilization techniques, sex control, incubation, and larval rearing to stock size. These systems address reproduction of select stocks of specific fish (mostly salmonids) whose growth performance has been scrupulously tested and approved under controlled culture conditions.

Cryopreservation.—Standard procedures for storage and transportation of gametes and embryos have been successfully developed. Monosex sperm from select broodstock of several species can now be stored and provided for opportune use in later spawning seasons. This enables sequential spawning from select broodstock throughout the year under natural environment or culture conditions. This process will permit the development of technologies for year-round seed supplies of appropriate broodstock of temperate and tropical species, taking timely advantage of the favorable grow-out conditions with lessened dependence on uncertain seed supplies that may be obtained only during climatic spawning seasons (Harvey and Carolsfeld 1993). The International Fisheries Gene Bank is perhaps the most recent development in networking tropical cryopreservation of gametes derived from select aquaculture species in China, Brazil, Venezuela, and even for endangered stocks of North American species.

Controlled sex differentiation.—Starting with Guerrero's (1975) initial work using 17α-methyltestosterone for regulating production of monosex tilapia, procedures are now further refined for the production of sexually sterile fish. Such fish are being used in intensive aquaculture systems where genetically altered fish or exotic species are commercially grown adjacent to areas with wild stocks. Methodologies including induction of female triploidy have successfully produced sterile stocks as well as monosex females of new strains. Such techniques will be applied to certain strains of transgenic salmon, which are much faster growing than wild stocks. This process avoids any possibility of cultured stocks breeding with natural or feral populations if, through accidents or floods, they escape from ponds or enclosures to the natural environment. The main benefit of sexually sterile fish for aquaculture food production is that the metabolizable energy of the fish is used for somatic growth and diverted from gamete production. This can eventually result in more efficient growth and bigger fish for marketing or greater yields from inputs in a given period.

Growth stimulation.—The use of peptides and proteins for enhancing growth and feed conversion of salmon aims to improve production of intensive systems by reducing feed costs. Increasing the growth rate of fish has, for example, been achieved by administering bovine placental lactogen to fish in amounts that effectively enhance growth. This technique, developed by Donaldson et al. (1993), is the subject of a U.S. patent with the Monsanto Company. It will be particularly important for fish farming of salmonids, tilapia, catfish, and carp, and applicable to fish of any age. It can be administered by slow release injection and in the diet.

Transgenics.—The development of recombinant DNA methodologies to produce fish with altered "geneconstructs" has improved the growth–size characteristics of cultured fish and in the future may be used to manipulate their reproduction, disease resistance, and tolerance to environmental conditions. Initial successes have been achieved with salmonids, carp, and loaches. Selected over several generations, transgenic fish will eventually be produced with desired market characteristics, much improved growth, feed conversion, nutritive value, and flesh quality. Therefore, in the future industry, mass-cultured fish may have such inherent qualities similar to select beef and poultry products.

Issues of special concern, which are being carefully addressed, include the safety of such fish for human consumption, the interaction of such fish with wild species (in reproduction, food competition or habitat displacements), and the public perception and consumer acceptance of modified organisms.

Conclusions

In summary, biotechnologies now being developed for potential commercial application with salmonids will produce fish that grow rapidly, never mature sexually, and therefore provide an optimal fish for culture conditions that is incapable of breeding with wild stocks in natural water bodies. Such technologies, when they are subsequently applied to other candidate tropical species of high aquaculture performance, can revolutionize intensive food-fish aquaculture as well as stock enhancement of water catchment reservoirs in those tropical "aquaculturally advanced" developing countries that are already the largest producers. Further, there are 64 cyprinid and 43 cichlid food species already being cultured, and only the most suitable candidates can be then directly targeted.

There is excellent opportunity for university researchers of developed countries to concentrate effort on high-value tropical fish species and contribute to the species survival programs. These efforts are being advocated for many tropical environments where there is evident threat of disappearance of certain endangered fish species because of environmental degradation, industrial pollution, pesticides, and agricultural chemicals. Many of these "ornamentals" are non-food species, but several are of the same genera of locally significant food species. They thus offer a challenging opportunity for university scientists to engage in significant overseas research partnerships. Teams of multidisciplinary personnel within a university complex will have a greater opportunity to solve breeding and culture problems by using laboratory-controlled aquaria than by traveling to remote field sites. In this way, scientists can help clarify difficult physiological aspects of reproduction and nutrition of promising indigenous tropical food fish for aquaculture.

Epilogue

At the outset of this review, it was indicated that in 1974 a scholarly World Bank group had determined the key global problems and set scientific goals and priorities. Some of these have been vigorously pursued in some regions through international collaboration of aquaculture researchers. Twenty years later, challenges facing aquaculture are more formidable, world food needs more urgent, but opportunities are perhaps greater. Accordingly, aquaculture systems may yet sustainably provide a more significant portion of the world's food-fish requirements and may yet achieve the FAO projected output of 20 million mt by the year 2010.

Aquaculture is rooted in the distant past, is valiantly serving the present, and we hope will contribute significantly in the future to assure the world's population an adequate supply of fish for human consumption. The future of aquaculture will depend on policy makers worldwide to courageously pursue and purposefully accomplish the scientific, management, and environmental goals that have been already clearly defined.

References

ADCP (Aquaculture Development and Coordination Programme). 1976. Aquaculture planning in Asia. Food and Agriculture Organization of the United Nations, ADCP/REP/76/2, Rome.

Allsopp, W. H. 1973. Aquaculture in Southeast Asia. International Development Research Centre, 015e, Ottawa.

Bhukaswan, T. 1977. The fisheries of Ubolratana in the first 10 years of impoundment. Pages 195-205 in Symposium on the development and utilization of inland fishery resources. Food and Agriculture Organization of the United Nations, Indo-Pacific Fishery Commission Proceedings 17(3), Rome.

Cheong, L. J. 1993. The ornamental aquatic industry in Singapore. In Aquararama '93, conference proceedings. Department of Primary Production, Singapore.

Coche, A. G., M. Vincke, and B. Haight. 1994. Aquaculture development and research in sub-Saharan Africa. Food and Agriculture Organization of the United Nations, Committee for Inland Fisheries of Africa, WPA/93/Inf. 3, Rome.

Donaldson, E.M. [et al.]. 1993. Aquaculture biotechnology in Canada including the development of transgenic salmon. Pages 47-57 in Aquatic Biotechnology and food safety. Organization for Economic Cooperation and Development, Paris.

FAO (Food and Agriculture Organization of the United Nations). 1976. Report of the FAO technical conference on aquaculture. Kyoto, Japan, 26 May–2 June 1976. FAO Fisheries Report 188, Rome.

FAO (Food and Agriculture Organization of the United Nations). 1984. Report of the FAO world conference on fisheries management and development. Rome, 27 June–6 July 1994. FAO, Rome.

FAO (Food and Agriculture Organization of the United Nations). 1992. Review of the state of world fishery resources. Parts 1: The marine resources. Part 2: Inland fisheries and aquaculture. FAO Fisheries Circular 710 Rev. 8.

FAO (Food and Agriculture Organization of the United Nations). 1993a. Aquaculture production 1985–1991. FAO Fisheries Circular 815 Rev. 5.

FAO (Food and Agriculture Organization of the United Nations). 1993b. Agriculture towards 2010—Overview of fisheries. Pages 175-195 in Conference Document C93/24, FAO, Rome.

Guerrero, R. D., III. 1975. Use of androgens for the production of all-male Tilapia aurea (Steindachner). Transactions of the American Fisheries Society 104:342-348.

Harvey, B., and J. Carolsfeld. 1993. Induced breeding in tropical fish culture. International Development Research Centre, UDC:639.3, Ottawa.

Jhingran, V. J., and V. Gopalakrishnan. 1994. A catalogue of cultivated aquatic organisms. FAO (Food and Agriculture Organization of the United Nations) Fisheries Technical Paper FIRI/T130, Rome.

Kapetsky, J. M. 1994. A strategic assessment of warm water fish farming potential in Africa. Food and Agriculture Organization of the United Nations, CIFA (Committee for Inland Fisheries of Africa) Technical Paper 27, Rome.

New, M. B., A. G. J. Tacon, and I. Csavas. 1993. Farm-made aquafeeds. FAO (Food and Agriculture Organization of the United Nations)–RAPA (Regional Office for Asia and the

Pacific), ASEAN (Association of Southeast Asian Nations)–Commission of the European Communities, Rome.

Prendergast, A. F. [et al.]. 1994. Searching for substitutes—canola. Northern Aquaculture, May:15-20.

UNCED (United Nations Conference on Environment and Development). 1992. Agenda 21 Rio declaration. United Nations, A/CONF. 151/4, parts 1-4, New York.

UNDP (United Nations Development Programme), NORAD (Norwegian Agency for Development Cooperation), and FAO (Food and Agriculture Organization of the United Nations). 1987. Thematic evaluation of aquaculture. FAO, Rome.

WB (World Bank) and FAO (Food and Agriculture Organization of the United Nations). 1974. Aquaculture research programme. World Bank, DDR:IAR/74/25, Washington, D.C.

World Bank, United Nations Development Programme, Commission of the European Communities, and Food and Agriculture Organization of the United Nations. 1992. A study of international fisheries research. World Bank, Policy and Research Series, 19, Washington, D.C.

Sea Ranching of Atlantic Salmon with Special Reference to Private Ranching in Iceland

ÁRNI ÍSAKSSON

Abstract.—This paper defines sea ranching and differentiates it from salmon enhancement activities occurring in most salmon-producing countries. Private and semi-private ranching of salmon can be considered aquaculture, as the purpose is to produce high-quality fish for world salmon markets. International production statistics for ranching demonstrate the dominant position of Pacific salmon (*Oncorhynchus* spp.) in sea ranching, primarily in Japan and Alaska. Status of Atlantic salmon (*Salmo salar*) ranching is discussed with a special focus on the development of private ranching in Iceland. This includes production statistics and discussion of the social, genetic, and ecological problems associated with this development.

The strategies used in ranching, including site selection and salmon stock, as well as release and recapture methods, tend to vary considerably between countries and areas, depending on the species used and the ecological and political framework. The absence of any sea fishery for salmon within Iceland's 200-mile (320-km) territorial limits, as well as private ownership of rivers, has facilitated the development of large-scale private ranching with the primary aim of producing high-quality salmon for the market. The industry, in response to market demands, emphasizes the production of large 1-sea-winter salmon, approaching 3 kg in mean weight. Genetic selection experiments suggest that marine survival and average weight at return can be improved through family selection. It is fairly clear that ranching of Atlantic salmon will always be somewhat small-scale compared with salmon farming, primarily because of ecological and economic limitations. The ranching industry must thus promote quality rather than quantity. The large commercial operations in Iceland are still developmental, with release and recapture methods and smolt production routines being generated. Periods of low marine survival in the late 1980s and early 1990s have further compounded the situation. As most of the large-scale operations in Iceland have gone into liquidation, the future of private ranching of Atlantic salmon seems highly uncertain.

Ocean or sea ranching is the practice of releasing young fish into the marine environment and allowing them to roam and grow in the wild until maturation. The term is most commonly used for the release of salmon smolts or fry, which are ready to migrate from freshwater into the sea and subsequently return after 1 to 3 years as mature fish to the same freshwater location. In the case of Atlantic salmon (*Salmo salar*), returning fish are mostly classified as grilse or 1-sea-winter (1SW) and 2-sea-winter (2SW); older salmon are rarely observed in ranching operations. Salmon ranching can further be categorized depending on salmon species, harvest strategies, and political structure.

- Private ranching is defined as large-scale releases of salmon smolts by private companies with the intent of harvesting all the salmon upon return at the release site. Genetic selection of the stock to improve performance is logical and desirable, and all the salmon are harvested at the release site. This activity is currently confined to Iceland.
- Semiprivate ranching is used in Japan and Alaska with Pacific salmon (*Oncorhynchus* spp.) where cooperative companies of fishers are releasing salmon to enhance local fisheries. Genetic selection of stock is possible, and the fish are harvested both in mixed stock fisheries and close to the release site.

- Enhancement is here used to define all other activities, such as releases of fry and smolts by governments or companies for mitigation or restoration purposes. Use of indigenous stock is often mandatory. This would include public releases in the Pacific and Atlantic oceans and the Baltic Sea.

International Aspects

International sea ranching is primarily based on Pacific salmon of the genus *Oncorhynchus*, of which there are 7 species in the Pacific, including steelhead trout (*Oncorhynchus mykiss*). The greatest contributions to commercial ranching come from species with the shortest freshwater rearing cycle, that is, chum salmon (*O. keta*) and pink salmon (*O. gorbuscha*), primarily in Japan and Alaska. These operations, producing over 250,000 metric tons (mt) of salmon annually (Ísaksson 1994), can be considered true ranching as they are conducted by semiprivate organizations in order to enhance their own commercial fisheries and are thus non-river based. These species are released at a size of 1-2 g a few months after hatching. All ranching operations in the Atlantic are based on Atlantic salmon and the conspecific Baltic salmon (*S. salar*), which are released as 1- or 1-year smolts at a size of 20–50 g.

This paper deals primarily with sea ranching of salmon

as an aquaculture venture (i.e., a method to produce food through various harvest strategies). It discusses sea ranching of Atlantic salmon with a focus on current production and the status of sea ranching in Iceland. It also reviews the major ecological and political problems encountered in private ranching programs and the ranching strategies envisioned for the Atlantic salmon ranching industry. Finally, the potential of sea ranching is compared with salmon farming with respect to future development and economics.

Ranching of Atlantic Salmon

There are considerable enhancement activities in various countries bordering the Atlantic Ocean. Total releases of Atlantic salmon smolts into the Atlantic in Europe for enhancement purposes, however, constitute less than 2 million smolts, with Ireland releasing the bulk, close to 1 million smolts. About a million Atlantic salmon smolts are released on the east coast of North America, fairly equally divided between Canada and the United States. Iceland is the only country actively involved in commercial ranching, releasing close to 6 million smolts annually in recent years.

Sweden and Finland are conducting considerable salmon enhancement activities in the Baltic Sea, releasing over 5 million smolts annually (Ackefors et al. 1991). This activity can be considered a prime example of public enhancement activities in Europe and is upholding a sizable mixed-stock commercial salmon fishery in the Baltic main basin, with serious consequences for the remaining wild stocks (Eriksson and Eriksson 1993). The issue is further confused by a large participation in the fishery by nations that do not contribute to release operations.

Private Ranching in Iceland

In Iceland, all salmon and trout fishing rights are controlled by owners of lands adjacent to rivers and lakes, who by law are obliged to form a fisheries association to manage the resource. The fisheries associations have been involved in enhancement activities for decades, including both fry and smolt releases and construction of fish ladders. In the early 1960s, the Institute of Freshwater Fisheries established the Kollafjördur Experimental Fish Farm, which was instrumental in promoting fish culture activity and started experimenting with salmon ranching in 1965. This activity, and the fact that harvest of Atlantic salmon in the sea has been prohibited by law since the 1930s, has laid the foundation for private ranching in Iceland.

Although there have been experimental releases since the mid-1960s, commercial ranching only started in the mid-1980s, peaking with the release of 6 million smolts in 1991. The proportion of ranched salmon in Icelandic salmon catches has thus increased from less than 20% in 1980 to more than 80% in the early 1990s (Figure 1). Most ranching activity takes place on Iceland's west coast; it is of minor importance in other areas.

The largest commercial facility operating in Iceland, established by Silfurlax, Inc., is a release facility located at Hraunsfjördur in western Iceland (Figure 2). Smolts are released from seawater pens in midsummer after a 2- to 3-month adaptation in freshwater and subsequently in seawater rearing pens. Recaptures are performed through an efficient seining process in estuarine areas within 100 m of the river mouth. Most recaptures take place during June through August, with a peak in July. In 1993, the Hraunsfjördur facility recaptured about 100,000 salmon,

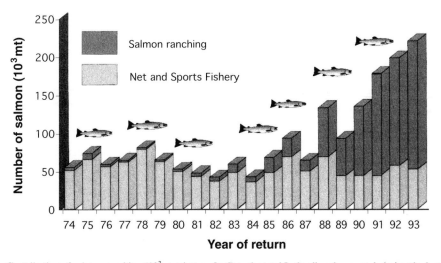

FIGURE 1.—Contribution of salmon ranching (10^3 metric tons [mt]) to the total Icelandic salmon catch during the last 20 years.

FIGURE 2.—A diagram of the Hraunsfjördur ranching facility, showing the smolt-rearing cages and the dam, which creates the freshwater lake. The trapping of adults is conducted within 100 m downstream of the highway bridge.

about 50% of the total Icelandic salmon catch of 206,000 salmon (650 mt).

The high proportion of ranched salmon in the Icelandic salmon catch has raised questions regarding the interaction of ranched and wild salmon populations. Considerable information has been gathered on the environmental and ecological problems associated with private salmon ranching and the improvements needed for its viability.

Problems Related to Ocean Ranching

There are various problems related to ranching, which in many cases are beyond the rancher's control. The success of ranching depends on the release of high-quality smolts at the right time into an oceanic environment that favors the survival and growth of the fish. The salmon rancher can often deal successfully with the rearing and release aspects of the ranching process, but the marine phase of the salmon's life cycle is entirely subject to the whims of nature. This section highlights some of the major problems related to ranching, with a special emphasis on problems observed in private Atlantic salmon operations in Iceland.

Ecological Constraints

Smolt quality and release techniques.—Probably the most important factor under human control is smolt quality and the success of smoltification, which determines the success or failure of an ocean ranching venture. Release time and techniques are also critical factors. It is by no means certain that a successful small-scale pilot project will be as successful after scaling up to commercially viable size. The pure logistics of releasing 2 to 3

million Atlantic salmon smolts over a 1-month period and harvesting 200,000 salmon over a 2-month period can be overwhelming. This scaling up of ranching to commercial sizes has often been difficult and can, in most cases, be considered a new stage in experimentation.

Interannual variation in ranching potential.—There is good evidence that ranching potential varies highly from year to year, which is reflected in survival and growth of ranched salmon in the sea. This is probably more prominent in polar and temperate areas, which are on the borderline of salmon distribution. It is well known that warm oceanic currents flowing northward, called El Niño, create unfavorable conditions for salmon in temperate areas of the Pacific. Similarly, polar currents have created difficult conditions for ranched salmon in Iceland (Ísaksson 1991), but most prominently for wild salmon stocks on the north and east coasts of the country (Antonsson et al. 1993).

A model (Figure 3) was constructed from data obtained at the Kollafjördur Experimental Fish Farm in Iceland in the 1980s (Ísaksson 1994). When the warm Gulf Stream flows north of Iceland and oceanic conditions are relatively favorable (as reflected in the left side of Figure 3), return rates are high, the ranched fish are larger, and most of the salmon return after 1 year in the sea, having a fairly even male-to-female ratio in the grilse population.

Conversely, if polar currents dominate and the Gulf Stream does not affect the north coast of Iceland (as reflected in the right side of Figure 3), the return rates are low and the grilse small in size. There is also a delay in maturation over to the second year, and males tend to dominate in the returning grilse. This condition is, in fact, frequently observed in wild salmon populations on Iceland's north and east coasts (Scarnecchia 1984).

Similar findings have been reported for the western

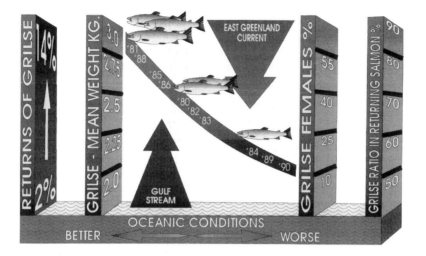

FIGURE 3.—A descriptive model showing the effects of oceanographic factors on return rates and various population parameters of Icelandic ranched Atlantic salmon (*Salmo salar*). The figure describes only general trends.

Atlantic. Friedland et al. (1993) suggested that salmon habitat in the Labrador Sea and Denmark Strait has been reduced in recent years and is especially critical in late winter. It is likely that this information mostly holds for arctic areas, and ranching conditions might be more stable in subarctic areas, especially in the grilse component. However, 2SW salmon from those areas frequently go to arctic feeding areas (e.g., west Greenland).

Carrying capacity of the ocean.—Some information from the Pacific indicates that larger numbers of feeding salmon in the ocean reduces the growth rate of several salmon species. Peterman (1984) established a reduced growth rate of sockeye salmon (*O. nerka*) in the Gulf of Alaska in years of high abundance. Similarly, record runs of sockeye to the Fraser River in Canada in recent years were accompanied by unusually small-size returning adults (J. C. Woodey, Pacific Salmon Commission, Vancouver, British Columbia, Canada, pers. comm.). Eggers et al. (1991) demonstrated reduced average size of pink salmon since the early 1970s in the wake of increased ranching. Similar observations have been reported for Japanese hatchery chum salmon.

Since salmon catches in the Atlantic are minute (5,000–8,000 mt) compared with those in the Pacific (600,000 mt), it seems unlikely that salmon populations in the Atlantic suffer from overcrowding. It must be borne in mind, however, that salmon production might in some years be limited by other fish species occupying the same oceanic niche, as well as by thriving populations of marine mammals.

Genetic effects from straying salmon.—Many biologists fear that continuous straying of reared and ranched fish into rivers may be detrimental to wild stocks, which have adapted to a specific environment for thousands of years.

Ranched and reared salmon, in contrast, have adapted to the rearing environment at least through a part of the life cycle and might thus be unsuited for life in the wild. Long-term genetic mixing of those salmon with wild salmon might consequently be detrimental to wild stocks.

A number of workshops and conferences have in recent years dealt with this issue, compiling information on detrimental impacts and focusing on research necessary for conclusive answers (Thomas and Mathisen 1993). Meanwhile, for precautionary reasons, it is considered important to prevent large-scale straying of ranched and reared salmon into pristine salmon rivers.

Icelandic enhancement and ranching operations have yielded a great deal of practical information on the straying of ranched salmon into rivers and between ranching stations, as well as the straying of wild fish into ranching stations and between rivers. Some of the findings are summarized in the following sections.

Straying of ranched salmon.—There is considerable information available in Iceland on the straying of ranched salmon to other ranching stations and into rivers. The riverine information, however, is only based on the screening of the sports catch for coded-wire tags, but little information is available on the proportion of strays in the escapement or their subsequent spawning success.

Straying varies from year to year, appearing to be higher in years of low return. Thus, there might be a common factor during imprinting, possibly related to smoltification, that affects both survival and homing precision. The number of strays from one ranching station to other stations can be on the order of 10–15% of total numbers of microtagged salmon returning in high stray years, but only half of that in years of low straying rates. The number of observed strays of ranched salmon into wild salmon rivers has been in the range of 2–4% of the total number of tagged salmon returning, being higher in rivers close to the ranching stations. Incidence of straying seems to be much higher in rivers flowing directly into the sea than in tributaries in complex river systems.

Straying of wild salmon.—Straying of wild salmon between rivers is assumed to be low, but considerable straying of microtagged salmon—tagged as wild migrating smolts—has been observed between river systems in Iceland, even those geographically far apart. These observations are limited to the detection of microtags in the sports catch; very limited information is available on strays in the escapement.

Since 1988, wild smolts have been tagged in the rivers Ellidaár on Iceland's southwest coast, Midfjardará on the north coast, and Vesturdalsá on the northeast coast. Salmon caught in the sports fishery have subsequently been screened for microtags although the screening process is systematically monitored only in some key rivers. During this period, 27,262 wild smolts were microtagged; 840 tagged salmon were caught in the rivers, excluding escapement. During this period, 13 wild tagged salmon were reported from ranching stations and 13 wild fish reported from nonnative salmon rivers in various areas. Of those strays, the majority were found in neighboring rivers, but significant numbers of strays seem to be returning to distant rivers.

Catches of wild salmon at ranching stations.—The occurrence of wild salmon at ranching stations has been of major concern as there is a reason to suspect that this phenomenon is related to harvest strategy as well as the location of the ranching station in relation to major salmon rivers. Estuarine traps might be catching more strays from wild salmon rivers than would occur in a freshwater trap, and a ranching station located in the migration path of wild salmon would catch more wild salmon than a station located at the bottom of a long fjord. Ranching experience, on the other hand, has shown that ranched salmon are reluctant to enter freshwater except during freshets, which can occur infrequently. The resulting salmon are colored because of maturation, and thus are practically unfit for export market. Such delays further result in greater strays from the ranching site to neighboring rivers.

Estuarine traps are thus of great importance for the salmon rancher. They procure a steady supply of bright salmon throughout the season and have the added benefits of reducing straying to salmon rivers, preventing genetic effects. The negative effects might be some catches of strays from wild salmon populations, which causes controversy between ranchers and river owners. A successful solution is the greatest challenge facing the Icelandic freshwater management system today.

Fish diseases.—Diseases and parasites originating in reared or ranched populations can, in theory, spread to wild populations, primarily through straying. Certain diseases that are transmitted through eggs to progeny can be magnified in ranched populations if infected smolts are released in great quantities. Bacterial kidney disease (BKD), which is found in most countries bordering the Atlantic, is a prime example.

During the last 30 years, BKD has periodically been observed in ranched populations in Iceland but has been curtailed and kept successfully in check by disinfecting salmon eggs and discarding eggs from infected ranched females. These procedures have allowed gradual but successful clean-ups of ranching stations.

In the summer of 1995, furunculosis (*Aeromonas* sp.) was observed for the first time in an Icelandic salmon river. It was a minor epidemic related to warm and dry weather. At the same time, it was observed in a nearby ranching station, which eventually resulted in total loss of broodfish and disinfection of the rearing facility. Although no further outbreaks have been observed, this one-time epidemic has resulted in stiffer controls for transporting both wild and ranched live salmon.

Pollution.—Marine pollution can be a threat to ranching, which needs to produce top-quality clean and healthy salmon. Organic residues can move up through the food chain to predators such as salmon and cause flesh contamination. Good examples of such a development are the salmonid populations in the North American Great Lakes and to a lesser extent in the Baltic Sea. Such problems have not been observed in populations feeding in the open Atlantic.

Political Constraints

The political framework for ranching is set by the laws of individual countries as well as international laws and treaties. Most of the issues focus on the questions regarding who should be permitted to ranch and where the salmon shall be harvested.

Public vs. private ownership.—In many countries, rivers and lakes are publicly owned, and no individual or company is permitted to utilize a public resource for a ranching operation. This is the case in most salmon-producing countries, especially in the Pacific rim and North America. In this case, the ranching or enhancement operations are run by governments to compensate for losses due to hydroelectric power development and habitat degradation. Enhancement operations in North America are a prime example. Some semiprivate ranching operations run by trade-based cooperatives have been permitted under this regime (e.g., in Alaska and non-river-based private operations in Oregon).

In some European countries, rivers—as well as the fishing rights—are in private ownership. This is the case in the United Kingdom, Norway, and Iceland. This system encourages private ranching, provided that the private use of marine resources is not restricted and the ranched salmon are not heavily harvested by coastal fisheries. Iceland is the only country in the North Atlantic that has forbidden salmon fisheries within its territorial waters for decades and has the proper political framework for the development of private ranching.

Mixed-stock fisheries.—Fisheries on mixed stocks of wild salmon have been a great management challenge for decades and probably the greatest cause for the decline of many small salmon stocks, which cannot tolerate the fishing pressure exerted on larger stocks. Similarly, it is certain that wild salmon stocks would be seriously affected by large-scale ranching if the returning salmon were harvested in a mixed-stock fishery. Such a ranching strategy is doomed to fail. Ranched salmon should thus be harvested only in a terminal fishery after they have separated from wild salmon stocks.

International migrations.—Salmon migrate long distances across the ocean and have frequently been harvested on oceanic feeding grounds by the host country. In the Atlantic, the fisheries at west Greenland and the Faroe Islands, which mostly harvest 2SW salmon, are well-known examples. Ranching schemes based on 2SW salmon would have difficulty enduring such harvests, but the private ranching operations in Iceland have focused on fast-growing 1SW grilse, which do not migrate to distant areas. The fisheries in west Greenland and the Faroe Islands are furthermore subject to quotas negotiated by the North Atlantic Salmon Conservation Organization, which in recent years have been purchased by private interest groups represented by the North Atlantic Salmon Fund.

Ranching Strategies

Ranching plans and strategies for Atlantic salmon in Norway and Iceland have been presented (Hansen and Jonsson 1994; Ísaksson 1994; Jónasson 1994). Hansen and Jonsson (1994) concluded that the potential for sustainable ranching in Norway was limited and it should be done primarily for enhancement purposes in large rivers that are threatened or have lost their salmon stocks. Harvest should be primarily through a sports fishery.

In Iceland, ranching has been developed at sites with relatively small freshwater flows. Very few large rivers are available for private ranching as they sustain a very valuable sports fishery for Atlantic salmon. Private ranching in Iceland must thus be adapted to small rivers, and release technology and harvesting methods must take this

into account. A freshwater lake or a lagoon is frequently an integral part of the ranching site in order to accommodate rearing cages for short-term adaptation of smolts to the site (Figure 2).

Practical experience and experimentation in salmon ranching in Iceland during the last 30 years have led to ranching strategies that are being recommended for a viable ranching plan. These strategies relate primarily to the practical aspects of ranching, such as the initial selection of a ranching stock, selective breeding of the stock for certain traits, and release and harvest strategies.

Selection of a Stock

Early experiments indicated that a single ranching stock could be used in a large geographic area in western Iceland and that release sites were a practical alternative to a rearing and ranching facility (Ísaksson and Oskarsson 1986). Jonasson (1994) compared the return rates for 3 wild stocks with those of the Kollafjördur ranching stock, and compared the salmon stock by release site interaction for those same stocks when they were released at 3 different locations. The results confirmed that the Kollafjördur ranched stock, developed for several decades at the Kollafjördur Fish Farm, had the best performance in ranching, with a higher return rate than the wild stocks tested. No significant interaction between salmon stock and release site for return rate was detected, suggesting that genetic selection can be based on families within one salmon-ranching stock in western Iceland. As a result of these findings, wild local stocks have been mostly excluded from commercial sea ranching in Iceland.

Genetic Selection in Ranching

The length of the generation interval (ranching cycle) is of great importance in a selection program. The shorter the interval, the greater the progress over a certain time span. The ranching cycle depends on the age of the smolts used in ranching as well as the age of the returning adults. In a hatchery program, the length of time from parent spawning to the return of progeny can vary from a minimum of 3 years to a maximum of 6 years depending on the age of smolts and the returning adults.

Since 1987, an experimental genetic selection program has been conducted by the Institute of Freshwater Fisheries and the Kollafjördur Fish Farm (Jonasson 1993, 1994). The program has been based on family selection and the release of 1-year smolts of several salmon stocks at various ranching localities. The program has demonstrated that survival and growth can be improved through selective breeding in both the freshwater and the marine phases (Jónasson 1994).

Future Ranching Plans

In the future, Icelandic ranching programs will be based mostly upon grilse (1SW), which at present have a mean weight of 2.6 kg. The return rates of grilse are much higher than those of 2SW salmon, but in the past, grilse have not been harvested to any extent in the Faroes and West Greenland fisheries. Assuming that grilse feed to a greater extent within Icelandic territorial limits, they will be protected by the Icelandic ban on salmon fishing in the sea. There are also indications that 3- to 5-kg salmon are in greater demand on the international market than larger salmon.

Jonasson (1995) presented a breeding plan for sea ranching. He concluded that the most economical genetic gain was to select for grilse biomass (per 1,000 smolts released), which ensures a 3-year generation interval and thus a more rapid selection progress than if 2SW salmon were selected.

Release and Recapture Strategies

The largest commercial ranching operations in Iceland conduct salinity adaptation of smolts prior to release. This seems to be the most practical method when dealing with large numbers of smolts, seems to ensure successful seawater migration, and precludes unwanted delay and possible abstinence of smolts from migration. In Icelandic ranching operations, salinity adaptation of smolts has given returns comparable to conventional releases from freshwater, but no significant benefits in survival have been observed (Ísaksson and Óskarsson 1986). Similar findings were reported by Hansen and Jonsson (1986).

It became apparent in the late 1980s that recaptures in large-scale ranching operations would have to take place in the estuarine areas of the freshwater outflow. Early attempts to build suitable fish ladders or other trapping facilities in small rivers or outflows from rearing stations demonstrated that salmon were reluctant to enter, especially during low flow periods, and that straying to other areas increased significantly (Ísaksson 1982; Ísaksson and Óskarsson 1986). Fish that entered did so mostly during freshets and were frequently discolored, indicating a long estuarine stay prior to upstream migration. Those conditions are well known in many salmon rivers during midsummer drought conditions.

The estuarine trapping now performed at most ranching stations has, on the other hand, procured bright, high-quality ranched salmon throughout the season and probably significantly reduced straying to other areas and rivers. Although controversial among Icelandic river owners, estuarine trapping may be the only practical way to harvest large numbers of salmon, procure high-quality product for the market, and prevent a major exodus of ranched salmon into Icelandic rivers.

Ranching as an Alternative to Salmon Farming

Salmon Ranching as a Substitute for Salmon Farming

It is fairly certain that sea ranching will not replace salmon farming in areas suitable for that activity. Salmon farming, in which salmon are contained throughout their life cycle, has the advantage of being able to supply fresh, high-quality salmon throughout the year, particularly when there is no supply of wild salmon from the Pacific and Atlantic oceans. In recent years, there have been great improvements in salmon farming through selective breeding and advances in disease control. These improvements have made the industry more economical, despite considerable reductions in salmon prices.

Ranching, however, has certain promotional advantages over farming that have resulted in 20% higher market price of ranched Atlantic salmon in recent years. These advantages need to be stressed in the marketing of ranched salmon:

- Ranching has a clean image compared with farming. The salmon are fed only during the juvenile stages; they feed on natural food through most of their life cycle. They are thus free of the antibiotics and disinfecting chemicals frequently used in salmon farming. Cage farming of salmon is also known to cause local marine pollution, which is nonexistent in ranching.
- Ranched salmon have to migrate long distances; thus, they have stiffer muscles and a different texture of flesh from farmed salmon.
- Ranched salmon are intentionally released from a location with the intent of harvesting the fish upon return at the release site. This seems to secure minimal straying. Fish escaping from sea cages often home to the rearing site, but upon maturation they tend to stray at random into nearby rivers.

Ranching also has its drawbacks, some of which are related to environmental concerns:

- The total production of salmon from ranching is entirely dependent on the total number of smolts released and resulting sea survival. Since sea survival is highly variable, the total production of ranched salmon tends to be unpredictable. Furthermore, sea survival tends to be less than 10%, making smolt production and release capacity a major factor limiting production.
- Harvest of ranched stocks with wild stocks in coastal fisheries is of major concern. Since total harvest of the ranched stock is one of the major objectives, there could be public pressure to increase fishing

effort in traditional coastal fisheries, which could be detrimental to wild stocks. Total separation of the ranched population from wild stocks should thus be a primary aim in the harvest strategy.

- Since straying of ranched salmon into salmon rivers is assumed to have a negative impact, the harvest strategy should ensure total recapture and preclude straying as much as possible.
- Ranched salmon return during a short period in the summer, which affects marketing and the price of the product.
- In certain parts of the world, there are indications that releases of salmon can reach or surpass the carrying capacity of the oceanic environment (Eggers et al. 1991). There is no information on this issue in the Atlantic, but future expansion of ranching in the north Atlantic could give rise to concerns both locally as well as internationally.

From the foregoing, it is fairly clear that Atlantic salmon ranching will be relatively small scale compared with salmon farming. The small production, however, could be promoted as a healthful product and thus command considerably higher prices than the mass-produced farmed salmon.

Economy of Ranching

On the basis of 1994 salmon prices, the total cost of ranching 100 Icelandic smolts was calculated as US$100. The cost per kilogram of returning salmon averaging 3.0 kg is thus US$3.30 assuming 10% return rates, and US$6.60 at 5% return rates. The going price per kilogram of gutted ranched salmon onboard a transport plane has been US$5.30. This gives the salmon rancher a price of US$4.10 for each kilogram of ungutted salmon.

Icelandic ranching operations thus need approximately 8% return rates to break even, and even higher returns to show a profit. Since the ranching operations have recently been operating below 5% returns, clearly none are generating profit. The larger commercial ventures in Iceland must thus be considered developmental projects; their viability is questionable, and some have already closed down. Release and recapture methods for large releases are still being developed, as well as smolt production routines. Since smolt price is a relatively high factor in the cost of ranching, there will be great benefits from any reduction in smolt production costs. Breeding programs are also expected to contribute significantly to the economy of ranching in the years to come.

References

Ackefors, H., N. Johanson, and B. Wahlberg. 1991. The Swedish compensatory programme for salmon in the Baltic: an action plan with biological and economic implications. International Council for the Exploration of the Sea, Marine Science Symposium 192:109-119

Antonsson, Th., G. Guobergsson, and S. Guojónsson. 1993. Possible causes of fluctuation in stock size of Atlantic salmon in northern Iceland. International Council for the Exploration of the Sea, Council Meeting 1993, Anacat Committee M:10.

Eggers, D. M., R. L. Pelz, B. G. Bue, and T. M. Willette. 1992. Trends in abundance of hatchery and wild stocks of pink salmon in Cook Inlet, Prince William Sound, and Kodiak, Alaska. In O. A. Mathisen, and G. T. Thomas, editors. Biological interactions of natural and enhanced stocks of salmon in Alaska. Interim report from the workshop held in Cordova, Alaska, November 10-15, 1991, JCFOS 9201, Juneau Center for Fisheries and Oceanic Sciences, University of Alaska, Juneau, Alaska.

Eriksson, T., and L.-O. Eriksson. 1993. The status of wild and hatchery propagated Swedish salmon stocks after 40 years of hatchery releases in the Baltic rivers. Fisheries Research 18:147-159.

Friedland, K. D., D. G. Reddin and J. F. Kocik. 1993. The production of North American and European Atlantic salmon: effects of post-smolt growth and ocean environment. International Council for the Exploration of the Sea, Council Meeting 1993, Anacat Committee M:13

Hansen, L. P., and B. Jonsson. 1986. Salmon ranching experiments in the river Imsa: effect of day and night releases and of sea water adaptation on recapture rates of adults. Institute of Freshwater Research, Drottningholm, Report 63:47-51, Lund, Sweden.

Hansen, L. P., and B. Jonsson. 1994. Development of sea ranching of Atlantic salmon towards a sustainable aquaculture strategy. Aquaculture and fisheries management. Pages 199-214 in D. S. Danielssen, B. R. Howell, and E. Moksness, editors. An international symposium: sea ranching of cod and other marine fish species. Aquaculture and Fisheries Management Supplementary Issue 25.

Ísaksson, Á. 1982. Sea ranching of salmon in Iceland, present status. In C. Eriksson, P. O. Larsson, and M. P. Ferranti, editors. Sea ranching of Atlantic salmon. A European Cooperation in Scientific and Technical Research (COST) 46/4 workshop in Lisbon, 26-29 October 1982.

Ísaksson, Á., and S. Óskarsson. 1986. Returns of comparable microtagged Atlantic salmon (Salmo salar L.) of Kollafjördur stock to three salmon ranching facilities. Institute of Freshwater Research, Drottningholm, Report 63:58-68, Lund, Sweden.

Ísaksson, Á. 1991. Atlantic salmon—present status and perspectives of sea ranching. In T. N. Pedersen, and E. Kjörsvik, editors. Sea ranching—scientific experiences and challenges, Proceedings from the symposium and workshop October 21-23 1990. Norwegian Society for Aquaculture Research, Bergen.

Ísaksson, Á., 1994. Ocean ranching strategies, with a special focus on private salmon ranching in Iceland. Nordic Journal of Freshwater Research 69:17-31

Jónasson, J. 1993. Selection experiments in salmon ranching: genetic and environmental sources of variation in survival and growth rate in fresh water. Aquaculture 109:225-236

Jónasson, J. 1994. Selection experiments in Atlantic salmon ranching. Estimations of genetic parameters for important

traits in the freshwater and sea ranching period—realized
response for increased return rate of grilse. Doctoral thesis.
Agricultural University of Norway, Ås, Norway.

Jónasson, J. 1995. Salmon ranching, possibilities for selective
breeding. A report of the Nordic Council of Ministers, Nord,
Stockholm

Peterman, R. M. 1984. Density-dependent growth in early life
of sockeye salmon (*Oncorhynchus nerka*). Canadian Journal of Fisheries and Aquatic Sciences 41:1825-1829

Scarnecchia, D. L. 1984. Climatic and oceanic variations affecting yield of Icelandic stocks of Atlantic salmon (*Salmo
salar*). Canadian Journal of Fisheries and Aquatic Sciences
41:917-935

Thomas, G. L., and O. A. Mathisen. 1993. Biological interactions of natural and enhanced stocks of salmon in Alaska.
Fisheries Research 18:1-17.

The Growth of Salmon Aquaculture and the Emerging New World Order of the Salmon Industry

JAMES L. ANDERSON

Abstract.—The aquaculture of salmon has become one of the most significant influences in the salmon industry. Pen-raised salmon aquaculture has moved from virtual nonexistence in the late 1970s to composing over 30% of global harvest in the 1990s, and it is still growing. Aquaculture in the form of salmon enhancement or salmon ranching has also become the dominant source of "wild" salmon in many areas of the world, such as Japanese chum (*Oncorhynchus keta*) runs, the Columbia River, and pink (*O. gorbuscha*) runs in Prince William Sound, Alaska. Aquaculture of salmon has influenced profound changes in the marketing and trade of salmon as well as salmon fisheries management. This paper attempts to document the importance of salmon aquaculture, its increase over the past 2 decades, and how it has changed the salmon industry worldwide.

The origin of salmonid aquaculture dates back to the late 1700s in Europe (Folsom et al. 1992). As shown in Table 1, the first hatchery propagation of Pacific salmon (*Oncorhynchus* spp.) was developed in Canada in 1857 (Bardach et al. 1972). Salmon hatchery techniques were adopted in the USA soon after 1857 and were introduced to Japan in 1877, when the first national hatchery was built in Chitose, Hokkaido Island. However, it was not until the 1950s that hatchery-based enhancement programs were introduced on a significant scale. The Japa-nese Aquatic Resources Conservation Act, enacted in 1951, stimulated the growth of chum (*O. keta*), pink (*O. gorbuscha*), and cherry (*O. masu*) salmon ranching in Japan (Nasaka 1988). Salmon enhancement programs were also growing in the USSR, the USA, and Canada. However, prior to 1960 in the USA and Canada, most of the salmon harvest came from natural stocks. Most of the growth of enhancement programs in North America occurred during the 1970s and 1980s.

The following sections present a global perspective of the substantial transitions in the salmon industry that began to emerge in the 1970s and continue today.

TABLE 1.—Milestones in the salmon aquaculture industry.

1857	First hatchery propagation of Pacific salmon
1950s-1960s	USSR, Japan, USA, and Canada enhancement programs
1960s	Norwegian salmon aquaculture emerged
1974	Private, for-profit salmon ranching starts in Oregon
After 1976	Japan chum enhancement increases rapidly
1979	Norway, USA, Canada, Chile, Japan, and Scotland have emerging salmon farming industries.
Late 1970s-1980s	North American and Japanese enhancement programs grow significantly
1980	World farmed salmon production accounts for about 1% of world salmon supply
1983	World farmed salmon production exceeds world wild and ranched chinook salmon harvest
1986	World farmed salmon production exceeds combined world wild and ranched chinook and coho salmon harvest
1990	World farmed salmon production exceeds combined world wild and ranched chinook, coho, and sockeye salmon harvest
1991	World farmed salmon production exceeds Alaskan salmon harvest (all species)
1992	World farmed salmon production accounts for ~32% of world salmon supply; all U.S. private, for-profit salmon ranching has failed
1990s	Farmed salmon will constitute an increasing share of world supply; increasing market development with farmed salmon as the leader; Atlantic salmon dominates pen-raised production. Increasing criticism of salmon enhancement programs. Source: Food and Agriculture Organization of the United Nations (FAO 1991a, 1993).

Growth of Salmon Enhancement and the Emergence of Salmon Farming: 1970–79

By the beginning of the 1970s, the USSR led the world in the stocking of pink and chum salmon (Bardach et al. 1972). Japan was not far behind, and with the enactment of the 200-mile-limit fishing zones and other constraints to high-seas salmon fishing, Japan stepped up its aquaculture efforts. By 1980, the hatchery-based salmon (mostly chum) harvest was 74,397 metric tons (mt), representing more than 45% of Japan's total salmon supply.

In 1971, Alaska created the Fisheries Rehabilitation, Enhancement, and Development Division, and, in 1974, authorized nonprofit hatcheries. In 1973, Alaska created the first comprehensive limited entry program in the USA, and 1975 was the first year of fishing under limited entry (Orth 1981). The limited entry fishery contributed to interest in nonprofit hatcheries and the first nonprofit hatchery harvests occurred in 1977 (Orth et al. 1981). Hatchery production was also becoming a greater factor in the lower 48 states, and by the end of the 1970s, much of the harvest of coho (*O. kisutch*) and chinook (*O. tshawytscha*) salmon in Washington, Oregon, and California was primarily dependent upon hatchery-released salmon. For example, U.S. coho hatchery production in the Oregon Production Index area (south of Ilwaco, Washington, through California) accounted for less than 10% before 1960, approximately 53% by 1969, and, in 1979, accounted for about 75% of the total production (Oregon Department of Fish and Wildlife 1982).

In the USA, private salmon ranching was attempted in California (only one salmon ranch received a permit) and Oregon. Anadromous, Inc. (started in 1974; controlling interest purchased by British Petroleum), and Oregon Aqua-Foods (started in 1974; purchased by Weyerhaeuser in 1975) were the most significant operations in Oregon under way by 1980 (R. Mayo and C. Brown, The Mayo Associates, Seattle, Washington, unpubl. ms.).

While public salmon enhancement programs were growing, private pen-raised salmon began to emerge throughout the world. In 1965, A/S Mowi planned production of Atlantic salmon (*Salmo salar*) in sea enclosures on the Norwegian coast, which was influenced by early trials conducted with rainbow trout (*O. mykiss*) by the Vik brothers. In 1969, the Grønvedt brothers began growing salmon on the Island of Hitra, Norway, in floating net pens (Edwards 1978). By 1972, there were five farms producing a total of 46 mt in Norway, and by 1980, there were 173 farms producing a total of 4,300 mt (Heen et al. 1993).

Although Norway took the lead in the production of pen-raised salmon, the industry was also developing elsewhere. In Japan, Nichiro Fisheries Company started the culture of sockeye (*O. nerka*), chinook, chum, and pink salmon. By 1973, Nichiro had focused on pen-raised coho salmon, modeled after Norway's use of eggs imported from Washington and Oregon (Nasaka 1988). By 1980, Japan reported production of 1,855 mt of pen-raised salmon (Japan Marine Products Importers Association 1977–88).

In Scotland, Marine Harvest, Ltd. (a subsidiary of Unilever at the time), started operations in 1968 and had its first harvest in 1972 (Marine Harvest International, Inc., 1994). Growth was slow through the 1970s, and by 1980, production was approximately 600 mt (Heen et al. 1993).

The western USA took the lead in the development of pen-raised salmon in North America. This began in 1969 with National Marine Fisheries Service experiments at the Manchester Field Station in Washington State. In 1971, Ocean Systems, Inc. (later Domsea, a subsidiary of Campbell Soup Co.), started coho and chinook cage systems and began producing farmed salmon (Sylvia 1989). By 1980, western U.S. salmon production had reached an estimated 391 mt. In 1972, the pen-raised salmon industry started in British Columbia with surplus eggs from a government hatchery (Folsom et al. 1992) and remained essentially undeveloped, producing only 39 mt by the end of 1979 (Heen et al. 1993).

In 1978, the first significant pen-raised salmon aquaculture operation started in New Brunswick, Canada, on Deer Island and, by 1979, it had produced 6.3 mt. In Maine, several salmon culture operations were attempted in the 1970s. In 1970, Maine Salmon Farms began producing coho at a pen site in an estuary of the Kennebec River; however, this company failed in the late 1970s. Fox Island Fisheries, which started production in 1973–74 in Vinalhaven, Maine, was probably the first truly marine salmonid operation on the U.S. east coast; however, it went out of business in 1979 (see Bettencourt and Anderson 1990 for a more thorough discussion of the history of pen-raised salmon aquaculture development on the U.S. east coast). Despite these failures, by the early 1980s, the U.S. and Canadian east coasts were in position to experience reasonable growth.

Throughout the 1970s, several experiments with private salmon ranching were undertaken in Chile. The first commercial hatchery was the Sociedad de Pesqueria Lago Llanguihue, Ltd., which began operation in 1975. In 1979, cage culture operations started with Nichiro Chile, Ltd.'s, efforts to raise coho salmon. In 1979, operations by the Sociedad Pesquera Mytilus, Ltd., also began producing coho (Mendez and Munita 1989).

By the end of the 1970s, only Norway had established a pen-raised salmon industry of any significance. In 1980, world salmonid mariculture production was approximately 13,321 mt (4,778 mt of Atlantic salmon, 2,371 mt of Pacific salmon, and 6,172 mt of pen-raised trout, primarily *O. mykiss*). In addition, Japan was greatly expanding its hatchery-based chum fishery at this time. As the 1980s began, world aquaculture producers had developed the potential for tremendous growth.

The Rise of Pen-Raised Salmon Farming: 1980–88

The 1980s ushered in the rise of pen-raised salmon aquaculture. Although worldwide pen-raised salmon accounted for slightly more than 1% of total salmon supply in 1980, the technology for pen-raised salmon was reasonably well developed in Norway, and the industry was poised for rapid growth. Between 1980 and 1987, world production of pen-raised salmon (led by Norway) increased over thirteenfold (Figure 1). In addition to Norway, Scotland, Chile, Canada, and the USA, noticeable production began in Ireland, the Faroe Islands, New Zealand, and Australia.

In the early 1980s, the North American salmon industry, notably the Alaskan industry, was relatively indifferent to the ramifications of the developments in Norway and elsewhere around the world. Although many in Norway believed that worldwide production of farmed salmon would eventually exceed Alaska's entire production by the end of the 1980s, few in the U.S. salmon

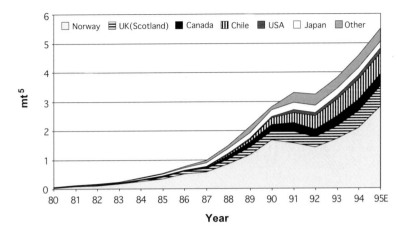

FIGURE 1.—World farmed salmon production by country, 1980–95 (1995 is an estimate). Source: Folsom et al. (1992), 1980–90 Pacific and Atlantic salmon; Food and Agriculture Organization of the United Nations (FAO) (1991b, 1994), 1991–92 Pacific salmon; Kontali Analyse A/S, No. 3 (1995a), 1991–95 Atlantic salmon and 1993–95 Pacific salmon.

industry seriously considered these claims. In fact, by 1990, farmed salmonid (trout and salmon) production of approximately 314,688 mt exceeded Alaska's entire salmon harvest. In 1991, the Alaska Seafood Marketing Institute (ASMI) stated that the "farming explosion last decade was unpredictable" (ASMI 1991, p. 15). In the early 1980s, the primary concern of those in the traditional salmon fishery focused on fisheries management and policy and, in the USA as the 1980s progressed, on activities to develop nonprofit ranching operations.

During the late 1970s through the mid-1980s, the Alaskan salmon industry moved a greater proportion of harvest away from canning to frozen salmon. During the second half of the 1970s, Japan reduced its high-seas fishing activity as a result of international agreements and 200-mile limit policies. Since 1976, as a result of these policies, Japan's dependence on imports increased. As Japan and Europe demanded more salmon (mostly frozen) from the USA, domestic production shifted from canned to fresh and frozen (Figure 2). Typically, over 70% of the Alaskan harvest was canned in the early 1970s. However, by the mid-1980s, the proportion of salmon for canning had dropped to the 30% range. During much of the 1980s, most fresh and frozen salmon from the USA was exported, primarily to Japan. Given the emphasis placed on the exportation to Japan, the Alaskan industry did not undertake major efforts to develop markets in the USA, which created an opportunity for the farmed salmon industry.

In the mid-1980s, imports to the USA accelerated rapidly (Figure 3), led by farmed salmon from Norway. Initially, the primary market was "white tablecloth" restaurants in the northeastern USA, but markets were soon developed throughout the country (Riely 1986). Although the USA was the dominant trader in the world salmon

market in the early 1980s, top-quality, fresh salmon from Norway and other regions began to displace wild salmon from Europe and most of the eastern USA by the latter half of the decade. The ability to produce sufficient quantities of fresh, farmed salmon year-round for large markets was achieved by Norway and other countries by

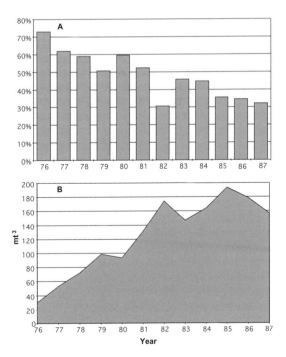

FIGURE 2.—(A) Percentage of Alaska salmon harvest that is canned; (B) Japanese imports, by weight, of frozen U.S. salmon. Source: (A) National Food Processors Association (1976–87), (B) Japan Marine Products Importers Association (1977–88).

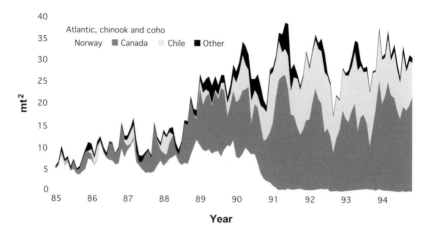

FIGURE 3.—United States imports of fresh and chilled Atlantic, chinook, and coho salmon, 1985–94. Source: National Marine Fisheries Services (1985–94).

the second half of the 1980s. As a result of considerable marketing efforts by Norway, and the growing economies in North America, Japan, and Europe, farmed salmon was absorbed by the market at relatively high prices (Figure 4).

By the mid-1980s, pen-raised salmon aquaculture was rapidly being established in Scotland, Ireland, Canada, Chile, and other regions. By 1983, total world farmed salmon (all species) production exceeded the world wild and ranched chinook salmon harvest. In 1986, total world farmed salmon production exceeded the combined world wild and ranched chinook and coho harvest.

In the USA, however, the industry met with a complicated and unclear regulatory environment coupled with frequent opposition from adversarial user groups. For

example, in June 1987, Alaska imposed a temporary moratorium on private, for-profit, farmed salmon and trout, which eventually became permanent in 1988. Although reasons given for this included environmental concerns, spread of disease, pollution issues, and genetic degradation of native stocks, other prominent motivating factors for the permanent moratorium were economic, such as market competition and concern about multinational corporations controlling the industry. In contrast, nonprofit aquaculture (enhancement) was growing rapidly in Alaska despite similar biological and genetic issues. For example, by 1988, hatchery-based pink salmon represented approximately 86% of Prince William Sound's harvest, up from less than 3% in 1979 (Brady and Schultz 1988) (Figure 5).

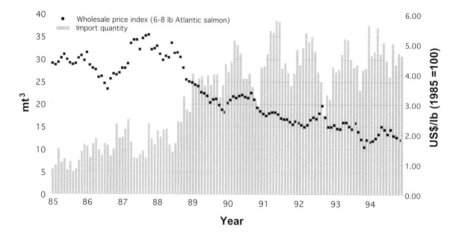

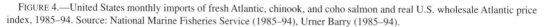

FIGURE 4.—United States monthly imports of fresh Atlantic, chinook, and coho salmon and real U.S. wholesale Atlantic price index, 1985–94. Source: National Marine Fisheries Service (1985–94), Urner Barry (1985–94).

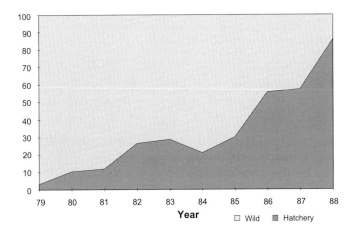

FIGURE 5.—Percentage of hatchery vs. wild pink salmon in Prince William Sound, Alaska, harvest, 1979–88. Source: Brady and Schultz (1988).

On both the east and west coasts of the USA, environmentalists (frequently local property owners), along with members of the commercial fishing sector, were often instrumental in stopping aquaculture projects. In Oregon, salmon ranching was heavily protested, primarily by the commercial fishing sector, even though private hatchery returns contributed substantially to the salmon harvest. This was a significant factor in the demise of these operations in the early 1990s. As early as 1983, researchers indicated that if Oregon's private salmon ranchers could not receive compensation from commercial and sport harvesters—or if restrictions were not placed on commercial and sport harvest—of privately ranched salmon, the industry would not survive beyond the early 1990s (Anderson 1983).

One of the few areas in the USA that generally favored salmon aquaculture in the early 1980s was the Eastport–Lubec region in Maine, partially owing to a high unemployment rate and the decline of the herring fishery. The most significant operation to emerge in this area was Ocean Products, Inc. (OPI), which began operation in 1982 with smolts from a Canadian hatchery and continued operation in its second year with 100,000 smolts acquired from the U.S. Fish and Wildlife Service. Ocean Products, Inc., soon became the largest salmon farm in the USA (Anderson and Bettencourt 1992).

As the pen-raised salmon industry emerged in Washington and Maine, there frequently was local resistance. In addition, new and changing regulations regarding such issues as discharge, marine mammals, navigation, disease control, feed additives, and migrating birds raised the cost and continued to erode the competitiveness of U.S. operations. The USA accounted for only about 2–3% of world production by the end of the 1980s, despite

the fact that it developed much of the extant hatchery technology and nutritional requirements for salmon and trout.

Price Declines and Restructuring in the Salmon Industry: 1988–93

The years 1988 and 1989 were remarkable for the salmon industry. Beginning in 1988, the farmed salmon industry increased production substantially, which contributed to downward pressure on prices (Figure 4). Falling prices were first observed in Europe late in 1987 and in the USA by mid-1988. In 1989, record supplies of farmed salmon (45% higher than 1988 levels), in conjunction with a record wild and ranched salmon harvest, led to an all-time record supply of salmon (Figure 6). A world supply of more than 1 million mt contributed to declining salmon prices worldwide. By the late 1980s, the salmon industry had become a truly year-round and globally competitive industry. By 1988, farmed salmon held the dominant market share of fresh and frozen salmon in Europe. In the USA, imports of fresh salmon more than doubled between 1988 and 1989. Even in Japan, pen-raised salmon accounted for approximately 90% of fresh imports and 11% of frozen imports (Kusakabe 1992). While imports in 1989 accounted for about 41% of Japan's supply, over 5% was derived from its domestic pen-raised salmon industry, and nearly 45% came from its hatchery-based chum fishery (Figure 7). By 1989, pen-raised salmon harvest accounted for over 20% (23% including pen-raised trout) of world production (Figure 6), and over 40% of world trade (Figure 8).

Over the past several years, bankruptcies, divestitures, and producer concentration have been commonplace in

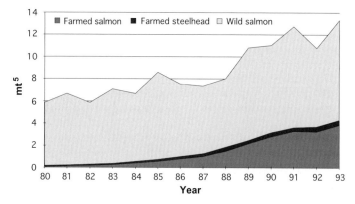

FIGURE 6.—Annual world harvests of wild and farmed salmon, 1980–93. Source: FAO (1991b, 1994), 1980–92 farmed and wild salmon; Alaska Department of Fish and Game (1994), 1993 wild salmon; Canada Department of Fisheries and Oceans (1994), 1993 wild salmon; Kontali Analyse A/S, No. 2 (1995b), 1993 farmed salmon.

the salmon industry. In the USA, price declines in 1989 precipitated a petition from the Coalition for Fair Atlantic Salmon Trade (FAST), led by OPI, which alleged that Norwegian producers had received countervailable subsidies[1] and were dumping[2] salmon in the USA, materially damaging the domestic industry. The U.S. International Trade Commission made a preliminary ruling on September 26, 1990, that Norwegian farms were dumping salmon. The final ruling was made on February 25, 1991, indicating that the Norwegians were selling below fair market value. This resulted in a countervailing duty of 2.27% and an antidumping duty that ranged from 15.65% to 31.81%, depending upon the company. The magnitude of these duties caused Norway to become uncompetitive in the U.S. market. As a result, Norway's share of imports sank to less than 5% by March 1991, and it has not changed much since then (Figure 3). From the time the petition was filed until the ruling, OPI sank into severe financial difficulty and was ultimately purchased by Connors Brothers, a subsidiary of George Weston, Ltd. (Weston also owns BC Packers).

A failed salmon freezing program that attempted to support prices, the U.S. duties levied on Norway, and similar actions in Europe resulted in the bankruptcy of many Norwegian firms as well as Norway's Fisheries Sales Organization. Between 1988 and 1990, bankruptcies, closures, or consolidations caused the number of

farms in British Columbia to decline from 150 to 118 (Folsom et al. 1992). Bankruptcies, divestitures, and consolidations were also common in Ireland and Scotland. For example, in October 1992, Unilever PLC sold Marine Harvest, the UK's largest salmon farm, to a U.S. firm, Marifarms (later renamed Marine Harvest International, Inc.).[3] The Scottish operations were restructured to increase efficiency and reduce cost (Marine Harvest International, Inc., 1994). In Maine, in addition to OPI's sale to Connors Brothers, Mariculture Products, Ltd., failed in 1992 and, in 1995, Maine Pride failed. Maine Pride's facilities were acquired by the Canadian firms International Aqua Foods, Ltd., Stolt Sea Farm, Inc., and Connors Brothers.

Despite declining prices, there was still tremendous growth in the farmed salmon industry. In 1990, world farmed salmon (all species) production exceeded the combined world wild and ranched chinook, coho, and sockeye harvest. Additionally, in 1991, world farmed salmon production exceeded the Alaskan salmon harvest (all species). Between 1988 and 1993, production of pen-raised salmon more than doubled (Figure 1). By 1991, pen-raised salmon accounted for over 50% of world trade (Figure 8), and by 1992, it composed approximately 32.5% (35% including pen-raised trout) of total world salmon harvest. By 1993, costs and other husbandry considerations resulted in Atlantic salmon becoming the preferred species for cage culture operations. Pen-raised Pacific salmon production declined as Canada and Chile increasingly switched to Atlantic salmon, and by 1993,

[1] To offset "material injury" to the U.S. industry caused by imported products tht receive certain subsidies from foreign governments, the U.S. government instituted countervailing duties, which are authorized under Section 701 of the Tariff Act (19 U.S.C. 1671).

[2] The antidumping provisions of the Trade Act (19 U.S.C. 160 & 1673) were developed to offset "material injury" to the U.S. industry caused by unfair price discrimination and below-cost sales.

[3] Marine Harvest International, Inc. (MHI), was traded on the American Stock Exchange until it was purchased by Booker plc in 1994. In addition to being the largest producer of Atlantic salmon in both Scotland and Chile, MHI is also involved with shrimp culture in Ecuador.

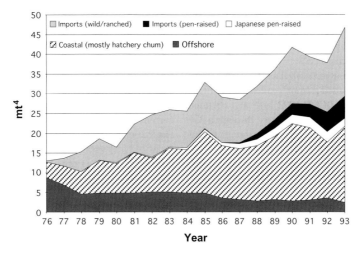

FIGURE 7.—Japanese salmon supply, 1976–93. Source: Japan Marine Products Importers Association (1977–88), Japan Ministry of Agriculture, Forestry, and Fishery (1977–94).

both British Columbia and Chile produced more Atlantic than Pacific salmon.

As Norwegian production was essentially eliminated from the U.S. market in 1991, Canadian and Chilean farmed salmon more than filled the gap. The growth of the global farmed industry and several record-breaking harvests in the wild and ranched salmon fishery (Figure 6) contributed to significant downward pressure on prices throughout the early 1990s. In the late 1980s and early 1990s, farmed salmon could be found nearly everywhere in the USA. It has been available daily in Seattle, Washington, for several years and is even commonly sold in Anchorage, Alaska.

The pressure on price since 1989 resulted in increased efforts to market salmon. Led by the farmed salmon industry, many efforts have begun to expand and broaden the market. New products such as portioned salmon fillets, microwaveable entrées, salmon medallions, salmon ham, and marinated salmon products have been introduced. Smoked salmon suppliers have been trying to broaden the appeal of smoked products. Brand naming is being attempted through gill tags and other labeling, and through enhanced service approaches. However, the vast majority of farmed salmon is sold fresh, whole, head on. Recently, Chile, British Columbia, the International Salmon Farmers Association, and others have all been working on generic marketing campaigns in the USA.

The emphasis on marketing in the farmed salmon sector has had a major influence on the wild salmon industry. A strategic planning report prepared by the Alaska Department of Commerce and Economic Development (ADCED 1993, p. 12) stated that "concerted action is urgently needed for the Alaskan salmon industry to regain its leadership position in the global marketplace."

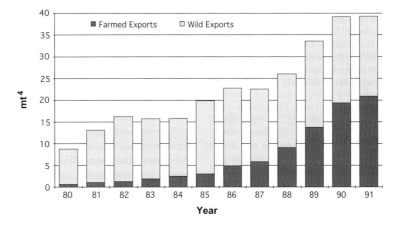

FIGURE 8.—Farmed share of world fresh and frozen salmon exports, 1980–91. Source: FAO (1991b, 1993).

as feeds from agriculture and manure from animal husbandry. However, fish culture is given a lower priority and is less accessible to these inputs, and large fish farms established in the late 1950s to raise table fish were usually in a deficit state, mainly because of the poor quality and high costs of feeds, thus resulting in low returns.

For such constraints to be overcome, it was evident that by combining fish culture with agriculture and animal husbandry activities, savings on feed could be attained, more employment opportunities and income could be generated, and a much more favorable and less-polluted environment could be established. An economically self-sufficient operation may be attained in which feeds and fertilizers are produced for its own use, thus obviating energy and feeds and reducing expenditures. Today, integrated fish farming practiced in pond fish culture has been further extended into open water bodies, such as lakes and small reservoirs.

Integrated fish farming in China is quite diversified. Apart from the simple to complex forms of integration, it is usually based on local socioeconomic conditions, availability of natural resources, climate, and geographic location. The most simple and popular type of integration is the combination of fish and crop farming. Other commonly practiced forms of integration are to combine pigs, ducks, or chickens with fish culture.

A rather direct and popular model of integrated fish farming practiced in China is the pig–grass (or vegetable)–fish integration. Under such a model, a portion of pig excreta is applied into the pond, most of it as fertilizer for the growth of high-yielding terrestrial grass or vegetation, such as rye grass, sudan grass, elephant grass, and Chinese cabbage. The grass is fed to the major stocked herbivorous fish species (grass carp). The water is then fertilized by the excreta of the grass carp, which produces the live feed for filter-feeding fish species. The silt of the pond is then used to fertilize the grass grown on land. It is reported that feces of 1 kg of grass carp may support 0.2–0.5 kg of filter-feeding silver and bighead carp through the fertilization of the water body. More complex integrated fish farming models are also practiced, such as cow–vegetable–mushroom–earthworm–duck–fish, fish–mulberry–sericulture and the aquaculture–agriculture–industry models.

Lake Fish Culture

China has about 2,800 lakes with areas exceeding 1 km^2, 124 lakes larger than 100 km^2, total lake area of 75,610 km^2 (Tu 1988), and total lake water area of 7,425,580 ha, of which 1,869,853 ha is suitable for aquaculture. By 1990, 615,770 ha were being used for aquaculture.

Extensive lake fish culture activities commenced in the 1950s when shallow lakes near towns and cities in the intermediate and lower reaches of the Yangtze River were stocked with fingerlings supplied by state fish farms. This practice was extended into medium and large lakes from the 1960s to the 1970s. Lake capture fisheries that previously relied on natural reproduction were transformed into extensive culture-based fisheries. However, the construction of water conservancy facilities, building of dikes to reclaim land for agriculture, industrial pollution, and intensified fishing resulted in a decline in lake capture fisheries.

Measures to enhance propagation and culture techniques and to conserve natural resources were proposed according to environmental conditions, productivity, and major constraints to fisheries in each lake. Management practices that were implemented consisted of extensive stocking, regulating fishing periods, controlling the number and types of fishing boats and gear, and establishing spawning environments. Although such measures contributed to increased production in open water bodies to some extent, fish production in large lakes seldom exceeded 40–50 kg/ha because of low primary production.

To further increase fish production in lakes, semi-intensive cage and pen fish culture was proposed. It has been proposed that 80% of the lake area be used for resource propagation, 15% for pen fish culture, and 5% designated as conservation area where no fishery activities are allowed. Cage and pen culture in lakes started in the 1970s. In the late 1970s, from experiences introduced from abroad, pen culture was carried out in many lakes in the Yangtze River Basin with encouraging results. The area of pen culture in lakes increased from 31,100 ha in 1985 to 70,000 ha in 1990. The average yield from present-day pen fish culture is 435 kg/ha. High yields of 12,000 and 25,380 kg/ha have been obtained through stocking of large-sized fingerlings and intensified feeding regimes. The size of pens varies from 0.15 to 0.73 ha; large pens, up to 1.86 ha, have also been reported.

Pen fish culture in lakes has also adopted an integrated approach that consists of two major components in the system: aquatic plants (*Trapa natans, Ipomaea aquatica*, and *Euryale ferox*) and fish (grass carp and blunt snout bream). Other species that have been stocked to a lesser extent are the common, black, silver, and bighead carp. This has been demonstrated as a successful combination in lakes in the lower reaches of the Yangtze River delta. For easy access, such aquatic vegetation is cultured near the pens, which also reduces the impact of waves on the pens. It is envisaged that pen fish culture has high potential in the management of large shallow lakes.

Reservoir Fish Culture

Compared with other fish culture practices conducted throughout China's history, mass-scale reservoir fish

- inappropriate management practices due to lack of funds in research, lack of technical expertise at grass-root levels, and the means and capability to cope with adverse situations; and
- pollution in coastal waters due to discharge of untreated industrial wastewater and domestic sewage.

Imports and Exports of China's Aquaculture Produce

The imports and exports of China's aquaculture products are not listed separately from other fishery products. Exports of fishery products in the 1950s and early 1960s consisted mostly of frozen fish exported to the former Soviet Union and Eastern Europe. By the 1980s, apart from the traditional markets of Hong Kong, Macau, and Japan, China's export of fishery products had entered into the countries of the European Community, northern Europe, and North America. To the present day, Hong Kong and Macau continue to be the major outlets for China's fishery products, representing about 60–62% of the total, followed by Japan with 35%.

During the 1950s, major export fishery products were traditional low-value, monotonous products such as live pond fish, frozen fish and dried cuttlefish, jellyfish, and salted fish. In the 1960s, exports consisted of frozen fresh fish, brine fish, frozen fillets, dried shrimp, and miscellaneous live aquatic products (swamp eel, loach, softshelled turtle, mitten crab, and sneakhead). By the 1980s, exports included live marine products (grouper, seabream, live eel, blue crab, lobster, and clams), glass eel, frozen clam meat, fish roe, and agar. The sources of these products are mainly from aquaculture and the processing industry. The export of the fleshy shrimp, mainly of capture species, started in the late 1950s. By the late 1970s, however, cultured shrimp constituted the bulk of shrimp exports.

The two major imports of fishery products continue to be fish meal from Peru and Chile, representing about 70–80% of the total, and frozen fish, mainly from North Korea, representing 15–25% of the total.

Faced with the pressure of a high population and the available terrestrial natural resources, domestic consumption of fishery products in the future shall no doubt continue to prevail in China, as the present-day per capita fish consumption in China is only 9.3 kg annually. Exports of fisheries products shall therefore be concentrated on high-value species group with high-value returns.

Conclusions

In the course of fisheries development from the 1950s to the 1990s, especially in the area of aquaculture, China has become the leading nation in world fisheries production. China is presently the sole nation where aquaculture production has surpassed capture fisheries, thus demonstrating that aquaculture can substantially increase production through rational use of available resources. However, this development and expansion was not uneventful.

In the 1950s, emphasis on aquaculture was proposed but was not effectively implemented for more than 2 decades. During this time, undue emphasis on agriculture, centralized monopoly on purchase and marketing, and egalitarianism in distribution all seriously hampered initiative by individuals, and the advantages of aquaculture could not be developed to any substantive extent. At the same time, intensifying marine capture fisheries resulted in overexploitation and depletion of natural fish stocks along the coast. It was under such circumstances that rational use of resources was put on the agenda for the future development of China's fisheries. Subsequent economic reforms completely restructured the management system.

In the future, apart from increasing the per unit yield in existing freshwater pond fish culture and developing aquaculture in underdeveloped areas, expansion of pond fish culture is possible only on lands unsuitable for agriculture. Moreover, to increase freshwater pond fish culture will require more intensified culture practices, with an integrated approach, combined with the overall renovation or reconstruction of existing ponds. With such limitations in pond fish culture, open-water bodies such as lakes and reservoirs hold high potential. Further use of these existing water bodies through stock enhancement as well as cage and pen culture holds promising prospects.

To overcome the shortage of feeds, China's aquaculture should continue to develop with an integrated approach by incorporating fisheries, agriculture, animal husbandry, and rural sideline occupations into one production system with available local resources.

Mariculture in China is directed to a few major cultured species in shrimp, mollusks, and seaweed. The expansion of the shrimp culture industry is most notable, and results are significant. But such development is not without consequences: poor management practices at the grass-roots level, water-quality problems, and shrimp diseases have resulted in heavy losses in certain areas of shrimp culture along the coast. Apart from environmental issues, such losses are believed to be due to lack of technology, expertise, and the application of standard shrimp culture techniques. More intensified research, training, and extension is urgently needed if China is to maintain its momentum in shrimp culture and prevent further deterioration.

Environmental issues that often are beyond the scope of aquaculture are also exceedingly important. Rapid

population growth and increased urbanization and industrialization have brought about increased socioeconomic prosperity, but have also greatly increased the discharge of industrial effluents and untreated domestic sewage. It is not surprising that red tides along the coast have occurred more frequently in recent years. Discharges of large quantities of untreated industrial and domestic wastes directly into rivers have become one of the major causes for adverse environmental conditions.

The seven large river systems that drain into the Pacific Ocean receive 22.4 billion m^3 of wastewater annually, 89.2% of which consists of industrial effluents. In 1992, the total wastewater discharge throughout the country was 36.65 billion mt (excluding discharges from township and village enterprises). The injection of such large amounts of waste is far beyond the self-purifying capability of the rivers.

The pollution of rivers and their associated water bodies will not only continue, it will intensify. It is estimated that by the turn of the century, the seven large river systems that drain into the Pacific will receive approximately 61.1 billion mt of wastewater annually. Although the water quality is currently acceptable, strict control measures accompanied with more investments in pollution control are required to curb the further deterioration of inland and coastal waters.

References

Anonymous. 1988. The "rice-azolla-fish" system. Food and Agriculture Organization of the United Nations, RAPA (Regional Office for Asia and the Pacific) Bulletin Vol. 4, Rome.

Chamberlain, G. W. 1997. Sustainability of world shrimp farming. Pages 195-209 in E. K. Pikitch, D. D. Huppert, and M. P. Sissenwine, editors. Global trends: fisheries management. American Fisheries Society Symposium 20, Bethesda, Maryland.

Edwards, P., et al. 1986. Pilot small-scale crop, livestock, fish integrated farm. Asian Institute of Technology research report 184, Bangkok, Thailand.

FAO (Food and Agriculture Organization of the United Nations). 1993. Aquaculture production 1985–1991. FAO Fisheries Circular 815, Rev. 5, Rome.

NACA (Network of Aquaculture Centres in Asia). 1989. Technical manual 7. Food and Agriculture Organization of the United Nations, NACA, Bangkok, Thailand.

Tu, F. J., et al. 1987. China fishery divisions. Zhejiang Science and Technology Publishing House, Hangzhou, Peoples Republic of China.

Sustainability of World Shrimp Farming

George W. Chamberlain

Abstract.—Shrimp farming has become a multibillion-dollar business that has attracted attention in many developing countries as a means of generating profits, local employment, and hard currency. Annual world production of farm-raised shrimp grew rapidly and steadily from a negligible level in 1980 to 799,000 metric tons (mt; live weight) in 1991. However, from 1992 to 1994, production sputtered between 789,000 and 891,000 mt, as increases in some regions were offset by declines in others. Although official Food and Agriculture Organization of the United Nations data are not yet available for 1995 and 1996, projections indicate that production declined because of a worsening disease epidemic in Asia. Although production hurdles exist, diseases have been the direct cause of recent production declines. Despite serious efforts to control them, diseases continue to plague the industry. The first sign of large-scale problems occurred in 1988, when Taiwan's production dramatically dropped. The cause of the widespread mortality was poorly understood. Another crippling decline occurred in 1993 in China. Much of Asia is now struggling with rapidly spreading viral epidemics. In some cases, extensive complexes of shrimp ponds have been abandoned owing to insurmountable disease problems. Nearly 20 shrimp viruses have been identified thus far, but the most threatening during the mid-1990s have been white-spot virus in Asia and Taura Syndrome virus in the Americas.

An increasing number of environmental and social issues have been documented as the shrimp aquaculture industry has grown and intensified. Regulations are tightening to address issues such as mangrove destruction, waste discharge, and overdevelopment. Concerns about the environmental sustainability of shrimp farming have become more pronounced in light of large-scale production failures. Disease and environmental issues are forcing shrimp farmers to reevaluate traditional management practices, which rely heavily on external resources such as healthy wild shrimp, clean estuarine water, and a large adjoining ecosystem to assimilate wastes. Farmers are being encouraged to rely less on uncontrollable resources and more on the fundamental disciplines of sanitation, animal health, genetics, nutrition, and sound management. This difficult transition is similar to that which traditional animal husbandry underwent long ago—from wild animals in a natural setting to domesticated animals in a more controlled setting. Promising new techniques are being developed to reduce water requirements, disinfect ponds, diagnose diseases, breed for disease resistance, improve feed efficiency, and predict the carrying capacity of ecosystems. These tools will pave the way toward greater production efficiency and a new phase of environmentally sustainable growth.

During the 1980s, world landings of shrimp from capture fisheries reached a plateau of approximately 1,900,000 metric tons (mt) per year while shrimp aquaculture began a period of rapid growth. World production of farm-raised shrimp grew dramatically from less than 84,000 mt in 1982 to 891,000 mt in 1994 (National Marine Fisheries Service [NMFS] 1992; Food and Agriculture Organization of the United Nations [FAO] 1996). However, widespread crop failures and environmental concerns have raised questions about the sustainability of shrimp aquaculture. Sustainability was defined by Hulse (1993) as "satisfying present needs without prejudice to the needs of the future." The following issues illustrate the major concerns challenging the shrimp farming industry.

Disease Outbreaks in Selected Countries

Taiwan

One of the early leaders in production of farmed shrimp was Taiwan, where 10,000 ha of small, family-owned, intensive farms produced 115,000 mt of *Penaeus monodon* in 1987. However, devastating shrimp mortality in 1988 dropped Taiwan's production to only 44,000 mt (NMFS 1992). The causes of mortality were attributed to industrial pollution, bacterial and viral diseases, and recirculation of pond effluents among farms. Repeated efforts to revive the industry in Taiwan resulted in collapses in 1992 and 1993 despite attempts to switch from *P. monodon* to alternate species such as *P. japonicus* and *P. penicillatus* (Chua 1993). In 1994, shrimp production in Taiwan was estimated to be 25,000 mt (Rosenberry 1994), and most Taiwanese shrimp farmers had switched to marine fish culture.

China

The 1988 crop failure in Taiwan had little impact on the world supply of shrimp because the shortfall was offset by rapidly increasing production from China. During the 1980s, China's production grew at an average rate of 80%/year. By 1987, China had become the world leader, with an annual production of 153,000 mt. In 1988, production increased to 199,000 mt, and some observers

expected China's production to continue climbing to 400,000 mt/year. However, production became erratic during the next 4 years owing to worsening water quality, red tide blooms, and sporadic disease problems in the primary shrimp production area on the Gulf of Bohai. This deterioration was attributed to increasing domestic, agricultural, and industrial pollution as well as self-pollution by organic material from the shrimp farms (Rosenberry 1990; Infofish 1994).

In 1993, China's production plummeted to 88,000 mt, down from 207,000 mt the previous year (Figure 1). This collapse caused the first decline in world production of farmed shrimp, and it resulted in a worldwide market shortage and record-high prices. Initially, the widespread mortality was blamed on pollution and red tide blooms. Researchers later discovered the causative viral disease, now known as white-spot virus based on the symptomatic presence of pinpoint to 1-mm whitish spots on the cuticle of some shrimp species (Huang et al. 1995). White-spot virus is a highly pathogenic disease that causes up to 100% mortality 2–3 d from onset. Epidemic mortality was first observed in China in the southern province of Fujian around June and July 1993. The virus moved north to the province of Zhejiang and Jaingsu, eventually reaching the Gulf of Bohai (Rosenberry 1994). Mortality affected farms in all kinds of systems and stocking densities ranging from 4 individuals per m^2 to more than $25/m^2$. In an attempt to manage risk, farms have decreased stocking densities to $3–4/m^2$ and switched species from *P. chinensis* to *P. japonicus*, which is more lucrative in the live market to Japan. Despite continual efforts to revive the industry, 1996 shrimp production in China was estimated as 80,000 mt, and China was a net importer of shrimp (Rosenberry 1996).

White-spot virus quickly spread from China to other Asian countries where it has been causing an explosive pan-Asian epizootic during 1994–97, affecting the following species: *P. monodon*, *P. chinensis*, *P. japonicus*, *P. indicus*, *P. merguiensis*, *P. penicillatus*, *P. indicus*, *Trachypenaeus curvirostris*, and *Metapenaeus ensis* (Nakano et al. 1994; Lightner et al. in press).

Thailand

Thailand has a long history of low-density shrimp production in coastal ponds. However, beginning about 1985, intensive culture techniques were introduced from Taiwan. Results were very encouraging, and intensive culture spread quickly. Rapid expansion was stimulated by the tax-free status of shrimp farming in Thailand and the role of major feed companies such as the Charoen Pokphand (C. P.) Group in providing a complete umbrella of support to small producers. The C. P. Group (1991) provided statistics on farms operating under different management regimes in 1991 (Table 1).

Most intensive farms in Thailand are small, family owned and operated, and have less than 2 ha of water surface. The typical intensive pond is 0.3–0.5 ha in surface area, 1.5–1.8 m deep and is equipped with aeration (Chamberlain 1991). It is managed with relatively high rates of water exchange, and its crop is fed 5 times/d. Yields range from 5–12 mt ha^{-1} $cycle^{-1}$.

Typical operation of intensive ponds relies on continuous exchange of organically loaded pond water with clean water from the estuary or ocean. The rapid proliferation of shrimp farms often results in recirculation of waste water among neighboring farms. This creates a stressful environment for shrimp, which is expressed as reduced rates of growth, food conversion, and survival. Recirculated waste water can also transmit diseases among farms and create an opportunity for outbreak of epidemics.

The problem of recirculating waste water was exemplified by a shrimp farming region in the northern Gulf of Thailand near Bangkok. Approximately 5,000 ha of intensive ponds were constructed in that area in a 2-year period. This was an unfortunate choice of sites because it receives much of the waste water from Bangkok through runoff from four rivers. Sediment deposition has created mud flats that extend into the gulf for 10–15 km at low tide. On the incoming tide, much of this sediment is resuspended and carried into shrimp ponds. Thus, farms in the area suffered high mortality and virtually the entire 5,000-ha development was abandoned. The industry simply relocated to new areas south of Thailand and growth has continued (K. Lin, Asian Institute of Technology, Bangkok, Thailand, pers. comm.).

In late 1992 and early 1993, the Thai shrimp farming industry began reporting widespread mortality in which one of the external symptoms was a light yellowish, swollen cephalothorax. The syndrome, known as yellow-head disease,

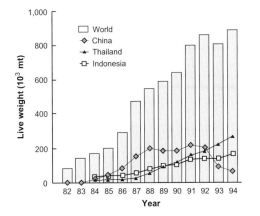

FIGURE 1.—World production (live weight, 10^3 metric tons [mt]) of farm-raised penaeid shrimp, 1982–94 (FAO 1996).

TABLE 1.—Statistics for size, number and productivity of Thai shrimp farms by management type, 1991 (C. P. Group 1991).

Management type	Number of farms	Area (ha)	Production (mt)
Extensive	2,587	22,000	7,555
Semi-intensive	3,909	25,163	14,274
Intensive	7,739	29,377	71,665

caused up to 100% mortality in 3–5 d in some farms. As a result, during the first quarter of 1993, shrimp production in Thailand dropped for the first time in 5 years.

Thai researchers identified the pathogen as a rod-shaped cytoplasmic virus with a single piece of ssRNA as its genome (Boonyaratpalin et al. 1993; Wongteera-supaya et al. 1995). They conducted bioassays with many pond organisms and determined that the natural carriers of the virus are small brackish-water shrimp, *Palaemon styliferus* and *Acetes* sp., which often reside in shrimp ponds in Thailand. They also determined that the virus is viable outside the host in seawater for 72 h.

On the basis of this information, a massive educational effort was instituted by the Thai Department of Fisheries. Farmers were instructed not to feed trash fishes, which often include palaeomonid shrimp. Filter socks were installed on inlet water systems to prevent entry of small shrimp. Farmers were instructed to hold incoming water in the reservoir for 72 h before use. If an outbreak did occur in a given pond, farmers were instructed to warn neighboring farms not to pump for 72 h after the infected pond was drained. After the initial widespread outbreak in 1992, subsequent outbreaks have been more sporadic and less acute. It is uncertain whether the subsidence of virulence was caused by management efforts or by changes in either the viral pathogen or the shrimp host.

White-spot virus, first observed in China in 1993, appeared in southern Thailand around November 1994, coincident with the rainy season. Mortality was severe until the end of January. Many researchers thought this virus would diminish like yellow-head virus. However, it appeared again in 1995 and in 1996, again during the rainy seasons.

Measures used in Thailand to prevent the spread of white-spot virus include careful screening of pond intake water to remove viral carriers, disinfection of pond water with chlorine or formalin, recirculation of pond water to minimize use of new water, and testing of postlarvae for white-spot virus using sensitive molecular diagnostics. In addition, the Thai government is attempting to reduce disease transmission by installing large offshore pumping stations in major shrimp farming regions. These pumping stations are designed either to draw clean oceanic water to the shore-based ponds or

to pump effluent from the ponds to the offshore site. Despite all these measures, shrimp production in Thailand is estimated to have dropped by 25–40% during 1996, and the long-term impact of white-spot virus is still unclear. If the white-spot epidemic expands to the Western Hemisphere, it could cause severe mortality in farm-raised *P. vannamei* and *P. stylirostris*, which are known to be susceptible.

Ecuador

Ecuador is by far the largest shrimp producer in the Western Hemisphere. Shrimp exports are the third largest income earner in Ecuador after petroleum and bananas. About 80,000 ha of shrimp ponds are concentrated in the Gulf of Guayaquil area, and another 50,000 ha are spread along the coast. Ecuador's annual production increased steadily through the 1980s to a peak of 113,000 mt in 1992.

Beginning in late 1989, shrimp mortality occurred in farms located on the Guayas River estuary during an extended drought, which caused salinity to rise and nutrients to concentrate in the estuary. These unusual conditions favored bacterial growth, especially pathogenic *Vibrio* spp. Bacterial infection may have been facilitated by gregarine parasites, which break the epithelial lining in the intestine. *Vibrio*-infected shrimp swim in a disoriented fashion near the water surface. This behavior attracted large flocks of sea gulls above ponds with severe infections. Hence the syndrome was named the "Gaviota" (Spanish word for sea gull) syndrome. The epizootic peaked in severity in early 1990, then persisted sporadically until early 1992 (D. Lightner, Dep. Veterinary Science, University of Arizona, Tucson, USA, pers. comm.).

A number of treatments were attempted to combat the Gaviota syndrome. Antibiotics and anticoccidials were added to feed in order to reduce bacterial and parasite infections. Some farms also tried treating their ponds with molasses to provide an alternate energy source to favor beneficial bacteria. None of the treatments were completely effective, but the onset of El Niño in 1993 brought heavy rains, which flushed the Guayas River estuary. Thus, the Gaviota syndrome disappeared.

In 1992, shrimp farms located near the Taura River about 25 km south of Guayaquil reported mortalities reaching 80–90% in some ponds. The disease, named Taura Syndrome (TS), dissipated for a few months, then returned in 1993 as a major epidemic affecting most of the farms in the Gulf of Guayaquil and some farms in northern Peru. Ecuadorian shrimp production declined from 113,000 mt during 1992 to 85,000 mt during 1993. Many shrimp ponds in TS-impacted areas were converted to tilapia monoculture or polyculture with shrimp (Chamberlain 1994).

The TS virus initially was considered to be caused by water-borne fungicides from the banana industry. However, it later was diagnosed as a viral disease (Brock et al. 1995). Retrospective histological studies showed that the TS virus was present in Ecuador as early as September 1991 (D. Lightner, Dep. Veterinary Science, University of Arizona, Tucson, USA, pers. comm.). In mid-1993, the TS virus was found in wild postlarvae collected near the mouth of the Gulf of Guayaquil, Ecuador, and in wild adults collected off the Pacific coast of Honduras, El Salvador, and Chiapas in southern Mexico (Lightner et al. in press). During 1994 to 1996, the TS virus quickly spread to all shrimp-producing countries in the Western Hemisphere, except Venezuela, where few farms import broodstock or larvae. Transmission of the disease is not completely understood, but both a winged aquatic insect and sea gulls have been documented to carry viable viral particles in their gut (Lightner et al. in press).

Despite the continuing presence of the TS virus, high market prices during 1995 drove Ecuador's shrimp industry to record production of 115,000 mt (Rosenberry 1996). By mid-1996, many growers reported gradually improving survival rates. Some speculate that Ecuadorian shrimp populations may be developing resistance to the TS virus. This pattern is not yet evident in other countries affected by this virus.

Philippines

As of mid-1996, white-spot virus had not been reported in the Philippines. However, production of shrimp is plagued by luminescent *Vibrio* spp., which are present almost all year. Farmers are treating it with an antibiotic, furizolidone, in the feed. Typical production results are not as good as earlier years: feed conversion ratio (FCR) of 1.9–2.2, survival of 40%, postlarval (PL) cost of $25–30/thousand shrimp, size at harvest of 31–32.5/kg. Many shrimp farms are switching to tilapia or milkfish culture (J. Vargus, First Philippine Holdings Corporation, Manila, pers. comm.).

United States

With the recognition of the devastating effects of viral pathogens, several attempts have been made to use viral-free stocks for cultivation. Marine Culture Enterprises, a commercial operation in Hawaii, stocked *P. stylirostris* in their super-intensive greenhouse-covered raceways. After a devastating outbreak of infectious hypodermal and hematopoietic necrosis (IHHN) virus in their research and development facilities in 1983, the farm instituted rigorous sanitation and quarantine procedures. However, a second outbreak of IHHN virus occurred in 1987, causing serious losses that ultimately resulted in farm closure (Mangiboyat 1987).

The species *P. vannamei* is more resistant to IHHN virus than *P. stylirostris*. Nevertheless, IHHN virus infects *P. vannamei* and causes "Runt Deformity Syndrome" (RDS) in which a portion of the affected population displays reduced growth, highly variable size classes, and sometimes reduced survival. Severe cases of RDS can reduce the productivity and profitability of farms by 30–50% (Kalagayan et al. 1991).

The U.S. Marine Shrimp Farming Program began isolating populations of specific-pathogen-free (SPF) *P. vannamei* in early 1988 as a seed source for the U.S. shrimp farming industry. The offspring from these populations, referred to as "High-Health" shrimp, performed well in commercial farms in the USA during 1991 and 1992. During 1993, High Health stocks were supplied to 400 ha of commercial ponds in Ecuador, but survival averaged less than 15% owing to an outbreak of the TS virus (Pruder et al. 1995). During 1995 and 1996, most shrimp farms in the USA also were hit by an outbreak of the TS virus despite their use of High-Health *P. vannamei*. Survival rates ranged from 15–30%. Thus, the SPF approach does not seem practical when the farms using those stocks are unable to control entry of pathogens (Lightner 1996).

Risk of Disease Transmission

Diseases are a primary limiting factor for shrimp farming today, and the risk of disease losses is likely to worsen as the industry continues to grow. Nearly 20 shrimp viruses have been identified thus far (Lightner et al. in press). Many of these began as localized pathogens but quickly spread to new regions (Lightner and Redman 1991). Once a new disease becomes established in wild aquatic populations, there is little prospect of extracting it, and it becomes another management hurdle for local farmers. With regard to highly virulent shrimp viruses, this has been an insurmountable hurdle in some cases. It is in the common interest of shrimp farmers and resource regulators to prevent entry of exotic diseases.

In Asia, the most serious viral diseases are caused by white-spot and yellow-head viruses. Both viruses have been shown to infect American penaeids (e.g., *P. vannamei, P. stylirostris, P. setiferus, P. aztecus,* and *P. duorarum*) and to cause serious disease (Lightner in press). White-spot virus also causes mortality in nonpenaeid, freshwater and marine crustaceans, including *Macrobrachium* spp., *Procambarus clarkii,* and *Ocronectes punctimanus.* White-spot virus infects but does not cause significant disease in a variety of crab and spiny lobster (Chang et al. in press; Wang et al. in

press). In Latin America, the most serious viral disease is the TS virus. The susceptibility of Asian species to the TS virus is unknown. It is, nonetheless, critical to prevent the establishment of Asian viruses in the Western Hemisphere and vice versa.

The most common means of shrimp disease transmission has been direct transfer of infected animals. Uncontrolled shipments of broodstock or postlarvae from infected farms or wild populations have quickly expanded the range of several viral diseases (Lightner and Redman 1991; Lightner 1996). Once a disease enters a new area, it can spread quickly through recirculation of effluents from or to neighboring farms; through live carriers such as crustaceans, birds, or insects; or through contamination of equipment, vehicles, or people.

Farmers observing the onset of a serious infection in ponds commonly conduct an emergency harvest to avoid a total loss. Infected shrimp, which pose no threat to human health, are processed and marketed through normal channels of distribution. Viruses are known to remain viable in frozen tissue for an extended time. Viable whitespot and yellow-head viruses from Asia have been detected in frozen shrimp at retail seafood outlets in the USA (Lightner et al. in press). Thus, an insidious vector for worldwide transmission of shrimp viruses is through international trade in frozen shrimp.

There are many ways in which viral material in frozen shrimp could reach susceptible populations. For example, a major portion of the shrimp entering the USA from Asia is not in ready-to-eat form. All wastes produced in the USA during processing of those shrimp at breeding plants, restaurants, retail outlets, and home kitchens are potential disease vectors. Another vector is undersized, frozen shrimp, which is sold in the USA as bait for recreational fishing (Lightner 1996).

In November 1995, white-spot virus and a possible co-infection of yellow-head virus were detected in a population of *P. setiferus* at a south Texas shrimp farm (Lightner et al. in press). The diseases had not reappeared as of January 1997 (P. Frelier, Dep. Veterinary Pathobiology, Texas A&M University, College Station, Texas, pers. comm.). A second case of white-spot virus, which again appeared with a possible co-infection with yellow-head virus, was detected at the Waddell Mariculture Center (Bluffton, South Carolina, USA) during January 1997. The origin and extent of the latter infection were unknown at the time of this writing (D.V. Lightner, Dep. Veterinary Science, University of Arizona, Tucson, Arizona, USA, pers. comm.). White-spot viral infections were also diagnosed in North American crayfish, *Procambarus* spp. and *Ocronectes punctimanus*, being held at the U.S. National Zoo (Lightner et al. in press). The cases illustrate the serious risk of introducing epidemic Asian viruses to the Western Hemisphere and vice versa. Tighter regulations are needed to assure proper disposal of wastes from frozen imported shrimp.

Other Risk Factors

Wiley (1993) categorized and prioritized the risks facing shrimp aquaculture from the viewpoint of an insurance company, Sedgwick, James, Ltd. The most important factor influencing the success of a farm, in Wiley's view, was the competency of personnel and management. Security, biophysical environment, and farm design were also considered key. Wiley recommended that sites with excess nutrient enrichment and pollution should be avoided. Preferred sites should be located in predominately rural areas with no surrounding industrial influences and reasonable space between farms. The only topic in which Wiley (1993) listed a specific value was stocking density. Sedgwick, James, Ltd. considers production levels above 8 mt/ha to be very risky and Wiley indicated that no insurance would be granted to a farm exceeding this production level. Shrimp price volatility was also considered a substantial risk to the shrimp producer. Futures and options are now available to protect growers from price fluctuations. The Minneapolis Grain Exchange initiated this program in July 1994, but the degree of acceptance and use of this program by the commercial industry is uncertain.

Environmental Issues

Discharge of Organic Wastes

Traditional shrimp pond management involves continual input of clean water and discharge of waste water. Water exchange is thought to improve water quality by flushing out waste products, avoiding excessive eutrophication, and maintaining a healthy plankton bloom. In extensive systems, it also may have value in adding additional forage prey. As a function of stocking density, average daily water exchange rates typically vary from 1–5% in extensive systems to 25–30% in intensive systems (Clifford 1985).

Nutrient budgets for intensive and semi-intensive shrimp ponds indicate that only 6–24% of the nitrogen and 4–13% of the phosphorus input is incorporated into harvested shrimp (Briggs and Funge-Smith 1994; Robertson and Phillips 1995). The remainder is retained in the pond water and sediments. These nutrients are exported to the environment during routine water exchange, harvesting, and sediment disposal. However, the receiving water body can assimilate only a limited quantity of nutrients, known as the critical load, before water quality and ecosystem diversity begin to change. Robertson and Phillips (1995) estimated the amount of mangrove area needed to assimilate the total nitrogen and phosphorous load generated by shrimp ponds as a result of water exchange, drainage during harvesting, and sediment disposal was 2–3 ha for each ha of semi-intensive pond area and 7–22 ha for each ha of intensive pond area.

In some cases, shrimp farming developments have clustered in certain regions, resulting in nutrient discharge well beyond the critical load of the receiving body. The scenario of overdevelopment often begins with successful results from a single pioneer farm in a given region. This leads to rapid, unplanned development of additional farms. As neighboring farms pump and discharge water to a common estuary, water quality begins to deteriorate, shrimp become stressed, and disease organisms are transmitted among farms. This type of development is vulnerable to mass mortality of shrimp, and hence is not sustainable.

One approach used to avoid such problems is to regulate the quantity and quality of effluents from existing farms. However, experience has shown that a proactive approach can be more effective by avoiding problems before they arise. This involves integrating aquaculture resource allocation within a broader system of resource planning that considers the needs of a variety of resources and resource users. For example, Canada has developed an application system for leasing coastal areas to salmon growers that involves evaluating the capability of the proposed site to assimilate the organic load of the farm. In addition, a minimum spacing of 3-km is required between salmon farms located within a single enclosed water body (Black and Truscott 1994).

Mangrove Destruction

The rapid growth of shrimp farming worldwide has resulted in the construction of new ponds in many coastal areas. The preferred environment for pond construction is salt flats, which are relatively unproductive and easy to develop. In the early days of shrimp farming, mangrove areas also were considered suitable sites for pond construction on the presumption that this was the environment where shrimp occurred in nature (Fegan 1996). However, experience has shown that mangrove areas make poor sites for shrimp ponds because their acid sulfate soils become extremely acidic (pH 3–4) when dried. Most farms now prefer to use land above the intertidal zone because it is more accessible to heavy equipment, more manageable, and more productive. Farm developers disturb the mangrove area only to construct an inlet canal for access to estuarine water. In some countries, even minor conversion of mangrove areas requires mitigation. Mitigation is an agreement between the devel-

oper and the concerned regulatory body in which the developer compensates for loss of mangrove habitat by creating a similar habitat nearby.

Large-scale destruction of mangrove areas can have serious ecological and social consequences. Mangroves are critically important as highly productive nursery areas for estuarine species, habitat for birds and mammals, buffer zones against storm events, stabilizing forces against soil erosion, and sources of revenue for poor coastal communities (Bailey 1988). Nevertheless, in the early rush to capitalize on the profitability of shrimp farming, local policy often overlooked or even encouraged conversion of mangrove areas to ponds (Bailey 1988). For example, in the Philippines, the Bureau of Fisheries and Aquatic Resources listed mangrove areas as "swamplands available for development" until 1984 (Primavera 1995).

Shrimp farming is rarely the main cause of mangrove destruction. It is estimated to have destroyed less than 5% of the global mangrove resource by 1988, but locally the impact may be more severe (Phillips et al. 1993). Aquaculture pond construction is estimated to have destroyed 20% of the mangrove forest in some parts of Ecuador (Snedaker et al. 1986). In Thailand, 34% of the cleared mangrove area is used for aquaculture ponds (Network of Aquaculture Centres in Asia and the Pacific [NACA] 1994). In Vietnam, 240,000 ha of coastal ponds already have been developed, largely in mangrove areas (C. P. Group 1994). In Indonesia, most of the 300,000 ha of habitat being used to culture shrimp was mangrove forest (Macintosh 1996).

Most shrimp-producing countries now recognize the value of mangrove areas and have regulations in place to protect them. However, they often lack the resources to monitor and enforce those regulations (Bailey 1988; Macintosh 1996). A good example is the "informal" shrimp farms in Ecuador (Fay 1995). These are usually small, poorly funded operations established without permission in intertidal mangrove areas because ponds can be constructed there without heavy equipment and operated with minimal funds and technology. In contrast, "formal" shrimp farms are larger and better capitalized units that are licensed by the government and required to avoid mangrove areas.

Several U.S. environmental groups have threatened to boycott imported farm-raised shrimp if mangrove destruction by the international shrimp farming industry is not stopped (Woodhouse 1996). The issue of mangrove destruction is complex and is the topic of considerable debate. It is complicated by the many different types of mangroves, which vary in their commercial, physical, and ecological value. Those mangrove areas along the coastal fringe are valued the most because they are thought to function as key nursery areas for the offshore fishery and as buffer zones against storms and erosion. The bulk of mangroves, which are located inland, are often assigned a lower ecological value (Hambrey 1996).

Social issues are also contentious. In most countries, legal ownership of mangrove areas is claimed by the state, even though local communities often depend heavily on exploitation of those resources. The transformation of such a multiple-user resource into a private aquaculture property can create social conflicts (Bailey 1988; Sebastiani et al. 1994).

In order to protect and manage mangrove resources more carefully, Hambrey (1996) recommended that coastal areas should be categorized relative to aquaculture into three zones: those unsuitable for development based on their high value for alternative uses, those highly appropriate for development, and other areas which may or may not be suitable. In those areas judged highly appropriate for aquaculture, Hambrey's (1996) economic analysis concluded that shrimp farming should be encouraged because of its high economic value relative to other competing uses "but only if the risks of failure can be reduced."

Pongthanapanich (1996) used linear programming to estimate the combination of uses that would yield the maximum net present value for mangrove areas in Thailand. The analysis excluded those mangrove areas within Thailand's conservation zone, which is kept as a natural forest. Pongthanapanich (1996) concluded that 61% of the area outside the conservation zone should be conserved in its natural condition, 17% should be designated for wood concessions, 10% for reforestation, 12% for shrimp farming, and none for plantations of rubber and palm oil.

Saltwater Intrusion

In some cases, the salt water used in shrimp ponds has impinged on neighboring agricultural areas such as rice fields. Salinization of agricultural soils makes them unproductive for most crops. This can occur through salt water seepage into the water table or through salinization of freshwater used for irrigation. Salinization of groundwater can affect scarce drinking water supplies in coastal villages. Salinization of shrimp ponds becomes an issue if ponds fail and require conversion back to agricultural production.

Increased Use of Therapeutants

The increased incidence of disease has led to increased use of various therapeutants, including antibiotics, anticoccidials, copper compounds, quarternary ammonium compounds, iodine, formalin, potassium permanganate, and malachite green. Although such treatments can be valuable tools in a health management program, they are sometimes used inappropriately or in excess. This results in ineffective treatment, financial loss, and potential contamination of natural waters and shrimp with residues. Clearly, improved methods are needed to diagnose, prevent, and treat shrimp diseases.

Other

Other environmental issues, which are not addressed in this paper, include dependence on wild postlarvae and reproductive adults, excessive pumping of groundwater, and land and water use conflicts.

Promising New Developments

The technology of shrimp aquaculture is advancing rapidly in the fundamental disciplines of animal health, sanitation, genetics, nutrition, and pond management. The challenge is to apply these new tools to overcome the hurdles now facing the industry.

Animal Health

Disease diagnosis.—Successful animal health programs are based upon appropriate prophylactic and diagnostic measures—rather than therapy—to detect early disease symptoms. Diagnostic technology has advanced rapidly in recent years with the application of methods such as DNA probes, enzyme linked immunosorbent assays, mono- and polyclonal antibodies, and polymerase chain reactions (PCR) (Bruce et al. 1994; Carr et al. 1996; Lightner et al. in press). Molecular techniques are especially critical for rapid, accurate, and sensitive diagnosis of viral diseases. The availability of these techniques has allowed researchers to study the biology and transmission of viral diseases to better understand means of managing them.

Very little tissue is required for molecular techniques, so they can be applied on a nondestructive basis to broodstock by hemolymph sampling or removal of a single pleopod. They also can be used to screen postlarval populations. In Thailand, many farms and government laboratories are now performing routine analysis of 5- to 8-day-old, hatchery-reared postlarvae for whitespot virus by using PCR followed by a dot blot test (T. Flegal, Dep. Biotechnology, Mahidol University, Bangkok, Thailand, pers. comm.).

In addition, sophisticated diagnostic tools are becoming available at the farm level through the use of commercially available DNA probes and diagnostic kits for viral (DiagXotics™, Inc., Wilton, Connecticut, USA) and bacterial pathogens (e.g., Radikit™, Disease Section, Freshwater Fisheries Research Centre, Melaka, Malaysia).

Thus, a test for IHHN virus, which previously required a 6-week bioassay followed by histology in a well-equipped lab, can now be accomplished in a simple laboratory in a matter of hours. Even faster tests are on the horizon. Human health researchers at the University of California (Berkeley, California, USA) recently developed polydiacetylene films that can detect target pathogens or their toxins instantly. The films, which are composed of highly ordered crystalline arrays coupled to antibodies, undergo mechanical disruption upon contact with the target antigen, which causes an instant color change (Coghlan 1996).

Therapeutants.—Necrotizing hepatopancreatitis (NHP) was diagnosed as the cause of high shrimp mortality rates in Texas ponds, which began in 1988 (Frelier et al. 1993). Since then, NHP has been diagnosed in Peru, Venezuela, Ecuador, and Panama (Lightner et al. 1992; Lightner and Redman 1994; P. Frelier, Dep. Veterinary Pathobiology, Texas A&M University, College Station, Texas, pers. comm.). The disease can be controlled through use of the antibiotic oxytetracycline in the feed. Oxytetracycline is not a U.S. FDA-approved medication for shrimp. However, shrimp farmers in the USA are able to use it through an FDA permit for an Investigational New Animal Drug (INAD). There has been no sign of resistance to this antibiotic (Frelier et al. 1994). However, treatments for other bacterial infections, such as *Vibrio* spp., are less satisfactory, and these bacteria are prone to develop resistance to the treatment. To avoid needless expense and danger of bacterial resistance to antibiotics, antibiotic treatments must be limited and judicious. They should not be prescribed for prophylactic use.

Immune enhancers.—Invertebrates cannot be vaccinated in the strict sense of the word because they lack an antibody-mediated immune response. However, they do respond to a variety of nonspecific immune enhancers. Itami et al. (1989) reported a reduction in mortality from 78.9% in controls to 29–36% in *P. japonicus* treated with formalin-killed *Vibrio* sp. Itami and Takahashi (1991) demonstrated that a reduction in mortality could be achieved by microencapsulating killed *Vibrio* cells and feeding them to larval *P. japonicus*.

Polysaccharides are being used to enhance the shrimp immune system. Sung et al. (1994) showed that beta glucans improved disease resistance and growth of *P. monodon* challenged by *Vibrio vulnificus*. Further research showed that beta glucans stimulated *P. monodon* hemocytes to increase production of reactive oxygen species, which are important in microbiocidal activity (Song and Hsieh 1994).

Other potential immune enhancers include elevated vitamins, selenium, and astaxanthin (Tacon and Kurmaly 1996). There is still much to learn about invertebrate immunity. Hemolymph lectins of shrimp are also recognized for their role in causing agglutination of foreign proteins (Fragkiadakis and Stratakis 1995). Recent research about inducible antibacterial peptides and primitive cytokines may lead to important disease-resistance treatments in the future (Beck and Habict 1996).

In pigs, probiotics have enhanced growth through colonization of the colon by microflora that block pathogenic microorganisms (Russell et al. 1996). A similar concept has been applied in shrimp larval culture tanks to control pathogenic bacteria. The probiotic concept involves intentionally seeding sterilized seawater tanks with beneficial bacteria to reduce the opportunity for colonization of pathogenic bacteria (Garriques and Arevalo 1995). This is an exciting alternative to using antibiotics for control of pathogens in larval culture systems.

Stress.—During viral epidemics, often certain farms seem to operate with minimal disease losses. Much of this advantage is attributed to low stress. For example, in a trial conducted in Thailand, shrimp were injected with a sublethal dose of white-spot virus and their growth in aquaria was compared to controls (T. Flegal, Mahidol University, Bangkok, Thailand, pers. comm.). There was no difference in growth or survival between the infected and control shrimp until they were subjected to stress, such as low dissolved oxygen (DO) and extreme pH. With stress, the infected shrimp demonstrated signs of disease and experienced high mortality. Control shrimp suffered only a temporary setback in growth. Thus, reducing the risk of disease outbreak in farms will require that culture systems be operated under optimal environmental conditions. Excessive stocking densities lead to accumulations of metabolic end products, to stress, and ultimately to heightened opportunity for disease.

Genetics

Genetic selection has been the cornerstone of animal and plant husbandry for many years. Specialized companies have been breeding chickens for meat consumption since about 1950. Over that time, the growth rate of chickens has increased by three- to fourfold while food conversion has simultaneously improved by 30%. In a study designed to evaluate the relative contribution of genetics versus nutrition to the improvement in broiler growth in the last 40 years, genetics was found to account for approximately 85–90% of the contribution while the remainder was attributed to nutrition (Haverstein et al. 1994). Genetic background has also been shown to influence disease resistance potential and immunocompetence in chickens (Ruff and Bacon 1984; Puzzi et al. 1990).

Atlantic salmon (*Salmo salar*) and carp (*Cyprinus* spp.) have demonstrated considerable improvement in disease resistance through genetic selection (Gjedrum and Fimland 1995). Invertebrates also have genetic potential to increase disease resistance. Gaffney and Bushek (1996) studied the resistance of oyster (*Crassostrea virginica*) to two parasites, *Perkinsus marinus* and MSX. They demonstrated that oysters originating from areas chronically infested by *P. marinus* were more resistant to infections with that parasite than oysters originating from non-affected areas. Furthermore, selective breeding studies increased resistance to either MSX or *P. marinus*. However, oysters selected for resistance to MSX had reduced resistance to *P. marinus*. Similarly, rainbow trout with improved resistance to *Aeromonas* sp. had significantly higher mortality when infected with *Vibrio* sp. (Fevolden et al. 1992).

Laboratory selection.—Early research on genetic disease resistance in shrimp is yielding encouraging results. The U.S. Marine Shrimp Farming Program is using specific pathogen free (SPF) *Penaeus vannamei* as a platform for selective breeding research (Carr 1996). Three series of trials have been conducted in which 60 full-sib families (30 maternal half-sib families) have been reared in each trial to a size of approximately 1 g. Specimens from each cross were distributed to cooperating laboratories to determine susceptibility to viral diseases based on *per os* challenge. Results indicated less than 1% survival for all families exposed to either white-spot or yellow-head virus. However, there was a high degree of variation in response to challenge by the TS virus. For example, mean family survival in the second trial ranged from 10% to 76.7%. This indicates significant heterozygosity and heritability ($h^2 = 0.22$). On the basis of this information, Carr (1996) anticipated a gain of 5–10% per generation in disease resistance to the TS virus. He reported no correlation between growth performance and resistance to the TS virus. Thus, selection for TS virus disease resistance should not affect growth performance.

Farm-level selection.—Farm-level breeding programs have reported improved viral resistance simply by rearing shrimp in the presence of viral pathogens for several generations. This strategy is based on mass selection of survivors versus non-survivors in areas with heavy disease pressure. It suffers the disadvantage of low selection intensity during periods of low disease pressure. Also, there is an associated risk of spreading pathogens with the transfer of stocks. Various groups have reported success with this approach:

- Weppe et al. (1992) reported that a population of *P. stylirostris,* called SPR43™, which was reared in captivity for more than 20 generations, is resistant to IHHN virus.
- Ricoa Agromarina, C.A. (Maracaibo, Venezuela) is marketing a population of *P. stylirostris* called Super Shrimp™, which was reared in captivity more

than 7 years and is resistant to IHHN virus (Wilkenfeld 1996).

- Aquamarina de la Costa (Caracas, Venezuela) reports that incidence of deformities thought to be caused by IHHN virus in captive populations of *P. vannamei* have decreased from initial levels of 35–45% to current levels of 3–10% due to rigorous selection over several years (Rosenberry 1995).
- Laboratory challenge tests at University of Arizona have confirmed field results indicating improved resistance to the TS virus of *P. vannamei* reared in captivity for 2–3 generations (D.V. Lightner, Dep. Veterinary Science, University of Arizona, Tucson, Arizona, pers. comm.).

Gaffney and Bushek (1996) suggested the following steps in implementing a farm-level selection program: (a) drawing founders from diverse sources, including areas with a long history of exposure to the pathogen; (b) culturing animals at known sites of heavy disease pressure; and (c) using appropriate breeding plans to increase disease resistance without compromising other production traits. This process may require years, but it can be accomplished with presently available technology.

Natural selection.—In addition, natural selection processes may be gradually increasing resistance of wild shrimp populations to certain viruses such as IHHN virus in Mexico, yellow-head virus in Thailand, and TS virus in Ecuador (Lightner 1996). The rate at which disease resistance develops would be expected to be a function of selection pressure. In other words, wild shrimp populations that are concentrated in areas strongly impacted by the disease would be expected to adapt faster than those which are more widely dispersed and include animals from noninfected regions.

Other approaches.—Disease resistance genes can be moved from one species to another through hybridization. At this point, only a few interspecies hybrids have been produced with penaeid shrimp, and these have had very low hatching rates. Another approach is to move the disease resistance genes from one species to another through genetic engineering. In mice, transgenic populations have been produced with disease resistance to specific pathogens. Similar research is contemplated for shrimp (Mialhe et al. 1995).

Pond Management

Water exchange.—Traditional shrimp pond management relies on relatively high rates of water exchange to maintain water quality, regulate plankton density, and introduce supplemental food organisms. This practice restricts the development of shrimp farming in areas with limited water availability, and it imposes a serious risk of disease introduction. It also has environmental and financial implications. Shrimp pond effluent represents a significant source of eutrophication for receiving waters. Briggs and Funge-Smith (1994) estimated that 22% of the total input nitrogen and 7% of the total input phosphorus of an intensive shrimp pond are released during routine water exchange. In addition, water exchange can cause mortality of impinged and entrained organisms during pumping. Financially, water exchange is costly in terms of pumping, maintaining predator screens, and removing sediments from supply canals and ponds (Boyd 1992; Peterson and Daniels 1992).

Until recently, little systematic research had been conducted to evaluate the need for water exchange in aerated systems. Browdy et al. (1993) demonstrated that daily water exchange in the range of 10–100% had little impact on growth or survival of *P. vannamei* in fiberglass tanks as long as acceptable DO levels were maintained. Allan and Maguire (1993) found that water exchange rates of 0–40% did not significantly affect performance of *P. monodon* stocked in plastic-lined pools at 20-40/m².

Hopkins et al. (1993) found that daily water exchange could be reduced from 25% to 2.5% in ponds stocked at 44/m² with no reduction in shrimp growth or production. In the absence of water exchange, biological oxygen demand (BOD), unionized ammonia, and nitrite of pond water tended to increase with increasing density. At a density of 22/m², shrimp performance was not hindered by lack of water exchange, but at densities of 44 and 66/m², water exchange was required to avoid mortality. Hopkins et al. (1993) cautioned that the assimilative capacity of ponds with zero water exchange and 20 hp/ha of aeration is approximately 70–140 kg of feed ha⁻¹ d⁻¹. Hopkins et al. (1995) demonstrated that fixed daily rations of 68–136 kg/ha could produce 5.8–8.2 mt/ha of *P. vannamei* with zero water exchange and 20–40 hp/ha of aeration.

Little systematic research on water exchange has been conducted in non-aerated ponds, where water exchange is used to control plankton density and thereby regulate DO levels. Nevertheless, with the onset of the TS virus in Ecuador, many farms reduced daily exchange rates from typical levels of 5–20% to 1–4% with no negative effects (L. Anderson, Morrison International, Guayaquil, Ecuador, pers. comm.).

Disinfection of inlet water.—Chlorine disinfection of pond water is being used by more farms in Asia as means of operating in areas affected by virulent pathogens and their carriers. The process involves treating pond water with approximately 30 ppm of hypochlorite solution (60% concentration), allowing a reaction time of 2–3 d, dissipating chlorine residue with aeration, and then using the pond for culture (C. P. Group 1994). In such disinfected systems, water exchange is reduced to a minimum because replacement water needed during the cycle must be disinfected in a separate pond before introduction into the culture pond. Boyd (1996) cautioned that it is not possible to recommend a standard chlorine dose applicable to all waters. The appropriate chlorine dose for disinfection depends upon the chlorine demand and pH of the water.

Harvest drainage.—Briggs and Funge-Smith (1994) estimated that 13% of the total input nitrogen and 3% of the total input phosphorus of an intensive shrimp pond are released during harvest drainage. Schwartz and Boyd (1994) found that 50% of the nitrogen, phosphorus, and BOD discharged during drainage of channel catfish (*Ictalurus punctatus*) ponds is released in the last 15–20% of effluent. The environmental impact of this waste can be greatly reduced by using sedimentation ponds to receive the final portion of harvest water (Chanratchakool et al. 1995).

Pond sediments.—Shrimp ponds can accumulate considerable quantities of sediment from suspended soil particles in inlet water supplies and from erosion of the pond walls and bottom. Two intensive shrimp ponds in Thailand accumulated an average depth of 7.5 cm of loosely consolidated sediment in a single 4-month production cycle (Boyd 1992). Such high rates of sedimentation can quickly diminish the working volume of ponds. Boyd (1992) recommended use of sedimentation ponds to remove sediments from incoming water before it is used in shrimp ponds. Erosion of ponds is caused mainly by aerators positioned around the periphery of the pond and by wind-driven waves. Smith (1996) estimated that erosion in aerated ponds for *P. monodon* accounted for 198 kg ha⁻¹ d⁻¹ of sediment from pond walls and 90 kg ha⁻¹ d⁻¹ from the pond bottom (Smith 1996).

The quantity of organic material that accumulates on the pond bottom from uneaten feed, organic fertilizer, shrimp excrement, and dead plankton increases in direct proportion to stocking density (Boyd 1992; Clifford 1994). This organic material enriches the sediment and generally results in a much faster accumulation rate. Briggs and Funge-Smith (1994) estimated that 31% of the nitrogen and 84% of the phosphorus input in an intensive shrimp pond was transferred to the sediments. To avoid inhibiting effects of accumulated sediment on shrimp performance, many farmers wash the sediments out of the pond following harvest (Clifford 1994). This practice can deteriorate the quality of the receiving water body. Sediments should either be dried and spread evenly over the pond bottom between cycles (Boyd 1992) or removed from the pond when wet and dried in a sedimentation pond. Chanratchakool et al. (1995) recommended that the black anaerobic layer be flushed out of the pond with a pressure washer and pumped into a sedimentation pond. Sandifer and Hopkins (1996) recommended that sludge be pumped from the ponds weekly during the production cycle, settled, dried, and used to improve agricultural land.

Polyculture.—Shrimp ponds generate a large amount of plankton, which is a concern in terms of effluent but a potential opportunity in terms of polyculture (Shpigel et al. 1993; Chanratchakool et al. 1995; Sandifer and Hopkins 1996). Promising polyculture species are bivalve mollusks such as oysters, clams, and cockles and filter-feeding fish such as tilapia and mullet. Seaweed such as *Gracilaria* sp. can also be reared in effluents to strip nitrogen and phosphorus from the water column.

Nutrition

Continual advancements in shrimp nutrition are reducing the cost and improving the efficiency of shrimp feeds. This is partly related to a better understanding of nutrient requirements, ingredient digestibility, attractants, pigments, health additives, and feed processing methods (D'Abramo et al. 1997). However, it is also due in large part to more efficient management of feeding methods, particularly the use of feeding trays to estimate the daily consumption rate of shrimp (Chanratchakool et al. 1995).

Low protein feeds.—One of the most promising new developments in shrimp nutrition is the use of low protein feeds in conjunction with zero water exchange. Research by Hopkins et al. (1995) indicated that *P. vannamei* reared in intensive ponds with zero water exchange performed as well with 20% protein feed as with 40% protein. Analogous results were also reported by Israeli researchers working with tilapia (Avnimelech et al. 1992). This approach uses the ecology of the pond environment to maximize the efficiency of feed (Kochba et al. 1994). In conventional shrimp ponds, only 6–24% of the nitrogen and 4–13% of the phosphorous input are incorporated into harvested shrimp while the remainder is exported to the environment in the form of effluents or sediment (Briggs and Funge-Smith 1994; Robertson and Phillips 1995). In ponds with zero water exchange, there is an opportunity to use the microbial community to recycle a portion of these wastes for later consumption in the form of enriched detritus or plankton.

The threshold carbon to nitrogen (C:N) ratio of food for zooplankton and bacteria is approximately 10:1 (Anderson 1992). Above this threshold, nitrogen is limiting and below it carbon is limiting. Typical feeds for intensive shrimp culture have a crude protein composition of 35–45% and a C:N ratio of 3–4:1. Bacteria are unable to efficiently use the waste from these feeds because they are limited by insufficient organic carbon. Avnimelech et al. (1992) demonstrated that the C:N ratio could be balanced by providing a carbohydrate supplement to the pond or by including a carbohydrate diluent in the feed. The lack of water exchange is important to give bacteria and other microorganisms an opportunity to colonize wastes without being continually flushed out.

Shrimp and tilapia apparently use bacteria-laden detritus and zooplankton as a secondary food source, which improves feed efficiency. Moss et al. (1992) found that effluent from a shrimp pond enhanced growth of *P. vannamei* 89% more than inlet water to the pond. Bombeo-Tuburan et al. (1993) found that organic detritus was the most important food source in penaeid stomachs. Some researchers attribute the nutritional value of detritus to microbially mediated digestion of refractory organic matter and others suggest that it is due to contribution of important nutrients (Moss et al. 1992; Harris 1993). Shpigel et al. (1993) found that nutritional value of particulate matter also was enhanced by attached benthic diatoms.

Both Hopkins et al. (1995) and Avnimelech et al. (1992) reported a 50% reduction in feed cost by using minimal water exchange and low-protein feeds, but both groups relied on high rates of aeration. Avnimelech et al. (1992) used aeration rates of 200 hp/ha, and Hopkins et al. (1995) used rates of 20–40 hp/ha. Apparently, aeration is essential to prevent sedimentation and release of growth-retarding anaerobic metabolites such as hydrogen sulfide, nitrite, and methane. Aerobic decomposition leads to the clean end-products of carbon dioxide and water. The potential of achieving high production rates and low FCRs using low protein feed and little or no water exchange looks very promising. Future research should focus on optimizing pond design to reduce aeration requirements.

Site Selection and Predictive Models

Ecological models are being developed to help predict the carrying capacity of estuaries based on tidal velocities, DO levels, and nutrient loads. Strutton et al. (1996) developed an expression for calculating the flushing time of coastal inlets based on mean depth, net evaporation rate, salinity of the open ocean, and salinity of the inlet. The state of South Carolina, USA, uses a dilution model to determine whether the effluent discharge from a proposed shrimp farm would cause a significant change in the water quality of the receiving stream (Hopkins et al. 1993). Ward (in press) developed a model of the assimilative capacity of a Honduran estuary with respect to oxygen-demanding constituents. His model consists of a hydrodynamic component that estimates the water velocity in the tidal receiving body based on varying degrees of farm development, and a mass transport component that predicts the sag in DO in the receiving body as a function of farm area. Such models, coupled with routine monitoring, could allow planners to regulate existing shrimp farming areas and avoid overdevelopment of new areas.

Legislation

A major effort has been undertaken in Thailand to prevent a recurrence of the failures in Taiwan, China, and the upper Gulf of Thailand. The following is a summary of recent shrimp farming legislation (NACA 1994):

- All shrimp farms must register with the Department of Fisheries.
- Farms larger than 8 ha must submit the farm's conceptualized design and layout to the Department of Fisheries for approval before construction. The design must include a waste water oxidation pond no smaller than 10% of the total pond surface area. The BOD of effluent should not exceed 10 ppm. Pond effluent from the final portion of a pond harvest must be passed through the sedimentation pond before discharge in order to reduce the load on public waters.
- Farmers are no longer allowed to wash their sediments into public waters following each harvest. They must dry the sediments and dispose of them in another manner.
- Biological treatment, such as cultivation of oysters, mussels, and seaweed in the sedimentation pond, is recommended.

Conclusions

Shrimp farming has been a multibillion-dollar income earner for the top-producing countries; however, the distressing pattern of rapid expansion followed by dramatic decline has raised questions about the sustainability of current shrimp farming practices. Worsening viral epidemics are the primary factor limiting shrimp production worldwide. By 1996, white-spot and Taura Syndrome viruses were documented in virtually every shrimp-producing country in Asia and the Americas, respectively. This is a multifaceted problem that must be addressed by a multifaceted approach.

Clearly, improved systems for disease management are needed if shrimp farming is to recover. Fisheries authorities and aquaculturists should work together to limit exposure to new diseases by restricting imports of live shrimp. Selective breeding for disease resistance shows promise as a means of adapting to viruses already established in the local environment.

White-spot and yellow-head viruses are known to be entering the Western Hemisphere through imports of frozen shrimp, and the first cases of white-spot and yellow-head viral infection of American penaeids have already been documented. Tighter regulations are needed to prevent introduction of exotic diseases in frozen shrimp.

Major environmental and social issues associated with shrimp farming include eutrophication of estuaries, destruction of mangroves, salt-water intrusion, discharge of chemicals and therapeutants, collection of wild postlarvae and reproductive adults, introduction of exotic diseases, and social conflicts concerning land and water use. Proper resource planning, allocation, monitoring, and enforcement can prevent many of these problems.

New pond management technology is being developed to greatly reduce water requirements, minimize loading of receiving waters, and reduce the cost of feed. Shrimp aquaculture is in a state of transition to a more controlled, efficient, and environmentally sustainable form, which will position it for substantial growth in the next century.

References

Allan, G. L., and G. B. Maguire. 1993. The effects of water exchange on production of *Metapenaeus macleayi* and water quality in experimental pools. Journal of the World Aquaculture Society 24:321-328.

Anderson, T. R. 1992. Modeling the influence of food C:N ratio, and respiration on growth and nitrogen excretion in marine zooplankton and bacteria. Journal of Plankton Research 14:1645-1671.

Avnimelech, Y., S. Diab, and M. Kocha. 1992. Control and utilization of inorganic nitrogen in intensive fish culture ponds. Aquaculture and Fisheries Management 23:421-430.

Bailey, C. 1988. The social consequences of tropical shrimp mariculture development. Ocean & Shoreline Management 11:31-44.

Beck, G., and G. S. Habicht. 1996. Immunity and the invertebrates. Scientific American 275:60-66.

Black, E. A., and J. Truscott. 1994. Strategies for regulation of aquaculture site selection in coastal areas. Journal of Applied Ichthyology 10:294-306.

Bombeo-Tuburan, I., N. G. Guanzon, Jr., and G. L. Schroeder. 1993. Production of *Penaeus monodon* (Fabricius) using four natural food types in an extensive system. Aquaculture 112:57-65.

Boonyaratpalin, S., K. Supamataya, J. Kasornchandra, S. Direkusarakom, U. Ekpanithanpong, and C. Chantanachookhin. 1993. Non-occluded baculo-like virus the causative agent of yellow-head disease in the black tiger shrimp *Penaeus monodon*. Gyobo Kenkyu (Fish Pathology) 28:103-109.

Boyd, C. E. 1992. Shrimp pond bottom soil and sediment management. Pages 166-181 *in* J. Wyban, editor. Proceedings of the special session on shrimp farming, Orlando, Florida, 22-25 May 1992, USA. World Aquaculture Society, Baton Rouge, Louisiana.

Boyd, C. E. 1996. Chlorination and water quality in aquaculture ponds. World Aquaculture 27:41-45.

Briggs, M. R. P., and S. J. Funge-Smith. 1994. A nutrient budget of some intensive marine shrimp ponds in Thailand. Aquaculture and Fisheries Management 25:789-811.

Brock, J. A., R. Gose, D. V. Lightner, and K. W. Hasson. 1995. An overview on Taura syndrome, an important disease of farmed *Penaeus vannamei*. Pages 84-94 *in* C.L. Browdy and J.S. Hopkins, editors. Swimming through troubled water, Proceedings of the Special Session on Shrimp Farming, Aquaculture '95. World Aquaculture Society, Baton Rouge, Louisiana.

Browdy, C. L., J. D. Holloway, C. O. King, A. D. Stokes, J. S. Hopkins, and P. A. Sandifer. 1993. IHHN virus and intensive culture of *Penaeus vannamei*: Effects of stocking density and water exchange rates. Journal of Crustacean Biology 13(1):87-94.

Bruce, L. D., R.M. Redman, and D. V. Lightner. 1994. Application of gene probes to determine target organs of penaeid shrimp baculovirus using an in situ hybridization. Aquaculture 20:45-51.

Carr, W. H. 1996. Advances in the genetics of species resistance to diseases. Proceedings of Camaronicultura '96 Foro Internacional, Mazatlan, Mexico, 1-3 August 1996. Organized by Banco de Mexico, Mazatlan, Mexico.

Carr, W. H., J. N. Sweeney, L. Nunan, D. V. Lightner, H. H. Hirsch, and J. J. Reddington. 1996. The use of an infectious hypodermal and hematopoietic necrosis virus gene probe serodiagnostic field kit for the screening of candidate specific pathogen-free *Penaeus vannamei* broodstock. Aquaculture 147:1-8.

Chamberlain, G. W. 1991. Shrimp farming in Indonesia, I: growout techniques. World Aquaculture 22:12-27.

Chamberlain, G. W. 1994. Taura Syndrome and China collapse caused by new shrimp viruses. World Aquaculture 25:22-25.

Chang, P. S., C. F. Lo, Y. C. Wang, and G. H. Kou. In press. Detection of white spot syndrome associated virus (WSSV) in experimentally infected wild shrimp, crabs and lobsters by in situ hybridization. Aquaculture.

Chanratchakool, P., J. F. Turnbull, S. Funge-Smith, and C. Limsuwan. 1995. Health management in shrimp ponds. Aquatic Animal Health Research Institute, Department of Fisheries, Kasetsart University Campus, Bangkok, Thailand.

Chua, T. E. 1993. Environmental management of coastal aquaculture development. Pages 74-101 *in* R. S. V. Pullin, H. Rosenthal, and J. L. Maclean, editors. Environment and Aquaculture in Developing Countries. International Centre for Living Aquatic Resources Management Conference Proceedings 31.

Clifford, H. C. 1985. Semi-intensive shrimp farming. Pages IV-15–IV-42 *in* G. W. Chamberlain, M. G. Haby, and R. J. Miget, editors. Texas shrimp farming manual. Publication of Texas Agricultural Extension Service, College Station, Texas.

Clifford, H. C. 1994. El manejo de estanques camaroneros. *In* J. Zendejas, editor. Camaron +94: Seminario internacional de camaronicultura en Mexico, Mazatlan, Mexico, 10-12 February 1994. Purina S.A. de C.V., Mexico City, Mexico.

Coghlan, A. 1996. Dipstick spots disease in minutes. New Scientist 149:9.

C. P. (Charoen Pokphand) Group. 1991. Techniques for feeding black tiger shrimp. C. P. Aquaculture Business Division, Technical Information Document, Bangkok, Thailand.

C. P. (Charoen Pokphand) Group. 1994. The closed recycle culture system for black tiger shrimp. C. P. Shrimp News 2(5):2-3.

D' Abramo, L. R, D. E. Conklin, and D. M. Akiyama. 1997. Crustacean nutrition. Advances in World Aquaculture, Volume 6. World Aquaculture Society, Baton Rouge, Louisiana.

FAO (Food and Agriculture Organization of the United Nations). 1996. Time series on aquaculture—quantities and values. FAO Fishery Information, Data, and Statistics Unit (FIDI), Aquacult-PC, Release 8494/96, April 1996.

Fay, R. R. 1995. Environmental study of the Cayapas-Mataje area for the purpose of recommending management strategies, province of Esmeraldas, Ecuador. Report prepared for Camara Nacional de Acuacultura de Ecuador by Geochemical and Environmental Research Group, Texas A&M University, College Station, Texas.

Fegan, D. F. 1996. Sustainable shrimp farming in Asia: vision or pipedream? Asian Aquaculture 2:22-24.

Fevolden, S. E., T. Refstie, and K. H. Roed. 1992. Disease resistance in rainbow trout (*Oncorhynchus mykiss*) selected for stress response. Aquaculture 104:19-29.

Fragkiadakis, G., and E. Stratakis. 1995. Characterization of hemolymph lectins in the prawn *Parapenaeus longirostris*. Journal of Invertebrate Pathology 65:111-117.

Frelier, P. F., J. K. Loy, and B. Kruppenbach. 1993. Transmission of necrotizing hepatopancreatitis in *Penaeus vannamei*. Journal of Invertebrate Pathology 61:44-48.

Frelier, P. F., J. K. Loy, A. L. Lawrence, W. A. Bray, and G. W. Brumbaugh. 1994. Status of necrotizing hepatopancreatitis in Texas farmed shrimp, *Penaeus vannamei*. Pages 55-58 *in* USMSFP 10th Anniversary Review, Gulf Coast Research Laboratory Special Publication No. 1, Ocean Springs, Mississippi.

Gaffney, P. M., and D. Bushek. 1996. Genetic aspects of disease resistance in oysters. Journal of Shellfish Research 15:135-140.

Garriques, D., and G. Arevalo. 1995. An evaluation of the production and use of a live bacterial isolate to manipulate the microbial flora in the commercial production of *Penaeus vannamei* postlarvae in Ecuador. Pages 53-59 *in* C. L. Browdy and J. S. Hopkins, editors. Swimming through troubled water, Proceedings of the special session on shrimp farming, Aquaculture '95, World Aquaculture Society, Baton Rouge, Louisiana.

Gjedrum, T. and E. Fimland. 1995. Potential benefits from high health and genetically improved shrimp stocks. Pages 60-65 *in* C. L. Browdy and J. S. Hopkins, editors. Swimming through troubled water, Proceedings of the special session on shrimp farming, Aquaculture '95, World Aquaculture Society, Baton Rouge, Louisiana.

Harris, J. M. 1993. The presence, nature and role of gut microflora in aquatic invertebrates: A synthesis. Microbial Ecology 25:195-231.

Haverstein, G. B., P. R. Ferket, S. E. Scheideler, and B. T. Larson. 1994. Growth, livability, and feed conversion of 1957 vs 1991 broilers when fed "typical" 1957 and 1991 broiler diets. Poultry Science 73:1785-1794.

Hopkins, J. S., C. L. Browdy, P. A. Sandifer, and A. D. Stokes. 1995. Effect of two feed protein levels and feed rate combinations on water quality and production of intensive shrimp ponds operated without water exchange. Journal of the World Aquaculture Society 26:93-97.

Hopkins, J. S., R. D. Hamilton II, P. A. Sandifer, C. L. Browdy, and A. D. Stokes. 1993. Effect of water exchange rate on production, water quality, effluent characteristics, and nitrogen budgets on intensive shrimp ponds. Journal World Aquaculture Society 24:304-320.

Huang, J., S. Xiaoling, Y. Jia, and Y. Conghai. 1995. Baculoviral hypodermal and hematopoietic necrosis—study on the pathogen and pathology of the explosive epidemic disease of shrimp. Marine Fisheries Research 16:7.

Hulse, J. H. 1993. Agriculture, food, and the environment. Food Research International 26:455-469.

Infofish. 1994. China's shrimp crop failure: the implications. Infofish International 94:42-43.

Itami, T., Y. Takahashi, and Y. Nakamura. 1989. Efficacy of vaccination against vibriosis in cultured kuruma prawns *Penaeus japonicus*. Journal of Aquatic Animal Health 1:238-242.

Itami, T., and Y. Takahashi. 1991. Survival of larval giant tiger prawns *Penaeus monodon* after addition of killed *Vibrio* cells to a microencapsulated diet. Journal of Aquatic Animal Health 3:151-152.

Kalagayan, G., D. Godin, R. Kanna, G. Hagino, J. Sweeney, J. Wyban, and J. Brock. 1991. IHHN virus as an etiological factor in runt-deformity syndrome of juvenile *Penaeus vannamei* cultured in Hawaii. Journal of the World Aquaculture Society 22:235-243.

Kochba, M., S. Diab, and Y. Avnimelech. 1994. Modeling of nitrogen transformation in intensively aerated fish ponds. Aquaculture 120:95-104.

Lightner, D. V. 1996. The penaeid shrimp viruses IHHNV and TSV: Epizootiology, production impacts and role of international trade in their distribution in the Americas. Revue Scientifique et Technique Office International des Epizooties 15:579-601.

Lightner, D. V., and R. M. Redman. 1991. Hosts, geographic range, and diagnostic procedures for the penaeid virus diseases of concern to shrimp culturists in the Americas. Pages 173-196 in P. DeLoach, W. J. Dougherty, and M. A. Davidson, editors. Frontiers in shrimp research. Elsevier Science Publishers B.V., Amsterdam, Netherlands.

Lightner, D. V., and R. M. Redman. 1994. An epizootic of necrotizing hepatopancreatitis in cultured penaeid shrimp (Crustacea: Decapoda) in northwestern Peru. Aquaculture 122:9-18.

Lightner, D. V., R. M. Redman, and J. R. Bonami. 1992. Morphological evidence for a single bacterial etiology in Texas necrotizing hepatopancreatitis in *Penaeus vannamei*. Diseases of Aquatic Organisms 13:235-239.

Lightner, D. V., R. M. Redman, B. T. Poulos, L. M. Nunan, J. L. Mari, and K. W. Hasson. In press. Status of the major virus diseases of concern to the shrimp farming industries of the Americas: known distribution, hosts and available detection methods. Proceedings of the Fourth Symposium on Aquaculture in Central America, Tegucigalpa, Honduras, 20-23 April 1997. World Aquaculture Society, Baton Rouge, Louisiana, USA.

Macintosh, D. J. 1996. Mangroves and coastal aquaculture: Doing something positive for the environment. Aquaculture Asia 1:2-8.

Mialhe, E., E. Bachere, V. Boulo, and J. P. Cadoret. 1995. Strategy for research and international cooperation in marine invertebrate pathology, immunology and genetics. Aquaculture 132:33-41.

Mangiboyat, A., Jr. 1987. Kahuku shrimp farm afflicted with another virus outbreak. Pacific Business News (September 21). Honolulu, Hawaii, USA.

Moss, S. M., G. D. Pruder, K. M. Leber, and J. A. Wyban. 1992. The relative enhancement of *Penaeus vannamei* growth by selected fraction of shrimp pond water. Aquaculture 101:229-239.

Network of Aquaculture Centres in Asia and the Pacific (NACA). 1994. Environment and aquaculture study concludes. NACA Newsletter 11:1-11.

NMFS (National Marine Fisheries Service). 1992. World shrimp culture. Volume 1: Africa, Asia, Europe, Middle East, North America. Office of International Affairs, National Marine Fisheries Service, Silver Spring, Maryland.

Nakano, H., H. Koube, S. Umezawa, K. Momoyama, M. Hiraoka, K. Inouye, and N. Oseko. 1994. Mass mortalities of cultured Kuruma shrimp, *Penaeus japonicus*, in Japan in 1993: Epizootiological survey and infection trials. Fish Pathology 29:135-139.

Peterson, J., and H. Daniels. 1992. Shrimp industry perspectives on soil and sediment management. Pages 182-186 in J. Wyban, editor. Proceedings of the special session on shrimp farming, 22-25 May 1992, Orlando, Florida, USA. World Aquaculture Society, Baton Rouge, Louisiana, USA.

Phillips, M. J., C. K. Lin, and M. C. M. Beveridge. 1993. Shrimp culture and the environment: lessons from the world's most rapidly expanding warmwater aquaculture sector. Environment and aquaculture in developing countries. ICLARM Conference Proceedings 31, Manila, Philippines.

Pongthanapanich, T. 1996. Economic study suggests management guidelines for mangroves to derive optimal economic and social benefits. Aquaculture Asia 1:16-17.

Primavera, J. H. 1995. Mangroves and brackishwater pond culture in the Philippines. Hydrobiologia 295:303-309.

Pruder, G. D., C. L. Brown, J. N. Sweeney, and W. H. Carr. 1995. High health shrimp systems: seed supply—theory and practice. Pages 40-52 in C. L. Browdy and J. S. Hopkins, editors. Swimming through troubled water. Proceedings of the special session on shrimp farming, Aquaculture '95. World Aquaculture Society, Baton Rouge, Louisiana, USA.

Puzzi, J. V., L. D. Bacon, and R. R. Dietert. 1990. A gene controlling macrophage functional activation is linked to the chicken B-complex. Animal Biotechnology 1:33-45.

Robertson, A. I., and M. J. Phillips. 1995. Mangroves as filters of shrimp pond effluent: predictions and biogeochemical research needs. Hydrobiologia 295:311-321.

Rosenberry, B. 1990. Chinese shrimp farming and its problems. World Shrimp Farming 15:1-2.

Rosenberry, R. 1994. World shrimp farming 1994. Annual report. Shrimp News International, San Diego, California, USA.

Rosenberry, R. 1995. Resistant strains of *vannamei* and *stylirostris* in Venezuela and Colombia. Shrimp News International 20:4-5.

Rosenberry, R. 1996. World shrimp farming 1996. Annual report. Shrimp News International. San Diego, California, USA.

Ruff, M. D., and L. D. Bacon. 1984. Coccidiosis in 15.5 congenic chicks. Poultry Science 63:172-173.

Russell, T. J., M. S. Kerley, and G. L. Allee. 1996. Effect of fructooligosaccharides on growth performance of weaned pig. Journal of Animal Science 74(Suppl. 1):61.

Sandifer, P. A., and J. S. Hopkins. 1996. Conceptual design of a sustainable pond-based shrimp culture system. Aquacultural Engineering 15:41-52.

Schwartz, M. F., and C. E. Boyd. 1994. Effluent quality during harvest of channel catfish from watershed ponds. Progressive Fish Culturist 56:25-32.

Sebastiani, M., S. E. Gonzalez, M. M. Castillo, P. Alvizu, M. A. Oliveira, J. Perez, A. Quilici, M. Rada, M. C. Yaber, and M. Lentino. 1994. Large-scale shrimp farming in coastal wetlands of Venezuela, South America: causes and consequences of land-use conflicts. Environmental Management 1:647-661.

Smith, P. T. 1996. Physical and chemical characteristics of sediments from prawn farms and mangrove habitats on the Clarence River, Australia. Aquaculture 146:47-83.

Snedaker, S. C., J. C. Dickinson III, M. S. Brown, and E. J. Lahmann. 1986. Shrimp pond siting and management alternatives in mangrove ecosystems in Ecuador. Final report. Office of the Science Advisor, U.S. Agency for International Development.

Song, Y.-L., and Y.-T. Hsieh, 1994. Immunostimulation of tiger shrimp (*Penaeus monodon*) hemocytes for generation of microbiocidal substances: analysis of reactive oxygen species. Developmental and Comparative Immunology 18:201-209.

Shpigel, M., J. Lee, B. Soohoo, R. Fridman, and H. Gordin. 1993. Use of effluent water from fish ponds as a food source for the Pacific oyster, *Crassostrea gigas* Thunberg. Aquaculture and Fisheries Management 24:529-543.

Strutton, P. G., J. A. T. Bye, and J. G. Mitchell. 1996. Determining coastal inlet flushing times: a practical expression for use in aquaculture and pollution management. Aquaculture Research 27:497-504.

Sung, H. H., G. H. Kou, and Y. L. Song. 1994. *Vibriosis* resistance induced by glucan treatment in tiger shrimp (*Penaeus monodon*). Fish Pathology 29:11-17.

Tacon, A. G. J., and K. Kurmaly. 1996. Nutrition and health management in shrimp culture. Page 397 *in* Book of Abstracts, 1996 Annual Meeting of the World Aquaculture Society, 29 January–2 February 1997, Bangkok, Thailand. World Aquaculture Society, Baton Rouge, Louisiana, USA

Wang, Y. C., C. F. Lo, P. S. Chang, and G. H. Kou. In press. White spot syndrome associated virus (WSSV) infection in cultured and wild decapods in Taiwan. Aquaculture.

Ward, G. In press. Hydrographic limits to shrimp aquaculture in El Golfo de Fonseca. Proceedings of the III Simposio Centroamericano Sobre Camaron Cultivado, Tegucigalpa, Honduras, 26–29 April, 1995. Asociacion Nacional de Acuicultores de Honduras, Choluteca, Honduras.

Weppe, M., J. R. Bonami, D. V. Lightner, and AQUACOP. 1992. Demonstacion de altas cualidades de la cepa de *P. stylirostris* (SPR43) resistente al virus IHHN. Pages 29-232 *in* Memorias Congreso Ecuatoriano de Acuicultura. CENAIM, Guayaquil, Ecuador.

Wiley, K. 1993. Environmental risk assessment in shrimp aquaculture. Infofish Magazine (February):49-55.

Wilkenfeld, J. 1996. Aspectos operativos en laboratorios de produccion de postlarvas. Presentation at Camaronicultura '96 Foro Internacional, Mazatlan, Mexico, 1-3 August 1996. Organized by Banco de Mexico, Mazatlan, Mexico.

Wongteerasupaya, C., S. Sriurairatana, J. E. Vickers, A. Akrajamorn, V. Boonsaeng, S. Panyim, A. Tassanakajon, B. Withyachumnarnkul, and T. W. Flegel. 1995. Yellow-head virus of *Penaeus monodon* is an RNA virus. Diseases of Aquatic Organisms 22:45-50.

Woodhouse, C. 1996. Farms avoid new U.S. turtle curb on shrimp imports. Fish Farming International 23(5):24.

Allocating Fishing Rights

Efficiency and Distribution Issues During the Transition to an ITQ Program

LEE G. ANDERSON

Abstract.—The transition to an individual transferable quota (ITQ) fisheries management program can be partitioned into three phases. The first phase is the actual implementation, which includes the initial allocation of quota and the structuring of regulations and institutions under which the system will operate. The second phase is the period in which the market for quota trading develops and the participants have the opportunity to make short-run changes in their operations as facilitated by an ITQ system, including trades in quotas. The third phase is when participants can make long-term changes through modifications in harvesting and processing capital equipment. There is the potential for developing new market channels in the second and third phases and for analytical purposes it may be useful to distinguish between those which are tied to capital investments and those which can be carried out with the existing capital stock. There will be different efficiency and distribution effects in each of these phases. The types of effects likely to occur, how they will differ in the various phases, and potential basic or fine-tuning adjustments in policy to mitigate potentially undesirable results are described.

Although the basic concept of individual transferable quota programs (ITQs) is relatively simple, the manner and timing with which they will achieve conservation and efficiency benefits and the distribution of efficiency gains are quite complex. (For a detailed discussion of ITQs, see Mollett 1986 and Neher et al. 1989. For a more detailed discussion on the transition to ITQ programs using traditional economic models, see Lindner et al. 1992; also, author's unpublished manuscript.) Among other things, the type and timing of the efficiency effects depend upon the amount and type of capital in the processing and harvesting sectors, the existing regulation program, and the current status and biological variability of the fish stocks. The purpose of this paper is to describe the types of changes likely to occur during the transition to a fully functioning ITQ program.

To set the stage, the first section is focused on the operational incentives for the processing and harvesting sectors under traditional and ITQ regulation programs. The main points will not be new to readers familiar with fisheries economics. However, the framework for discussion is different and uses several somewhat restrictive assumptions, but it has been designed to focus attention on elements of industry structure that are important in ITQ management. This focus will provide the basis for predicting the timing and the types of conservation and efficiency gains that are described in the next section (Timing of Efficiency Gains and Distributional Effects). The initial discussion also uses somewhat restrictive assumptions concerning the status of the stock and the malleability of capital. The implications of relaxing those assumptions is the subject of the next two sections (ITQs in Overfished Stock and Non-Malleability). Although distribution effects are covered throughout the paper, the next section (Other Distribution Issues) summarizes

and expands the coverage of this topic. The final section presents a summary and general conclusions.

Industry Structure and Regulation

Basic Operational Incentives

In order to study the workings of a commercial fishery, we must understand the conditions that influence the harvesters' and processors' choice of capital equipment and the level of operation. This can be described using the break-even conditions in each sector.

At the outset, it will be useful to define three different types of fish prices:

P_f = the price received by processors or wholesalers for final fish products,

P_p = the price processors pay for raw fish, and

P_{ex} = the raw fish price received by the boat.

The distinction between the last two may seem artificial because under traditional fishing arrangements they are the same. However, as will be described below, they are conceptually different in ITQ programs and, further, this difference is critical.

Although most economic analyses of commercial fishing assume homogeneous producers, in the real world individual processors and harvesters are often dissimilar owing to differences in capital, skill, location, and other factors. These differences are explicitly considered in this discussion. The basics of the operation of participants in the processing sector can be explained using the following expression:

$$\alpha P_f - APC - P_p \geq 0 \tag{1}$$

where α is the average recovery rate of final product from the raw fish, which is a function of the type of capital,

213

and APC is the average processing cost (total processing costs including normal profits divided by output), which is a function of the type of capital and the output level.

Average variable processing costs (total variable costs divided by output) is an important variable in the context of non-malleable capital (see Non-Malleability). Assuming constant marginal processing costs, firms will strive to minimize APC by operating at full capacity so that fixed costs will be spread over the largest possible level of output.

The maximum amount a given processor can afford to pay for raw fish (hereafter the maximum bid price) can be determined by solving expression (1) as an equality. This maximum bid price will only be valid for the quantity of raw fish that will allow APC to be minimized. Those processors whose maximum bid price is below the current market price for raw fish cannot afford to operate in the long term. Those whose maximum bid price is above the market price will be earning intramarginal rents; that is, they are earning more than enough to cover all their costs. The marginal producer will operate where expression (1) holds as an equality, and will just cover all costs. Processors will tend to organize their investments in plant and equipment and their annual level of output such that they can have the highest possible maximum bid price. This will put them in the best possible position to purchase raw fish to keep their plants operating. At the same time, they will be hoping that they can buy fish at a price lower than their maximum possible bid price.

The operation of participants in the harvesting sector can be explained in terms of following expression:

$$P_{ex} - AEC/CPUE \geq 0 \qquad (2)$$

where AEC = average cost of effort, and CPUE = catch per unit effort. Note that in order to make the distinction between the effects of stock (CPUE) and effort cost on the average cost per unit of harvest, we must define effort cost in terms of a standardized unit of effort. The quotient is the average cost per unit of output. Assuming constant marginal cost of effort, boats will try to minimize AEC by producing at full capacity.

The minimum a harvester can afford to receive for a unit of fish (hereafter minimum reservation price) can be determined by solving expression (2) as an equality. Analogously, this minimum reservation price will only be valid for that level of output where AEC is minimized. Harvesters whose minimum reservation price is above the market price cannot afford to operate, while those for whom it is below the market price will earn intramarginal rents. Analogous to processors, harvesters will try to keep their minimum reservation price as low as possible so as to increase their chances of finding a buyer

for what they catch and to make intramarginal rents if they can.

Overall equilibrium in a non-ITQ fishery will occur in the following situation:

$$\text{Maximum bid } P_p \text{ of marginal processor} = \\ \text{minimum reservation } P_{ex} \text{ of marginal harvester} \quad (3)$$

This expression will determine the equilibrium raw fish price and the number of operators in each sector. Processors will use expression (1) to determine their maximum bid price for raw fish, and harvesters will use expression (2) to determine their minimum reservation price for raw fish. If the existing price of raw fish is below the maximum bid price of a particular processor, the processor will choose to operate, and if that price is above the minimum reservation price of a particular boat, the harvester will choose to operate. Entry of processors and harvesters will cease when the maximum bid price of the marginal processor equals the minimum reservation price of the marginal harvester. The marginal operators will earn normal profits and all others will earn intramarginal rents.

In the long run, operators in both sectors will tend to invest in new capital equipment if this will allow for a reduction in APC or AEC. This reduction in cost will allow processors to increase their maximum bid price and harvesters to lower their minimum reservation price, which will allow the operators to increase their intramarginal rents per unit of output and perhaps to increase their market share.

While these expressions do not capture all of the intricacies of the two sectors, they do provide the basis for a heuristic description of how the sectors work and the important parameters. The market channels and overall final product demand are important in the determination of P_f. Technology and the level of output determine APC, AEC, and α. The condition of the stock determines CPUE. Regulations can affect, directly or indirectly, any of these items. The main focus of this discussion is to show how traditional regulations affect these parameters and then to show what changes can be expected with the introduction of ITQs.

Traditional Fishing

In order to describe the effects of ITQs, we need to describe in more detail the process through which an open-access fishery will reach a situation represented by expression (3). The real issue centers on the types of things that will change the parameters of expressions (1) and (2) such that the minimum reservation price of vessels is pushed up and the maximum bid price of processors is pushed down.

Open access.—Open access leads to overfishing and

economic waste (Anderson 1986). While the analysis herein cannot demonstrate the results of open access as elegantly as some of the more formal models, the basic point is made. The relationship between the economic parameters and the state of the fish stock is the driving force for overfishing. Anything that raises the maximum bid price for processors or lowers the minimum reservation price for boats will change the incentives for entry to the fishery. These changes will lead to a situation where expression (3) will not hold. To the extent that entry occurs, increased effort will reduce CPUE, which will directly affect expression (2) for all boats and help move the system to a new equilibrium. As an example, a new technology that increases the product recovery rate, α, will increase the maximum bid price. This will encourage entry, which will have a tendency to decrease CPUE. In simple terms, with no regulation, anything that improves profitability in harvesting and processing will have an adverse effect on the fish stock.[1]

At the same time, changes in market or stock conditions that change any of the parameters in expressions (1) and (2) can potentially lead to changes in the industry structure where condition (3) applies. For example, the change in the recovery rate described above could lead to a reduction in fleet size if the decrease in CPUE pushes the minimum reservation price for some boats above the maximum bid price for the marginal processor. In addition, the distribution of the intramarginal rents could change, especially if only some of the processors were able to achieve the improvement in recovery rate.

Open access regulation.—The relationship between profitability and the condition of the stock is at the heart of traditional regulation. Administrators try to reduce pressure on the stock by making it less profitable to fish. For example, a gear restriction will affect the average cost of producing effort, AEC, for those participants who use that particular gear. According to expression (2), their minimum reservation price will go up. Those whose reservation prices are pushed higher than the market price will not be able to continue fishing. Effort will go down and pressure on the stock will be reduced. In one sense, however, such a program will sow the seeds for its own destruction. To the degree that CPUE goes up as a result of the decrease in effort, the minimum reservation price of the remaining boats will go down. This will tend to increase effort, which will push the CPUE back down again. Also, given sufficient time, the affected boats can

adjust their technology and operating procedures to minimize the increased cost of the gear restrictions. Such adjustment will reduce the biological efficacy of the regulation; in the long run, the effect of the gear restrictions on reducing fishing mortality will be less than it was originally. And there is a non-symmetrical effect here: While the gear restrictions increased the minimum reservation price of only boats with that gear, the increase in CPUE lowered the minimum reservation price of all boats.

Because gear restrictions tend to become less effective over time, regulators may implement a new round of gear restrictions, which will start the cycle one more time. The history of open-access management is replete with examples of tighter and tighter regulations. The results are short-term improvements in stock size (which tend to be reduced over time as participants change operating procedures or introduce new technologies) and inefficiencies in production.

The same conclusion follows for area or season closures. Season closures reduce the number of days a boat can operate, which means it must spread its fixed costs over less output. This will increase AEC and raise the minimum reservation price. It can also have the same effect on processors, and they will reduce their maximum bid price. This will force some of the marginal producers to stop fishing, which will increase CPUE. Given time, however, boats and plants can adjust their activities so that they can produce more output in the open season and new participants may find it profitable to enter if the initial increase in CPUE is high enough. There will be improvements in stock size, which tend to be lost over time, and permanent inefficiencies in production.

A total quota is an open-access regulation that, if properly enforced, can lead to long-term improvements in the stock. However, total quotas provide incentives for inefficiencies. A properly enforced quota can increase CPUE. All else being equal, this will reduce the cost of taking a unit of fish (AEC/CPUE), which will lower minimum reservation prices. However, with the total quota, the ability to produce at the lowest cost is not very useful once the quota is reached. This leads to a race for fish. In order to get the fish, boats will gear up to harvest as fast as possible. This will increase AEC. Gearing up for the race for fish will cease when expression (2) holds for the marginal participants. Quotas can also have effects on the processing sector. They have to gear up to handle fish in shorter amounts of time, which will affect their costs as well. The new equilibrium with the quota will be achieved when changes in other parameters offset the long-term changes in CPUE such that expression (3) holds again. It is important to note that the change to the new equilibrium will affect the number of processors and harvesters that can participate in the fishery. Different types of regulations can affect dissimilar participants in

---

[1]Throughout this discussion, CPUE will be used as a measure of the health of the stock. This is very simplistic because such things as distribution of cohort sizes and fishing mortality during critical life periods can also be important. However, given the abstraction of the discussion, it will suffice, especially if the results are interpreted correctly.

different ways. Therefore, the choice of regulations will affect how many participants will operate, which ones will operate, and the amounts of the intramarginal rents that will be received.

ITQ Regulation

Individual transferable quota programs build on the ability of the total quota program to achieve conservation goals. They also change the incentives facing participants, and so there is no race for fish. This can reduce some of the biological problems such as discard and by-catch, which frequently accompany derby fisheries, and will tend to improve the profitability and efficiency of the harvesting and processing sectors.

As indicated above, a successful quota can maintain CPUE at the desired level. (Optimal quota size and, hence, optimal CPUE are considered exogenous, and so this discussion is focused on the ability of ITQs to keep catch within that limit and to maximize the net gain from harvesting it.) Improvements in CPUE improve profitability in the harvesting sector. However, with an ITQ program, harvesters can obtain the "right" to take or catch a unit of fish through market mechanisms rather than by building a faster and bigger boat to win the race.

For discussion purposes, assume the ITQ fishing rights are given to third parties who are neither harvesters nor processors. (This assumption is for pedagogical purposes only. It will make the description of effects more clear, but is not an endorsement of this type of ITQ allocation.) These third parties will hire vessels to harvest their catch and will sell it to processors. They can make the most profit from their ITQs by maximizing the spread between the P_{ex} they pay the boats and the P_p they receive from processors. Alternatively, vessels will have incentives to lower their minimum reservation prices so that they will have a better chance of being hired to harvest fish. This means they will try to keep AEC as low as possible. At the same time, processors will have incentives to increase their maximum bid price so that they will have a better chance of getting the fish. In terms of this model, they will try to find better markets to increase P_f, and they will choose capital and operation procedures so as to increase α and decrease APC. The market price that processors will pay for raw fish, P_p, will depend on the total quota and the distribution of maximum bid prices and capacities of processors. It will tend toward the P_p, which causes expression (1) to hold for enough firms that the quota can be processed. Those processors for whom it holds as an inequality will earn intramarginal rents even if they do not own ITQs.

Similarly, the market price received by boats, P_{ex}, will depend upon the distribution of minimum reservation prices and capacities of harvesters. It will tend toward the P_{ex}, which causes expression (2) to hold for enough boats that the quota can be harvested. Again, some of the boats may earn rents.

The market process does not equalize P_p and P_{ex}. Instead there is a tendency for the difference to be maximized subject to the economic and biological realities of the fishery. This difference is the annual market price of the right to harvest one unit of fish:

$$P_{ITQ} = P_p - P_{ex} \qquad (4)$$

The above conclusions are not changed if ITQs are owned by harvesters or processors. A harvester with quota will have incentives to sell raw fish at the highest possible P_p and to keep the cost of harvest as low as possible. Similarly, a processor with quota will hire boats to harvest the fish at the lowest possible P_{ex}, and will have incentives to maximize profits from processing and marketing.

No matter who owns the quota, in the long run harvesters and processors will at least make normal returns and those with special attributes may earn intramarginal rents. The size of rents earned by ITQ owners will depend upon the maximum bid price of the marginal processor and the minimum reservation price of the marginal harvester. For the most part, the allocation of ITQs will not have an effect on the amount or timing of efficiency gains, but only on who receives the gains. However, see the following discussion of non-malleable capital.

The focus of this discussion on the technical operation of an ITQ program tends to mask the potential distributional effects. It is important to note that the number and identity of the processors and harvesters and the amounts of intramarginal rents obtained will likely differ under open access (expression [3]) and ITQs (expression [4]).

Timing of Efficiency Gains and Distributional Effects

For purposes of analyzing the efficiency and distribution issues in the transition to an ITQ program, the preceding discussion can be summarized as follows. Under open-access or traditional regulation, a commercial fishery will tend toward a situation where expressions (1), (2), and (3) will hold. All elements in expressions (1) and (2) can be affected by the open-access race for fish or the type of regulation. For example, the race for fish caused by seasonal closures may reduce the quality of the product so that P_f is lower than need be. Processors try to arrange their activities so that they maximize profits and, in doing so, arrive at a situation where they have the highest possible maximum bid price for raw fish (see expression [1]). This means they will be in the best position to bid for raw fish, and if they can purchase for a lower price, they will be earning intramarginal rents. To the extent that current regulations affect α, P_f, or APC, efficiency in the processing sector will be adversely affected and maximum bid prices will be lower than they otherwise could be. In summary, expression (1) will be brought to zero at the margin because firms will try to maximize their bid price.

Because of the race for fish, there is a different equilibrating mechanism for expression (2). Harvesters will be motivated to arrange their activities so that they can catch fish before the season is closed or before others can with the given set of regulations. They can afford to spend more for the race until AEC/CPUE is equal to the maximum bid price of the marginal producer. In summary, under open-access or traditional regulation, expression (2) will be brought to zero by increases in AEC/CPUE.

An ITQ program will tend towards a situation where expressions (1) and (2) hold and expression (4) is maximized. It will change the process in two ways. First, the motivation of harvesters will change such that they will try to minimize AEC/CPUE so that they can have the lowest possible minimum reservation price for raw fish. This will set them in the best position to sell their harvesting services if they do not receive quota, and it will maximize the profits from their ITQ if they do receive an initial allocation. Second, the direct or indirect adverse effects on the elements of expressions (1) and (2) can be eliminated.

The annual efficiency gains from an ITQ program are represented by the price of the ITQ in expression (4). One way to look at the transition to an ITQ program is to consider how the various elements can and are likely to change over time. While the expression focuses on efficiency, distribution issues can be described as well.

The elements of expressions (1) and (2) can be broken down into final market, processing, harvesting, and stock effects (Table 1). Given the previous discussion, the price of an ITQ will be as indicated by the equation in the footnote for Table 1. The table also contains a summary of the types of changes likely to occur to the various elements in the short, medium, and long run. The terms—short, medium, and long run—are relative and will differ according to the nature of the fishery; the short run could last 6–12 months and the long run could be as much as a decade.

The remainder of this section describes these changes in more detail. To keep things simple, assume that an ITQ program has been implemented on top of a system that maintains a safe catch level such that stock rebuilding is not an issue. (The case with stock rebuilding will be discussed in the next section.) In all cases, the predicted results are quite general. The actual results will depend upon the peculiarities of the case involved. The specifics of the regulation program in effect prior to the ITQ program and the size, composition, and the type of capital (malleable or non-malleable) in the two sectors will be especially important in determining the types and timing of gains.

In the short, medium, and long run, producers can potentially find better markets for the final product. While they may try to do so in traditional regulation programs, such programs sometimes preclude certain marketing options. The most striking example of this is the Australian southern bluefin tuna (*Thunnus maccoyii*) ITQ program. Prior to the ITQ program, the race for fish meant that the bluefin tuna were caught at small sizes and in such condition that their best use was for canning. With the advent of ITQs, market incentives encouraged harvest at larger sizes and better onboard treatment such that the fish could be sold on the Japanese raw market at

TABLE 1.—Economic factors influencing the ITQ price in the short, medium, and long run.[a]

Source of gain	Timing		
	Short run	Medium run	Long run
Final market, P_f		Market development	Market development
Processing α, APC	Regulation removal Effects of initial allocation	Operational changes Quota trades and contracting Regulation removal, minor retooling	Operational changes Quota trades and contracting Regulation removal, investment New opportunities, investment
Harvesting, AEC	Regulation removal Effects of initial allocation	Operational change Quota trades and contracting Regulation removal, minor retooling	Operational change Quota trades and contracting Regulation removal, investment New opportunities, investment
Stock, CPUE	Low TAC Same CPUE Elimination of race for fish	Low or possible increasing TAC Improving CPUE Elimination of race for fish	Higher TAC Higher CPUE Elimination of race for fish

[a] $P_{ITQ} = (\alpha P_f - APC) - \left(\dfrac{AEC}{CPUE}\right)$

a significant increase in market value. The Canadian halibut (*Hippoglossus stenolepis*) ITQ program is another good example. With the elimination of the derby fishery, virtually all of the product went to the fresh rather than frozen market, which significantly increased the market price and eliminated much of the processing and storage costs. This increase in P_f obviously increased the ITQ price. While the gains from improved marketing only appear as a single row in the Table 1, the potential for such gains may be quite large. Indeed, the gains here may come more quickly and be larger than gains from increased efficiency in harvesting and processing.

In the short run, firms may have little opportunity to make changes in capital and operating procedures, and so there are likely to be few gains. Positive benefits might be possible from the removal of certain types of restrictive regulations. For example, the surf clam (*Spisula solidissima*) ITQ program replaced a vessel moratorium and a limit on the number of fishing days per boat. Owners of quota were immediately able to reduce the number of boats they used. The least effective boats were retired, and this reduced maintenance costs and allowed for the replacement of high-cost vessel days with low-cost vessel days. As a result, AEC was reduced, which produced a small wedge between P_p and P_{ex}. Whether such opportunities are possible in other fisheries depends upon the nature of the regulations and the possibility to make rapid changes when they are removed.

If the initial allocation of ITQs is drastically different from existing production levels of current participants, there may be efficiency losses in the short run. The ITQ owners themselves may not be geared up to harvest, and it will take time for trades to take place or for contracts to be established so that the fish can be harvested and processed.

In the medium run, removing regulations might still yield some gains. Certain types of restrictions affect operational behavior such that it takes time to adjust to their removal (e.g., restrictions on power winches or dredge width). Harvesters in an ITQ system may find it profitable to replace old winches or dredges, depending upon the cost savings and the age of the existing equipment. While this will not involve major investment such as a new boat, the retooling will take time. Given the refitting capacity of existing boat yards, it might take several years before the full benefits are achieved.

Two other types of related changes are also possible in the medium run for both harvesters and processors. First, as a result of quota trading or long-term contracting, firms can increase their level of output such that they can take advantage of economies of scale and, hence, reduce AEC or APC. Also, because of the increased flexibility of the ITQ program, firms may be able to make small changes in their operations that increase efficiency.

For example, scheduling at processing plants may be improved such that average storage costs fall.

In the long run, the same types of changes that are possible in the short and medium run can still occur. However, the long run also opens up the possibility of new investments. Some opportunities may be possible because of the removal of regulations (e.g., things that could not be fixed by simple refitting). For example, boat length limits and days at sea limits will both result in vessels that may not be the most efficient overall. An ITQ program will produce incentives to replace these less efficient boats.

In addition, an ITQ program may result in the design of more efficient boats and deck equipment. Owners and naval architects design boats and gear to maximize production based on existing and likely regulations. With restrictive regulations, certain possibilities and avenues of research are not considered or explored: There is no sense spending research and development efforts in areas where they will not prove useful. With the removal of these restrictions, such barriers to research and development are lifted. This could possibly lead to efficiencies that would not be predictable prior to the ITQ program.

In summary, the economic gains from an ITQ program are represented in annual terms by the annual price of an ITQ. The types of changes that will lead to efficiency gains depend upon the elements in the equation, and the timing of the changes depends upon the time periods in which the various elements can change.

There will be distribution effects of these efficiency gains, and who will gain and who will lose will depend upon the initial allocation of quota *and* the type of capital and the operating procedures of the participants in the two sectors. The first point is obvious and is frequently made in the literature. The second point is a little more subtle. One way to make this distinction as sharp as possible is to separate rents earned from the ITQs and intramarginal rents earned by processors or harvesters. Those who receive ITQs will receive an annual rent per unit of quota equal to P_{ITQ}, and as explained previously, this rent is likely to increase over time.

Whether they receive ITQs or not, all participants will also be potentially affected by the change in industry operations. Prior to the institution of ITQs, the industry would be operating in response to expressions (1), (2), and (3) in the ways described previously. Marginal producers would tend to make normal profits and others would tend to make intramarginal rents. With regulations that restrict flexibility of operations, those participants whose vessels, operational procedures, or personal skills enable them to work well under the restrictions are likely to be among those who earn intramarginal rents.

With the institution of ITQs, the number and composition of both sectors may change. In addition, the individuals who receive intramarginal rents and the amounts they earn can change. It is not possible to describe exactly how this will happen because it will vary from fishery to fishery depending upon the circumstances. It is possible, however, to describe the nature of the changes.

With ITQs, there will be pressures for processors to increase their maximum bid price and for harvesters to decrease their minimum reservation price. Those who are best able to do so will be best suited to continue in the industry and will be most likely to earn intramarginal rents. For example, those participants who were able to operate well under the restrictive regulations may not be the ones who can survive when the regulations are removed. Also note that one way to increase maximum bid price or reduce minimum reservation price is to increase yearly output. This will spread fixed costs over more output and may result in economies of scale with respect to variable costs. (This does not mean that production will necessarily always end up in the hands of few producers. For one thing, there are limits to the efficiencies of increasing size. In addition, in some cases, the lowest costs may come from smaller boats.) Therefore, there will be a tendency for the number of processors and harvesters to decrease. In summary, the amounts and the recipients of intramarginal rents will change, regardless of whether they receive initial allocations or not.

Some believe that the initial recipients of ITQ will receive all the potential rents. This is only true, however, if all firms are identical in both current make up and potential investment possibilities, and anything that can improve efficiency in either sector is known with certainty. These are very restrictive assumptions indeed.

The simple argument that the initial recipients will receive all rents is as follows: Assume that with an ITQ program, P_p will tend to $21 while P_{ex} will tend to $15. Therefore, the initial recipient will receive a right that has an annuity value of $6/year. If the owner sells it for the present value of annuity flow, the buyer will just receive normal profits from fishing.

However, given the differences in skills and capital assets among potential participants, the situation is likely to be more complex. To keep the story simple, assume there are two harvesters with minimum reservation prices of $15 and $10, respectively, and the former receives an initial allocation. If P_p is likely to remain at $21, the most that the first individual can make will be $6/year per unit of quota. However, the second individual can make up to $11/year per unit of quota. If the second individual desires to buy quota from the first, the sale price will be somewhere between $11/year and $6/year depending upon their relative bargaining power. The closer it is to $6, the more of the gains from the ITQ program the new owner will capture.

This discussion oversimplifies things to make the point. The case for gains going to purchasers may be stronger in real-world situations. Once an ITQ system is started, it is unlikely that new vessels will be constructed on speculation. The owners will want to obtain sufficient ITQs (or long-term contracts) before building the boat. The prospective builder will know that its P_{ex} will be $10, but current ITQ owners will not. The market price for ITQs will be $6, and the new boat owner will be in a good position to buy at that price and to receive some of the gains that result from the ITQ program. In fact, it is these sorts of gains that will keep an ITQ fishery dynamic.

ITQs in Overfished Stock

If ITQs are established as part of a stock rebuilding program, the efficiency and distribution effects described previously will be altered somewhat. The main difference will be due to the reduction in fishing mortality that will be necessary in the early years.

In the previous case, it was assumed that the TAC associated with the ITQ program was roughly equal to current harvest. This was the case for the surf clam and ocean quahog (*Arctica islandica*) fishery and will be the case for halibut and sablefish (*Anoplopoma fimbria*), two of the three ITQ fisheries in the USA. With overfished stocks, however, initial TACs will be less than current harvests. Since ITQs are normally issued on a percentage of TAC basis, this means that, initially, each ITQ owner will have less fish to harvest. Over time as the stock rebuilds, TACs will presumably be increased and, depending on the current level of overfishing, may even surpass current catch levels.

Even with lower output levels, there may still be the potential for savings owing to the removal of restrictive regulations. The key question centers around how participants will react to a production path with low harvest levels during rebuilding. They may engage in short-term leases or contracting to allow the catch to be harvested and processed with fewer boats and plants. The efficiencies from the less restrictive regulation will apply to fewer boats and plants, and so the gains will be less than they otherwise would be. Alternatively, owners may elect to operate more boats and processors than necessary, although at lower levels of operation, in order to maintain their labor force and keep market channels open until TACs increase. This will allow for more potential savings from more operators, but the inefficiencies of low-level output may counteract some or all of the savings.

Serious distribution effects will occur, especially if the cutbacks are severe and long-lasting. Even those operators who are included in the ITQ allocation will suffer short-term reduction in intramarginal rents. However, in the long run, producers who receive ITQs will reap the benefit of the current investment in the stock. This is one way that ITQs are different from other management regimes. If cutbacks lead to bankruptcies under traditional regulations, current producers may be gone when the stock is rebuilt. The gains will go to the new participants when the fishery is re-opened, not to those who suffered from the reduction in fishing mortality. With ITQs, however, the original participants will have received a significant portion of the eventual economic gain from rebuilding. The medium- and long-term effects are likely to be the same in this case, but it will just take longer to achieve. There is no sense in investing in new vessels until the stock recovers. In addition, the rebuilt stock may exhibit characteristics that call for different types of harvesting capital and operating procedures than did the overfished stock. However, it will take time for stock growth, biological research, and architectural research and development to come together to design and build a new boat.

Non-Malleability

One issue that has received relatively little attention in the ITQ literature is the effect of the malleability of capital on the distributional effects of an ITQ program. While a complete analysis of this problem is beyond the scope of this paper, some important conclusions are described.

Malleability refers to the ease with which plants and equipment can be switched from one use to another. Some harvesting equipment can be used in a number of fisheries on a day-to-day basis, others can be changed but only with substantial time in dry dock, and some are suitable for only one use. Similarly, processing equipment can be single or multiple purpose. Since processing equipment is normally immobile, the ability to move to other uses is also dependent on its current location relative to the other uses.

The main effect of non-malleability can be summarized as follows. If processing or harvesting equipment is non-malleable and there is excess capacity, the distribution consequences go beyond merely who receives initial allocations of quota. This can be demonstrated using a simple example: Assume a fishery with perfectly mobile capital in processing and harvesting that is 100% overcapitalized owing to an open-access regulation program. Assume an ITQ program is instituted and, for whatever reason, the rights are distributed such that 50% of the quota is given to half the processors and the other 50% is given to half the harvesters. Those that receive quota can arrange it such that their equipment can harvest and process the TAC. They can remain in the fishery and obtain the rewards from the program. Those that do not receive quota shares must take their capital and

use it elsewhere in the economy. They do not receive a share of the rents, but at least they can still earn a return from their equipment for the rest of its productive life even though they must switch to another use.

Now consider the same case, except assume that the capital is completely non-malleable. Again, producers who receive quota can receive the gains from the program by using their equipment. Those who do not receive shares are in a different position. Essentially the value of their capital has fallen to zero. Given the race for fish, they were able to harvest or process such that they received a return for the use of their equipment. With the elimination of the race and with the institution of a market-allocated quota, they can earn nothing from the equipment and, hence, its zero value.

To carry the argument a step further, assume that the participants who do not receive quota decide not to give up but are willing to enter the market to bid for product. Focus for the moment on a processor. Recall that processors normally use expression (1) to determine what they can bid for raw fish to keep their plant going. To remain successful in the long run, the most they can afford to pay is the difference between the returns from processing a unit of fish and average total processing cost, where total costs include a normal return on investment. In this case, however, at the extreme, the plants will only consider variable processing costs when determining their maximum bid price. Anything they earn over the variable costs is money in their pocket that they would not earn if they did not operate. As the processors bid for fish to keep their plants going, the price will approach a point where only variable costs are being covered. The firm may be operating, but again there is no (or a greatly reduced) return on the equipment. The process can be reversed if overcapacity is eliminated as some of the plants reach the end of their productive life. Non-malleability and overcapacity are joint requirements for the loss of capital values. This reduction or loss of return will also happen to vessel owners who must bid for the right to harvest. This means the price of ITQs will go up as processors bid for fish and harvesters bid for the right to harvest. Some of the lost capital values of processors and harvesters are transferred to ITQ owners.

In summary, with non-malleable capital, firms that do not receive quota not only lose out on the gains from the program, but the capital value of their equipment can be lost or diminished. How much they will lose depends upon the remaining useful life of their equipment and time period over which overcapacity will continue.

It is important to recall that overcapacity causes many biological and economic problems in a fishery. There are great advantages to reducing it. The point is that the amount and the distribution of the costs of doing so can vary according to the malleability of capital. These are important social issues, and they will be extremely important to industry participants when ITQs are being proposed. One way to address these problems is to design the initial allocation scheme such that participants are compensated for the loss in the capital value of their equipment with the capital value of the ITQs they receive.

Other Distribution Issues

Many of the distributional issues of implementing ITQs have been reviewed in the context of the preceding discussion. In this section, I discuss two distributional topics that have not been covered: (1) The potential effects of an ITQ program on the share system of crew remuneration and how this can affect the distribution of net gains between boat owners and crew members, and (2) issues relating to sharing the gains of fisheries management between ITQ owners and the general public.

Crew Share

In most fisheries, the crews are paid a share of the trip proceeds rather than a fixed wage (Sutinen 1979, Anderson 1982). There are two reasons for this arrangement. First, it distributes the risk of good and bad trips between the boat owner and the crew. Second, because the amount caught on any one trip depends in some degree on the skill and exertions of the crew, the share system can be used to attract good crew members to a given boat and to provide incentives for extra effort during a trip. As will be demonstrated, the exact percentage of the net proceeds that goes to the crew depends upon the relative sizes of the opportunity cost of labor, the annual fixed cost of owning and maintaining the boat, and the relative bargaining power of crew members and owners. Since ITQs will provide incentives to change the types of boats used and the way they are operated, they will also tend to change share rates and the relative rents earned by the two groups.

The following definitions will be used in developing a simple model that can show how the share rate and the relative rents can be affected by an ITQ program:

s = the percentage share of net proceeds received by the crew; $(1-s)$ = the percentage received by the boat owner,

LC = the opportunity cost or minimum reservation wage of the crew for a representative trip. It depends upon the number and types of workers used,

NLC = the non-labor cost of operating a vessel on a representative trip,

FC = the fixed cost of owning and maintaining a boat apportioned on the basis of a representative trip, and

Q = the expected harvest from a representative trip.

If the crew is to earn the minimum amount necessary to cover its opportunity costs, the share rate, s, must be such that the following equation holds:

$$s[P_{ex}Q - NLC] = LC \qquad (5)$$

Solving for P_{ex} produces an expression for the minimum P_{ex} that will cover labor opportunity costs for various share rates for this hypothetical vessel:

$$P_{ex} = [LC/s + NLC]/Q \qquad (5')$$

This break-even equation for the crew is plotted as the curve CBC (Figure 1). If the share rate goes as high as 1 (i.e., all of the net proceeds go to the crew), P_{ex} can get as low as $[LC + NLC]/Q$. As s approaches zero, P_{ex} must approach infinity if the crew is to break even. All combinations of P_{ex} and s above and to the right of the curve represent points where the crew will earn an income higher than their reservation wage. That is, the crew will be earning rents.

If the boat is to earn the minimum amount necessary to cover fixed costs, the share rate must be such that the following holds:

$$(1 - s)[P_{ex}Q - NLC] = FC \qquad (6)$$

By solving for P_{ex}, this becomes

$$P_{ex} = [FC/(1 - s) + NLC]/Q \qquad (6')$$

This equation shows the combinations of P_{ex} and s that allow the boat owner to cover fixed costs. This boat break-even equation is plotted as the curve BBC (Figure 1). If the share rate gets as low as zero, (i.e., all of the net proceeds go to the boat owner), P_{ex} can get as low as $[FC + NLC]/Q$. As s approaches 1, P_{ex} must approach infinity if the boat owner is to cover fixed costs. All com-

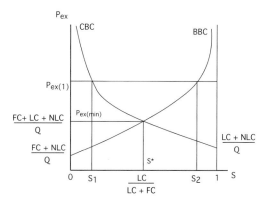

FIGURE 1.—Relationship between crew share (S) and break-even ex vessel price for crews (CBC) and boat owners (BBC).

binations of P_{ex} and s above and to the left of the curve represent points where the boat owner earns rents.

The intersection point of these curves is the lowest P_{ex} that the boat can receive and still cover all costs, and the share rate that must be used with that P_{ex} if the net proceeds are to be distributed such that both crew members and the boat owner will break even. Although the algebra is somewhat messy, the values of P_{ex} and s at the intersection are indicated in the figure. The reservation price is average total cost. This is no surprise as it follows directly from the preceding discussion. However, the share rate that must hold is illuminating. It is the ratio of labor opportunity costs to the sum of labor opportunity costs and fixed costs. This means that there will be a tendency for the share rate on any vessel to equal a value that is a function of the amounts and types of capital and labor used. Therefore, to the extent that ITQs lead to changes in the way capital and labor are used, they will have a tendency to change the share rate.

There is more to be learned from this figure, however. With heterogeneous vessels, each will have a different set of break-even curves. In the competition to sell fish, P_{ex} will be pushed down to the point where the marginal boat is just able to cover costs. This boat will have no choice as to where to set the share rate if both the owner and crew are to break even. The analysis is different for the non-marginal boats, which will be able to earn rents because their reservation price is lower than the market ex vessel price.

For example, assume that the boat represented by Figure 1 is non-marginal and the market price is $P_{ex(1)}$. The boat will be earning rents that will be distributed according to the size of the share rate. For example, if the share rate equals s_1, the crew will just cover their opportunity costs and all the rents will go to the boat. Alternatively, if the share rate equals s_2, all of the rents will go to the crew, and the boat owner will just cover fixed costs. At share rates between s_1 and s_2, the rents will be distributed between the boat owner and the crew. The closer the share rate is to s_1, the greater the proportion of the rents that will go to the boat, and vice versa.

It is not possible, at this level of abstraction, to determine the final share rate. It will depend upon the bargaining skills and strengths of the two groups and perhaps on the customs for particular types of boats or in different fisheries or ports.

ITQs create situations where the parameters of the CBC and the BBC (Q, FC, LC, and NLC) are likely to be changed. This will result in a downward shift of the two curves. The degree of movement will differ for different types of boats. While the reservation P_{ex} will shift down, the share rate at the intersection could increase or decrease. Similarly, the curves for newly constructed boats will have different shapes from those of existing boats.

The same forces that cause the curves to shift down are also likely to decrease the market ex vessel price. This could lead to situations where the share rate for particular vessels will have to change in order to ensure that both the owner and the crew break even.

In any event, the total amounts of rents and their distribution will change. For example, a boat that was relatively efficient under a race-for-fish program could have been providing rents to owner and crew. However, if it becomes the marginal boat under an ITQ program, the rents to both will disappear. This is an example of the distribution effects that could follow from the efficiency incentives to build more cost-effective boats.

The nature and extent of these distributional changes will depend upon the composition of the fleet and the bargaining process that determines the share rate, and either or both of these can be affected by the types and timing of operational changes and investments encouraged by ITQs.

Returns to the Public

Because ITQs create quasi-property rights out of what is normally considered to be property of the entire nation, there are often questions about how the rents from ITQs should be distributed with respect to ITQ owners and the general public. The discussion will be somewhat limited and focused on issues raised in other sections.

In the first place, there are several views on whether the public should be reimbursed when fisheries are privatized. Some would argue that fisheries are like oil and gas resources, which the government regularly leases or sells for private exploitation. Others see a subtle difference in that the petroleum resources to be leased have normally not been previously used, but given the existing laws, fish stocks have been used by certain individuals for many years. According to this view, these previous users have earned at least a preferential right to the stocks. Ultimately, it is a political question, but one which has economic overtones.

Two interrelated questions seem particularly relevant. First, are the basic incentives of ITQ regimes affected by programs that extract payments from owners? Second, if payments are to be collected, what is the best way to do it?

The first question is a form of the principal–agent problem and is quite important to the entire philosophy of ITQs. The notion behind granting property rights is that, since owners can claim the present value of all gains and are responsible for the present value of all costs related to the use of their property, there will be incentives for both biological conservation and economic efficiency. But how will these incentives be changed if some of the gains are to be transferred from the "owner" of the ITQ back to the public?

The question is especially relevant if the transfer is to take place through an annual fee. Consider once again the breakdown of the price of ITQs on the top of Table 1. The value that will be generated from an ITQ program depends upon cost, price, and production elements in harvesting and processing and the CPUE. One of the goals of the quota in an ITQ program is to maintain or improve CPUE. Any increases in value that stem from changes in CPUE are therefore directly related to the management program. Improvements that stem from changes in the other parameters are in a sense only indirectly related to the program. ITQs provide incentives to make the appropriate investments and operational changes to develop new markets, increase recovery rates, and lower production costs. The incentives to make these changes could be adversely affected by an improperly structured or overly aggressive fee collection program. New Zealand's stated intent to drive the price of ITQs to zero by collecting all rents raises questions in this regard.

This is not to say that fee programs are inappropriate or will limit the gains from an ITQ program, because there certainly are elements of Ricardian rent in the optimal use of fisheries. In addition, since the price of ITQs will be determined by the returns to the marginal producers and harvesters, a fee that collects some of the gains related to market price will not destroy all incentives. This situation just indicates that fee programs must be designed with care.

Elsewhere, I have argued that because of the difficulties of measuring the rents from ITQ programs, and in order to reduce acrimony in the management process, it might be wise to collect rents based on *ad valorem* fish fee (Anderson 1994). If these fees are to be collected, it is also important that the types and amounts of fees be announced early in the program development and not be changed without due consideration. If ITQ owners become weary of capricious changes in taxes, the value of the right will be diminished and some of the beneficial incentives will be destroyed.

These problems can be avoided if the ITQs are initially allocated through an auction system. Prospective owners will pay a one-time fee to obtain the right and they will base their bid on the present value of the profits that can be earned taking into account the potential gains through investments in plants or boats. As far as revenue generation is concerned, this has some merits because those individuals who have access to better information concerning possible future investments will be able to make higher bids.

Alternatively, as concluded above, a considerable portion of the gains from ITQ programs will probably be a long time in coming because harvesters and processors will make capital investments and it will be difficult to predict the gains because the types of investments may not be known at the outset. Therefore, auctions based on current information may not return the same present value of revenue as a properly designed fee system.

Summary and Conclusions

While this paper has covered a wide range of topics, the main points can be summarized as follows.

- The potential efficiency gains from implementing an ITQ program will depend upon the existing regulation program, the number and types of harvesting and processing firms, and the state of the fish stock prior to its implementation.
- The status quo regulation program will most likely have an effect on the choice of capital equipment and operating procedures of participants in both sectors, and this will have an effect on the nature of the gains from implementing an ITQ program.
- The full gains of an ITQ program will probably not be achieved for several if not many years, as it will take time for participants to change their capital structure and operating procedures. Further, it will be difficult to predict exactly the size of the expected gains if current regulation programs have restricted or misallocated research and development activities.
- Although ITQ programs provide incentives to solve conservation and efficiency problems, with heterogeneous fleets and processors there can be changes in the amounts of intramarginal rents earned by the various participants.

References

Anderson, L. G. 1982. The share system in open-access and optimally regulated fisheries. Land Economics 58(4):435-49.

Anderson, L. G. 1986. The economics of fisheries management. Johns Hopkins University Press, Baltimore.

Anderson, L. G. 1994. Rents, rentals, and cost recovery in fisheries management. Pages 201-210 *in* R. H. Stroud, editor. Conserving America's fisheries. National Coalition for Marine Conservation, Savannah, Georgia.

Huppert, D. D., L. Anderson, and R. Harding. 1992. Consideration of individual fishing quotas in the North Pacific groundfish trawl fishery. Page 85-91 *in* Report prepared under contract to NOAA, NMFS. Silver Spring, Maryland.

Lindner, R. K., H. F. Campbell, and G. Beven. 1992. Rent generation during the transition to a managed fishery: the case of the New Zealand ITQ system. Marine Resource Economics 7:229-248.

Mollett, N. 1986. Fishery access control programs worldwide. Alaska Sea Grant Report Number 86-4. Fairbanks, Alaska.

Neher, P. A., R. Arnason, and N. Mollett. 1989. Rights based fishing. NATO ASI Series E, Applied Sciences. Dordrecht: Kluwer Academic Publishers.

Sutinen, J. G. 1979. Fisherman's remuneration systems and implications for fisheries development. Scottish Journal of Political Economy 26:147-62.

The Icelandic Individual Transferable Quota System: Motivation, Structure, and Performance[1]

RAGNAR ARNASON

Abstract.—This paper provides a brief description of the evolution, structure, and economic performance of the individual transferable quota (ITQ) system in the Icelandic fisheries. The ITQ fisheries management system in Iceland was instituted gradually over a period of 15 years: The system was initially imposed in the herring (*Clupea harengus*) fisheries in 1976 and, subsequently, in the capelin (*Mallotus villosus*) fishery in 1980 and the demersal fisheries in 1984. Since 1990, all Icelandic fisheries have been subject to a uniform system of ITQs. The system, however, is still subject to some dispute and, consequently, further modification and change. The key steps in the ITQ system's evolution were initially taken in response to financial crises in the respective fisheries. More recently, however, the fishing industry has agreed to a significant improvement in the fisheries management system without being threatened with the alternative of a financial disaster. The passing of the comprehensive ITQ fisheries management legislations in 1990 is a case in point. While a definitive study of the economic impact of the ITQ system is not available, the indications are that there has been a substantial economic improvement in the fisheries.

Individual transferable quotas (ITQs) constitute one of the most promising approaches to improved fisheries management. Within the framework of analytical models, it is possible to show that an appropriately designed ITQ system is capable of producing full economic rents in fisheries (e.g., Arnason 1990). It should not be forgotten, however, that these analytical models represent only an approximation to the economics of actual fisheries. In addition, they generally ignore the social environment within which the fisheries operate. Therefore, studying the socioeconomic conditions that allow the actual introduction of an ITQ system and determine the subsequent course of the fishery is of great practical importance.

This paper considers the ITQ system in the Icelandic fisheries. It describes the origin, evolution, and the current structure of the system and presents indicators of its economic impact. The paper is composed roughly of the following sections: a background description of the Icelandic fisheries, an outline of the origin and evolution of the ITQ fisheries management system, a description of the structure of the current ITQ system, and an assessment of the economic performance of the ITQ system.

The Icelandic Fisheries: A Descriptive Background

The most important Icelandic fishery by far is the demersal or groundfish fishery. In recent years, this fishery

has usually generated between 75% and 80% of the total value of all fisheries catches combined. The most important demersal species are haddock (*Melanogrammus aeglefinus*), redfish (*Sebastes* spp.), saithe (pollock *Pollachius virens*), and, in particular, cod (Atlantic cod [*Gadus morhua*]). Pelagic fisheries based exclusively on capelin (*Mallotus villosus*) and herring (Atlantic herring [*Clupea harengus*]) normally account for about 50% of the total catch volume and about 10–15% of the total catch value. In addition to demersal and pelagic fisheries, there are significant shrimp (northern shrimp [*Pandalus borealis*]), Norway lobster (*Nephrops norvegicus*), and Icelandic scallop (*Chlamys islandica*) fisheries. The history of the catches is illustrated in Figure 1. A more detailed numerical description of these fisheries and their relative importance is provided in Table 1. The fishing fleet measures about 120,000 gross registered tons (grt). It consists of the following four main vessel classes.

Deep-Sea Trawlers

Deep-sea trawlers are relatively large fishing vessels usually between 200 and 1,500 grt and 40 and 80 m in length. They are engaged almost exclusively in the demersal fisheries, employing bottom and occasionally midwater trawl. Because of their size, the deep-sea trawlers have a wide operating range and are able to exploit practically any fishing ground off Iceland. The two main types of deep-sea trawlers are conventional or fresh fish trawlers, and freezer trawlers. The fresh fish trawlers conserve their catch by refrigeration. Each fishing trip usually lasts for about 5–15 d. In recent years, there has been a trend toward freezer trawlers. Currently, there are about 30 freezer trawlers and about 75 conventional

[1]This paper is a slightly modified and updated version of one published in the journal Marine Resource Economics (Arnason 1993, 8:201-218).

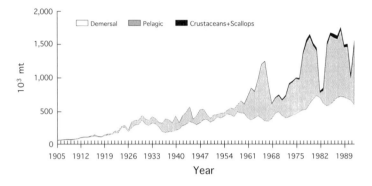

FIGURE 1.—History of the Icelandic fisheries by species categories, 1905–92 (10^3 metric tons [mt]).

trawlers. The deep-sea freezer trawlers are generally considerably larger than the conventional ones. The fishing trips usually last between 20 and 45 d.

Specialized Purse Seiners

From 200 grt and up, purse seiners primarily engage in the capelin fishery. Most also participate in other fisheries, particularly the deep-sea shrimp fishery and the herring fishery. The specialized purse seiners usually follow the capelin schools over great distances and land their catches where it is most convenient.

Multipurpose Vessels

Multipurpose vessels cover a wide size range. The typical multipurpose vessel is smaller than those previously discussed. Some, however, are quite large (i.e., >200 grt). The multipurpose fleet is, for the most part, not specialized with respect to fishing gear or fishery. Most of the multipurpose fleet is designed as gillnetters or longliners although it is technically capable of employing trawls and purse seines as well. The geographical range is limited for the smaller multipurpose vessels, which are normally confined to 1- to 3-d fishing trips exploiting fishing grounds relatively close to their home port.

The Part-Time Fleet

The numerous part-time fishing vessels have sizes up to 12 grt although most are under 10 grt. These vessels are typically owner-operated and employed on a seasonal basis. This fleet employs handline, gillnets and longlines. Depending on the fishery, the crew size is one to three persons. As the smaller of the part-time fleet (i.e., vessels <10 grt) were not subject to vessel quota restrictions until 1990, this component of the part-time fleet has mushroomed in recent years.

Further details about the Icelandic fishing fleet (Table

2) show that the average age of the fishing fleet is rather high. This may reflect the effects of more restrictive fishery management measures and official efforts in recent years to halt new investment in the fishing fleet.

The Origins and Evolution of the Fisheries Management System

Until the introduction of the vessel quota system in the demersal fisheries in 1984, most Icelandic fisheries were what may be characterized as common property fisheries. (Owing to the introduction of vessel quotas in the herring and capelin fisheries a few years earlier, the common property nature of these fisheries had been eliminated; however, these fisheries only accounted for about

TABLE 1.—Icelandic fisheries: catch and value data.

Species	Average catch 1981–90 (1,000 mt)	Estimated catch values[a] (US$ million)	Estimated MSY[b] (1,000 mt)	Estimated MSY values[a] (US$ million)
Demersal				
Cod	356.0	367.8	400.0	412.7
Haddock	55.6	79.9	60.0	86.3
Saithe	68.3	41.8	85.0	52.0
Redfish	98.1	96.9	90.0	88.9
Other	92.3	102.0	75.0	82.9
Total	670.3	688.4	710.0	722.8
Pelagic				
Herring	67.6	10.4	110.0	16.9
Capelin[c]	662.1	42.2	750.0	47.8
Total	729.7	52.6	860.0	64.7
Crustaceans				
Shrimp	24.1	42.9	30.0	53.4
Lobster	2.4	11.6	3.0	14.5
Total	26.5	54.5	33.0	67.9
Shellfish				
Scallop	13.3	5.8	14.0	6.1
All	1,439.8	801.3	1,617.0	861.5

[a]At 1990 catch prices.
[b]Maximum sustainable yield.
[c]MSY estimate represents the Icelandic share.

TABLE 2.—The Icelandic fishing fleet, decked vessels, December 1992. NA = data not available. Source: Anonymous (1992a, b).

	Number	Total tonnage (1,000 gross registered tonnes [grt])	Average age (years)
Deep-sea trawlers	107	56.756	15.6
Standard	79	NA	NA
Freezer	28	NA	NA
Purse seiners	44	20.626	23.2
Multipurpose fleet	328	39.681	25.3
>200 grt	53	14.199	26.7
111–200 grt	91	14.120	24.0
51–110 grt	103	8.091	28.0
13–50 grt	128	3.271	22.9
0–12 grt	427	3.333	13.0
All	906	120.396	

10% of the value of Icelandic fisheries). First, until the extension of the fisheries jurisdiction to 200 miles (320 km) in 1976, the Icelandic fisheries were essentially international fisheries. Large foreign fishing fleets featured prominently on the fishing grounds, taking almost half of the demersal catch. The extension of the fisheries jurisdiction to 200 miles all but eliminated foreign participation in the Icelandic fisheries. However, the initial management measures taken in the demersal fisheries following this extension consisted mostly of total quotas, limited access, and effort restrictions. Consequently, they did not alter the common property nature of these fisheries for domestic fishers, who were still forced to compete for shares in the catch. Therefore, not surprisingly, the development of the Icelandic fisheries in the post-war era closely followed the path predicted for common property fisheries (e.g., Gordon 1954), exhibiting increasingly excessive fishing capital and effort compared with the reproductive capacity of the fish stocks (Figure 2). The value of fishing capital employed in the Icelan-

dic fisheries increased more than twelvefold from 1945 to 1983 while real catch values only tripled. Thus the growth in fishing capital exceeded the increase in catch values by a factor of more than four. This means that in 1983, the output/capital ratio in the Icelandic fisheries was less than a third of the output/capital ratio in 1945.

This long-term decline in the economic performance of the Icelandic fisheries did not go unnoticed by the fisheries authorities. In fact, over the years, various measures were taken in an attempt to reverse this trend. However, before the extension of the exclusive fishing zone to 200 miles in 1976, effective management of the fisheries, especially the demersal ones, appeared impracticable because of the large foreign fleets on the fishing grounds. For this reason, fishery management prior to the 200-mile extension was minimal.

With the de facto recognition of the exclusive 200 mile fishing zone in 1976, the situation was dramatically changed. Since that time, the Icelandic fisheries have come under gradually increasing management, culminating in a uniform ITQ system in practically all fisheries since 1990 (Table 3). A more detailed review of the evolution of the ITQ fisheries management system in individual Icelandic fisheries follows.

The Herring Fishery

In 1969, owing to an alarming decline in the herring stocks, an overall quota was imposed on this fishery. Since this did not halt the decline in the stocks, a complete herring moratorium was introduced in 1972. In 1976, when fishing on the Icelandic herring stocks was partly resumed, it was obvious that the whole fleet could not participate. Hence, an individual vessel quota system with limited eligibility was introduced. Vessel quotas, however, were small and, in 1979, by a ministerial decree and with industry support, transfers of quotas were permitted between vessels eligible to participate in the herring fishery. In 1988, the vessel quota system in the

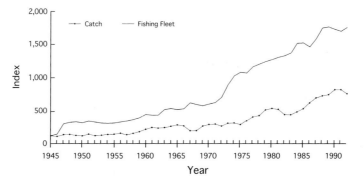

FIGURE 2.—Fishing capital and catch values (fixed prices), 1945–92.

TABLE 3.—Key steps in the evolution of the ITQ management system: a chronological overview.

1975	Individual vessel quotas applied to herring fishery
1979	Vessel quotas made transferable in herring fishery
1980	Individual vessel quotas applied to capelin fishery
1984	Individual vessel quotas applied to demersal fisheries
1985	Effort quotas option introduced to demersal fisheries
1986	Vessel quotas made transferable in capelin fishery
1988	Transferable vessel quotas in all fisheries; effort quota option retained
1990	Complete uniform system of transferable vessel quotas in all fisheries

herring fishery became part of the general fisheries vessel quota system.

The Capelin Fishery

The capelin fishery, which became very big in the 1970s, was subjected to limited entry and individual vessel quotas for license holders in 1980, a time when the stock was seriously threatened with exhaustion. The arguments for this arrangement were similar to those used previously in the herring fishery. The positive experience with the vessel quota system in the herring fishery also proved a convincing argument for adopting a similar system in the much more important capelin fishery. In 1986, the capelin vessel quotas became partly transferable. In 1988, the capelin vessel quota system became a part of the general fisheries vessel quota system with fairly unrestricted transfers of quotas.

The Demersal Fisheries

Following the extension of the exclusive fishing zone to 200 miles in 1976, the major demersal fishery (the cod fishery) was subjected to an overall catch quota. The annual quotas recommended by the marine biologists soon proved quite restrictive and thus difficult to maintain. Hence, individual effort restrictions, taking the form of limited allowable fishing days for each vessel, were introduced in 1977. However, since new entry remained possible and the demersal fleet continued to grow, the annual allowable fishing days had to be reduced from year to year. At the beginning of the individual effort restriction regime in 1977, deep-sea trawlers were allowed to pursue the cod fishery for 323 d only. Four years later, in 1981, the number of allowable fishing days for cod had been reduced to 215. This system was obviously economically wasteful. Consequently, in 1984, following a sharp drop in the demersal stock and catch levels, a system of individual vessel quotas was introduced. Initially, the Icelandic Parliament, the Althing, passed legislation to this effect for 1 year only. In 1985, because of generally favorable results of the individual quotas, the system was extended

for another year. However, an important provision was added: Vessels preferring effort restrictions could opt for that arrangement in place of the individual quota restriction. This system was extended largely unchanged for an additional 2 years in 1986. In 1988, the Althing passed a general vessel quota legislation for all Icelandic fisheries to be effective for 1988–90. In 1990, a complete, uniform vessel quota system for all fisheries, the Fisheries Management Act, was legislated, abolishing the limited effort option in the demersal fisheries as well as a few other loopholes.

The Shrimp and Scallop Fisheries

The inshore shrimp and scallop fisheries are relatively recent additions to the Icelandic fisheries. These fisheries were largely developed during the 1960s and 1970s and have, practically from the outset, been subject to extensive management consisting primarily of limited local entry as well as overall quotas. In recent years, there has also been a strong movement towards vessel quotas in these fisheries. With the fisheries management legislation passed in 1988, the deep-sea shrimp fishery, the only remaining significant Icelandic fishery not closely managed, was also subjected to vessel quotas. The management of the shrimp and scallop fisheries is now part of the general ITQ system according to the general fisheries management legislation of 1990.

As may be inferred from the preceding descriptions, fishery management in Iceland has evolved more by trial and error than by design. In most countries, and Iceland is no exception, there is a strong social opposition to radical changes in the institutional framework of production and employment. A great deal of this opposition seems to derive from traditional values and vested interests rather than rational arguments. Therefore, in Iceland, it was probably unavoidable from a sociopolitical point of view to pass through an evolutionary process during which various management methods were tried in different fisheries. The knowledge and understanding gained from these experiments was probably crucial for the eventual acceptance of an efficient fisheries management system.

At the same time, it should be noted that the key steps in the evolution of the fisheries management system have usually only been taken in response to crises in the respective fisheries due to a sudden reduction in stock levels. Thus, management of the herring fisheries started in 1969 in response to an imminent collapse in the herring stocks. Similarly, the management of the capelin fishery and the current management of the demersal fisheries were implemented in the early 1980s in response to a perceived danger of a corresponding collapse in these fisheries.

This pattern reflects, above anything else, the reluctance of members of the fishing industry to accept changes in the traditional organization of the fisheries. Only when faced with a disaster in the form of a significant fall in income due to fish stock reductions or a drop in the world market price for fish products have the interest groups been willing to consider changes in the institutional framework of the fisheries.

The passing of the comprehensive fisheries management legislation in 1988 and, even more so, in 1990 constitutes a break with this pattern. For the first time, the fishing industry agreed to a significant strengthening of the fisheries management system without being threatened with a financial disaster as the alternative. This must, I think, be attributed to the fact that the potentially immense benefits of the vessel quota system were becoming apparent to most of the participants in the fisheries.

The Current ITQ Fisheries Management System

The Icelandic ITQ fisheries management system was instituted at different times and in somewhat different forms in the various fisheries. As mentioned previously, it was made uniform by fisheries management legislation passed in 1990.

The essential features of the current ITQ system are as follows: All fisheries are subject to vessel catch quotas. The quotas represent shares in the total allowable catch (TAC). They are permanent, perfectly divisible, and fairly freely transferable. They are issued subject to a small annual charge to cover enforcement costs.

The ITQ system was superimposed on an earlier management system designed mainly for the protection of juvenile fish. This system, involving certain gear, area, and fish size restrictions, is still largely in place. The ITQ system has not, in other words, replaced these components of the earlier fisheries management system. Further details of the ITQ system in the Icelandic fisheries follow.

Total Allowable Catch

The Ministry of Fisheries determines the TAC for each of the most important species in the fisheries. This decision is made on the basis of recommendations from the Marine Research Institute, which the Ministry of Fisheries has followed quite closely.

Currently, 10 species are subject to TACs and, consequently, individual quotas. These include six demersal species (cod, haddock, saithe, redfish, Greenland halibut [*Reinhardtius hippoglossoides*], and plaice [*Hippoglossoides platessoides*]), two pelagic species (herring and capelin), and two species of crustaceans (shrimp and

lobster). In addition to these, several exploited species are not currently subject to TACs. This means that the corresponding fisheries can be pursued freely. For the most part, these species are subject to relatively light fishing pressure or appear primarily as bycatch of other fisheries. Most are also commercially negligible. The 10 species subject to ITQ restrictions account for well over 90% of the total value of the Icelandic fisheries.

Permanent Quota Shares

Each eligible vessel is issued a permanent quota share in the TAC for every species for which there is a TAC. These permanent quota shares may be referred to as TAC shares.

Initial Allocation of Permanent Quota Shares

The initial allocation of TAC shares to individual vessels varies somewhat among fisheries. In the demersal, lobster, and deep-sea shrimp fisheries, the TAC shares were normally based on the vessel's historical catch record during certain base years. In the demersal fisheries, this usually equaled the vessel's average share in the total catch during the 3 years prior to the introduction of the vessel quota system in 1984. There are noteworthy exceptions to this rule, however. For example, a demersal vessel not operating normally during 1981–83 because of major repairs or because it entered the fleet after 1981 could have its calculated share adjusted upwards. Also, during 1985–87, it was possible to modify the TAC shares by temporarily opting for effort restrictions instead of vessel quotas and demonstrating high catches during this period. In the herring and inshore shrimp fisheries, the initial TAC shares were equal and the same applied to the capelin fishery except that one-third of the TAC shares were initially allocated on the basis of vessel hold capacity.

Annual Vessel Quotas

The size of each vessel's annual quota in a specific fishery is a simple multiple of the TAC for that fishery and the vessel's TAC share. For instance, if the vessel's TAC share is 1% and the TAC is 100,000 metric tons (mt), then the vessel's annual quota would be 1,000 mt.

Divisibility and Transferability

Both the TAC shares and the annual vessel quotas are transferable and perfectly divisible. This means that any fraction of a given quota may be transferred to another vessel.

The TAC shares are transferable without any restrictions whatsoever. Transfers of annual vessel quotas, on the other hand, are subject to some restrictions. Annual vessel quotas are freely transferable between vessels within the same geographical region. Transfers of annual quotas between geographical regions are subject to revision by the respective fishers' unions and the local authorities. The rationale for this stipulation is to stabilize local employment in the short run. In practice, however, it appears that few interregional transfers are actually blocked.

Apart from this, transfers of quotas only are subject to registration with the Ministry of Fisheries. The particulars of the exchange, including price, are not registered.

Restricted Access: Fishing Licenses

All commercial fishing vessels must hold valid fishing licenses. Moreover, a fishing license is a prerequisite for being allocated a quota. Fishing licenses are issued only to vessels already in the fishery in 1990 and their replacements provided they are deemed comparable in terms of fishing power. The fishing licenses are not transferable.

Thus, in addition to the ITQ system, the Icelandic fisheries are subject to restricted access. One impact of a well-designed ITQ system is to provide the socially appropriate incentive for disinvestment (investment) in the fishing fleet. The fishing license stipulation clearly adds a deterrent to increasing the number of fishing vessels.

Exemptions from the ITQ System

There are two minor exemptions from the current ITQ system, both in the demersal fisheries. The first concerns longline demersal fisheries in mid-winter. More precisely, 50% of the demersal catch of vessels employing longline during November through February each winter is exempt from quota restrictions. The reason for this exception is primarily to support regional employment during this period.

This exemption means that vessels employing longline from November through February can exceed their quota allocation by 50% of their catches during these 4 months. Before allocating annual ITQs to vessels, the fisheries authorities set aside a certain fraction of the TAC to allow for this.

Second, hook-and-line vessels under 6 grt may elect to be exempted from quota restrictions, in which case they are subjected to limited fishing days. This effort limitation is adjusted downward if the total catch of this part of the fleet exceeds a certain volume.

Quota Fees

The annual vessel quotas calculated in the previously described manner were initially issued by the Ministry of Fisheries free of charge. However, according to the Fisheries Management Legislation of 1990, the Ministry of Fisheries is to collect fees for catch quotas to cover the cost of monitoring and enforcing the ITQ regulations. The law imposes an upper bound on this fee currently amounting to 0.4% of the estimated catch value. This percentage is probably more than sufficient to cover the extra costs of operating the ITQ system compared with the costs of the previous fisheries management system.

The Icelandic ITQ system exhibits most of the crucial features of the ideal ITQ system as discussed in the theoretical literature (e.g., Arnason 1990). It is important to realize, however, that there are certain aspects of the Icelandic ITQ system that deviate from the theoretical ideal and almost certainly subtract from its economic efficiency.

First, as discussed earlier, there are certain exemptions from the ITQ system. These exemptions are admittedly relatively minor compared with the total volume of the fisheries. Nevertheless, being exempt from the ITQ regime, the vessels in question are essentially engaged in the traditional competition for catch shares, with the associated economic waste.

Second, in the Icelandic ITQ system, the ITQs are closely associated with fishing vessels. More precisely, only people who own fishing vessels with a valid fishing license can hold quotas. In addition, the total quota holdings must not exceed the fishing capacity of the vessel in question (although this particular stipulation actually seems to be loosely interpreted and enforced). The set of potential holders of ITQs is thus severely restricted. This clearly subtracts from the ability of the quota market in effecting the most economically beneficial allocation of quotas.

Third, the holders of TAC shares must harvest at least 50% of their TAC share every second year to retain the share. This stipulation is designed to obstruct speculative quota holdings. However, in so doing, it reduces the efficiency of the quota market and may induce the fishing industry firms to maintain more fishing vessels seaworthy than would be optimal.

The Performance of the ITQ System

The main purpose of the vessel quota system is to improve the economic efficiency of the fisheries. The Icelandic fisheries are biologically very productive and should be able to generate high economic rents. However, until the adoption of the vessel quota system, comparatively low rents were generated in the industry. In fact, during the years preceding the introduction of the vessel quota system in the various fisheries, annual losses were often quite substantial.

The Herring Fishery

When the herring fishery was resumed in 1975, at the end of a 3-year fishing moratorium, a system of individual catch quotas was imposed. Because of the generally favorable experience with this system, the quotas were made perfectly divisible and transferable in 1979. In 1990, the herring fisheries management system was incorporated, largely unchanged, into the comprehensive fisheries management system for the Icelandic fisheries.

The ITQ system in the herring fishery has been very successful. Since 1975, the herring stock biomass has tripled and catches have increased fivefold (Figure 3). Fishing effort,[2] on the other hand, has not increased. In fact it has declined by some 20%. This means that the technical efficiency in the herring fishery is now more than 6 times higher than it was at the outset of the vessel quota system in the fishery 19 years ago.

---

[2]Fishing effort is here defined as the application of fishing capital to the fishery and is measured as a multiple of vessel tonnage and days fishing.

Since the introduction of the individual quota system in 1975, catch per unit effort has clearly increased (Figure 4). This development is especially pronounced after the quotas were made transferable in 1979.

The Capelin Fishery

An individual vessel quota system was introduced in the capelin fishery in 1980. In 1986, the quotas were made transferable. In 1990, the capelin management system was incorporated into the overall Icelandic fisheries management system.

The capelin is a short-lived species and the fishery, which depends on a single cohort each year, is very volatile. Since the introduction of the vessel quota system in 1980, in spite of rather dramatic fluctuations, there has been no discernible trend in annual catch levels (Figure 5).

While the catches have remained largely unchanged, the capelin fleet has been substantially reduced. At the outset of the capelin ITQ system in 1980, 68 capelin purse-seiners were operating. At the end of 1992, there were only 39. This change represents a reduction in

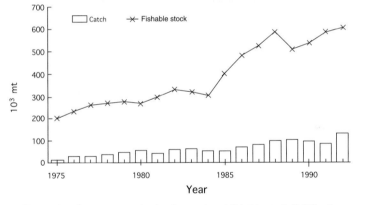

FIGURE 3.—Summer spawning herring catch and fishable stock (1,000 mt).

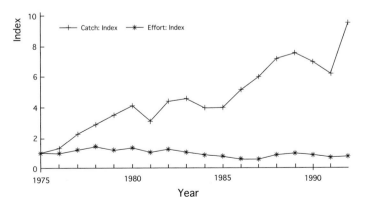

FIGURE 4.—Summer spawning herring catch and effort.

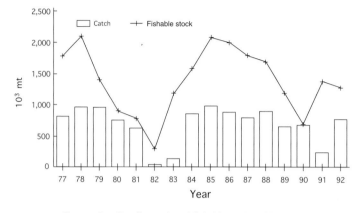

FIGURE 5.—Capelin catch and fishable stock (1,000 mt).

vessel numbers of over 40%. Capelin fishing effort, measured as the multiple of vessel tonnage and days of fishing, has also been substantially reduced (~35%). Thus, there are strong indications that the efficiency of the capelin fishery has been substantially increased since the introduction of the vessel quota system. The development of the catch and the fishing effort fleet size in terms of tonnage is illustrated in Figure 6.

The Demersal Fisheries

The demersal fisheries are by far the most important Icelandic fisheries, accounting for over 75% of the total value of the catch. These fisheries were subjected to ITQs in 1984. However, already in 1985, an optional effort alternative became available. The system was subsequently under almost continuous revision until the adoption of the uniform fisheries management system based on ITQs in 1990.

The trend in fishing capital and fishing effort.—We have seen that one of the reasons for the dissipation of economic rents in the Icelandic fisheries has been overinvestment in fishing capital and excessive fishing effort. Therefore, one of the tests of the efficacy of the vessel quota system is the development of fishing capital and aggregate fishing effort since the introduction of the system.

When the vessel quota system was introduced in 1984, the previous growth in the value of aggregate harvesting capital halted (Figure 7). In fact, fishing capital contracted in both 1984 and 1985. This was the first time since 1969 that the value of the fishing fleet actually decreased. In the preceding 15 years, this capital value had grown at an annual rate of over 6%. Thus, at present, the vessel quota system seems to have generated beneficial results. (The years 1982–84, however, were periods of heavy losses for the fishing industry; therefore, the halt in investment in 1984–85 can hardly be attributed exclusively to the vessel quota system.) In 1986, on the other hand, investment in fishing capital resumed at a high rate. However, this resumption of investment should not be interpreted as a failure of the vessel quota system as such. After all, the increase in the value of fishing capital since the inception of the ITQ system has amounted to just over 3% annually. Moreover, most of the investment since 1986 can be explained by factors extraneous to the ITQ system.

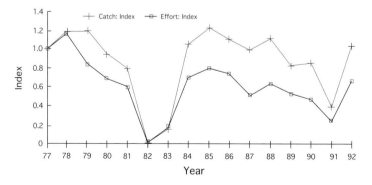

FIGURE 6.—Capelin fishery catch and effort.

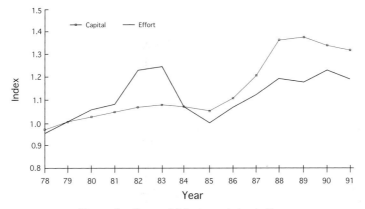

FIGURE 7.—Demersal fisheries capital and effort.

First, a good deal of the investment in fishing capital from 1986 onwards has consisted of the installation of freezing equipment and the corresponding modifications of several deep-sea trawlers.[3] In other words, this part of the investment is in fish processing capital employing new and profitable techniques. Second, a part of the investment was in specialized trawlers for the emerging and very valuable deep-sea shrimp fishery, which was not subject to vessel quotas until 1988. Third, by the mid-1980s, a significant fraction of the deep-sea trawler fleet was due for replacement. As 1986 and 1987 were unusually profitable for the harvesting sector, many firms took the opportunity to replace their aging vessels. Fourth, this period saw a very significant investment in small vessels (<10 grt) that were not subject to the vessel quota system. Last but not least, the effort quota option in the demersal fisheries, introduced in 1985, undermined the efficiency incentives of the ITQ system, inducing many vessel owners to upgrade or replace their vessels. During 1986–90, less than 50% of the demersal catch was taken under the ITQ system. The effort quota option was abolished in 1990 and a significant reduction in fishing capital occurred in that year. More importantly, this reduction continued in 1991 and 1992; from 1990 to 1992, the value of fishing capital was reduced by almost 15%.

The course of demersal fishing effort tells a similar story. Fishing effort in the demersal fisheries, measured as a multiple of fleet tonnage and days at sea, dropped by some 15% in 1984 (Figure 7), the first year of the vessel quota system, and by an additional 6% in 1985. From 1986 to 1990, on the other hand, fishing effort increased considerably, primarily owing to the widespread selection of the ill-advised effort quota option within the ITQ system during 1986–90. Another important expla-

nation for the increase in fishing effort in 1989 and 1990 is the decline in the demersal fish stocks without a commensurate reduction in the TACs, thus requiring more fishing effort to fill the catch quotas. Although exact statistics are not available, demersal fishing effort in 1991 apparently declined somewhat again.

The main question, however, is not whether fishing effort has been reduced from its 1983 level. The crucial measure of the impact of the vessel quota system is the difference, if any, between the actual fishing effort in 1984–90 and the fishing effort level that would have prevailed during that period had the vessel quota system not been introduced.

Clearly, it is not at all straightforward to predict the course of fishing effort under the earlier management regime. However, in an attempt to provide a partial answer to this question, a simplistic model has been used to explain the path of fishing effort under the two different management regimes. Essentially, a simple trend model describing the path of fishing effort under the two management regimes was specified. Somewhat more precisely, it was hypothesized that during the 13-year period (1978–90), fishing effort evolved over time according to the following relationship:

$$e(t) = (a \cdot D_1 + b \cdot D_2) \cdot \exp[(c \cdot D_1 + d \cdot D_2) \cdot t]$$

where e(t) = fishing effort in year t,
    t = years measured from 0 to 13,
D_1 and D_2 = dummy variables for the two management regimes,
    a and b = intercepts under the two management regimes, and
    c and d = growth rates of effort under the two management regimes.

Thus, $D_1 = 1$ during the years of restricted effort (e.g., 1978–83) and 0 thereafter, and $D_2 = 0$ in the years preceding 1984 and 1 thereafter.

---

[3]In 1983, there were three freezer trawlers; in 1990, there were 28.

The hypothesis that there is no structural break in the evolution of fishing effort between the two management regimes is resoundingly rejected. The relevant test statistic is c^2 (2 df) = 58.9. Similarly, the growth of fishing effort under the ITQ regime is significantly lower than under the previous fisheries management regime. The relevant test statistic is c^2 (1 df) = 6.9.

Employing this estimated relationship, we may predict the fishing effort assuming that the vessel quota system had not been introduced in 1984 and compare this with the actual fishing effort observed (Figure 8). According to estimation results (Figure 5), fishing effort under the earlier fisheries management system would, in all likelihood, have continued to increase at a high rate after 1983. Judging from these estimates, the vessel quota system appears to have reduced total demersal fishing effort by some 34% compared with the expected fishing effort under the previous management system. The financial benefits of this kind of effort reduction are very substantial.

In interpreting these results, however, readers should be mindful of the extreme simplicity and mechanistic nature of the underlying model. They should also realize that the observed path of fishing effort under both fisheries management regimes is a result of all the operating conditions of the fishery. In particular, TAC restrictions and the ratio between the TAC and the fishable biomass level are probably more important determinants of overall fishing effort than the ITQ system itself. It is another matter that ITQs may make it easier to impose restrictive TACs and to enforce them.

More direct estimates of economic benefits.—We now turn to more direct estimates of the economic benefits generated under the vessel quota system. Unfortunately, little research has been done in this area and the available information is, consequently, rather scant.

From a theoretical point of view, the economic benefits of a vessel quota system should include the following items.

A reduction in fishing effort: Under the vessel quota system, competition between vessels for a limited stock of fish is eliminated. Consequently, the fishing firms will attempt to catch their vessel quota with minimum fishing effort. It is important to realize, however, that aggregate fishing effort will not necessarily be reduced if the TAC is excessive. Given the size of the fish stocks, each TAC requires a certain minimum fishing effort. If the TAC is set high relative to the size of the fish stocks, aggregate effort may actually increase under an ITQ system.

Reduced cost of fishing effort: Having secured private ownership of a certain volume of catch under the vessel quota system, the fishing firm can concentrate on taking that catch with minimal costs.

Improved quality of the catch: Being bound by its vessel catch quotas, the fishing firms can increase revenues only by improving the quality of this catch.

In a study carried out in 1985, the National Economic Institute (Reykjavik, Iceland, unpubl. rep.) attempted to estimate the benefits of reduced fishing effort and improved quality of the catch in the demersal fisheries for the year 1984. The conclusion was that the benefits of reduced fishing effort amounted to some US$14 million and improved quality of the catch to some US$6 million. The total number, US$20 million, is about 8.5% of the value of the demersal fisheries in that year. These results were confirmed in a less comprehensive study done in 1987 (Althing 1987).

Quota values.—Yet another way to approach the problem of estimating the rents generated in the demersal fisheries as a result of the vessel quota system is to look at quota values. As the catch quotas are transferable, a market for quotas has developed. In this market, quotas are exchanged for other valuables such as money. Hence,

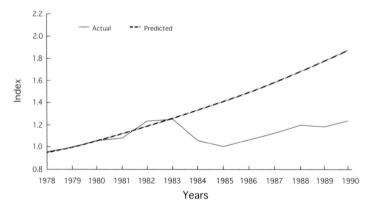

FIGURE 8.—Actual vs. predicted effort in the demersal fisheries.

applying standard economic theory and assuming that the market for quotas is reasonably effective, the value of the fishery should equal the total value of quotas.

There are two quota values to consider. One is the annual quota value, which is the multiple of the annual quota price and the corresponding total volume of annual quotas. This measure may be more formally characterized as the annual rental price of TAC shares. The other is the permanent quota value. This is the current market value of the TAC shares themselves. This, presumably, reflects the present value of expected annual quota values in the future. Alternatively, it measures the present value of expected future profits of using the TAC shares for fishing in the future. Clearly, the permanent quota value depends on the expected future path of biomass, TACs, fish prices, fishing technology, and fishing costs. It is the market evaluation of the future value of the fishery.

In what follows, we estimate the value of the fishery on the annual basis using annual quota values. This, as discussed previously, should provide a reasonable estimate of the economic rents generated annually in the fishery. Notice, however, that this value will not necessarily equal bookkeeping results. Quotas will be bought on the market at a price up to their marginal variable profits. Fixed costs (e.g., those associated with harvesting capital) are irrelevant for these transactions. Therefore, a firm may buy quotas at a high price without being able to cover fixed costs.

The extent of the demersal quota market is considerable (Table 4). In 1984, the first year of the vessel quota system, about 11% of quotas held by the fishing industry were exchanged. Since 1986, this percentage has fluctuated between 20% and 30%. The fourth column of Table 4 shows that, between 1986 and 1990, less than one-half of demersal catch was taken under the ITQ system. This reflects the widespread use of the limited effort option during these years. Hence, although the fraction of outstanding quotas being traded in 1986–90 had more than doubled compared with the initial year, 1984, the actual

quantities exchanged were not any greater than in 1984–85.

The economic rents generated in the demersal fisheries can be inferred from the quota values (Table 5). Judging from the fourth column of Table 5, it appears that the total value of outstanding quotas evaluated at the midpoint of the given price range was some US$46 million in 1984 and US$245 million in 1990. However, these numbers almost certainly underestimate the true value of demersal catch rights. The reason is that they ignore the value of the non-tradable catches, which are mostly taken under effort quotas. If all the demersal catch is evaluated at the vessel quota prices, we obtain the valuation in the last column of Table 5. According to this column, the economic rents generated in the demersal fisheries are considerable and constitute a significant fraction of the gross earnings of the fisheries. Thus, in 1989 the demersal quota values exceeded one-quarter of total earnings in the demersal fisheries.

These estimates, however, must be interpreted with great care, especially during the latter part of the period. During these years, demersal catches were good and fish prices extremely good. For this reason, the quota prices for 1986–90 were probably higher than would otherwise have been the case although considerably lower than in a fully efficient equilibrium. More importantly, it must be realized that one of the first effects of a reasonably complete ITQ system is to make excessive fishing capital commercially redundant. This means that its market price falls drastically, the opportunity cost of its use is reduced, and the market value of catch quotas is correspondingly increased. This, however, is a short-term effect that will be reversed in the long run when the level of fishing capital reaches a new equilibrium.

How does the economic performance of the demersal vessel quota system compare with optimal economic use? Comparing the above quota price valuation of fisheries rents with the maximum attainable rents according to an empirical study of the Icelandic demersal fisheries (see Arnason 1984) makes it apparent that the realized economic rents in 1990 were well over 50% of the maximum

TABLE 4.—The extent of the demersal quota market (NA = information not available).

Year	Total quota transfers		Vessel quotas as fractions of total demersal catch
	1,000 mt	% total vessel quotas	
1984	64.1	11.2%	98.1%
1985	81.0	19.2%	76.8%
1986	52.2	24.4%	36.0%
1987	40.1	21.1%	28.1%
1988	61.4	23.7%	43.8%
1989	72.2	29.8%	42.5%
1990[a]	78.9	22.6%	68.2%
1991	NA	NA	95.9%

[a] The new complete ITQ system took effect on Sept. 1 of this year.

TABLE 5.—Economic rents in demersal fisheries: quota price valuation.

Year	Annual quota price range US$ million/mt		Total quota values US$ million	
	Cod	Other	All quotas	Whole fishery
1984	55–87	24–40	36–57	36–57
1985	84–126	54–72	36–53	51–72
1986	127–176	79–109	23–32	66–91
1987	206–259	104–131	35–44	104–131
1988	208–277	154–205	49–65	108–144
1989	262–349	157–209	62–83	143–189
1990	428–514	256–308	151–182	222–267

attainable ones. Previous reservations concerning the probable upward bias of the quota values as estimators of sustainable rents in 1990 must, however, be stressed. Nevertheless, it is clear that significant benefits have been generated by the demersal vessel quota system. As fishing effort has been reduced only slightly, these benefits must primarily be attributed to reduced harvesting costs per unit of catch and higher quality of the catch. In fact, there is ample evidence that this has occurred.

The Problem of Discards

Discarding of catch or high-grading is an often-cited problem with ITQ systems, especially in mixed fisheries (e.g., Rettig 1986, Squires and Kirkley 1991). The Icelandic demersal fisheries are certainly mixed fisheries. Nevertheless, there is little evidence of increased discarding under the ITQ system. According to measurements published in a recent report by a government commission (Nefnd um mótun sjávarútvegsstefnu 1993), demersal discards range from 1% to 6% of total catch volume depending on gear and vessel type. Moreover, according to this report there has been no detectable increase in discards since the introduction of the vessel quota system in 1984. Since 1993, however, the cod quotas have been drastically reduced and the quota market prices have skyrocketed and there have been growing rumors that discards of cod have increased. This, if true, would in fact conform with the economic theory of discards (Anderson 1994; Arnason 1994).

Conclusion

Versions of the ITQ fisheries management system have been in operation in Icelandic fisheries since 1979. The evidence on the performance of this system, although somewhat mixed, is generally favorable.

The introduction of the ITQ system in the herring fishery appears to have resulted in a dramatic increase in efficiency. In the capelin fishery, the ITQ system also appears to produced substantial economic benefits although less striking than in the herring fishery. In the demersal fisheries, the evidence is less conclusive: The fishing fleet has increased and aggregate fishing effort has contracted only slightly; however, various indicators, including quota values in the demersal fisheries, strongly suggest that significant rents are being generated by the system.

When interpreting the development of the demersal fisheries since 1984, it is important to be mindful of the imperfectness of the ITQ system employed during the early part of the period. This applies especially to the widespread use of the effort quota option during 1985–90. In fact, as shown in Table 4, the fisheries management system was only partially an individual quota system in these years. For this reason, the upward trend in harvesting capital and demersal fishing effort during this period is hardly evidence of the failure of the ITQ system. In fact, as previously explained, since the abolition of the effort quota option and the consolidation of the ITQ system in 1990, demersal fishing capital has declined substantially. Finally, there is little evidence that the vessel quota system has increased discarding of catch.

References

Althing. 1987. Parliamentary bill for fisheries management 1988–1991. Annexes III and IV. Icelandic Parliament Session 110.

Anonymous. 1992a. Utvegur, Fiskifélag Íslands. Fisheries Society of Iceland, Reykjavik.

Anonymous. 1992b. Sjomannaalmannak, Fiskifélag Íslands. Fisheries Society of Iceland, Reykjavik.

Anderson, L. G. 1994. An economic analysis of highgrading in ITQ fisheries regulation programs. Marine Resource Economics 9:209-226.

Arnason, R. 1984. Efficient harvesting of fish stocks: the case of the Icelandic demersal fisheries. Ph.D. dissertation, University of British Columbia.

Arnason, R. 1990. Minimum information management in fisheries. Canadian Journal of Economics 23:630-653.

Arnason, R. 1993. The Icelandic individual transferable quota system: a descriptive account. Marine Resource Economics 8:201-218.

Arnason, R. 1994. On catch discarding in fisheries. Marine Resource Economics 9:189-207.

Gordon, H.S. 1954. Economic theory of a common property resource: the fishery. Journal of Political Economy 62:124-142.

Nefnd um mótun sjávarútvegsstefnu. 1993. Skýrsla til Sjávarútvegsráoherra. Reykjavík, Iceland.

Rettig, R. B. 1986. Overview. In N. Mollett, editor. Fishery access control programs worldwide: Proceedings of the workshop on management options for the north Pacific longline fisheries. Alaska Sea Grant Report 86-4, Fairbanks, Alaska

Squires, D., and J. Kirkley. 1991. Production quota in multiproduct Pacific fisheries. Journal of Environmental Economics and Management 21:109-126.

The Political Economy of ITQs

Rögnvaldur Hannesson

Abstract.—Fish resources worldwide are coming under increasing pressure. This generates incentives to claim exclusive use rights. Some social scientists seem to favor communal rather than private rights. It is argued that the differences between these rights are more apparent than real, as both imply exclusion of some users. Individual transferable quotas (ITQs) are use rights that are becoming more widespread. Surprisingly, the initiative for these rights has come from from public authorities rather than the industry, even though politicians or public servants do not have strong incentives to establish such rights. ITQs are, nevertheless, unlikely to be established without the support of a critical mass in the industry. This support requires that the industry gets a share of the rent of the resource. To what extent this will be fair and compatible with the public interest is discussed. Contrary to "ordinary" industries, rents in the fishing industry will be eroded not through healthy competition but through unnecessary costs, since these rents reflect a pure natural scarcity. Any lasting gain in efficiency will therefore mean lasting rents, which will be captured by the first generation of rights holders unless some suitable measures are being taken. If accrual of rents to private individuals is considered unacceptable, it is neccessary to identify a community that is broad enough to be a worthy receiver of rents and narrow enough for the rents to make a difference, since rents are the driving force for achieving economic efficiency. Despite possibly substantial gains by those who are grandfathered into a closed industry, there is widespread opposition among boat owners and fishers to exclusive use rights such as ITQs. Uncertainty, distributional conflicts, and ideology, may explain such opposition.

Of Forests and Fish

In the early 1840s, a liberal newspaper called the *Rheinische Zeitung* was published in Cologne, Germany. The paper soon acquired a young and resourceful editor whose name was Karl Marx. In one of his articles, he offered the following observations: "If every violation of property, without distinction or more precise determination, is theft, would not all private property be theft? Through my private property, do not I deprive another person of this property? Do I not violate his right to property?" (McLellan 1973).[1]

Of particular interest in this context is that Marx was writing about exclusive use rights over what used to be a free-access resource. The public had by custom gathered dead wood for their stoves and fireplaces from the forests of the Rhineland. By the time Marx was editing the Rheinische Zeitung, the demand for wood had increased to such an extent that the wood collectors were much resented, and the great majority of the court cases in the Rhineland dealt with the theft of wood. A law was proposed empowering forest keepers to assess damages and impose fines in order to ease the burden on the courts. Marx argued that the general public had common rights in the forests by tradition and that the state should defend those rights.

This story illustrates two points. First, as open-access resources come under increased pressure, exclusive use rights are likely to develop. If this does not happen, the resources will be depleted, possibly beyond recovery. Second, such exclusive use rights must ultimately be sanctioned by the state to ward off a challenge by those who are unwilling to accept being excluded.

There are obvious parallels between what happened in the Rhineland in the 1840s and what is happening now in world fisheries. Improved technology and increased demand for fish have increased the pressure on fish stocks to the point where it is impossible to accommodate all those who wish to harvest. Exclusive use rights, such as individual quotas and fishing licenses, are already beginning to develop but are still highly contested. What form should such rights take? Who should get them?

---

[1]Quoted from McLellan (1973), p. 56. The *Rheinische Zeitung* was not popular with the Prussian authorities of its day, and the censor vetoed many articles (the Rhineland was awarded to Prussia at the Vienna conference, after the Napoleonic wars). In 1843, the newspaper was banned. Prior to that, Marx once wrote the following, possibly to placate the censor or his wealthy bourgeois financiers: "The 'Rheinische Zeitung', which cannot even concede theoretical reality to communistic ideas in their present form, and even less wish or consider possible their practical realization, will submit these ideas to thorough criticism" (quoted from McLellan, op. cit., p. 54). Marx wrote a series of articles on the proceedings on the new forest law in the Rhineland parliament and argued forcefully for respecting the traditional rights of the wood collectors. The original text of these articles can be found, inter alia, in "Karl Marx, Werke, Artikel, Literarische Versuche bis März 1843," Volume I:1 in Karl Marx/Friedrich Engels Gesamtausgabe, Dietz Verlag, Berlin 1975. The papers on the wood question occupy pages 199-236. Those who take the trouble to look up this volume will be rewarded by some beautiful love poems Marx wrote to his fiancé, Jenny von Westphalen, as he was studying philosophy under Hegel in Berlin.

What is the role of the state in establishing and enforcing such rights? Since exclusive use rights for terrestrial resources have a much longer history, it is worthwhile to consider these briefly before turning to exclusive use rights in fishing.

The Development of Property Rights

Property rights undoubtedly arise from human greed, and they amount to an institutionalization of the rule "This is mine and not yours." Or, to use Marx's language, one person's property deprives another of the same property. Undoubtedly, also, establishing and enforcing private property rights often have amounted to theft or robbery. Farmers have been forcibly deprived of common pastures by greedy landlords ultimately relying on the coercive power of the state. A famous example is the clearances of the Scottish Highlands. In a process that nowadays might be called "ethnic cleansing," the Highlanders were driven off the land by brute force and replaced by sheep (Galbraith 1977).

Yet it is a challenging thought that a social institution arising from such ignoble origins can in fact be of great benefit for the common good. Introducing property rights is not a zero sum game; it is not just a question of depriving somebody else of property. Clearly defined and enforced property rights are productive in the sense of making possible surplus production over and above what would be obtained without property rights. Nowhere is this perhaps clearer than in the context of natural, common-pool resources where the absence of any exclusive use rights can lead to irreversible depletion. The challenge is to harness the productive nature of exclusive use rights for the common good—to combine the exclusive nature of property rights with a fair distribution of their benefits.

With respect to the development of property rights, it is possible to discern two schools of thought whose difference nevertheless may be more apparent than real: the "property rights school" and the "common property school." Both share the view that exclusive use rights are likely to develop when the pressure on an open-access resource has reached the point where the benefits of enforcing these rights outweigh the costs. The schools part company in their view of what form these rights can take.

The property rights school sees use rights as individualistic and vested in persons, including persons "de jure." A classic article written within this genre is Demsetz (1967), in which are cited examples such as territorial rights to what used to be common hunting grounds in Quebec. Libecap (1989) discusses how miners in the American West agreed among themselves to establish individual property rights to settle their competing claims

and managed to have these rights recognized by lawmakers and the courts.

The common property rights school points out that use rights often are held in common by a clearly defined group. Members of a group, identified by location, lineage or ethnicity, see themselves as the rightful users of a given resource. Together, they exclude others from access to the resource while developing among themselves allocation rules and use rights that may or may not be egalitarian and democratic. Ostrom (1990) cites numerous examples where users of a common resource have developed systems of common use rights.

For the purpose of comparison with fisheries, perhaps the most interesting case is the system of common grazing rights on alpine pastures in Switzerland (Stevenson 1991). As population pressure in the Swiss villages increased and threatened the carrying capacity of the alpine pastures, rules were devised at the village level on how to limit the use of the pastures and how to allocate the rights to use them. In some cases, the rights were limited to the descendants of those who had settled in the village before a certain time. In other cases, each landowner in the village got grazing rights in proportion to his landholding. From the fisheries perspective, it is particularly interesting that in yet other cases the Swiss grazing rights are allocated by individual and transferable grazing quotas. These rights are very similar to the individual transferable quotas (ITQs) used in some fisheries.

Scholars of the common property tradition do, in my view, overemphasize the distinction between individual and communal property rights. Both types of rights would appear to have a common origin in human greed; both are an application of the rule "This is mine and not yours," or in the common property case "ours and not yours." The reason why use rights have taken the form of common rights and not individual rights often appears to depend on technical circumstances such as economies of scale; grazing land, for example, is likely to be more efficiently used by letting a suitably sized herd of animals roam over a large area rather than by fencing it in and having smaller herds graze on each lot. In the case of the Swiss alpine pastures, economies of scale in pasture use, herding of animals, and processing of milk are arguments for a common use of the pastures. The use rights do in some cases, however, come fairly close to being individual rights. This is particularly true of the individual transferable grazing quotas, which can be held by non-farmers and leased to anyone who can use them.

The notion that the form that use rights can take is a practical matter in large measure decided by technical circumstances emerges very clearly from Libecap (1989). He sees use rights to resources as arising from attempts by a group of individuals to regulate the use of a resource for their own benefit and to seek legal protection of the arrangements that best suit their interests. The successful examples he cites involve both individual and common use rights. Which of these prevails is conditioned by circumstance. Mineral rights are private, vested in one individual or company. Property rights to mineral deposits were defined instrumentally, as in following an ore vein and not the boundaries drawn on the land above. Competing claims arising when ore veins met could be settled by relatively transparent and simple rules. By contrast, an efficient use rights system for oil fields requires definition of fieldwide use rights because of the migratory nature of the oil underground and incomplete information about the dynamics of extraction. When fields are unitized, leaseholders share in the profits that can be generated on the field under a unified management. This can be characterized as a system of use rights held in common by the lessees on the field. The parallel with the ITQ system in fisheries is close. The difference is that the "quota," or profit share, is not related to the quantity produced by each lease. The optimal use of a field depends not only on the total quantity of oil extracted but also on the location of extraction. It may therefore be optimal not to produce at all on certain leases, or to use them for injecting water to increase the pressure in the field and enhance the recovery of oil. A similar situation could arise where young and old year classes of a fish population occupy different areas. It would probably make more sense to take the older and bigger fish and so harvest in locations where these are to be found.

Exclusive Use Rights in Fisheries

Exclusive use rights to fish resources are for the most part a recent phenomenon. Overfishing of the most valuable stocks has been with us for a good portion of this century, but technological and institutional factors have long prevented the emergence of property rights as a way of rationing these scarce resources. The single most important impediment was international law. Until the late 1970s, most of the important fish stocks of the world were accessible outside any national jurisdiction and were, therefore, effectively open-access resources. As pressure on the stocks increased, some coastal states claimed extended fishing limits to secure exclusive fishing rights for their citizens or whomever else they authorized. This can be seen as a first development towards individual use rights, as such rights would be impossible without first clarifying which state jurisdiction would provide the ultimate sanction. After a long political struggle and sometimes armed confrontation, the national jurisdiction over fishing was extended from 3 nautical miles (~5 km) to 12 nautical miles (~19 km), enclosing fiords and bays, in the late 1950s, and then to 200 miles

(320 km) in the late 1970s. Some important fish stocks are still accessible outside 200 miles, and recent improvements in technology have made them more so. This has prompted some coastal states to demand binding and enforceable regulations of fishing outside 200 miles that would conform to fisheries regulations inside 200 miles. (In 1993, the United Nations convened a conference on straddling and highly migratory fish stocks to deal with this issue. The conference ended in August 1995 with an agreement on a convention on fishing outside 200 miles.)

Once the question of national jurisdiction over the area where fish stocks are located has been settled, the ground is clear for erecting the necessary rights structure for a sensible harvesting of the stocks. What kind of structure would be adequate? A first point to note is that territorial access rights would be inadequate unless defined to include the entire habitat of a fish stock. There is a quite large volume of anthropological literature dealing with territorial use rights in traditional fisheries (e.g., Ruddle and Johannes 1985). These rights are, in most cases, insufficient to deal with stock externalities in fisheries owing to the wide migrations of fish. The territorial use rights may, however, have dealt successfully with crowding externalities arising from competition for the best fishing spots. Clearly, such arrangements can be useful as far as they go. The parallel of access rights in fisheries with those for oil fields is immediate: optimal use requires that the entire field be owned or controlled by one authority. Optimal or near-optimal solutions have been obtained in cases where there are only a few companies or consortia holding rights to the field, but fragmented ownership has typically resulted in competitive and wasteful extraction.

The migrations of fish usually are so extensive that defining exclusive territorial rights that cover the entire habitat of a fish stock will probably not be seen as politically acceptable. In cases where two or more states share the habitat of a stock, it would not even be possible except by an unlikely consent of all states involved. For this reason, and the fact that individual fish cannot be identified and labelled unless captured, individual rights to the fish resources themselves are highly unlikely or impossible.

Individual use rights can, on the other hand, be defined with respect to harvesting. There are two principal ways of doing this: as a right to harvest a certain quantity (which can be defined indirectly, such as a share in a total allowable catch [TAC]) or as a right to use certain harvesting equipment (a boat of a given specification). In some cases, rights have been defined with respect to both.

The choice between these two types of harvesting rights is a practical one rather than one of principle. Both limit the access to resources but in a slightly different way. Each has its own advantages and disadvantages, and circumstances of time and place will determine which is to be preferred. Boat licenses are easier to monitor than fishing quotas. The problems with boat licenses are due to an uncertain and changing relationship between a license and the quantity of fish that the license permits one to catch. It is notoriously difficult to define a unique relationship between a boat license and the resulting harvesting capacity in all but the most simple and static circumstances. A boat license gives fishers an incentive to increase the use of factors not specified by the license and to economize on the use of other factors. This becomes easier as boats are replaced and their design changed. Attempts to avoid this pitfall risk trapping the fishery in a technological straitjacket and forgoing the benefits of technological change.

To what extent such exclusive rights can be said to amount to individual property rights depends on the degree of transferability and length of tenure. For example, nontransferable fishing quotas that are reallocated every year are a fairly tenuous set of rights, but they would probably amount to rights beyond that 1-year allocation since courts would look for precedence and equal treatment if having to rule on future allocations. Permanent and transferable fishing quotas, on the other hand, would appear to come as close as is practically possible to individual property rights in the resource itself or, rather, in a share thereof.

The length of tenure and degree of transferability of fishing rights introduce a classic dilemma with respect to exclusive use rights: that of efficiency versus equity. Efficiency is best served by long and secure tenure. It facilitates long-term planning and gives rightholders a long-term interest in the resource. But the exclusive right also protects the profits of the rightholders; the more protective of profit, the more successful the rights system is from the efficiency viewpoint. This may be deemed undesirable on the basis of social equity. A limited tenure of rights would make it easier to address any inequities emerging by means of a new allocation.

If fishing rights are permanent, the resource will in effect become a shared property of those who hold the rights. Under this arrangement, the holders of the use rights probably would assume the full responsibility for managing the resource, from stock assessment through the setting of TAC and monitoring of individual quotas or licenses, much as is the case with the owners of share rights in the Swiss alpine grazing lands. The state would become redundant as a management authority, and its only role would be the ultimate upholding of the rule of law and the honoring of contracts. Such a system might in fact become self-enforcing; each holder of a fishing quota would have an incentive to keep his fellow quota holders under surveillance, as the erosion of the resource

base through cheating would diminish the value of his property. This mechanism is particularly likely to work in relatively homogeneous industries with few participants. It is probably important as well that the rents obtained from the fishery be relatively moderate. Otherwise, the claims by outsiders who want to get a share in these rents might become difficult to resist, particularly if the use rights are a recent phenomenon in what used to be an open-access industry. If this is the case, the only way to preserve the incentive structure inherent in the fishing rights would be to have the state, or a regional body, act as a rent collector and manager of a publicly owned resource in the interest of the public. I shall return to this point in the ensuing section on fishing rents.

The Origins and Purpose of ITQ Systems

Since the early 1980s, there has been a trend towards establishing fishing rights based on fishing quotas. These quotas, transferable or not, seem to have emerged in two quite different situations. Usually, they have been a delayed and tortuous response to a declining resource base and increasing economic difficulties in the fishing industry. In Iceland, individual quotas first emerged in the herring (*Clupea harengus*) fishery and soon became transferable after it became clear that the quantity allocated to each boat was too small to make much economic sense (Arnason 1995). In 1984, ITQs were introduced as a 1-year emergency measure in the cod (*Gadus morhua*) fishery and subsequently were developed and renewed until ITQs of an indefinite duration emerged in 1990. In Canada and the USA, the ITQ systems—those proposed and those in place—have emerged in response to a long history of economic and biological overexploitation of stocks. An exception in the USA is the wreckfish (*Polyprion americanus*) fishery where ITQs were introduced only a few years after the fishery developed, in response to rapid overcapitalization. The initiative appears to have come from fishery administrators, with fishers catching on to the idea after being skeptical initially (see Gauvin et al. 1994). In Norway, the use of vessel quotas is widespread, but there is hardly any transferability.

Australia and New Zealand are partial exceptions. The ITQ system in the New Zealand deep-water fishery was put in place before signs of overexploitation had emerged and, in fact, before much was known about the productivity of the resources. The reason appears to have been precautionary: both the industry and government officials were well aware of what could be expected to happen as a result of free access. The solution was undoubtedly facilitated by the fact that only a few companies were engaged in the deep-water fishery, some of which had a large share of the fishery. In Australia, the ITQ system seems to have been introduced in the south east trawl fisheries for much the same reason, as the knowledge of the resources appears to have been rather limited (Geen et al. 1993). But even in these two countries, some ITQs emerged in response to resource depletion. In the New Zealand inshore fishery, there was a perceived need to reduce catches. This was accomplished simultaneously with introducing ITQs. In Australia, a rapid and substantial decrease in vessel participation in the southern bluefin tuna (*Thunnus maccoyi*) fishery was achieved by an ITQ system.

In the late 1980s, Chile made plans to introduce ITQs countrywide, in part to enhance the overall efficiency of the economy. These plans met with considerable resistance in the industry, and in the end ITQs were imposed for a limited number of stocks considered to be threatened by overexploitation. These ITQs are allocated by competitive bids (see Peña 1995).

One noteworthy thing about the attempts, successful or not, to establish ITQ systems is that the public authorities more often than not seem to have played a major and decisive role. This is somewhat surprising. People already in the fishery would benefit from keeping out newcomers and so prevent a further erosion of their profits. The transferability of quotas would make it possible for some to buy out others and to take advantage of economies of scale. A quota allocation would make it possible to avoid costs that otherwise would be incurred in order to attain the highest possible catch in the short term, or a highest possible share of the given total catch if the fishery is controlled by an overall catch quota. In short, there would be some aggregate gains obtained from an ITQ system, gains that would become capitalized into a market value of the quotas and accrue, roughly speaking, to the first generation of quota holders. Those who eye the chance to become the first generation of quota holders would appear to have a very clear economic interest in trying to convince their governments that this would be a sensible solution to the perennial problems of fisheries management.

On this basis, both the property rights school and the common property school would predict that initiatives for limited access and exclusive use rights would probably come from fishers acting in self-interest, even if scholars in these two traditions might differ in their views of what sort of arrangements would be most desirable or likely to emerge. The industry, nevertheless, seems usually to have played second fiddle to government officials in proposing the ITQ schemes that have been put in place.

The active role played by government officials in initiating ITQ systems is somewhat surprising because neither they nor the governments they represent have a par-

ticularly strong incentive to do so. There are two reasons why governments might be interested in an ITQ system. First, such systems can be used to raise revenue by leasing or selling fishing quotas. This is rarely done. Second, ITQ systems increase the economic efficiency in the fishing industry and so the productivity of the economy as a whole. However, the aggregate benefits of this efficiency and productivity are usually not very great and, furthermore, they are widely dispersed. Thus, ITQ systems would not appear worthwhile pursuing against the majority opinion in the fishing industry under those circumstances. Nevertheless, government officials have initiated these systems for the purpose of reducing the sometimes gross economic waste that occurs in free-access fisheries; the overinvestment in harvesting capacity manifested by ever-shorter fishing seasons, the costly regulations to limit fishing effort to what the resources can take, and the threat to the viability of the resources themselves posed by the excessive harvesting capacity that might be difficult to adequately restrain. The fishing industry is not unique in this respect. In fact, the situation with respect to unitization of oil fields with fragmented ownership seems rather similar. The initiative to unitize has typically come from the regulatory authorities who have been concerned with the wasteful allocation of resources arising from competitive oil recovery (cf. Libecap 1989).

Not surprisingly, in no case has an ITQ system been pushed through without the support of a critical mass of the industry. This support has been obtained through allocating ITQs on the basis of historical rights—usually the quantity fished, but sometimes the level of investment has been taken into account as well. Limited entry shields those who are in the industry from a further erosion of rents through free entry and holds out a promise of higher rents as some leave the industry and sell their rights to others who can enlarge their scale of operations. Since the efficiency gains from ITQ systems are not likely be large enough to impress the general public, political support for ITQ systems must be sought from within the fishing industry. Without a sufficiently strong constituency in the industry, ITQ plans will probably be abandoned despite the promises they hold with respect to economic efficiency. Support may be forthcoming only if the constituents expect to get a capital gain from the ITQ system. Therefore, it is probably unavoidable that some of the fishing rents become capitalized into a market value of ITQs. To the extent that the government, representing the ultimate owner of the resources (the public), is regarded as the rightful rent collector, this is a price that in all probability must be paid for putting an ITQ system in place.

Is that a price worth paying? From the viewpoint of economic efficiency, the most important property of ITQs is that they provide incentives to cut waste by limiting investment and the use of manpower and other inputs to what is strictly needed to catch what the resources can support. The social benefit of this is the amount of goods and services that can be produced instead. For recreational fishing the curtailment of this "waste" would be a doubtful proposition; recreational anglers would not necessarily devote their time to any productive effort or buy shares on the stock exchange instead of their fishing rod in case they could no longer go fishing. More importantly, they have decided that using time and effort in this way provides pleasure largely or wholly unrelated to the value of the material production it might ultimately provide. However, there are no less clear and important cases at the other end of the spectrum. Most commercial fisheries are conducted solely for obtaining cash income. Such activity has little or no intrinsic value and often is capital-intensive. Every society has limited investment funds and needs to use them as efficiently as possible. Investment in fishing fleets is no exception; it needs to be limited to what yields a competitive return at the margin.

Many fisheries will fall somewhere between these two extremes. Small-scale fisheries conducted with simple and inexpensive capital equipment may have a certain element of recreational fishing. For fisheries with a large recreational element, the ITQs are hardly the most appropriate instrument. Individual quotas may be adequate for preserving the resource, but some equitable rule of dividing them between prospective recreationists would seem to be preferable to ITQs that would select the most "efficient" or affluent ones.

The Tradeoff Between Efficiency and Fairness

Over time, fishing rents become capitalized into market values of licenses or quotas held by private individuals. The role of rents provides a useful perspective on the tradeoff one may be forced to make between efficiency and fairness in limited entry schemes.

Fishing rents play a dual role. First, they are the result of enhanced economic efficiency. As unnecessary costs are cut, a difference emerges between aggregate revenue and aggregate costs. This is what we call rent. In a limited entry fishery, rent will not disappear but accrue as capital gains to those who have the privilege of being "insiders" (to the extent they did not buy their way in). Ultimately, the rent is due to the scarcity of the fish; that is why we need limited entry. Second, rents also serve as an incentive to achieve economic efficiency; no one would make an effort to cut costs or increase revenue without a hope of making a net gain. Most of us probably do not have too much difficulty with the latter of these roles; given that we can ignore the possibility that

fishing may serve other purposes than merely producing fish, we would probably find it an indisputable benefit that the fish be provided at the lowest possible cost in the sense of forgoing, as little as possible, other goods that we also cherish. It is the question of who gets these rents that gives us trouble. Is it really necessary that only a few boat owners get rich just by providing fish at the lowest possible cost?

Contrary to "ordinary" industries, which are characterized technologically by "constant returns to scale," this question will not resolve itself with time. In a constant-returns-to-scale industry, production units can be replicated endlessly, a process that will not stop until the market price has fallen to the lowest attainable cost of production. In such industries, there are no rents; they are eroded by competition. In an industry built on a limited, free-access resource, the rents would be eaten up by unnecessary costs. In a constant-returns-to-scale industry, they disappear because of expanded production.

Yet many, possibly all, constant-returns-to-scale industries have started out as industries with high rents and because of high rents. Somebody had some kind of monopoly, an exclusive technological knowledge, a patent perhaps, that made it possible to sell a product at a price that vastly exceeded the cost of providing it. But in the end it was impossible to hang on to the monopoly. The technology of IBM could be replicated and improved. In the beginning, the salmon farming industry was characterized by high rents, which took about 10 years to erode through increased production and international dissemination of technology. Even the ball-point pen started out as a luxury item protected (but not for long) by patents (Lipsey and Steiner 1972, pp. 287-88).

This scenario, however, will not happen in the fishery. Fishing technology may be replicated anywhere, but not fish stocks. Fish stocks are limited by the productivity of Mother Nature, and fishing must be limited in order not to exceed that productivity. If this is done through market mechanisms such as ITQs, the scarcity value of the fish resources—the fishing rent—will become capitalized in a market value of the quotas. This value will not be eroded over time because the scarcity of fish stocks will always be a fact of life. But more poignantly, unless the quotas are leased, sold, or taxed by the government, those who will get this value are the first generation of owners who got them for free. By contrast, for a constant-returns-to-scale industry, the argument can be made that high rents are an acceptable price for progress because they are transitory anyway; the nature of economic and technological progress is such that these rents will be self-defeating. This is not so in an industry based on a limited natural resource.

So, is the accrual of fishing rents to private operators,

or more precisely, the first generation of private operators, an unacceptable price to pay for economic efficiency in the fishery? This is a question for which the general answer is necessarily ambiguous. Many fisheries are not particularly profitable, and people who engage in them would not be made immensely rich even if they received the present value of future rents of a permanent but modest share in the catch. Such inequities as might result from this would at any rate appear modest compared with those resulting from ownership of land, minerals, financial assets, or scarce talent. Private ownership of fishing quotas would under such circumstances appear to be a moderate "price" to pay for ensuring sustainable and nonwasteful fishing.

When, on the other hand, it is deemed inequitable to let the fishing rents accrue to the industry, we need to look for a community that is "worthy" of receiving these rents and at the same time small enough that these rents make a difference so as to provide incentives to husband the resource appropriately. Vesting the ownership in institutions representing the coastal communities of which the fishing industry is a part is one option. Such institutions would have to be given the right to levy user fees or resource rent taxes to be used in the interest of the community. Furthermore, they would have to be organized in a way that provides incentives to spend the money wisely and in the interest of the community at large rather than wastefully in the pursuit of political goals or personal enrichment of public officials.

Why Fishers Are Against ITQs

Given that ITQ systems have the potential of providing capital gains for those who first get them, it is paradoxical that fishers often are less than enthusiastic about them and sometimes even outright hostile. Nevertheless, there are some valid reasons for this. Prominent among those reasons are uncertainty, conflicts over distribution of gains, and ideology. Furthermore, there is the human capacity for misunderstanding and wishful thinking, which should not be underestimated.

Uncertainty

Both risk aversion and risk loving could account for fishers' opposition to ITQs. Replacing uncertain catches with a certain catch quota appeals to economists who often take risk aversion for granted as a description of human preferences. Fishers may think otherwise. Accepting an ITQ (e.g., on the basis of previous catch history) means forgoing the opportunity to do better. If many fishers believe they will be able to do better than they did in the past, this could account for opposition against ITQs.

They may not realize that only a few, if any, are likely to make any gain from this. A free-access fishery is at best a zero sum game in the long run; one fisher's gain is another's loss, but to attain that result they will have incurred additional costs in racing for the fish. Gambling against odds is, however, a tendency that keeps a thriving industry going in every country, and a similar attitude is not unexpected in fishing.

A different type of risk is related to fishers' business prospects. An ITQ is an asset fishers can and may be forced to sell in times of economic difficulties. An ITQ program will raise the cost of entering the fishery. A fisher who has gone broke and wishes to enter the fishery again will, in addition to having to buy a new boat, have to buy a quota. In this scenario, risk aversion in fact would account for opposition to an ITQ scheme.

Again, these problems are not unique to the fishing industry. Similar problems are encountered as well in the unitization of oil fields (cf. Libecap 1989). Oil leaseholders, particularly small firms, have typically resisted unitization for two reasons. First, the negative externalities caused by small firms mainly affect other firms. Second, there are enormous uncertainties with regard to what future production profiles will look like, as these depend on largely unknown geological characteristics of the underground reservoir. Many leaseholders are tempted to believe that they will do better than what they are offered in a profit-sharing agreement under unitization. Both of these apply to fisheries. Fishing firms are typically small, and the negative effects their harvesting has on the fish stock are mainly external to the firm. Future fishing possibilities are highly uncertain because of uncertainty about migrations, growth, and recruitment of fish.

Distribution

The fact that an ITQ program will produce aggregate benefits means there is a potential to make everyone better off. In practice, it is difficult to attain such fine-tuned fairness. The actual circumstances will always be somewhat complicated. Consider an allocation of permanent ITQs on the basis of catch history, a method that would appear to go a long way towards being fair. Someone might just have bought their first boat and thus be "without history." For some, the engine may have broken down last year, or maybe the family needed extra attention. There will always be cases like this. The more generously they are treated, the fewer benefits there will be for others.

One distributional issue particularly worthy of note is the sharing of income between boat owners and hired crew. ITQs have usually been allocated to boat owners and not to crew, even if crew members that are not boat owners may have just as long a track record as the boat owners themselves. This situation is important because ITQs will probably affect the incomes of crew members. This may come about in two ways. First, as fishing quotas acquire value, this value will be a cost for those who have to buy or lease quotas. This cost is liable to be subtracted from the value of the catch before the shares of crew and boat owner are calculated. If the current contracts with crew members do not allow for this, it is indeed likely that they will be so amended, for it might not otherwise be profitable to hire the crew. Effects of this kind have recently started to appear in Iceland where this practice is explicitly forbidden in existing contracts with the fishers' union but where crew members have been willing to do this nevertheless in order to secure their employment. An additional reason why crew members' wages are apt to suffer is that the demand for crew will recede somewhat as rationalization proceeds in the fishery. Fishers who are not boat owners will, therefore, probably be unequivocally against ITQs.

Fishers, boat owners or not, likely will not get any share in the fishing rents if the quotas are bought by the processing industry or outside financiers. Boats will then be hired on a contractual basis, and neither boat owners nor crew will get any share in the fishing rent over and above their skill rents. Instead, the fishing rent will accrue to the quota owner, except for such rental fees or taxes that might have to be paid.

Ideology

Here, I deal briefly with ideology in the political sense and lifestyle issues. It is not surprising to find vigorous resistance to ITQs on grounds of ideology. ITQs amount to a privatization of the right to harvest fish. Since private versus public ownership of resources and means of production has been at the core of ideological disputes for more than a century, it is natural to expect ITQs to be controversial. Those who for ideological reasons are against private ownership and market-determined allocation processes will be against ITQs on principle. Many of those who are against ITQs have explicitly taken this stand because of opposition to giving away a common resource to private interests. Some of those, but probably not all, might be satisfied by an explicit statement to the effect that the resource itself is a public property and by using fees or taxes to divert the resource rent from the private holders of ITQs. Seen in this light, the ITQs are merely an instrument to achieve economic efficiency and not a vehicle to enrich a small group of "sea lords." There are, however, limits to how far once can go in confiscating the fishing rents without losing the necessary industrial support to put such a system in place.

ITQs will to some extent change the way fishers conduct their business. With ITQs, there will be no point in competing for the largest possible catch (provided the quotas can be enforced). However, that is a game many fishers appear to cherish. One can hear statements such as "We must not let fishing become a business," which many people see happening as a result of less competitive fishing under an ITQ system. What is a bit surprising is hearing such statements from people who are making a living from fishing, an endeavor whose success is not unrelated to the profitability of the enterprise.

Conclusion

As economic development proceeds, open-access resources typically come under increasing pressure. Sooner or later, a point is reached where it is impossible to accommodate all those who wish to use a resource unless its yield is reduced, possibly beyond repair. In response, exclusive use rights have often developed, sometimes in a violent and unfair manner, but sometimes by mutual consent among the parties involved.

Sometimes, not least with respect to fishery resources, the response has been public regulation. Such regulations have often been based on the premise that all claimants can be accommodated. To accomplish this, various restrictions have been imposed on the use of harvest equipment and technology. Such restrictions have usually succeeded in protecting the resource from destruction or severe depletion, but at the expense of diverting capital and manpower to unproductive use. This, no less than the tendency to deplete open-access resources, deserves to be called the "tragedy of the commons."

There are better ways, ways that avoid the waste typically associated with traditional effort regulations and that are, at the same time, reasonably fair. ITQs represent such an alternative. By "grandfathering," a high degree of fairness may be obtained with respect to those who harvest a resource at a certain time. Fairness across generations requires that the gains from less wasteful harvesting not be capitalized into a windfall gain for the generation that was grandfathered in. Some of this is probably unavoidable to ensure the necessary support from the fishers who are active at the time when an ITQ scheme is established, but a substantial part of the resource rent could be confiscated through user fees or special taxes. The revenue so generated would have to be channeled to a sufficiently narrow constituency in order to make a difference and provide incentives for efficient harvesting. Communities or areas that depend on the fish resources and in which the industry itself is a part would seem to be the most natural candidates. The fairness of this is open to debate, but quite often such communities are poorly endowed with respect to other natural resources and are otherwise disadvantaged.

References

Arnason, R. 1995. The Icelandic fisheries. Fishing News Books, London.

Demsetz, H. 1967. Toward a theory of property rights. American Economic Review 57:347-359.

Galbraith, J. K. 1977. The age of uncertainty. Houghton Mifflin Company, Boston.

Geen, G., W. Nielander, and T. F. Meany. 1993. Australian experiences with individual transferable quota systems. Pages 73-94 in Anonymous, editor. The use of individual quotas in fisheries management. Organization for Economic Co-operation and Development, Paris.

Gauvin, J. R., et al. 1994. Description and evaluation of the wreckfish (*Polyprion americanus*) fishery under individual transferable quotas. Marine Resource Economics 9:99-118.

Libecap, G. D. 1989. Contracting for property rights. Cambridge University Press, Cambridge, England.

Lipsey, R. G., and P. O. Steiner. 1972. Economics, 3rd edition. Harper and Row, New York.

McLellan, D. 1973. Karl Marx. His life and thought. Harper and Row, New York.

Ostrom, E. 1990. Governing the commons. Cambridge University Press, Cambridge.

Peña, J. 1995. Chilean fishing regulation: a historical perspective. University of Chile, Department of Economics, Working Paper 135, Santiago, Chile.

Ruddle, K., and R. E. Johannes, editors. 1985. The traditional knowledge and management systems in Asia and the Pacific. United Nations Educational, Scientific and Cultural Organization, Paris.

Stevenson, G. G. 1991. Common property economics. Cambridge University Press, Cambridge, England.

Additional References on Property Rights and Fisheries Economics

Anderson, L. G. 1986. The economics of fisheries management. Johns Hopkins University Press, Baltimore.

Bromley, D. 1991. Environment and economy. Property rights and public policy. Blackwell, Oxford.

Christy, F. T., Jr. 1975. Property rights in the world ocean. Natural Resources Journal 15:695-712.

Copes, P. 1986. A critical review of the individual quota as a device in fisheries management. Land Economics 62:278-291.

Cunningham, S., M. R. Dunn and D. Whitemarsh. 1985. Fisheries economics. An introduction. Mansell Publishing, London.

Eggertsson, T. 1990. Economic behavior and institutions. Cambridge University Press, Cambridge, England.

Edwards, S. F. 1994. Ownership of renewable ocean resources. Marine Resource Economics 9:253-273.

Hannesson, R. 1993. Bioeconomic analysis of fisheries. Fishing News Books, London.

Mollett, N., editor. 1986. Fishery access control programs worldwide. Alaska Sea Grant Report 86-4, Fairbanks, Alaska.

Neher, P. A., R. Arnason, and N. Mollett, editors. 1989. Rights based fishing. Kluwer, Dordrecht.

Allocation of Fishing Rights:
Implementation Issues In Australia

M. Exel and B. Kaufmann

Abstract.—The allocation of property rights in fisheries is considered by some to be a solution to the problems of overexploitation and excess capacity in fisheries. Unfortunately, there are many implementation issues in allocating rights that, if not properly addressed, endanger the use of rights assignment as a policy tool. This paper examines implementation difficulties encountered in the allocation of fishing rights in two Australia individual transferable quota (ITQ) fisheries, the south east fishery and the southern bluefin tuna fishery, and in one input-controlled fishery, the northern prawn fishery. A number of the specific problems identified relate to the allocation process, the lack of meaningful industry involvement, and inadequate preparation on the part of the management bureaucracy. The seemingly innate inability of government bureaucracies to manage fisheries effectively is argued to stem from a lack of direct accountability to user groups and the existence of non-management-related incentives. In Commonwealth-managed fisheries in Australia, the creation of a statutory management authority, the move to 100% recovery of management costs, and the continuing devolution of management responsibility to management advisory committees are seen as possible means of improving the allocation-of-rights process.

The primary objective of this report is to examine the implementation of individual transferable quotas (ITQs) in two Australian fisheries, the south east fishery and the southern bluefin tuna fishery. The intent is not to provide a bioeconomic evaluation of ITQ effectiveness in these fisheries, but to provide a frank and open discussion of ITQ implementation difficulties experienced and lessons learned. A secondary objective is to outline new policy initiatives being developed in Australia to address overexploitation and overcapacity problems that characterize many fisheries worldwide.

The south east fishery (SEF) and the southern bluefin tuna fishery are managed through ITQs. A number of difficulties were encountered in implementing ITQs, especially in the SEF. After we discuss ITQ implementation in each fishery, we outline various lessons learned from these programs.

South East Fishery

Background

The SEF is a multispecies fishery currently composed of approximately 118 otter board trawlers and Danish seiners. The value of the catch in 1993 was roughly AUS$61 million. Of the 96 species caught in the fishery, 16 species are under harvest quotas. In 1993, two quota species, orange roughy (*Hoplostethus atlanticus*) and blue grenadier (*Macruronus novaezelandiae*), represented 45% and 10%, respectively, of SEF landed value. The fishery is essentially composed of three distinct components— Danish seiners that target on whiting (*Sillago flindersi*) and flathead (*Platycephalus richardsoni*), offshore orange roughy trawlers, and multispecies inshore trawlers.

Pre-ITQ management arrangements included the creation of boat units that were assigned to each vessel in the fishery. The number of boat units allocated to a vessel was simply the sum of hull and engine units, which in turn were based on vessel hull size and engine power. Boat units were used as a proxy for harvesting capacity. It was thought that growth in fishing capacity would be constrained by limiting the total number of boat units in the fishery. Boat units were transferable and acquired value. (For an in-depth discussion of pre-ITQ management arrangements, see Geen et al. 1990.) Concern that the boat-unit system was not stemming capacity expansion and overexploitation resulted in a formal government review of management options for the SEF in 1989. In April 1990, the Commonwealth government announced that the SEF would be managed by ITQs. In January 1992, 20 months after the formal decision, ITQs were introduced into the fishery.

ITQ Implementation Issues

Even though quotas were not introduced until almost 2 years after the formal decision to proceed, preparation for ITQs was far from complete. Consultation with industry was inadequate and ineffective. Certain sectors of industry remained opposed to ITQs upon their introduction. Many substantial difficulties currently experienced in the SEF can be traced back to ITQ implementation failures.

The sections that follow outline various ITQ implementation difficulties experienced in the SEF. Specific issues include quota allocation and appeals, jurisdictional overlap, elimination of effort controls, the quota monitoring system, the compliance program, overquota harvest policy, and stock assessment and total allowable catch (TAC) setting.

Quota Allocation, Consultations and the Appeal Process

The quota allocation process has been subject to much criticism. Over 2 years after the introduction of ITQs, dissatisfaction over the allocation process in general and quota allocations in particular continues to jeopardize acceptance of the management regime.

Quota allocations had two components. First, quota shares in the 16 quota species were divided between the Danish seine and the otter board trawl sectors in proportion to each sector's share of historical catch. Second, the quota for each species was assigned to individual operators on the basis of each operator's catch history and boat unit holdings (see Geen et al. 1993 for a detailed explanation of the quota formula).

There were a number of layers to the consultation process. The ITQ Liaison Committee, composed of industry and government members, was established in August 1990. The committee met eight times from August 1990 to September 1991 to discuss implementation. In addition, a committee of state and Commonwealth officials was formed to discuss ITQ implementation. The Quota Implementation Team, composed solely of Commonwealth officials, was tasked with the day-to-day responsibility for ITQ implementation. Two rounds of port meetings were held in September 1990 and in November–December 1990, and approximately 80% of fishery operators attended at least one meeting.

A number of complaints were raised in relation to the consultation process and the allocation formula. For example, a number of industry operators argued that the consultation process was perfunctory. Industry was not permitted to see the management plan being developed, and final quota allocations were not shown to operators until a month prior to the introduction of ITQs. As the consultations were not concerned with whether ITQs should be introduced but how they should be implemented, a number of operators (who did not want ITQs) were not predisposed to engage in constructive consultations.

Operator dissatisfaction resulted in the establishment (at the request of the Minister of the Department of Primary Industries and Energy [DPIE]) of the Review Committee to review the ITQ implementation process. It also resulted in the following: appeals by operators to internal review process of the Australian Fisheries Management Authority (AFMA) and to the Administrative Appeals Tribunal (an independent body established to hear appeals against Commonwealth administrative decisions); application by two operators to the federal court to overturn the management plan; and an AFMA-initi-

ated independent review related to application of the allocation methodology.

The Review Committee issued their report in 1992 (AFMA, Burns Centre, Forest, ACT, Australia, unpubl. mimeo.), wherein they found that industry was not adequately consulted on either the allocation process or the allocation formula. The process of consultation and implementation was driven, first and foremost, by a ministerial deadline for ITQ implementation. The Review Committee noted "that this objective unfortunately was achieved at the expense of a sound consultative process in the latter stages." The committee also noted that industry was not even shown a draft of the management plan, even though it had been indicated previously by DPIE that this would be done.

In a decision by the federal court, the allocation formula was found to be "capricious and irrational," and the court declared the paragraph in the management plan containing the formula void. As a result of the reviews and the federal court decision, the quota allocation formula was changed in October 1992. In particular, the catch component of the formula was changed so that each operator's quota share was based on his or her catch share over the entire qualifying period (as opposed to an average of the sum of catch shares in each year of the qualifying period).

Notwithstanding the qualifying formula change over 2 years after the commencement of ITQs, a number of operators are still not satisfied with their quota allocations, and political and legal avenues are being explored to satisfy quota grievances. A recent Senate committee review of Commonwealth fisheries legislation (Commonwealth of Australia 1993) found "the evidence to be convincing that the quota allocation process in this fishery was seriously flawed from the beginning. The inequities which resulted will continue to provide a major obstacle to establishing a satisfactory management regime for this fishery unless addressed urgently."

The Senate committee further noted that the "restructuring of the south east fishery, in particular the change from a unitized fishery to one based on ITQs, provides a telling example of how the property rights issue has been mishandled. . . ." (As outlined earlier, prior to ITQs, capacity in the SEF was managed through boat units—hence the reference to "a unitized fishery.") It is also worth noting that because of uncertainties surrounding quota allocations, permanent quota transfers were not permitted until January 1994.

Jurisdiction considerations.—In general, the Commonwealth's jurisdiction over quota species extends from 3 nautical miles (~5 km) to the 200-mile (320-km) Australian fishing zone limit. State jurisdiction extends from the low-water mark to the 3-mile limit. Therefore, many of the SEF species managed under ITQs in Commonwealth waters are also managed by effort controls in state jurisdictions.

Mixed jurisdiction creates an enforcement loophole in the SEF ITQ system, as most operators holding quota are also endorsed to fish in state waters. Operators have an incentive to report catch of quota species harvested in Commonwealth waters as having been taken in state waters, where quotas do not apply. Clearly, some operators are misreporting their catch. For example, in 1990, 91% of the total redfish (*Centroberyx affinis*) catch (a quota species) was reported as being harvested in Commonwealth waters; however, by 1993, that share fell to 30%.

The ability of operators to declare quota species as being caught in other jurisdictions is a fundamental problem in the SEF ITQ system. Negotiations were undertaken with state governments to close the "state loophole," and it was thought that compatible management arrangements could be worked out prior to the introduction of ITQs. Unfortunately, the quota loophole was not completely closed, and over 4 years after ITQs were introduced, negotiations with states to achieve workable management arrangements are still not finalized. Segments of industry unhappy with quota allocations are lobbying certain state governments not to close the loophole until specific changes are made to the ITQ system (especially with respect to allocations).

Elimination of effort controls.—Prior to the introduction of ITQs, the following input controls were in place in the SEF: vessel length and mesh-size restrictions, boat unit restrictions, limits on the number of vessels, and a limit on the number of boat units in the fishery. However, orange roughy was managed by a competitive TAC, and gemfish (*Rexea solandri*) was managed under ITQs.

With the introduction of ITQs in 1992, boat units were no longer used as an input control. However, all of the remaining input restrictions have been maintained. Regulations on mesh size were kept to "protect young fish." Vessel length and number restrictions continue at the request of the majority of operators, who are concerned about the potential for an effort blow-out directed at nonquota species. The existence of nonquota species in the fishery has created unexpected difficulties in eliminating input controls on quota species.

Quota monitoring system.—As pointed out in Hannesson (1993), "the main drawback of quotas is the control needed to make them effective." Quotas increase the incentive for misreporting. If quota limits cannot be enforced, the benefits that should flow from ITQ management are likely to be lost. A number of difficulties have been experienced in developing and implementing an effective quota monitoring system for the SEF. Before outlining current monitoring problems, the operational aspects of the SEF quota monitoring are briefly discussed.

SEF quota monitoring costs are approximately AUS$0.5 million per year. Roughly 4.5 full-time-equivalent person years are needed to manage the monitoring system (including data entry). All monitoring costs are recovered from industry. The framework of the monitoring system is as follows. Operators landing quota species are required to contact a central reporting facility (essentially a private 24-hour paging service) 2 hours prior to landing and provide information on estimated landing time, port of landing, and estimated catch. The central reporting facility immediately notifies both AFMA and the appropriate fisheries enforcement officers of the landing notification. Once a landing notification is received, enforcement officers decide whether to be physically present for the unloading of catch on a discretionary basis.

Upon landing, operators must complete a catch disposal record form. Part B of the form shows the operator's determination of catch weight by species; it is forwarded to AFMA immediately. While operators are to provide an "accurate determination" of catch weight on part B, operators are not obliged to weigh catches. Part C of the catch disposal record form accompanies the fish and is completed by registered fish receivers (e.g., processors and exporters). Fish receivers specify on a weekly basis the catch weight by species on part C and forward the completed form to AFMA. The quota monitoring section located in AFMA enters the catch data from part C and decrements the operator's available quota.

Compliance program.—Most of the major difficulties with the monitoring system stem from weaknesses in compliance. In moving to ITQs in the SEF, a number of compliance difficulties were encountered. The "paper trail" associated with the above quota monitoring system was seen as the backbone of the compliance program. However, two factors greatly reduced the effectiveness of the quota monitoring system in detecting misreporting. First, the SEF is not yet managed under a formal management plan, and AFMA has no legislative basis to require processors to complete part C of the catch disposal record form when a formal management plan is not in place for a fishery. Second, the possibility of collusion between processors and fishers raised questions concerning the validity of part C when it is completed by processors.

Compliance difficulties have also been experienced with reported overquota harvests. Some fishers declared catch levels for certain species on part B of the catch disposal record form that exceeded their quota allocation—in other words, self-reported overquota harvest. However, current legal advice questions the reliability and admissibility of evidence taken from part B of the catch disposal record form.

An additional difficulty in the SEF relates to non-trawl harvests of quota species. Currently, only the trawl sector is under an ITQ system. The harvest of quota species by gillnet and hook in Commonwealth waters is managed by input controls. Since some trawl operators have gillnet and hook endorsements (and others can purchase such endorsements), it is possible for trawl-caught quota species to be reported as harvested by non-trawl methods (therefore, the harvest would not come off the operator's quota).

While not all quota monitoring and compliance difficulties can be foreseen prior to ITQ implementation, active participation by industry in developing the monitoring and compliance system may help minimize the number of unexpected and unwelcome surprises. In the case of the SEF, when industry raised monitoring and compliance concerns, the literal response from the then Australian Fisheries Service was, "Don't worry, we've got that covered." In reality, little was covered.

Overquota harvest policy.—If relative TACs do not reflect relative abundance (owing to the uncertainties inherent in stock assessment, or as a result of stock rebuilding strategies), overquota bycatch problems are likely to be a problem for quota species that are harvested together. In the SEF, a number of additional factors increased the possibility of overquota harvesting such as a freeze on permanent transfer of quota, lack of a formal lease market, leasing transaction costs, and a policy rule that allowed operators 15 days after the end of the month in which overquota fish were landed to obtain quota.

Australian Fisheries Service managers appear to have given little attention to the issue of overquota harvesting prior to ITQ implementation. Unfortunately, dumping of overquota catch is currently considered to be a serious problem for a number of species. In response to overquota difficulties, the South East Fishery Management Advisory Committee established a working group, with a majority of industry members, to consider overquota policy options. Options being considered include quota substitution (allow fishers to land overquota harvest of a particular species with unused quota of another species), deemed value (setting a surrender price for overquota harvests), voluntary surrender and quota carry over or carry under. For a more in-depth discussion of various overquota options, see Baulch and Pascoe (1992).

Stock assessment and TAC setting.—As with many aspects of ITQ implementation in the SEF, the initial stock assessment and TAC-setting process was ad hoc and rushed. For most species, TACs were not based on formal estimates of relative abundance but were based on average catch levels over the 1986–91 period. The use of average catch levels can hide yearly fluctuations in abundance, and this can in turn create credibility problems with the ITQ system.

No formal stock assessment process was established prior to commencement of ITQs in the SEF, and no direct funds were made available to the management au-

thority to undertake stock assessment. Although AFMA developed a research and stock assessment strategy in 1992 to help rectify the situation (AFMA, Burns Centre, Forest, ACT, Australia, unpubl. mimeo.), it would have been more productive if a greater amount of the stock assessment and TAC-setting groundwork had been completed prior to ITQ implementation. In particular, the cost of stock assessment, the impact of ITQs on the usefulness of catch and effort data, high-grading, and other stock assessment issues should have been more fully considered prior to introducing ITQs.

Southern Bluefin Tuna Fishery

Background

The value of the southern bluefin tuna fishery is approximately AUS$95 million. The fishery is based on the harvest of a single highly migratory species, southern bluefin tuna (*Thunnus maccoyii*). Southern bluefin tuna spawn south of Indonesia, and migrate southeast, passing south of the Australian continent. They are mainly harvested by Australian, Japanese, and New Zealand fishers, although catches by Indonesian, Taiwanese, and Korean operators are increasing. Australian, Japanese, and New Zealand catches of bluefin tuna are managed through the Commission for the Conservation of Southern Bluefin Tuna (formerly the Trilateral Management Group), with each country receiving annual quota allocations.

In 1984, Australia introduced ITQs into the Australian fishery. Before beginning discussions on ITQ implementation issues, it is useful to consider both the Australian pre-ITQ management regime and the forces that led to the introduction of ITQs. During the 1982 and 1983 trilateral meetings, concern was expressed that bluefin tuna were being overfished. In particular, declines in the spawning stock were seen as increasing the risk of recruitment failure. As well, as noted in Geen and Nayar (1989), the financial performance of many Australian operators had deteriorated in the early 1980s—the result of increasing capacity and reduced tuna prices. In 1984, in a response to biological and economic concerns, the Commonwealth government announced that the Australian bluefin tuna fishery would be managed under ITQs. A more in-depth background on the introduction of ITQs is available in Robinson (1986).

ITQ Implementation Issues

Quota allocation.—The quota allocation formula for bluefin tuna was based on catch history and investment (for specific details concerning the allocation formula; see Robinson 1986). To be eligible for quota, operators had to demonstrate that they had either taken at least 15 metric tons (mt) of bluefin tuna in any one year over a specified period or had made irrevocable financial commitments to the fishery (e.g., through the purchase or construction of vessels). Recreational fishers were not eligible to receive quota allocations. Approximately 220 fishers applied for bluefin tuna quota units; 140 satisfied the eligibility criteria and were allocated quotas ranging from 1.4 mt to 823 mt. It is also worth noting that 485 mt were not allocated to quota holders; instead, they were set aside to cover incidental bycatch arrangements and quota needs flowing from the appeals process.

Operators who caught more than 15 mt in any year over the qualifying period were not included in the formal quota scheme. To handle operators with small catch histories, a bycatch allowance of either 1 or 5 mt of bluefin tuna was permitted as a condition on their license. This would allow fishers who did not hold bluefin tuna quota to land accidental bluefin tuna bycatch. It was recognized that this bycatch condition would require amendment at a later date (as nonquota holders used the bycatch allowance as a means to target bluefin tuna), but it was not possible to arrive at an alternative solution at the time.

As has been noticed in other countries (Arnason 1993), leaving even small fleet sectors outside the quota system can create difficulties later—the 1- and 5-mt bycatch allowance was not an exception. In response to increasing catches by operators through the bycatch allowance, the bycatch arrangements were removed in 1989. Operators who had used the bycatch condition were allocated units and formally brought into the ITQ system. Elimination of the bycatch provision was not welcomed by all operators, and in 1994 it remained a contentious issue.

Even with extensive industry consultations, there were many operators who thought the formula was discriminatory. The use of investment information in the allocation formula was seen by some operators as unduly rewarding late entrants into the fishery at the expense of pioneers. Elimination of the bycatch condition was argued to penalize smaller operators. Allocation difficulties were compounded by the 30% decline in TAC that accompanied ITQ introduction.

Appeals.—After allocations were announced, an informal appeal group composed of representatives from each state fisheries management authority and one federal fisheries manager was established. While a large number of operators put forward submissions in this process, only three applicants were successful.

The next step in the appeal process was the Administrative Appeals Tribunal (AAT). Initially, 25 appeals were lodged with the AAT, with only five appellants being granted additional quota. However, subsequent appeals to the AAT, the ombudsman, or the courts continued for

another 5 years. The ITQ implementation process was not helped by the uncertainty surrounding such a long appeal process. However, sufficient quota was set aside to satisfy appeal requirements.

In 1989, legislative amendments eliminated the possibility of appeals with respect to the initial quota allocations. This guaranteed quota holders that their quota share would not be reduced through further appeals.

Setting TACs.—The TAC-setting process starts with a trilateral scientific group meeting around September each year. Until 1994, the scientific group made recommendations on appropriate catch levels; managers and industry representatives from Australia, Japan, and New Zealand met immediately after the scientific meeting to negotiate global catch limits and national allocations.

The process by which the TAC is set is evolving. Since 1994, there has been a growing split among industry, managers, and scientists concerning the status of bluefin tuna. After the 1993 trilateral scientific report raised concerns over stock status, managers, at the request of scientists, initiated a scientific review of specific biological uncertainties in the stock assessment analysis. For the first time, external scientists were employed by managers and industry to assist in the review. Managers also broadened the scientific committee to include industry and other user groups (including conservation groups) in the stock assessment process. In keeping with the stock assessment and TAC-setting processes outlined by Francis (1992), Lane (1992), Pearse and Walters (1992), Hilborn et al. (1993), and Lane and Kaufmann (1993), a more strategic, decision-making, TAC-setting framework is being examined that explicitly identifies management objectives, predetermines management decision-making rules, and incorporates risk and uncertainty.

Quota monitoring and enforcement.—Catch is monitored through the use of catch record forms, which are completed by the skipper (providing estimated catch weight) and processing establishments (stating actual catch weight). The skipper and processor forms are forwarded to AFMA, and the processor-verified catch weights are used to decrement quota.

The ITQ monitoring process has evolved over time. In particular, ITQ-induced changes in fleet and processing structure have impacted the quota monitoring system. Bluefin tuna are no longer harvested by purse seiners for the canning market. In 1994, roughly 48% of Australian quota was leased to Japanese vessels to harvest, and 23% of quota was caught and subsequently reared in cages (see following text). In the face of such structural changes, the "paper trail" component of the quota monitoring system remains essentially unchanged. However, additional procedures were introduced to deal with joint venture and farming operations.

Under joint venture arrangements, domestic quota

holders lease their quota to Japanese longliners that harvest Australian bluefin tuna quota inside the Australian fishing zone. In 1994, approximately 50 longliners were fishing for 2,400 mt of leased Australian bluefin tuna quota. Given that joint venture vessels do not land bluefin tuna in Australia, a number of changes in the monitoring program were made to monitor these vessels. These changes include compulsory pre- and post-fishing inspections of the holds, 15% observer coverage, random at-sea inspections, as well as daily position and catch reporting (via satellite transmission), daily catch-and-effort logbooks, and individual length and weight measurement of every tuna.

Cage rearing involves catching bluefin tuna by purse seine or pole and line, placing the fish into pontoons, towing the pontoons to safe anchorages, and transshipping them to cages. The fish are fed for 3 to 6 months, then harvested for the Japanese sashimi market. With respect to the monitoring of fish farm activity, underwater cameras were introduced in 1994 to count fish as they are transferred from towing pontoons to holding cages. Farming is a very recent change in the fishery and has moved from 200 mt in 1992 to approximately 1,200 mt in 1994.

Carry-over policy and minimum quota holdings.— Currently, there is no provision for the carry over of quota underruns or overruns from one season to the next. This can create planning difficulties for operators. The value of bluefin tuna varies according to size and fat content. During the beginning and middle of the season, it is relatively easy to take medium-sized fish. However, under appropriate oceanic current conditions, the larger, more valuable fish can be caught by longliners in the final month of the season (usually off New South Wales and Tasmania).

Over recent years, quota holders have been "caught out" with either a shortage or surplus of quota at the end of the season. The problem is greatest for fishers who are over quota, as the lease or purchase of quota at the end of the season can be very expensive.

There is growing pressure from industry to increase the flexibility of the ITQ system through the introduction of a carry-over and carry-under policy. Current industry proposals include a maximum carry over or carry under of 5% of the total quota. The international nature of the fishery and concerns over the status of the stock make the use of carry overs and carry unders a contentious issue in the fishery.

There are no restrictions on minimum quota holdings, other than that quota holdings must be a multiple of 1 unit. Minimum quota holdings were not introduced as it was considered inequitable to force operators that harvest small amounts of bluefin tuna as a bycatch to purchase a minimum quota holding. Also, it was thought that the expense associated with minimum holdings would only result in increased black-market activity (quota lease prices are currently about AUS$4,000/mt, so any minimum quota holding could create a sizable expense for smaller operators).

The Southern Bluefin Tuna Management Advisory Committee is considering a minimum lease level of 500 kg (for the first lease) to eliminate administrative expenses associated with processing small leases. There would still be no minimum limit on the purchase level of quota.

An additional issue related to minimum quota holdings is the potential for effort increases for nonquota species. Vessels that hold bluefin tuna quota and are endorsed to fish in another fishery could sell their bluefin tuna quota and increase effort in the nonquota fishery. This in fact happened—the rationalization of the bluefin tuna fishery led to effort expansion in the south east fishery. Similarly, it is possible for new operators to enter the bluefin tuna fishery by purchasing a small amount of quota, then using the newly acquired access right to target nonquota "bycatch" (such as albacore tuna [*T. alalunga*] and bigeye tuna [*T. obesus*]).

A per-trip possession limit is being considered in order to avoid the latter problem. This would set specific limits on the amount of bycatch that could be taken relative to the amount of bluefin tuna onboard. The fishery lends itself to this approach as it is a highly targeted, single-species fishery with little "unintentional" bycatch. Longliners catch only one or two other species regularly in any quantity, and the purse seine and pole-and-line operations are virtually single-species fisheries.

Distributional and adjustment considerations.—Under ITQs, the bluefin tuna fishery has moved from purse seining for the low-valued canning market to a longlining and pole-and-line fleet that satisfies the high-valued sashimi market. This structural shift created adjustment pressures and raised distributional concerns with respect to canneries and the purse seine fleet. However, a 75% decline in the TAC would have likely generated greater adjustment pressures.

Two factors helped to mitigate adjustment costs. First, the purse seine fleet developed a skipjack tuna (*Katsuwonus pelamis*) fishery, which had positive downstream implications for canneries. Second, infrastructure requirements related to cage rearing and joint ventures generated new economic benefits.

In an attempt to address distributional concerns, one state government retained approximately 200 mt of quota, which was leased annually to operators that agreed to harvest the tonnage in that state's waters. However, as a result of administrative complexity, debates over reallocations, and difficulties in determining lease prices, the state government finally sold the remaining quota to industry.

ITQ Implementation Lessons

Industry Acceptance of ITQs and Involvement in Implementation

Broad-based industry acceptance of ITQs and industry partnership in the implementation process are the most important prerequisites to the development of a successful ITQ regime. It is better to continue to manage with dysfunctional input controls than to introduce ITQs in a fishery where industry is strongly opposed to the regime.

Confine the Appeal Process

Even if industry is supportive of ITQs, an equitable quota allocation can be difficult to achieve. Appeals are likely to continue for a longer period than first anticipated. The issue of appeals should receive a great deal of attention prior to ITQ implementation, and appeal rights should be constrained to a predetermined time period if at all possible. As well, quota should be set aside to handle successful appeals.

Minimize Management Involvement in Quota Allocations

The allocation of quota in ITQ fisheries is notoriously tricky. If possible, fisheries managers should have no decision-making involvement in the allocation formula or allocation process. Fisheries managers may be called on to provide suggestions and input into the allocation process, but they have no comparative advantage in quota allocation decision making.

The management organization should be kept out of decision making in quota allocation for at least three reasons. First, because of their intimate involvement in the fishery, managers tend to have subjective preferences about who deserves quota. Second, for this very reason, industry often considers managers to be biased. Third, long after allocation is complete, any perceived inequities in quota allocation will make it difficult for managers to work with industry. The establishment of an independent allocation body is preferable to encumbering the management body with allocation decisions.

Ensure Adequate Planning Is Undertaken

New rights regimes are often introduced into fisheries that are experiencing overexploitation and capacity difficulties. As a result, a false sense of urgency frequently accompanies the introduction of ITQs. Worse, bureaucratic imperatives (e.g., meeting ITQ implementation deadlines) often take precedence over operational realities (e.g., determining acceptable allocations and establishing monitoring and surveillance systems).

A methodical approach that fully and openly examines the implications of ITQs has the best chance of success. For example, the cost of effective quota monitoring and compliance programs should be determined; compliance penalties and enabling legislation should be in place prior to implementation; quota allocation should be largely agreed upon; and issues related to high-grading, stock assessment, overquota harvest, and TAC setting should be settled prior to ITQ implementation. Industry should be fundamentally involved in the decision-making process. This may seem a tall order; however, ITQs may not be in the short-run interests of various scientific and fisheries management groups, and failure due to implementation mistakes may be used by such groups as evidence that ITQs do not work in principle.

One lesson that has become apparent in the SEF is that it can be quite difficult to correct mistakes that flow from the premature and unplanned introduction of any new management regime. After ITQs are implemented, new vested interests are created and some previously available management options are foreclosed. Once mistakes are made, the management authority can become frozen at particular points on the policy landscape, making implementation errors difficult to rectify.

Recent Policy Initiatives in Australia

In Commonwealth-managed fisheries (as well as in many state-managed fisheries), new policy initiatives are being developed to address overcapacity and overexploitation. Initiatives in Commonwealth-managed fisheries include creation of a statutory fisheries management authority (the AFMA), introduction of 100% recovery of fisheries management costs, establishment and/or revitalization of management advisory committees with strong industry representation, explicit identification in legislation of economic efficiency maximization as a management objective, and the identification of ITQs as the preferred management tool.

Establishment of AFMA

To achieve wider participation by industry, it may be necessary to change the institutional structures that deliver fisheries management services. This was the case in Australia. The Australian government policy statement (Commonwealth of Australia 1989) highlights the following benefits from moving fisheries management to a statutory authority from a government department:

> The structure of a statutory authority would enable the Government to effect its responsibilities in a flexible, open and less bureaucratic way. It would also allow greater community and industry participation in deter-

mining the appropriate management programs for Commonwealth fisheries than has been the case in the past.

The AFMA is a statutory authority legislated to manage Australian Commonwealth fisheries. The AFMA is headed by an eight-member board of directors. Board directors are selected on the basis of expertise in the following areas: commercial fishing, fishing industry operators other than commercial fishers, fisheries science, marine ecology, natural resource management, economics, or business management. The managing director, who is also a board director, is responsible for the day-to-day management of AFMA.

The AFMA's current budget is approximately AUS$18.5 million. Roughly 85 staff are employed, including management and overhead (financial management, systems, human resource management, etc.) staff.

The AFMA assumed responsibility for the management of Commonwealth fisheries in February 1992. Prior to the establishment of AFMA, Commonwealth fisheries were managed by the Australian Fisheries Service, a division of the DPIE.

There were a few teething problems in the creation of a statutory authority to manage fisheries. The first difficulty relates to the continued existence of a fisheries branch within the DPIE (which advises the minister, determines strategic policy, and undertakes international negotiations). The creation of two groups involved in fisheries issues, where there was previously one, has the potential to complicate the fisheries management process.

It is essential that AFMA and the Fisheries Policy Branch of the DPIE maintain a close and cooperative working relationship, and that their respective roles be clearly communicated to industry. Without effective cooperation, the AFMA could quickly become decoupled from the government decision-making process. Many of the individuals presently employed in the Fisheries Policy Branch were formally involved in fisheries management prior to the establishment of AFMA. The potential for conflict in such a situation should not be underestimated or overlooked.

A second challenge that AFMA had to overcome relates to the establishment of its own identity. The AFMA inherited a number of fisheries management difficulties, such as the SEF ITQ system. As noted in a recent Senate review of Commonwealth fisheries legislation (Commonwealth of Australia 1993),

Most of the decisions about the management of the SEF were taken prior to AFMA's establishment. As such, AFMA cannot be held responsible for the difficulties which exist in this fishery. However, they are required to develop and implement a solution, an extremely difficult task. Indeed, AFMA's Chairman was moved to comment that "the poisoned chalice was passed to us."

An additional identity difficulty relates to the fact that, initially, AFMA engaged many of the former DPIE fisheries managers. Therefore, there was an appearance of little immediate difference between AFMA and DPIE. However, it is interesting to note that at present less than 30% of AFMA's current staff were previously employed by the former DPIE fisheries managers.

The AFMA has established a management advisory committee (MAC) for each major Commonwealth fishery. Management advisory committees are the focal point for joint management/industry participation in fisheries management decision making.

Industry Involvement: Management Advisory Committees

A prerequisite to the development and implementation of cost-effective and workable fisheries management arrangements is meaningful industry participation in the management process. In Australian Commonwealth fisheries, industry participates as members on the AFMA Board, and on MACs. In some fisheries, industry is involved in day-to-day management through the appointment of industry-based MAC executive officers. Membership on MACs comprises an independent chairperson, an AFMA fisheries manager, and up to seven other members, who usually include one member from state fisheries organizations, one member from the research community, and up to five user-group representatives.

The MACs advise and make recommendations to the AFMA Board with respect to the development, monitoring, and amendment of management plans, and the implementation of management measures, such as closures, size limits, and TACs. The MACs are also involved in setting research priorities and coordinating the stock assessment process. They are heavily involved in the preparation of annual budgets for each fishery. Budgets include costs associated with surveillance and enforcement, overheads, logbook collection and processing, and day-to-day management activities. However, budgets must be approved by the AFMA Board.

Currently, most MACs are focused on the development of formal management plans. A management plan is a subordinate legislative instrument that is determined by AFMA and accepted by the minister. Management plans must include the following information:

- identification of management objectives and performance criteria,
- description of the fishing concessions to be used and determination of how they will be allocated, and
- identification of the rules governing those operating in the fishery.

Management plans allow for the creation of statutory

fishing rights (i.e., access rights defined in legislation), which exist for the length of the management plan.

Involving industry in management is not without its difficulties. Active participation in management decision making is frequently new to industry, and industry may not be prepared for the new role. Institutional structures within industry needed for broad-based consultations may not be in place. Existing power groups within industry understand how to influence management decisions through the political and bureaucratic process, and shifting this power structure may not be in the vested interests of some industry players.

Constructive dialogue between managers and industry is not easy to achieve. There is often profound distrust and suspicion between parties. It takes considerable time and effort on behalf of individuals from each group to make a new cooperative approach work effectively.

Cost Recovery

Under current Commonwealth cost-recovery policy, the commercial harvesting sector pays 100% of recoverable management costs in Commonwealth-managed fisheries. Recoverable management costs include the running costs of MACs, licensing, day-to-day management activity, ongoing costs associated with maintaining management plans, logbooks, surveillance, and quota monitoring. The commercial harvesting sector also contributes to the cost of fisheries research. Nonrecoverable management costs include enforcement of domestic fishing, a portion of recoverable costs in exploratory and collapsed fisheries, surveillance and deterrence of illegal foreign fishing activity in the Australia Fishing Zone, and Commonwealth-requested participation of AFMA in international forums (e.g., Organization for Economic Cooperation and Development [Paris] and the Food and Agriculture Organization of the United Nations [Rome]).

The mechanism by which management costs are recovered varies by fishery. For example, in the SEF, which is managed under ITQs, each operator's share of management costs is determined by the operator's share of quota for each of the 16 quota species. For example, in 1993–94, orange roughy represented 53% of the value of total quota, and therefore was allocated 53% of SEF management costs. Levies on individual orange roughy quota holders are in turn based on the proportion of orange roughy quota held by each operator.

The move toward 100% cost recovery in Commonwealth-managed fisheries is probably the most important factor in setting the dynamics in place for both increased industry participation and the continuing devolution of power to MACs. While cost recovery is not a prerequisite to increased industry involvement, it appears to be a powerful stimulant.

Summary

The failure of current fisheries management arrangements to solve overcapacity and overexploitation problems is receiving increasing worldwide attention in both the academic literature and popular press. Pressure is being placed on governments to implement policies that redress the failures of current management arrangements, and rights-based regimes such as ITQs are seen by many as a possible policy solution. However, the premature and ill-considered introduction of ITQs can jeopardize the acceptability and potential usefulness of ITQs as a fisheries management tool. Ironically, the greatest threat to ITQs is not the existence of scientific and management vested interests but rushed ITQ implementation.

Before ITQs are introduced into a fishery, a dispassionate and empirical analysis of the timing and magnitude of potential benefits and costs should be undertaken. Without such analysis, passion—rather than fact—is likely to remain the pervasive force in ITQ debates. The underlying premise of this paper is that resolution of current fisheries mismanagement problems does not require greater willpower on the part of governments to make the hard fisheries management decisions. Nor is it likely that ITQs or bigger and better input-control programs contain solutions to mismanagement. The decision-making role of government bureaucracies in the harvesting sector is substantial, and probably unequaled in comparison to other sectors in most industrialized economies. Unfortunately, public sector management has resulted in unfocused and costly scientific research, inflated management costs, economic waste, overexploitation, and social disruption. Public sector management of fisheries may be the problem, not the solution.

Cost recovery, devolution of management responsibilities, and significant involvement by industry and other user groups in management decision making are avenues worth exploring as replacements to the current command-and-control institutional framework. A case could be made that the management of Commonwealth fisheries in Australia is moving quickly in this direction.

References

Arnason, R. 1993. Icelandic fisheries management. Pages 124-144 in S. Cunningham, editor. The use of individual quotas in fisheries management. Organization for Economic Cooperation, Paris.

Baulch K., and S. Pascoe. 1992. Bycatch management options in the south east fishery. Australian Bureau of Agricultural and Resource Economics Research Report 92.18, Canberra.

Commonwealth of Australia. 1989. New directions for Commonwealth fisheries management in the 1990s. Canberra.

Commonwealth of Australia. 1993. Fisheries reviewed. Report of the Senate Standing Committee on Industry, Science,

Technology, Transport, Communications and Infrastructure. Canberra.

Francis, R. I. C. C. 1992. Use of risk analysis to assess fishery management strategies: a case study using orange roughy (*Hoplostethus atlanticus*) on the Chatham Rise, New Zealand. Canadian Journal of Fisheries and Aquatic Sciences 49:922-930.

Geen G., D. Brown, and S. Pascoe. 1990. Restructuring the south-east trawl fishery. Pages 129-141 *in* Bureau of Rural Resources, editor. Australian and New Zealand southern trawl Fisheries conference: issues and opportunities. Bureau of Rural Resources Proceedings No. 10, Canberra.

Geen G., and M. Nayar. 1989. Individual transferable quotas and the southern bluefin tuna fishery. Australian Bureau of Agricultural and Resource Economics, Occasional Paper 105, Canberra.

Geen G., W. Nielander, and T. Meany. 1993. Australian experience with individual transferable quota systems. Pages 73-94 *in* S. Cunningham, editor. The use of individual quotas in fisheries management. OECD, Paris.

Hannesson, R. 1993. Bioeconomic analysis of fisheries. Fishing News Books, Oxford.

Hilborn, R., E. K. Pikitch, and R. C. Francis. 1993. Current trends in including risk and uncertainty in stock assessment and harvest decisions. Canadian Journal of Fisheries and Aquatic Sciences 50:874-880.

Lane, D. E. 1992. Regulation of commercial fisheries in Atlantic Canada: a decision perspective. Optimum 23-3:37-50.

Lane D. E., and B. Kaufmann. 1993. Bioeconomic impacts of TAC adjustment strategies: a model applied to northern cod. Pages 387-402 *in* S. J. Smith, J. J. Hunt, and D. Rivard, editors. Risk evaluation and biological reference points for fisheries management. Canadian Special Publication of Fisheries and Aquatic Sciences 120.

Pearse, P. H., and C. J. Walters. 1992. Harvesting regulation under quota management systems for ocean fisheries. Marine Policy 16:167-182

Robinson, W. L. 1986. Individual transferable quotas in the Australian southern bluefin tuna fishery. Pages 189-205 *in* N. Mollett, editor. Fishery access control programs worldwide: proceedings of the workshop on management options for the north Pacific longline fisheries, Orcas Island, Washington, April 21-25, 1986. Alaska Sea Grant College Program report 86-4, University of Alaska, Fairbanks, Alaska.

Unraveling Rent Losses in Modern Fisheries: Production, Market, or Regulatory Inefficiencies?

JAMES E. WILEN AND FRANCES R. HOMANS

Abstract.—The H.S. Gordon paradigm has a long and prominent intellectual history and is certainly one of the most important and often cited articles in natural resource economics. However, modern fisheries have evolved in important ways beyond those envisioned by Gordon when he wrote his paper on rent dissipation in 1954. The most important impetus for change was the extension in jurisdiction in 1976, which gave coastal nations the legitimacy to exercise control over harvesting and other aspects of fisheries management by creating exclusive economic zones (EEZs). The creation of EEZs resulted in the establishment of new regulatory structures in most fisheries, converting formerly open-access commons into regulated open- (or closed-) access fisheries. Regulatory structures have had dramatic impacts on the evolution of technology, on markets, and on other aspects of the bioeconomics of modern fisheries. Recent adoptions of systems based on property rights, such as individual transferable quotas (ITQs), give us some insights into the types of impacts and the sources of rent losses that have emerged out of modern regulated fisheries. In particular, a surprising outcome from the adoption of ITQs in several fisheries has been that significant post-ITQ gains have emerged not from cost savings (as the Gordon paradigm would predict), but from revenues arising as fishers and processors were freed up from the tight season, gear, and area restrictions of old regulatory regimes. Whether this first wave of gains on the marketing side will be followed by additional cost savings from input reduction and consolidation is an open question, but it is evident that the regulations adopted after 1976 affected the market as much as or perhaps even more than they affected costs.

Economists and biologists have traditionally viewed the goals of fishery management from different perspectives. Biologists tend to focus on the health of the stock or biomass and view management objectives in terms of ensuring a safe stock level. Thus, biologists have tended to recommend policies principally aimed at preventing harvest levels that place the stock in danger of collapse. Economists have tended to focus on net economic returns from resource use and have promoted policies designed to achieve economically efficient, or rent-maximizing, outcomes (Scott 1955; Crutchfield 1965, 1979; Alverson and Paulik 1973). In the last 2 decades, the biologists' and economists' two perspectives have begun to converge. There are several reasons for this shift. First, the laws governing fisheries management in most countries have begun to be framed in a manner that explicitly incorporates economic and other goals into the policy-making process. The effect has been to tilt the management process away from one in which biological criteria dominate planning and actions toward one in which economic criteria are also important. Second, fisheries biologists have been exposed to new intellectual approaches that cast fisheries management in a decision-analysis framework, enabling rigorous incorporation of multiple goals including stock safety and economic returns in stochastic settings (Walters 1986; Hilborn and Walters 1992). Finally, fisheries managers increasingly face pressure from the industry over the financial consequences of regulatory changes in day-to-day management decisions. All these conditions necessitate the joint consideration of economic and biological factors in formulating fisheries management policy.

Given the growing importance of rent as a partial indicator of the success of the management process, one would expect that the concept of rent would have been refined and its measurement made more precise over the 2 decades since the formation of exclusive economic zones (EEZs). This has not happened. If one were to ask economists to define economic rents, one would end up with a manageably small number of relatively close definitions focusing on economic surplus. But if one were to ask how to measure or forecast rents in fisheries, one would end up with many conflicting suggestions.

This paper discusses rent generation in modern fisheries, with the aim of illuminating how potential rents from management changes might be forecast. The next section summarizes the intellectual origins of the concept of rent and the role played by the dominant paradigm developed by H. Gordon in 1954. The section that follows examines some modifications to the basic Gordon paradigm that incorporate features of today's fisheries, and the implications of these modifications for unraveling rent losses. Case studies of three different fisheries illuminate where rents seem to be emerging in rationalization schemes and what this may imply in general for predicting potential rents in fisheries yet to be rationalized.

Rents and the Gordon Paradigm

The literature in fisheries economics is dominated by

a simple theme—that open-access resource use dissipates the rent, or economic surplus potential, of fisheries. This theme has roots in a paper by Gordon (1954). Gordon compared an ordinary enterprise such as farming—where a legal owner directs the application of inputs to his land and extracts a payment for the services of the land and his entrepreneurship—with an open-access enterprise such as a fishery, where no such legal owner exists. Gordon's insight was to realize that in an open-access fishery, there is no proprietor or owner of the seabed and fisheries resources. The implication of this absence of ownership (or absence of property rights) is twofold. First, there is no entrepreneur present to direct the application of effort in a manner that yields the highest returns to the resource. Second, the extra uncaptured surpluses in the fishery, which normally would be withdrawn from the system by the owner, remain to attract more effort than would be the case under a circumstance with ownership. Thus, in Gordon's view, the absence of well-defined property rights in fisheries is the source of the problem because without an owner to extract surplus (rents), too many fishers enter and ultimately drive the harvest and biomass down to a level below what would be sustained if ownership rights to the seabed could be claimed and exercised.

This simple paradigm has been important to fisheries policy in several respects. First, by bringing attention to the rent losses inherent in open-access resource use, the paradigm brought economic efficiency and other social objectives into explicit consideration in the policy process. Second, by hypothesizing a rent dissipation mechanism, the Gordon model provides a conceptual structure with which to anticipate what might be happening in open-access fisheries. Third, by focusing on the role of incentives in motivating fishing behavior, policy options were broadened beyond the standard command and control restrictions that were used almost exclusively for regulation.

In many ways, however, the simplicity of the original Gordon (1954) open-access paradigm has been unfortunate. We would argue, in fact, that the simplicity of the original paradigm, coupled with economists' almost literal adoption of its premises, has set back the understanding of modern fisheries significantly. In particular, to sim-

plify his model, Gordon abstracted from four important features of fisheries. First, he basically ignored the connections between economic rents, harvests, and biology. His basic model linked rents and fishing effort but failed to include the feedback between effort and total harvests and biology. This omission was purposeful, because he did not believe that fishing could have more than a negligible effect on populations.[1] In hindsight, Gordon underestimated the strength of his own process; the tremendous response of fishing capacity to rents that occurred in the 1950s and 1960s drove several important fisheries to collapse.

The link between rents and entry, harvests, and population dynamics was eventually made by Smith (1968). In terms of impact on management philosophy, Smith's extension was too late; by then, another paradigm, Hardin's "tragedy of the commons," had become the dominant model in resource management literature (Hardin 1968). Hardin looked at the same process, but instead of focusing, as Gordon did, on the cause (individual incentives under open access), he focused on the effects on the resource itself. While Hardin's tragedy metaphor proved a popular device for drawing attention to resource degradation, it did little to illuminate solutions because it implied that degradation was inevitable and inexorable.

A second simplification that Gordon employed was the assumption that effort is unidimensional. This, too, had serious consequences for policy development in the late 1960s. In particular, the first limited entry programs were guided by an intellectual characterization of the rent dissipation problem as literally "too many boats chasing too few fish." As a result of this simple view, early management programs aimed directly at limiting boats. These programs were quickly found to be ineffective as fishers increased vessel capacity by altering other inputs (Wilen 1988). These problems might have been avoided if economists had developed a more general depiction of the open-access rent dissipation process in a multiple input setting.

The third abstraction Gordon employed was that the industry is assumed to operate in a completely open-access institutional setting, with no regulatory structure or other constraints on behavior. While this was probably an accurate depiction of most fisheries in the 1950s, it certainly was not after 1976, when coastal nations extended their resource jurisdictions to 200 miles (320 km). Economists have not responded to this new setting by adopting model structures that incorporate regulations in a meaningful way (see Wilen [1985] for a model of a regulated open-access fishery that incorporates both industry and regulatory behavior). As is discussed below, one cannot really predict how a fishery will evolve without a good understanding of the interrelationships between biology, fishing technology, regulations, and markets

---

[1]Cf. Turvey and Wiseman (1958), p. 77, which summarizes debates about fisheries policy at a 1956 symposium sponsored by the Food and Agriculture Organization of the United Nations. Gordon's views were summarized as follows: "Gordon said that the essential biological fact about the effects of fishing on stock numbers was that the reproductive capacity of fish was very, very high. Statistics supported the view that the size of a fish population was not related to the number of potential spawners. The effect of fishing was not on spawning but on average age."

(Homans [1993] provides an integrated model of industry and regulatory behavior and market interrelationships).

Finally, Gordon's fourth abstraction was that there is no feedback between entry, harvesting, and the market; thus, prices can be assumed given and independent of the open-access rent dissipation process. This was a natural assumption and one that allowed a simple analytical depiction of the process. However, in retrospect this assumption has led to an almost single-minded depiction of the open-access rent dissipation process as one of excess inputs, whose primary effects are to increase the costs of exploitation. Little attention has been given to the possibility that open-access processes might also affect product quality and the market side of the ledger. This is not to say that economists completely ignored the impact of open access on markets and vice versa. An early and particularly thorough analysis of open-access behavior is contained in Crutchfield and Zellner (1962), who developed an empirical model of price determination in the Pacific halibut (*Hippoglossus stenolepis*) fishery. Their model mainly focuses on price determination within the production period, and does not link this pricing to price determination during the marketing period between harvests.

Gordon's abstractions and the manner in which they have colored our view of fisheries have caused some major surprises in the face of evidence emerging from new fishery rationalization schemes. In particular, as the data begin to accumulate from individual transferable quota (ITQ) schemes, the results are not at all what the simple Gordon paradigm predicted. We should be seeing an unraveling or reversal of the rent dissipation process as depicted by Gordon. Specifically, as property rights have begun to emerge, we should be seeing a reduction of excess inputs with the consequent cost savings and generation of production efficiency rents. Surprisingly, in many cases what seems to be emerging are large rents from the marketing side of the ledger. As fishers have gained security of harvest rights, in many cases there have been more efforts expended to improve product quality and develop new markets, rather than saving inputs. In some cases, production costs have even risen in order to generate new, high-quality raw product, which earns higher prices in the market.

The next two sections discuss a new and broader paradigm appropriate to modern fisheries. This new view recognizes some factors that Gordon left out of his model, particularly the roles of and the interdependencies between the regulatory structure and the market. Contrary to what the Gordon paradigm suggests, the major inefficiencies in modern fisheries may be actually associated with marketing losses rather than cost inefficiencies. What is more important, these marketing losses seem to

be induced by the very nature of the regulatory structure that has evolved, a possibility the Gordon paradigm did not allow. When one considers a more realistic depiction of modern fisheries, which includes some of the factors Gordon ignored, the issue of identifying inefficiencies becomes considerably more complicated, involving technology, markets, regulatory structure, and industry entry and exit dynamics. This complexity is illustrated in the case studies section of this paper.

An Evolutionary View of Fisheries Institutions

The world has changed considerably since H. Gordon's seminal paper in 1954, which described the process of rent dissipation in a pure, open-access fishery. Since the extension of marine resource jurisdiction in 1976, virtually all coastal nations have had the ability, if not the resolve, to avoid the waste inherent in open access. As it turns out, the evolution toward rationalized systems has been slow. Most fisheries today are neither Gordon's pure open access nor anything approaching Scott's (1955) sole owner ideal. Rather, a hybridized system has evolved that is best defined as "regulated open access," or "regulated restricted access." By this, we mean that most fisheries retain biologically determined effort restrictions within an open-access framework or, in some cases, within limited entry programs. Thus, the inherent incentives are basically as Gordon described them although their impacts are generally stifled and modified by regulations.

If one looks at either the regulatory history of individual fisheries or at a large cross-section time-series of many different fisheries, an evolutionary process is apparent. In particular, most fisheries exploitation begins under pure, open-access conditions. Often, a fishery can remain in this situation for a long time because a new fishery usually begins as barely profitable. This may allow harvest levels to remain low enough to sustain the biomass at a reasonably safe level.

At some point, however—either because a market develops and increases revenues or because inputs increase as technology changes—rents rise, new vessels enter, harvests increase, and the biomass may be endangered. When this happens, if interested parties can overcome the transaction costs necessary to agree to stop overexploiting the resource, a regime shift occurs and some biologically motivated regulations are adopted. Often these are adopted after the fishery has been driven to a low level, necessitating a rebuilding period with fairly stringent controls.

In the next stage, which can be called regulated open access, the fishery proceeds under continuous adversarial pressures generated by increasing rents, increasing fishing capacity, and the resolve and ability of the new regulatory structure to protect the stocks. In this phase, access is open, and hence forces as Gordon described them drive entry and capacity growth. Potential rents drive the process (e.g., increasing real prices, technological change, and decreased opportunity costs), and as new entrants increase potential capacity, regulators must stifle the capacity in order to ensure stock safety.

Regulatory systems may use a variety of potential instruments such as gear restrictions, size and sex restrictions on harvests, and area and season closures. Often, fisheries in this stage start out employing biologically motivated restrictions such as closing the season during spawning periods or establishing minimum sizes. As rents emerge, because of technological change or real price increases, the pressure of excess capacity begins to threaten stock safety, and these original instruments begin to be employed to stifle effort. This stage may last for a very long time because there are an infinite number of ways for regulators to react to and mitigate growing effort when technology is flexible. Often, managers simply continue to tighten a single instrument, such as season length, as potential capacity grows. This has happened in the Pacific halibut, sablefish (*Anoplopoma fimbria*), and Alaska king crab (*Paralithodes camtschaticus*) fisheries. In other cases, the regulatory structure selects from among a suite of instruments, successively tightening one, then switching to another. For example, managers may shorten the season, then increase minimum mesh size or legally allowable size, then shorten seasons again, and so on. Most of the world's fisheries currently operate under a regulated, open-access regime.

Beginning in the 1960s, a modified version of the above institution emerged that might be called regulated restricted access. In regulated restricted-access systems, a limited-entry system overlays the types of controls used in the regulated open-access system. There are many prominent examples of this situation, almost all of which were adopted when open-access conditions threatened the safety of the biomass. These systems emerged to halt the entry process described by Gordon by also freezing numbers of vessels. Depending on how flexible technology is, these systems may or may not be effective, and they may last a long time without significantly eroding regulatory control. For example, Alaska's limited entry for salmon (*Oncorhynchus* spp.) has been in place for 20 years, and high license values attest to the existence of rents and the ability of the system to lock up capacity. However, these systems are ultimately vulnerable because the fundamental incentives to capture as large a share of potential rents as possible still exist. An unforeseen technological change can upset the delicate balance and send these systems into danger. For example, in British Columbia (B.C.), Canada, managers introduced a limited entry system in both their halibut and sablefish fisheries in 1980. By the early 1990s, however, the length of season for both of these fisheries contracted from 1 month to 1 week even with limited entry and higher allowable catches. In these cases, rapid technological change caused individual fishing capacity to grow, even though there were fewer vessels. Hence, regulated restricted-access fisheries ultimately face erosion, in which case regulators and fishers may look to other options.

The most recent option in this chain of institutional evolution has been a system based on property rights, notably ITQs (Neher et al. 1989). This type of system has become a viable option in many fisheries previously managed by traditional methods. Unlike any of the previously defined systems, which only control the symptoms of open-access behavior, property-rights systems fundamentally tackle the cause of the problem by altering basic incentive structures. These systems appear to be costly to set up and involve considerable uncertainties for existing participants and—equally important—for the existing management structure. If successful, however, these systems have the potential to end most problems associated with the rent dissipation process described by Gordon (1954) because they address the problem rather than its symptoms.

Unraveling Rents in Modern Fisheries Systems

If it is true that most modern fisheries are not as simple as Gordon depicted them, what are their characteristics and how do they alter his fundamental predictions? First, as argued previously, the nature of the regulatory structure critically affects the way rents are generated. For example, gear restrictions, size restrictions, season lengths, and many other biologically motivated regulations affect rent potential directly as constraints on input choices and fishing practices. In addition, these regulations affect the rent generation process indirectly because they affect subsequent rent-seeking behavior. For example, season length restrictions have a direct effect by shortening time on the grounds; but they also have an indirect effect by causing fishers to build vessels with higher quick-response capacity that will operate effectively in a derby setting.

Second, the regulatory structure is not static; it interacts with the industry and is responsive to investment, technological change, and so on. For example, in the B.C. halibut fishery, when circle hooks, fish shakers, and automated baiting equipment were first adopted, catch rates increased, regulators shortened seasons, fishers adopted the new technology at a faster rate, and seasons were shortened again. The significance of this cycle is that technology, costs, and regulations are dynamic and endogenous.

As a result, the technology and cost structure that we observe at a snapshot in time is unlikely to be anything close to efficient in the rent-maximizing sense. Instead, it reflects inefficient regulation-evading behavior on the part of the industry, as well as the whole evolutionary history of action and reaction between the industry and the regulatory structure. Third, these interactions among the industry, the regulatory structure, rents, and technology also affect and are affected by the market. To take a common situation, as seasons are compressed, ex vessel prices are lower than they otherwise would be because fish quality is poorer in a derby fishery, and because the final product must adapt to the short season.

These points about the connections among the regulatory structure, industry behavior, and markets in modern fisheries are particularly important for several reasons. If regulations reduce rents on the market side, overcapacity is mitigated somewhat because the potential profit incentive is not as strong as it would be otherwise. At the same time, it means that if we attempt to measure rent losses (or the other side of the coin, the potential gains from rationalization) by focusing only on costs, we may miss a large part of those losses. This observation may explain what seems to be happening worldwide in several adoptions of ITQs (Wilen and Homans 1992; Homans 1993). In particular, as old regulated open-access fisheries are freed from the regulatory constraints that have evolved to stifle effort, the first change is often new opportunities to generate products different from those produced under regulated open access. A common situation, for example, is for a fishery that has been backed into a short season derby to convert to a year-round fishery. In this situation, marketers and harvesters have new incentives to work together to maximize the value of the catch by fishing when demand is high and by developing and targeting niches in the market that never existed before. All of this is unimaginable within the Gordon paradigm, which ignored the market, the regulatory structure, and any interactions between them associated with rent dissipation.

In the next section, several case studies reveal some of the intricacies of rent generation in an expanded framework that accounts for regulations and market effects. These case studies are drawn from several fisheries recently converted to ITQ programs, and they collectively point to several conclusions. First, real systems that involve endogeneity and dynamic interactions among the industry, regulators, technology, biological factors, and the market are obviously complex. Understanding where rents are dissipated, and consequently where they might be released, is a difficult task that requires a serious look at the peculiarities of each situation. Often a single random event in the evolutionary history of the fishery can determine much of the entire subsequent path of rent dis-

sipation. Because of the evolutionary nature of technology, confining one's investigation to technologies and fishing practices that exist at a given moment is likely to be a misleading indicator of what might exist under a more rational management structure. Current technology, markets, and regulatory structures should probably be regarded as accidents of history rather than as an outcome of a rational optimizing plan. For the most part, trying to forecast future structures of technology, costs, prices, and markets using current structures requires caution and caveats, and may, in some cases, be futile.

Rent Dissipation and Rent Generation—Case Studies

British Columbia Halibut

As discussed previously, the north Pacific halibut fishery has a long history stretching back over a century. Its history reveals a process in which continued increases in rents generated (1) entry, (2) reductions in season lengths to 5 days in the late 1980s, and finally (3) limited entry in British Columbia and the recent move toward a rights-based system in the 1990s.

During the 1991 season, Canadian fishers decided to conduct their portion of the fishery under a vessel quota system rather than the usual "fishing derby." The U.S. fishery retained the short derby season system; hence, observers had a vivid comparison of operations under alternative incentive systems. During the brief U.S. fishery openings in 1991, there were reports that greater than 50% of the catch landed was never iced; during the biggest 1-day opening (in the first week of May), about a third of the fish were not even gutted. The British Columbians, in contrast, mostly chose to hold their quotas until after the May opening in the USA and before the final closing date of November. As a result, the Canadians obtained significantly higher prices in the first year of the program. *Fishermen's News* (June 1991) reported that ex vessel prices received in late May by B.C. fishers averaged US$1.10 higher than those received by their U.S. counterparts (who received about US$1.70). From the first year's data, this suggests a price "penalty" associated with open-access fisheries and its regulations of around 40%.

Since British Columbia adopted an individual quota system in 1991, there have been dramatic changes. Although it is too early to tell exactly what is happening to fishing costs, one study suggests that costs did not decline but rather rose slightly as fishers took longer trips and readjusted fishing practices (EB Associates 1993). The absence of significant reductions in inputs is due, in part, to the fact that quota trades and quota consolidation have been limited by design during the early stages of the program. Current restrictions allow limited leases but prohibit sales for the first few years, with the intent of allowing the industry to learn how the system will work before being freed up to quota trades. Anecdotal evidence suggests that fishers are still fishing for halibut as part of a larger complex of species targeted over the year, including salmon and other fish. Thus, quotas are still being filled by concentrated effort over periods that last less than the whole season, and full-time, year-round halibut fishers are not yet replacing part-time fishers.

There have been dramatic changes in processing and marketing, however. The B.C. industry has changed from one producing only frozen fish to one producing mainly for the fresh-fish market. New and higher-valued markets have developed for a year-round supply of fish, and marketers often work with fishers to spread out supply and avoid short-term market gluts. Surprisingly, the market is composed of a completely new set of brokers who are not affiliated with the old derby fishery. Research in progress suggests that processing during the derby regime was not particularly profitable and processors participated almost as a service to fishers who supplied the real bread and butter species, namely salmon. Hence, when the individual quota (IQ) fishery developed, a new group of aggressive and imaginative marketers took charge. Recent estimates suggest an ex vessel price premium of 60–70% in the IQ fishery, arising from this new market structure, better quality raw product, and perhaps more competition among handlers.

New Zealand Groundfish

In the New Zealand fisheries, the evidence for rent gains from improved marketing is equally startling. An important case is that of the snapper (*Chrysophrys auratus*), which is the mainstay of the northern New Zealand inshore groundfish fishery. This fishery has a long history of exploitation, much like that of the north Pacific halibut fishery. In the early part of the century, steam-powered trawlers caught snapper in vessels over 40 m in length. In the 1920s, Danish seine vessels came to dominate as gasoline and diesel power replaced steam vessels. In the 1970s, a new market for higher-quality, frozen, whole snapper opened in Japan. This market used more carefully handled fish caught on longlines, a more costly technology. The conversion to ITQs accelerated this conversion of technology and opened further avenues in which to increase revenues. One marketing innovation is a live-fish market, in which live snapper are packed in styrofoam containers with a water supply and whisked to market.

Early reports are that revenues *tripled* in some groundfish fisheries after the introduction of the ITQ system. This was made possible by a major shift in marketed products away from mixed batches of small, trawl-marked fish caught in compressed seasons with long tows. More recent changes in technology and practices favor longline-caught fish, or short-tow, trawl-caught fish, selected for size and market characteristics and spread over the whole year. Interestingly, both of these are evidence of switches to costlier technologies induced by the prospect of the marketing gains made possible by the changes in incentive structure. Thus, in contrast to what most economists would have expected, in this case unit costs actually rose after a more efficient regulatory scheme was adopted, but revenue increases more than compensated to increase profits. An *ex ante* analysis that attempted to predict rent gains in this fishery by examining production cost savings not only would have missed the most significant component of the rent gains, but also would have likely assumed that trawl technology would continue to be the dominant, if not sole, technology.

Australian Southern Bluefin Tuna

The bluefin tuna (*Thunnus maccoyii*) is another interesting ITQ case study involving a very valuable fish with high market value in Japan. These tuna have been harvested for many years in the south Pacific by fisheries from Australia, Japan, and other countries. Tuna spend roughly the first 8 juvenile years in coastal waters off Australia where they slowly migrate counterclockwise as they grow. After maturing, tuna enter the high seas and continue to grow and migrate over large areas. During the late 1970s, biologists warned of impending overharvests and recommended a catch reduction. In 1983, Australia and Japan both agreed to catch reductions. To implement them and to improve economic returns, Australia implemented an ITQ system in late 1984, the first such system in the country.

With the introduction of ITQs, some dramatic changes occurred in the bluefin tuna fishery. The most important change in fishing practices was geographical movement. With secure rights to specific quantities of tuna, fishers refrained from fishing small tuna in nearshore regions and moved off the continental shelf to target fish at the eastward edge of their life-cycle migration, where the fish are larger. Before ITQs, only 13% of the Australian catch was greater than 15 kg; within 2 years, more than 35% was of this larger size class. Targeting of the larger fish was a direct response to more lucrative prices paid by the Japanese in the sashimi market. Significantly, average prices rose in real dollars from AUS$988 per metric ton (mt) at the beginning of the program to over AUS$2,000 per mt in only 3 years (Geen and Nayar 1988). Hence, revenues more than doubled, largely because fisheries were able to tap new markets that were unavailable during the open-access phase.

With respect to fishing costs, Geen and Nayar (1988) reported that average variable costs have risen while average fixed costs have fallen because of consolidation and exit of excess capital. Variable costs have risen because fishers now travel at least 2 days to get to the edge of the shelf, as opposed to the short day trips that used to characterize fishing practices. Geen and Nayar estimate that fixed capital investment in the southern bluefin tuna fishery has been reduced by approximately 25% beyond what would have occurred without the program. Thus, on the whole, rents have emerged owing to removal of redundant capital and consequent savings, but these savings are offset somewhat by increased variable costs, and are swamped by rent generation from the marketing side. An analysis that attempted to project rent generation due to ITQs without including the marketing side changes wrought by new incentives would have substantially underestimated total rent gains and probably overlooked fishing practice changes induced by these gains.

Conclusions

As discussed in the introduction, the concept of rent is achieving increasing importance in fisheries management. First, rent plays a prominent role as the engine that drives fisheries into biological overharvesting or regulatory straitjackets. Second, prospective rents are the main indicator for gauging the potential success of changes in fisheries management policy. In view of these factors, it is important to understand exactly how rents are generated, how this process affects the nature of modern fisheries, and what might be expected from rationalized systems.

The Gordon paradigm has a long and prominent intellectual history and is probably one of the most important models in resource management literature. In important ways, however, modern fisheries have evolved beyond the structure Gordon considered. In particular, modern fisheries are heavily influenced by the nature of regulations in a manner that confounds the unraveling of resource rents. Regulations not only affect basic biological factors, they also affect rents, entry and exit behavior, technological choices, and markets. If that were not enough, regulations in turn are endogenous and dynamic in a fisheries system; hence, prediction becomes very problematic. Of particular importance is the fact that current technology, fishing practices, and product markets may have little in common with what might evolve over time, and even less in common with what might emerge under different institutional circumstances.

Individual transferable quotas and other institutional systems based on property rights create radically different incentives. Guessing what may emerge after these systems are introduced requires serious and careful consideration of factors unlikely to have been part of a fishery's history. At the very least, this suggests that even the most rigorous and sophisticated statistical and econometric analysis based on extrapolation of current circumstances may be wide of the mark (examples of sophisticated econometric analysis designed to project quota program impacts can be found in Squires and Kirkley [1991] and Squires [1990]). It suggests that methods not normally part of the purview of fisheries analysts might be particularly useful, including focus groups, Delphi and informed opinion methods, programming and simulation models, and calibration techniques. All of these methods require as basic input the qualitative and quantitative opinions of industry representatives, creative brainstorming about potential effects of programs, and judgment about the reasonableness of projections.

References

Alverson, D. L., and G. Paulik. 1973. Objectives and problems associated with managing aquatic living resources. Journal of the Fisheries Research Board of Canada 30:1591-1600.

Bell, F. H. 1981. The Pacific halibut. Anchorage, Alaska Northwest Publishing Company.

Crutchfield, J. A. 1965. The fisheries problem in resource management. University of Washington Press, Seattle.

Crutchfield, J. A. 1979. Economic and social implications of the main policy alternatives for controlling fishing effort. Journal of the Fisheries Research Board of Canada 36:742-752.

Crutchfield, J., and A. Zellner. 1962. Economic aspects of the Pacific halibut fishery. In Fishery industrial research, vol. 1. U.S. Department of the Interior, Bureau of Commercial Fisheries, Washington, D.C.

EB Associates. 1993. An evaluation of the British Columbia ITQ program. Consultants' report submitted to the Department of Fisheries and Oceans, Vancouver, British Columbia.

Fishermen's News. 1991. Newsnote (June):9.

Geen, G., and M. Nayar. 1988. Individual transferable quotas in the southern bluefin tuna fishery: an economic appraisal. Marine Resource Economics 5(4):365-388.

Gordon, H. S. 1954. The economic theory of a common property resource: the fishery. Journal of Political Economy 62:124-142.

Hardin, G. 1968. The tragedy of the commons. Science 162(3859):1243-1248.

Hilborn, R., and C. Walters. 1992. Quantitative fisheries stock assessment: choice, dynamics, and uncertainty. Chapman and Hall, New York.

Homans, F. R. 1993. Modeling regulated open access resources. Doctoral dissertation. Department of Agricultural Economics, University of California, Davis.

Neher, P., R. Arnason, and N. Mollett, editors. 1989. Rights based fishing. Kluwer Academic Publishers, Dordrecht, the Netherlands.

Scott, A. D. 1955. The fishery: the objectives of sole ownership. Journal of Political Economy 63:116-124.

Smith, V. L. 1968. Economics of production from natural resources. American Economic Review. 58: 409-431.

Squires, D. 1990. Individual transferable quotas: theory and an application. National Marine Fisheries Service. Administrative report LJ-90-16, Southwest Fisheries Center, La Jolla, Calif.

Squires, D., and J. Kirkley. 1991. Production quota in multiproduct Pacific fisheries. Journal of Environmental Economics and Management 21(2):109-126.

Turvey, R., and J. Wiseman, editors. 1958. The economics of fisheries. Food and Agriculture Organization of the United Nations, Rome.

Walters, C. J. 1986. Adaptive management of renewable resources. Macmillan, New York.

Wilen, J. 1985. Towards a theory of the regulated fishery. Journal of Marine Resource Economics 1:369-388.

Wilen, J. 1988. Limited entry licensing: a retrospective assessment. Journal of Marine Resource Economics 5:313-324.

Wilen, J., and F. Homans. 1992. Marketing losses in regulated open access fisheries. VIth International Institute for Fisheries Economics and Trade Conference, Paris. Institut Francais de Recherche pour l'Exploitation de la Mer, Issy-les-Moulineaux.

A Government Perspective on New Zealand's Experience with ITQs

PHILIP MAJOR

Abstract.—New Zealand has introduced a range of fisheries into individual transferable quota (ITQ) management since 1982 when explicit mechanisms to introduce transferable property rights were first used in New Zealand's deep-water fisheries. Pre-dating this, there were individual quota arrangements in the Bluff oyster fishery and the Ellesmere eel fishery. A comprehensive regime for quota management and individual transferable quota was introduced into New Zealand fisheries law in 1986.

This paper explores the mechanisms that were used to encourage fishers to accept the implementation of an ITQ management regime. The incentives principally were a buy-out mechanism, a quota appeal authority regime to review quotas of individuals, and a mechanism whereby government bought quota on catch reduction and sold quota on catch increases. The paper further proceeds to examine the incentives that have developed for fisheries conservation and for efficient fisheries economic use in the years since 1986. Discussion focuses on a range of issues. These include the development of claims for aboriginal title, the outcome of the various litigation and government negotiations on this issue, the need for further structural reform in the quota management system, and warnings to those who are proceeding to implement similar systems to ensure that policies are clear on the role of government and the role of industry in the management regimes that are established. The paper advises that the return, if any, to the Crown or governing body should be determined prior to establishing the management regime, and equally clear rules and policy should be established before introducing quota regimes with regard to the costs to industry and government.

I have been asked to address this conference from the perspective of a fisheries manager assessing the value of the quota management or individual transferable quota (ITQ) management system in operation. In speaking to similar audiences over the last 2 or 3 years, I have come to understand that New Zealand, which has embraced a quota management system, is now into a second generation of issues and problems that flow from such a system. Those of you who are considering such systems are grappling with problems that are associated with implementation. As a consequence, I focus on those issues related to implementation while pointing out the benefits and pitfalls encountered in New Zealand.

As a ministerial policy advisor and fisheries manager, I am particularly concerned with three primary objectives. The first is conservation, which is aimed at ensuring that stocks are able to replenish themselves in a manner that does not lead to a dislocation of the marine ecosystem. The second objective is to ensure that there is some economic efficiency in the commercial exploitation of fisheries and resources. The third objective is to ensure that mechanisms are in place for establishing a suitable balance among competing interests of groups who wish to have access to New Zealand's fisheries resources—commercial, recreational, and indigenous fishers and people who primarily seek to have the resource preserved in its natural state.

In New Zealand, fishery managers are concerned with the activity of commercial fishers and, to a lesser extent, the activity of recreational fishers. First, I briefly focus on some positive aspects of the quota management system in New Zealand. (Branson [1997] expands on these quite substantially.) The first aspect is that New Zealand's domestic fishing industry currently spends some US$3 million on research on the orange roughy (*Hoplostethus atlanticus*) fishery. Further, the industry has entered into communal contracts to regulate its own activity for the purposes of conservation and to ensure there is no overfishing in a range of subdivided marine areas.

In addition, New Zealand scallop fishers presently contribute several million dollars as the total cost of enhancing the Nelson Golden Bay scallop (*Pecten novaezelandiae*) fishery, and snapper (*Chrysophrys auratus*) fishers voluntarily pay for additional enforcement in their fishery. Snapper fishers also pay for research projects studying different effects of harvesting, which will lead to more efficient enforcement and better management practices for the restoration of the fishery.

In rock lobster (*Jasus edwardsii*) and abalone (*Haliotus iris*) fisheries, fishers pay over US$150,000 per annum for enforcement contracts to be carried out by the Crown. I doubt that there is anyplace else in the world where fishers voluntarily contribute substantial funds to enforcement agencies. Everywhere else in my experience, they seem to be doing their best to have such agencies underfunded and to ensure that they are unable to conduct their activities with strong enforcement presence. In New Zealand's squid (*Nototodarus gouldi*) fishery, when the government would not reduce quotas, the industry voluntarily set its own. There were two reasons for this voluntary action: first, people in the industry believed the resource needed a rest to recover, and second, there was a glut of squid on the world market.

Perhaps the most outstanding example of the effectiveness of New Zealand's fisheries regime on industry behavior is that fishers voluntarily chose not to increase the 50,000-metric-ton (mt) hoki (*Macruronus novaezelandiae*) fishery, which would have placed US$2.5 million directly into fishers' hands through joint venture arrangements. Instead, fishers chose to put the increase on hold, stating that this action would result in a greater number of fish of increased quality and size in the future. That is truly remarkable behavior for fishers anywhere in the world who are faced with the opportunity to increase their catch.

What led to this extraordinary behavior by fishers in New Zealand? Quite simply, it is that we have constructed a system in which people acting in their own self-interest have discovered that it benefits them to act in a manner that enhances the fishery. The reason for this behavior focuses on four elements in our management scheme.

1. Ownership: With ownership there is a capital value ascribed to an asset. Many fishers in the industry have had to purchase their quota, so now they have great respect for their invested capital.

2. Perpetuity: A quota is held in perpetuity, which reinforces the capital value and gives fishers the security to invest for the long term, giving them a vested interest in looking at the longer-term value of their asset. Consequently, they are starting to act with the capital value of their asset in mind rather than merely assessing their ability to obtain a cash flow from an annual catch.

3. Market in rights: By being able to trade fishing quota, fishers can obtain the parcel of fish that best suits their interests (i.e., the species they are liable to catch).

4. Enforcement: Again, the quota management system has changed the old game of outfoxing the government to catch more fish. It is now socially unacceptable to catch more than one's quota, as this constitutes stealing from one's mates rather than from the government. Further, with the increasing prominence of the conservation movement in New Zealand, there is a reinforcing effect of the value of environmental protection, which is leading to a discernible conservation ethic not previously found among fishers (or for that matter the general public).

At this point, I review the introduction of the ITQ scheme into New Zealand's fisheries. (I do not deal with it in detail as this is well covered in the literature by Clark et al. 1980.) There was an individual quota fishery in New Zealand for the oyster (*Tiostrea lutaria*) fishery prior to the declaration of the New Zealand exclusive economic zone (EEZ). However, after the introduction of the EEZ in New Zealand waters, there was a dramatic increase in fishing effort through joint venture arrangements. In 1982, New Zealand introduced a quota management system for nine species, which affected 12 companies. This was a relatively small management arrangement and the issues at the time were as follows:

- Method of allocation: Allocation was based on a formula of catch processing and capital investment in catching capacity.

- Tradability of quota among companies: While the government had approved this arrangement with maximum holdings of 35% for any one company or individual, the establishment of a market and accompanying registry was not initiated, with changes of ownership being notified to government resulting in cumbersome regulatory administration.

- Total allowable catch (TAC): Increases and decreases in TACs were not well handled at the time: A proportional arrangement and an arrangement for keeping a share for competitive fishing were elements of the process in place.

A final issue, not dealt with at the time, was that of record keeping to ensure that catch and quota allocation were properly accounted for. Nevertheless, for the first 2 or 3 years, the system worked quite well because the resource was not under any great pressure and the arrangements that New Zealand companies used to control their joint venture partners were adequate to ensure that overfishing did not occur.

In 1986, a much more comprehensive inshore structure was instituted, and the inshore and offshore components of the fishery were merged. This was a much more difficult change as the inshore fishery was overcapitalized and overfished. Consequently, new legislative mechanisms were introduced to encourage acceptance by industry. This is the situation many fisheries managers now face. The New Zealand fishery is large and complex, and has some 160 fisheries areas with quota.

The mechanisms used in New Zealand's regime are as follows. The first mechanism, designed to encourage many fishers to accept a change to a quota management regime, provided the government a means to purchase quota from fishers when a TAC reduction occurred and sell it to them when there were opportunities to increase TAC. This was designed to ensure that the fishers had a secure investment and could see that, in the event of severe decreases (which were anticipated at that time), they would be able to leave the fishery with dignity and invest in other areas of the economy. Equally, there was a benefit to the Crown, which recognized the potential for exponential increases to future TAC, enabling it to sell any recovered stock at a fairly significant return—a return that would have been greater than the cost of the reductions plus interest. In retrospect, this was not a good idea as the movements in quota were such that the government decided that the risk of the fishery should be borne by the participants and not by the Crown. Thus, 4 years after the introduction of the quota management system, New Zealand moved to a system of proportional quotas, which was coincident with the necessity for a large reduction in orange roughy catches. In moving to this system, the government then had to negotiate with industry for a compensation payment as a transitory mechanism for its accepting the risk of the fishery.

The next mechanism the government introduced as an incentive for moving from open access to a quota management system was to provide for a buy-back scheme. All fishers were issued a quota that was calculated according to their fishing history, adjusted for a commitment and dependence arrangement, at which stage they were offered the opportunity to tender a portion of their quota back to government. The government spent NZ$40 million on this program to buy quota from the fishery to reduce the TAC to levels that would allow the stocks to recover. This ties in to my previous point; the government believes that in the future, with the recovery of stocks, it will be able to sell quota at a price that would recover what it cost to reduce quota. A number of administrative reductions in quota were then made. The fishers who suffered these administrative reductions were guaranteed that when future increases in quota occurred, these reductions would be restored.

The next area the government addressed was bycatch. Fishers complained that it would be impossible for them to target the exact tonnages of their quotas and that arrangements had to be made for bycatch over and above the face value quota, which takes into account the fluctuations of stock and the inherent difficulty of targeting one species alone. As a consequence, arrangements were made for a 10% provision for over-catch, for under-catch being carried forward to the following year, for a bycatch trade-off where the primary species was surrendered or a sum of money could be paid in lieu, and for a leasing arrangement for quota.

In allocating catch histories, fishers argued that they should be entitled not just to their established history, but personal circumstances that deprived them of the opportunity to make a fair and reasonable catch over the qualifying years should also be considered. Consequently, the government permitted claims for commitment to and dependence in the fishery. Additionally, there was an administrative appeal system. The Minister of Fisheries established a body to hear cases, followed by a quasi-judicial body, known as the Quota Appeal Authority, to whom claims could be made. Beyond the Quota Appeal Authority, it was possible to go to the courts. The consequence of the administrative appeal system was that large quantities of additional quota were allocated in the fishery, which resulted in some reallocation of the original cuts, particularly in the snapper fishery.

The government also believed that the quota management system would make fishing extremely profitable, so fishers should pay a resource rental, which would include a return to the Crown to recover the costs of managing the fishery. The legislation supporting this mechanism was not as specific as it might have been. The resource rental was to be set annually and would involve a revolving fund for incoming management costs, and total allowable commercial catch reductions.

The final element of the 1986 legislation was enforcement, which was to be managed through a paper trail that followed the catch of fish on a vessel through the wholesaling process to retail or export. Documentation was made easier by the introduction of a goods and services tax in New Zealand, which meant that all companies had to keep trading records. While the paper trail would form the principal mechanism for investigating over-catches, it was to be supplemented by a range of other on-the-water and off-the-water observing operations.

The more astute of you and those who have visited New Zealand will have already recognized a range of fundamental flaws in the way the original system was established and the way that it differed from the theory that was developed by the original quota management system proponents. I will deal with these flaws in a moment. The reality is that over 6 years, the flaws in this system became apparent. Consequently, for the last 3 years, we have been reexamining the nature and extent of our quota management system. The nature of this reexamination has consisted of a number of expert reviews. However, this review process has become fraught with difficulties because of the wider public interests, the active involvement of conservationists, and the need clearly spelled out by government to have input from all stakeholders prior to final decision making.

Notwithstanding these difficulties, the review of New Zealand's quota management system was absolutely essential because of problems arising from the narrowness of the original established parameters. Areas of particular concern were as follows:

- bycatch issues;
- the nature of the fishing right;
- aboriginal title, both commercial and customary;
- recreational fishing access rights;
- conservation value rights;
- the return to the Crown (the state);
- the charging mechanism for management; and
- the mechanism that would be used to allocate quota in the future, bearing in mind that we had already

shifted from the prescribed tonnage quotas to a percentage quota.

The first issue—bycatch—is perhaps the biggest and most vexing. At this time, the mechanism for tracking catch against the quota leases and other fishing arrangements that have developed under the quota management system has become extraordinarily complex. Adding to the complexity is the need to track bycatch and overcatch, which must be counted in descending order. Consequently, quota holders and fishers are demanding a simpler system for administering quota holdings. This debate has spawned the concept of the annual catching entitlement (ACE), which is derived from the quota held by an individual, then bought and sold each year without having to be assigned to any individual owner.

The ACE led to a review of penalties in the fisheries, principally forfeiture. (In this case, forfeiture includes boats, quota, and other property, including cash, used in the commission of the offense.) There has been a move to ensure that forfeiture is either abandoned or is only used for the most excessive breaches of fisheries law; lesser breaches would be dealt with by a sliding scale of penalties. The issue is that the original market established for quotas did not meet the needs of fishers to rapidly trade quotas or fishing rights to ensure that their catches matched their holdings. It is believed that the ACE concept will provide a better market for trading fisheries rights. However, because ACE is divorced from ownership, an inherent conflict is set up in that ACE holders may not have the long-term interest of the fishery at heart, as individual owners would.

The second issue that arose is the nature and definition of the right allocated to quota holders. When the government initially allocated rights, it believed it was establishing a catching right in an area and that it could modify such rights by passing rules and regulations from time to time. However, these rights were not so readily modified. The courts ruled that people must be able to catch their fish in a reasonable manner, as they had done historically. The Crown therefore had to be extraordinarily careful in implementing modifying regulations. Consequently, the government has come to appreciate that the rights granted need to be very clearly defined—such as what is permitted in the water column, seabed extraction, general fishing areas, and size of animals that can be taken. My advice to anyone who is contemplating a rights-based system is to look very closely at all the interlinkages that may occur in terms of the scope and nature of the rights being established.

The third and probably principal issue that had to be addressed in the recent reviews of the quota management scheme is indigenous title. Initially, no account was taken of indigenous title. The consequence was that a high court imposed an injunction on the Crown, requiring it to consider indigenous title. This could occupy an entire paper in itself, but suffice it to say that claims relating to commercial harvesting rights have been settled with New Zealand's aboriginal people, the Maori. The range of customary fishing rights that will apply in the future is yet to be fully determined and defined, which must be done in full consultation with the Maori. The next issue to be considered was recreational fishing rights, which are undefined and regarded by recreational fishers as a priority in terms of the TAC allocation. Discussions held during the recent reviews resulted in an outpouring of anger from anglers at the suggestion that recreational fishing may be restricted or subject to licensing.

A further right, yet undefined, relates to conservation and the interests of those people who wish to see the resource preserved in a natural state. Conservation may be achieved either through marine reserves or through preservation of large numbers of a population. While such rights are based on the desire to preserve an ecosystem, there is also the perspective of divers and tourists who may wish to view these resources in their natural state. Aboriginal, recreational, and conservation rights do constitute a threat to the quota management system as it was originally established. Each group wishes to claim a priority, and recognition of these new rights would affect the perpetual nature of the existing fishing rights and any ongoing share of them. Therefore, it is essential in establishing quota management systems that a balance be established at the start between the four claims on rights: the aboriginal, the recreational, the conservationist, and the commercial.

Another principal issue that remained unresolved was whether there should be a monetary return to the Crown for the allocation of catch rights. As mentioned earlier, resource rentals did not clarify this matter. This has led to a huge debate on the issue. Obviously the industry argued strenuously against such a policy, while conservationists, and to a lesser extent recreationists, argued that there should be a fee for the environmental damage incurred while harvesting resources. This fee would also assign a value to modification of an ecosystem. Lately, the New Zealand government decided that there would be no resource rentals in the form of an access fee or a return to the Crown. The absence of the resource rentals, the Government stated, would create a climate for investment in the fishery and improve the certainty associated with fishing rights. It would also ensure that New Zealand was not hampered in terms of international competitiveness. If it becomes essential to obtain a return to the Crown rather than apply a charge to each quota holder, it would be better in the initial allocation process for the Crown to retain a small percentage, which it would lease out each year. This meets the return to the public although it does not meet the criteria required by conservationists for a payment for damage occasioned to the resource.

The question of a return to the Crown leads inexorably to the question of whether the Crown should charge users for resource management. From the beginning, it has been intended that the New Zealand commercial fishing industry should pay the costs of managing the resource. This intent generates two questions: Which costs are associated with management of the resource? What role do the commercial fishers or resource users have in terms of overseeing management costs and services?

At this time, the Crown has determined that the commercial fishing industry will be charged on an avoidable cost basis (there would be no costs if there was no commercial industry; thus, industry should pay all costs). Provision has been made for reasonably generous percentage deductions to apply to noncommercial fishing, especially recreational and Maori fishing rights, together with an allowance for government policy advice, which is a cost that lies with the Crown. In effect, the fishing industry will likely pay attributable costs rather than avoidable costs although there will inevitably be debate from the industry as to which charges really do belong with them.

In association with this problem, an additional decision has been made that gives industry wider influence in controlling the administrative costs of the quota management system. Similar decisions are yet to be made as to the way research and enforcement of the resource will be managed. In both cases, there is a desire by government to (1) ensure independence of action by the particular agencies to guarantee effective conservation and enforcement, and (2) assure that conflicts of interest are avoided.

The final question I address is resource allocation at the time the system was changed from open access to quota management. As explained earlier, New Zealand bent over backwards to ensure fairness and equity in providing an allocation to fishers. This accommodation worked dramatically against management of the system. For example, the Quota Appeal Authority has, by nature of its legal requirements, been exceedingly fair. As a consequence, additional quota has been issued to inside fisheries, which took us back to the catch limits established before the buy-out scheme and administrative reductions. Upon reflection, it would seem to be better to take a much harder line initially—to rely on an allocation based on a number of qualifying years and let the luck of the draw stand. Otherwise, like New Zealand, you could experience the prospect of administrative appeals continuing up to 8 years after the initial allocation occurred, followed by court appeals. In fact, in terms of the court appeals process, the government passed legislation to limit the grounds on which appeals were granted.

A further consideration in allocating new quota is whether to sell the quota upon issue. New Zealand determined that there should be no charge, initially, and that decision was sustained after the most recent review by the Minister of Fisheries. A range of fisheries management material (Anderson 1986; Johnson and Libecap 1982; R. Johnson, Dep. Agriculture Economics and Economics, Montana State University, Bozeman, Montana, USA, unpubl. ms.) on resource theory suggests that there should be some charge and suggests a range of mechanisms for tendering charges, either in total or in part, to provide some return to the state. Again, New Zealand found that it would be subject to a range of real challenges were it to adopt a tendering mechanism. The tendering issue, coupled with the allocation of 20% of new species to the Maori Fisheries Commission as part of the indigenous settlement arrangement, ensured that it was unreasonable for quota to be issued at a cost to the people who had fishing histories and whose quotas may be proportionately reduced to ensure that the Maori receive their quota under the settlement.

The final item that I refer to is aquaculture rights, which fall into the similar areas of conservation, recreational, and indigenous rights. A separate regime exists for allocation of aquaculture rights, or private enhancement rights; this will conflict and undermine the status and validity of wild fishery rights. Again, it is essential to (1) ensure that there is a spectrum of rights and mechanisms for managing them, and (2) avoid conflicts among competing rights or a hierarchy that usurps one set or the other. If this is not the case, we will see the benefits of perpetual ownership undermined.

Conclusion

I have briefly covered a large number of issues in this paper. However, if there is one piece of advice that I have for you, it is to try to get it as near right as possible the first time. As managers or fishers, you will inevitably be drawn into trying to introduce a quota management system on a staggered basis. This will only create more difficulties and more complexities. If you do seek a staged introduction, I suggest that you do it on a fishery-by-fishery basis, and make sure that the whole range of issues is covered in each fishery, rather than leaving the questions and points that I have covered open for dispute or debate at a later date. Once rights are implemented, it becomes very difficult to legislate, restrict, or minimize them. Legislation that makes such changes undermines the very benefits of the system that has been established. As a fisheries manager and a policy advisor to the Minister of Fisheries, I can categorically say that, notwithstanding the problems that we have encountered and the future problems that we will inevitably face, the right choice has been made in terms of developing the quota management system in New Zealand.

References

Anderson, L. G. 1986. The economics of fisheries management. Revised edition. Johns Hopkins University Press, Baltimore.

Branson, A. 1997. An industry perspective on New Zealand's experience with ITQs. Pages 270-273 in E. K. Pikitch, D. D. Huppert, and M. P. Sissenwine, editors. Global trends: fisheries management. American Fisheries Society, Bethesda, Maryland.

Clark, I. N., P. J. Major, and N. Mollet. 1988. Development and implementation of New Zealand's ITQ management system. Marine Resource Economics 5:325-349.

Johnson, R. N., and G. Libecap. 1982. Contracting problems: the case of the fishing. American Economics Review 72(5):1005-1022.

An Industry Perspective on New Zealand's Experience with ITQs

ANDREW R. BRANSON

Abstract.—The New Zealand fishing industry actively participated in planning and supporting government implementation of quota management (individual transferable quotas [ITQs]) for deep-water fisheries in 1983 and inshore fisheries in 1986. The 1986 implementation of ITQs enabled delivery of restructuring assistance as well as changes in overt fisheries policy. Major dispute occurred in 1990 with the change from direct to proportional quotas, and other, frequent regulatory and policy changes regarding ITQ management have resulted in an over-complex and administratively burdensome system. Legal claims by the indigenous Maori, who challenged the basis of property rights allocation, resulted in the Maori being significant holders of ITQ in the commercial fishery. Recreational and other users have no quantifiable property rights although fisheries law upholds their right to partake in fisheries. Despite problems, the New Zealand industry strongly supports the ITQ management system since it facilitates rationalizing individual fishing practice and business investment against a secure property right. Quota holders are strongly motivated towards collective action with regard to fisheries exploration, research, management, and enforcement, and management planning occasionally includes all the user groups. Examples of each of these effects are given. Current debate is focused on bringing the remaining fisheries within the ITQ management system and rationalizing the ITQ operation. Aquaculture concerns are seeking a similar secure property right. The government captures economic rent from the fishery by imposing a "resource rental" or special tax on quota holdings; this is opposed by industry as a disincentive to investment, but industry has expressed willingness to pay more actual costs of research and administration, given a more efficient, effective, and contestable delivery of services. Allocation rights policies for new fisheries continue to be a matter of debate at time of writing.

New Zealand implemented individual tradable quota management (ITQs) for its major deep-water fisheries in 1983 and for many inshore fisheries in 1986. Fishing industry leaders were active in planning and supporting moves by government to make such overt changes to fisheries policy based upon historical experience of both open-access and limited-entry fisheries management regimes.

Many procedural details covering the implementation of ITQs have already been described by Falloon (1993). Sissenwine and Mace (1992) have also described the way in which the implementation of output controls by way of quota management did not reduce the incidence of existing input controls on fisheries. They also described some of the additional administrative and record-keeping burdens placed upon industry and government as a result of ITQs. Some years later (1994), not all fisheries are yet managed by ITQs, and most fisheries still experience a range of both input and output controls.

The purpose of this short address is to summarize some problems and benefits experienced by the New Zealand fishing industry from the inception of quota management in 1983 to the present. I hope to demonstrate the reasons why the New Zealand industry is a staunch supporter of ITQs. A number of aspects referred to in this paper are also discussed by other speakers from New Zealand (e.g., P. Major; ITQ forum/discussion panel session by J. Mace and G. Clement).

Adoption of ITQs required a considerable change of culture by New Zealand's fishing industry. An abandonment of competitive fishing—where the first and apparently best fisher can achieve the highest catch at the expense of his peers—required a good deal of thought before it was accepted. Debates with regard to such a change in outlook took considerable time in New Zealand. The cultural change has occurred, and a stronger conservation commitment among fishers and businesspeople is one outcome.

The next obvious and major problem was to find a recipe acceptable to all concerned by which the initial quota allocation might be made. As with any other allocation of property rights, this issue was difficult, contentious, and hard fought by respective interests. Initial quota allocations were made in 1983 for a range of fish species taken in deep water. Allocations at this time were based upon such criteria as catch history and measures of investment and commitment to and dependence on the fishery. Subsequent allocations in 1986 for remaining deep-water and inshore species followed more formal procedures specified in revised legislation and were based upon established and demonstrated catch histories: quotas were allocated to those persons or companies who held fishing permits in the past and had thus demonstrated an effective catch history. No quota allocations were made to persons employed within a fisheries organization or involved only in processing and subsequent sale of fish but who had no history of actually catching fish.

Next in the order of necessary events was the appeals process, by which those who felt they were disadvantaged in the formal quota allocation might submit applications for more. An independent quota appeal authority was established whose processes proved lengthy and the criteria by which it operated liberal. Accordingly, a variety of arguments were advanced and received not only to demonstrate inaccuracy in government catch-history records, but also to show that individual fishers could have had, and should have had, higher histories had they not suffered vessel breakdown, illness, family distress, or other events. However justifiable the reasons for allocating additional quota to individuals who lodged successful appeals, the accumulation of quota increments had two effects. First, since quotas were at first based on weight rather than on percentage, it meant that the total allowable commercial catch (TACC) that had been established immediately became inflated to the extent of successful appeals. Second, where total quotas were subsequently reduced to reestablish the target total catch, individuals who had not lodged appeals or who were unsuccessful in appealing were obliged to carry a share of the proportional quota cuts. Nonetheless, the implementation of ITQs for a range of fisheries was successfully achieved. This was an administrative feat in itself, one which supplied a mechanism for restructuring the industry and, in particular cases, for amending and controlling its overall catch.

As implemented in 1983–86, the New Zealand ITQ system was very much a creature of its time. It was born of a concerted push from many people in the fishing industry to seek a process for effective restructuring and catch reduction in many fisheries. Since 1963, when a Commission of Inquiry recommended that all limited license fisheries should be decontrolled, there had been considerable multiplication of fishing permits, vessels, effort, and catch. Pressure increased within the inshore fisheries until the mid-1970s when fishers once again petitioned the government to control the number of participants in several fisheries, some of which again became controlled- or limited-entry fisheries from 1977 onward. Nonetheless, in a number of cases the industry continued to urge that there were too many fishers and too many dollars chasing too few fish. Such debates fueled the development of ITQ policies and their implementation during the 1980s.

Although the government during the 1980s demonstrated no interest whatsoever in supporting buy-back schemes to acquire and retire fishing vessels, it was persuaded that an alternative form of restructuring, first to allocate and then to retire fishing quota, could benefit both the resource and the fishing industry in the future. Accordingly, the first major benefit of ITQ implementation was the ability to implement processes whereby government might purchase, then retire, excess (weight-based) fishing quota to better achieve a more appropriate TACC in each target fishery. Government was satisfied that it might recoup its investment in this restructuring by instituting a fee or resource rental charged against quota ownership. Industry supported and endorsed this process—though it was later to regret supporting the overall resource rental concept.

At the same time that inshore fisheries were experiencing quota cuts, the deep-water fisheries were expanding and developing following the declaration of the 200-nautical-mile exclusive economic zone (EEZ) in 1979. For the last several years, all fisheries within New Zealand's management jurisdiction have been owned and exploited by way of ITQ held by New Zealand businesses and citizens. Fishing by foreign-licensed vessels under government-to-government arrangements diminished and ceased. Fishing by foreign vessels can still occur, but only under contractual arrangements with the New Zealand ITQ holder, effectively transferring a greater level of business control over New Zealand's fishing zone to those within the fishing industry.

The new ITQ management scheme created additional administration and record-keeping requirements, and new and severe penalties for breach of the rules. Upon conviction for an offense it might be possible, in addition to a fine, to lose one's whole catch, fishing gear, and fishing vessel, and to have one's quota confiscated. Repeat offenses might lead to disqualification from further involvement in the fishing industry.

The fishing industry and government management agencies came to a compromise with respect to some rules that control catch. Neither a particular TACC nor an individual quota was totally inviolate, and an underrun or overrun of 10% was permissible. This deficit or surplus could be carried forward into a future quota fishing year. Some additional rules sanctioned over-catch by payment of penalty fees to the government or allowed a trade-off of quota in one species for excess catch made in another.

These compromise rules were judged very necessary by many people in the fishing industry in order to encourage acceptance of this new fishing policy. However, the rules increased the burden of record keeping within industry and led to a good deal of friction with the government management agency since the two parties interpreted the rules differently for counting catch against quota. Despite frequent amendment and clarification of these rules and much effort from both sides, the rules remain very complex. There is still little agreement over annual quota balances between individual quota managers from industry and those responsible for administering individual quota balances from within the government agency.

On the positive side of the ledger, the ability to buy, sell, and trade in quota, coupled with the economic penalties for overcatch, has allowed the industry to evolve procedures to balance their catch with their quota holdings and supplied added incentives to avoid bycatch species.

When implemented using weight-based quotas, quota management offered a straightforward system whereby government carried the principal risk of change in allowable catch, making plain that it would sell for cash any additional quota created by a TACC increase and that it would repurchase for cash, at market rates, quotas in any fishery where a reduction in TACC was deemed necessary. There was a major argument between industry and government in 1989 when the government found that it was neither prepared nor able to repurchase expensive weight-based ITQ from industry and, thus, it decided to change the rules such that any change in total quota would be shared in proportion among ITQ holders. Most unpopular at the time, this particular action to substitute percentage-based quotas nonetheless strengthened the nature of the property right. It clearly transferred the risk of quota change from government to industry and effectively further strengthened industry's resolve to have a greater say in research, management, and establishment of quotas.

The rapid rate of change in fisheries policy within New Zealand—first, to implement ITQs and, second, to change and revise many of its rules and regulations—led to an over-complicated system that is expensive and difficult to administer within both industry and government agencies. The time spent in consultation and in preparation for, or defense of, litigation is a significant cost that must have a prominent place on the problems side of the balance sheet. Nonetheless, the fishing industry supported the changes that were implemented at the time, supported the change toward ITQ management, and implemented the significant restructuring and catch reduction necessary in a number of fisheries.

With the advent of the quota management system and the allocation of ITQs, there came many opportunities for rationalization of fishing activity and business investment. Fishers could and did combine their quotas and fish them onboard a single vessel, rather than employing two or more vessels as in the past. Fishers who had previously raced each day to the fishing grounds to secure the biggest share of the available catch could and did seek to take smaller hauls of fish and spread their catch throughout the available season, with obvious benefit to fish quality as well as to marketing and distribution. Fish processors could better plan their staff and distribution requirements with far more certainty. Those fishers who wished to retire from the fishery could do so with dig-

nity upon the sale of their fishing assets, and those who wished to expand their investment had a clear and obvious route by which to do so. Despite the warts and wrinkles of the new regime, the fishing industry in New Zealand became a strong supporter and advocate for quota management and for the advantages of ITQs.

The ability to buy, sell, exchange, and trade quotas has naturally resulted in some redistribution of fishing effort; this is most obvious within certain inshore fisheries. Over time, a number of fishers have sold their quota and left the fishery for pastures green, and the quota has shifted somewhat from the hands of fishers (remember that initial allocations were based upon catch histories made good) into the hands of new investors or fish processors and exporters. Whereas the original fishers may have had a reasonably consistent fishing pattern, exploiting on a regular basis their preferred and established fishing grounds, current quota holders are more likely to allocate their quota into areas of more favorable catch by the simple expedient of making contract fishing arrangements with alternative fishers. Some redistribution of fishing effort is therefore inevitable and brings the potential for gear conflict and spatial conflict among fishers now employed by those quota holders. To date, in the New Zealand fishery scene, such conflicts have not proven major.

An important effect of the quota management system was the influence it proved to have upon fishery exploration and development. It proved to be a marked disincentive for any individual quota owner to make major effort or investment in exploration or new development. The disincentive was that discovery of new stocks might lead to an increase in TACC, which would either be captured by the government—which might sell it to the highest bidder—or under the proportional quota regime, be distributed among all quota holders in that fishery.

While individual incentives are diminished, there is considerable positive encouragement for collective group action. A particularly good example of this exists in the Orange Roughy 3B Exploratory Company. All fishers holding ITQ in area 3B of the orange roughy (*Hoplostethus atlanticus*) fishery have collaborated to form a company whose shareholding reflects quota holding in the fishery. Over the years, they have engaged in a variety of projects. More of these activities are discussed by I.T. Clement in the transcript of the ITQ forum (this volume, p. 281-300).

A number of other examples of collective action among quota holders can be seen. In a prominent dredge scallop fishery, the government fisheries agency has operated an enhancement scheme to catch seed scallops (*Pecten novaezelandiae*) and distribute them to the seabed where they may be harvested for commercial gain. Quota holders in this scallop fishery have recently established a company whose shareholding is a mirror image of quota holding in the fishery. This company will take over and run, on a commercial basis, all activities of spat catching and enhancement. Quota holders in the paua (abalone, *Haliotis iris*) fishery are currently investigating the potential to create a paua quota holders company, and they will explore similar management initiatives. Quota holders in a dredge oyster (*Tiostrea lutaria*) fishery operate a similar oyster enhancement company with the aim of mutual (proportional) benefits.

The matter of enforcement and supervision of quotas has required a considerably different emphasis by the New Zealand Ministry of Agriculture and Fisheries. Reference has already been made to the additional layers of record keeping required to verify catch and cross-check the information against subsequent data covering fish sales, processing, and exports. A comprehensive paper trail exists that should, in theory, allow each fishing enterprise to be subject to a comprehensive audit, but the frequency of such audits is not great given available staff. Those involved in fisheries for rock lobster (*Jasus edwardsii*), paua, and snapper (*Chrysophrys auratus*) have each implemented processes to collect funds so that they may commission additional supervision and enforcement programs.

The New Zealand fishing industry finds it somewhat ironic that in recent times the government agencies that might receive these collected funds and implement enhanced enforcement and supervision programs to preclude poaching deem it constitutionally inappropriate to receive direct funding from some of the persons they may be obliged to supervise. It is a double irony that any industry group should actually volunteer to give extra funds to government only to discover that the government agencies find it difficult to accept such funds. Industry members are still of the view that death and taxes are among life's inevitabilities and we are both confident and fearful that the government agency will, in the near future, discover a way around this apparent inconsistency and charge for its services. (Events subsequent to 1994 have seen a move to full "cost recovery" for various services, including enforcement, and this confirms that we had good cause to be fearful.)

In addition to changes in the culture and activity of the fishing industry, there has also been a considerable change in the disciplines and programs required of fishery researchers. Many have had to be retrained from being fish biologists to becoming fish counters, and there is considerably greater emphasis on stock assessment techniques than in years gone by.

Similar influence has been exerted over industry members, who clearly now have an increasing incentive for investment in research so as to be assured that stock assessment information, on which decisions on TAC quotas are based, reflects the reality they see in their daily exploration of the fishery. An increasing number of individual companies now employ persons with qualifications and experience in fisheries science, and industry organizations, such as the New Zealand Fishing Industry Board, employ in-house stock assessment experts and contract with consultancies comprising international experts. An increased participation in and ownership of the products of research to establish optimal and sustainable yields from New Zealand's fisheries is a real cost, in financial terms, but offers real benefits to the industry in its future planning for the management and business of fisheries.

Within New Zealand's inshore rock lobster fisheries, several regional groups voluntarily contribute significant sums of money employing their own fisheries technicians to supervise onboard catch sampling programs and fishers' logbook schemes in order to contribute consistent data for supporting a better assessment of their fisheries.

A particular problem occurred for the government, and potentially for industry, as a result of the legal claim made by the indigenous people of New Zealand (the Maori) that fisheries property rights were theirs, guaranteed by treaty, and that government had no right to allocate such property rights to industry or anyone else. Debates on this score exhausted considerable effort and expense. The claims were settled, establishing for the Maori a clear position in the activity and business of fishing while avoiding wholesale confiscation of property rights from industry participants.

Recreational and subsistence fishers exert significant pressure on a number of fisheries, yet there is to date no quantifiable property right allocated to this category of fishery user. Neither, regrettably, are there accurate data to quantify their catch. While a realistic cap is placed on industry's ability to exploit fisheries by virtue of the existing quotas, considerable potential still remains for catch and effort increase by noncommercial fishers.

Cooperative management planning groups exist for the rock lobster and snapper fisheries; they involve industry, recreationists, conservationists, and government research and management agencies. These groups are working well to establish management plans within the overall context of ITQ management of commercial fisheries.

Fisheries management by means of ITQs has brought some interesting consequences for bankers and the investment community. In New Zealand to date, fisheries legislation decrees that ITQ cannot be registered as a security by way of mortgage or loan, which creates problems for both sides of any business deal. Since total quotas might be reduced or individual quota might be forfeit for offenses against the system, quota holders have often had to use alternative assets as security for finance. Nonetheless, the monetary value of quota as an asset in the books of each business is increasingly recognized, and we have learned that considerable care must be taken over the nature of publicity and information given with respect to quota change. Uninformed gossip or misinformation concerning the potential for cuts in fisheries quota can lead to disinvestment. To counter adverse consequences of this nature, both the fishing industry and the banking community now make an enhanced effort toward understanding stock assessment and TACC change. Despite misunderstandings at the hint of quota changes, there is increasing recognition by the banking community that ITQ leads to security and improved business.

Problems have continued with respect to resource rentals payable upon ownership of individual quota. In industry's view, the legislation passed at the outset clearly indicated that these rental payments should go into a revolving fund that would reimburse government for the costs of restructuring the industry and would then accumulate to pay the ongoing costs of fisheries research and management. Regrettably, the government saw things differently and set itself the target of capturing the majority, if not all, of the economic rent from the fisheries. This proved an obvious and blunt disincentive to investment. We are happy to report that in 1994, the New Zealand government indicated its intention to reverse this policy and do away with special resource taxes. The industry indicated its willingness to pay the direct costs of research, administration, management, and so on, but clearly expects to see such services delivered in a cost-effective manner with market competition between alternative service providers.

A review of the events affecting ITQ management over the last decade shows that the New Zealand fishing industry has experienced a major change in culture and finds itself in a considerably more complex world of fisheries management than it ever thought likely. Nonetheless, the fishing industry is confident in ITQs and the very real benefits they bring to fisheries conservation, to industry organization, and to business management. The way in which both industry members and others perceive the property right associated with ITQ has changed during the decade. The property has an increasing strength and value in that it can be defended against attack or threat by various sources, including pollution, reclamation, reserves, parks, and restrictive fisheries rules.

The industry looks forward with enthusiasm to see remaining fisheries brought within the same ITQ management policy. On the basis of the lessons we have learned in recent years, the industry shall be closely scrutinizing the further legislative amendments promised for late in 1994, after which we look for simplification of the system, some significant rationalization of its administration within government agencies, then a period of stability so that we may further build our business based on a secure ITQ.

References

Falloon, R. 1993. Individual transferable quotas. The New Zealand case. Workshop on individual quota management. Organization for Economic Cooperation and Development, Paris.

Sissenwine, M. P., and P. M. Mace. 1992. ITQs in New Zealand: the era of fixed quota in perpetuity. Fisheries Bulletin 90:147-160.

Summing Up:
An Overview of Global Trends in Fisheries, Fisheries Science, and Fisheries Management

Ellen Pikitch, Michael Sissenwine, Daniel Huppert, and Marcus Duke

The symposium, "Fisheries Management: Global Trends," comprised paper and poster presentations and panel discussions, many of which appear in these proceedings. Together, these constitute a comprehensive body of information that provides perspectives on the current situation of the world's fisheries. Both within and among regions, the world's fisheries are characterized by immense biological, geographic, economic, social, political, and cultural heterogeneity. Yet, amidst this background of tremendous diversity, global trends are evident. Here we attempt to synthesize the many contributions to the symposium and proceedings to elucidate the status and trends occurring within fisheries, fisheries science, and fisheries management.

Trends in Fisheries

The worldwide yield from marine fisheries leveled off during the 1990s, following 4 decades in which landings increased by over 300%. The rate of growth in marine catches has more or less steadily declined to near zero during this period, with most fishery resources now fully or heavily exploited. This trend indicates that we are at or near the limit of what the world's oceans can produce. The yield might be increased somewhat by recovering the estimated 25% of populations now believed to be overfished and depleted, by reducing waste, or by expanding the remaining few underdeveloped fisheries. However, the scope for an increase in yield is now judged to be very small, particularly for the traditionally important wild harvest fisheries. Unless overfishing is prevented, future yields will decline.

The interconnectedness of the world's fish-producing regions is strong, with over one-third of global production exchanged annually through international trade. Since 1970, there has been a dramatic change in the geoeconomic distribution of world fish production. The proportion of the catch produced by developing countries now accounts for more than 60% of global landings and continues to increase. Much of the catch by developing countries is from industrial fisheries, but the importance of small-scale fisheries to society, the economy, and human nutrition cannot be ignored. In fact, more of the fish consumed by the human population is produced by small-scale fisheries than by industrial fisheries. The perception of small-scale fisheries as marginal to the mainstream of fisheries activity is erroneous and needs to be changed.

Overcapacity is globally rampant, affecting small-scale and industrial fisheries in developing and developed countries alike. While there is little promise for increasing yields, the potential for substantially increasing net benefits from fisheries is enormous if overcapacity can be corrected. Harvest costs greatly exceed revenues from fishing, leading to an estimated global deficit that may be as high as US$60 billion. About US$300 billion of investment in the harvesting sector is not earning an economic return. Many (if not most) fisheries today are stressed economically, whereas they could produce tens of billions of dollars in rent.

Much of the overcapacity seen today stems from a history of open access to fisheries or from a breakdown of traditional limits to access, further stimulated by active development and subsidization. This overcapacity not only leads to the dissipation of potential economic rent but to wasteful and destructive fishing practices that ultimately could have irreversible effects. This is evidenced in such disparate realms as the Alaskan groundfish fishery in which large industrial vessels compete in a massive "race for fish"—engendering potentially avoidable bycatch and waste—and in small-scale, tropical, artisanal fisheries where the use of poisons and dynamite fishing is spreading.

While overcapacity plagues both industrial and small-scale fisheries alike, the nature of appropriate solutions to this problem differs. The remoteness of many small-scale fisheries occurring in isolated coastal communities where alternative employment opportunities are often lacking poses unique problems both in the developed and developing world. At the other end of the spectrum are the problems posed by the fleet of about 27,000 high-seas and distant-water fishing vessels. With extended jurisdiction and development of exclusive economic zone (EEZ) fisheries by coastal states, the trend toward less opportunity for these vessels is apparent. What is less clear is where this fishing capacity will be redirected and what effect it will have.

In the face of stagnating global fish supplies and the continuing growth in fleet capacity, the demand for fishery products will continue to grow because of human population growth, increases in wealth, and shifts in consumer food preferences. One analysis projects an increase in current global demand for fishery products of at least 39% by the year 2010. A requisite increase in supply to meet this demand is not likely attainable. This situation will continue to fuel overcapitalization and pressure to overfish. Unless there are controls, conditions will deteriorate.

Aquaculture is one method of closing the gap between demand and production of wild harvest fisheries. There has been astronomical growth in aquaculture production (much of it in freshwater), which now accounts for approximately 15% of the volume and 30% of the value of fishery production. But aquaculture is suffering from many growing pains, such as market gluts for some products, disease problems, and environmental damage (both caused by and adversely impacting aquaculture), and the limits to its growth and its ultimate potential to satisfy global demand for fishery production are as yet unknown. There are also serious concerns about allocation of limited, valuable space in coastal zones between aquaculture and other uses, and about the impacts of fish cultured for stock enhancement purposes on wild (particularly endangered) populations. The current dependence of aquaculture on fish protein as input (i.e., feed)—another factor limiting its growth—is controversial, with some arguing that fish are better used for direct human consumption.

Another potential means for increasing world fish production is to reduce both the amount of fish caught as bycatch and the amount of bycatch wasted. The trend in bycatch and discarding as a proportion of the global catch is unknown, but there is no doubt that concern about bycatch is increasing. Recent estimates indicate discards comprise about one-third of landings globally. This is a striking statistic considering that it represents just the "tip of the iceberg" of heretofore largely unaccounted for sources of mortality, which also include such factors as catch underreporting, recreational bycatches, fish killed following contact with fishing gear, "ghost fishing" of lost fishing gear, and unrecorded recreational and subsistence catches.

Recreational fisheries are increasingly important, particularly in developed countries. There is a need to develop new tools for allocating fishery resources between recreational and commercial users. For example, total allowable catch (TAC) and individual transferable quota (ITQ) management, which are commonly applied to commercial fisheries, may not be practical for most recreational fisheries. But when a fish species is targeted by commercial and recreational fisheries, management must find effective ways to regulate both sectors or resource conservation may be sacrificed in the absence of an effective allocation scheme.

Trends in Fisheries Science and Management

Recognition of the ubiquitous nature of the uncertainties that pervade all aspects of fisheries science and management has grown enormously. Concurrently, resource assessments are becoming increasingly sophisticated in accounting for uncertainty in the provision of management advice. Evidence indicates that managers both want and will use this information.

Uncertainty can be reduced, however, with regard to long-term management objectives and policy. Regulation by pre-agreed management procedures has considerable advantages over the more common regime of annual quota setting, and a trend of greater emphasis on the former is apparent. Reaching agreement on long-term policy is not an easy task, and it becomes increasingly difficult as the number of parties to a decision multiplies and the heterogeneity of their perspectives and circumstances increases. This is illustrated by the difficulty to define a multiannual framework for the Common Fisheries Policy (CFP) by Member States of the European Union. In this instance, while there is agreement on what constitutes a step in the right direction, there is concern about whether evolution of the CFP is proceeding quickly enough to avert more serious problems in the future.

One method for coping with uncertainty that has gained prominence is the development and testing of management procedures that are robust to alternative hypotheses about how stocks will behave in response to fishing. The scope of assessments have been extended beyond those that consider only uncertainty in parameter estimates for a single model, to examination of alternative single-species models, and even further to examining multiple models with varying specifications of stock structure or species interactions. Unfortunately, the management advice provided by different models is often qualitatively different, leading to the result that no management policy can be robust to all models. Weighting alternative hypotheses according to their relative plausibility has been suggested as an approach for dealing with this phenomenon.

Scientists are now beginning to recognize and understand large-scale temporal and spatial (decade long; over ocean basins) "regime shifts." These shifts are probably climate-driven and chaos-like phenomenon inherent to complex systems. Regime shifts are associated with major changes in ecosystem structure (e.g., species composition) and function (e.g., productivity). The collapse of North Atlantic cod stocks, poor ocean survival of Pacific salmon, and coherent variations in small pelagic fish populations (anchovy, herring) throughout the Pacific may all be part of regime shifts. How much fishing influences these shifts (or if they can be controlled by management) is not known. Variability on large time and space scales places new challenges on scientists to diagnose and predict regime shifts, and to develop management strategies that are robust in the face of such change.

In instances where there is potential for spatial and temporal diversification in management, an experimental approach may be a feasible means to reduce uncertainty about resource dynamics. The implementation of such an approach in an Australian multispecies trawl fishery proved to be both economically and scientifically viable, and was particularly successful at illuminating the effects of trawl-induced changes to benthic habitats.

Significant progress in understanding multispecies fisheries has been made on other fronts. For example, the North Sea multispecies modeling effort is the successful outcome of a 15-year series of theoretical models, model-driven data collection, parameter estimation for numerical models, and model verification against new independent data. The question now is, "Can these model results be translated into management advice and will managers respond?"

The quantity and quality of scientific information available to address fisheries management problems are also of concern. Unfortunately, in many parts of the world there is a trend towards deterioration in the quality of fisheries statistics, which clearly increases uncertainty. There is also concern that management measures, such as ITQs, that are designed to combat the pervasive problem of overcapacity may lead to further deterioration of fishery databases by increasing incentives to underreport, high-grade, or discard fish.

Economic factors are increasingly important to fishery managers, but the scientific information needed to assess and account for them is woefully inadequate. At present, there is little evidence of improvement nor is there sufficient funding to ensure good progress.

The importance of non-monitory values (e.g., from recreation, cultural uses, existence, biodiversity) is receiving increasing recognition. In fact, with the net value of commercial fisheries near zero or negative, it could be argued that non-monitory values are the most important. Unfortunately, even less is known about the social aspects of fisheries than about the economics.

There is a tendency to encourage members of the seafood industry, recreational fishing interests, conservation-

ists, and other stakeholders to participate in fisheries management. Many countries, such as Canada, New Zealand, Australia and Chile, have recently initiated consultative processes that are moving in the direction of U.S. fisheries management councils, which have been broadly empowered for nearly 20 years. At the same time in the USA, these councils have been subject to numerous criticisms, such as failing to adhere to conservation standards and favoring some user groups over others.

The trend toward broader involvement of stakeholders in the fisheries management process is accompanied by increasing controversy. A causal relationship is not implied, but it is possible. Controversies are fueled by politics, lobbyists, and litigation. The scientific basis for fisheries management is subjected to greater scrutiny, which undoubtedly improves the scientific basis for management up to a point. However, when alternative "scientific" views are posed and argued primarily as a tool to identify flaws that will undermine the management interests of opposing interest groups, scientific advice, the science profession, and fisheries management may all suffer.

The principle of "freedom of fishing" on the high seas and elsewhere has, appropriately, been eroding during the past 50 years. The area of high-seas management zones has declined as national jurisdictions have been extended. During this decade, significant progress has been made in establishing principles for international management of highly migratory and straddling fish stocks and for improving compliance with some high-seas management measures designed to combat problems caused by vessel reflagging. Progress to ensure that principles are put into practice is a current urgent challenge. In the face of the lack of international agreement on how to resolve high-seas fisheries issues, there has been a trend toward greater use of unilateral actions.

Within national EEZs, open-access fisheries predominated most industrial fisheries and fisheries of developed countries a few decades ago, but most of these fisheries, and certainly most fisheries in the USA, are now subjected to some form of controlled access. There is an increasing trend towards use of individual transferable quotas (ITQ), with some countries having embraced this form of management wholeheartedly although the overall impact of this method on worldwide fisheries is still small. Unfortunately, in the USA, the controversy over ITQs has led to a moratorium imposed by Congress.

Where ITQs have been used, there have been benefits and problems. The view presented about ITQs in New Zealand is extremely positive, whereas the view of ITQs in Australia and Canada is mixed. One positive aspect of ITQs that seems to be exceeding expectations is the improvement in market opportunities, quality, and price of fish products. On the other hand, initial allocation decisions to implement ITQs are inherently controversial and difficult. Of course, delaying these decisions does not make them any easier. Rules on rent extraction and quota consolidation are tough decisions that should also be addressed in the design of the management system. Once ITQs are implemented, enforcement, bycatch and high-grading, and data quality are difficult problems.

There is a trend toward imposition of user fees (or users pays, e.g., New Zealand, Australia, some Canadian fisheries, and even serious discussions in the USA). This is inevitable as government budgets are cut and use opportunities are restricted (no longer is the resource available to everyone, and some users are given individual rights). The trend toward "user pays" has also led to user demands for a greater say in how funds are spent for research and management. To the extent this increases accountability and efficiency, it is a positive trend. But the influence of paying users should not be allowed to jeopardize the long-term public interest, which is not necessarily the same as the users' interest.

Small-scale fisheries, particularly in developing countries, present some special problems, owing to factors including limited data collection, resource assessment, and fisheries monitoring and enforcement capabilities. Management approaches such as gear restrictions, closed areas or refugia, and community-based management may be particularly appropriate, and evidently these forms of management have a tradition of use in some areas. Community-based management may entail delegating management responsibility for a geographic area or a share of the overall allowable catch to a community. The community has responsibility for internal allocation decisions, monitoring, and enforcement. Allocating quota shares to communities is a hybrid form of management combining elements of community-based and ITQ methods. Community development quotas, which have been allocated to remote communities in Alaska, are an example of this approach.

Final Comment on Global Trends

Fisheries, fisheries science, and fisheries management are all changing, as is our experience with, and understanding of, the attendant causes and consequences. Further change is inevitable as technology advances, populations grow, and less developed countries strive to achieve the standard of living of developed countries. These changes are all associated with an evolution from the era when oceans and fisheries resources were considered so vast that they could not be damaged by mankind to a future of sustainable use, we hope. The challenge is to successfully manage the transition to more rational fisheries. The status quo is not an option.

ITQ FORUM AND PANEL DISCUSSIONS

Forum on Individual Transferable Quotas

Thursday, June 16, 1994

ORGANIZER: DANIEL HUPPERT, SCHOOL OF MARINE AFFAIRS.

We organized this Individual Transferable Quota (ITQ) forum with intention of giving participants in the symposium an expanded opportunity to learn about the experiences that our foreign and domestic colleagues have had with individual quota management systems. Further, we wanted this to be an occasion for extended interaction between speakers and the audience. In accordance with that goal we have assembled two panels. To open each panel session each panelist is asked to provide a brief response to the following focal question: What is the most important aspect or consequence of ITQ management from your perspective?

The first panel is chaired by R. Bruce Rettig, Agricultural and Resource Economics, Oregon State University, Corvallis, and it consists of the 10 speakers from the morning session on "Allocating Fishing Rights" as follows:

Lee Anderson, College of Marine Studies, University of Delaware, Newark

Ragnar Arnason, Faculty of Economics and Business Administration, University of Iceland, Reykjavik

Andrew Branson, Manager Technical Division, New Zealand Fishing Industry Board, Wellington

Martin Exel, General Manager of Southern Bluefin and Northern Prawn Fisheries, Australian Fisheries Management Authority, Canberra

Rögnvaldur Hannesson, Norwegian School of Economics and Business Administration, Bergen

Barry Kaufmann, Chief Economist, Australian Fisheries Management Authority, Canberra

Harlan Lampe, Las Vegas, Nevada. Formerly with Instituto Fomento de Pesquero, Valparaiso, Chile

Phil Major, Director, Fisheries Policy, Ministry of Agriculture and Fisheries, Wellington

John Pope, Lowestoft Fisheries Laboratory, England

James Wilen, Department of Agricultural Economics, University of California at Davis

The second panel is chaired by Richard Marasco, Director of the Resource Evaluation and Fishery Management Division, Alaska Fishery Science Center, National Marine Fisheries Service (NMFS), Seattle. Members of the second panel are drawn from the fishing industry, government management agencies, and environmental groups. The individuals participating are as follows:

George Clement, Director, Clement & Associates Limited, Tauranga, New Zealand

James Joseph, Inter-American Tropical Tuna Commission, La Jolla, California

Doug Hopkins, Environmental Defense Fund, New York

Jim Mace, SeaLord Products LTD, Nelson, New Zealand

Stuart Richey, Richey Fishing Co. PTY LTD, Tasmania

Bruce Turris, Canadian Department of Fisheries and Oceans, Vancouver, British Columbia

The narrative below is a slightly edited transcription of the Panelist's statements, comments and questions from the audience, and the panelist's responses. Questions from the audience are numbered in sequence of their occurrence. If we were able to determine the speaker's identity from the tape recording, the name of the questioner was included in parentheses.

Panel 1

R. Bruce Rettig, Chair

Rögnvaldur Hannesson.—Since I am from Norway, I will make some comments on what has happened there, or rather on what has not happened. Some years ago, the Norwegian Fisheries Administration became convinced that ITQs in some form was the appropriate way to go in fisheries management. They prepared a white paper. A critical portion of the industry turned down the proposal. Why? Uncertainty was a major part of it. The cost of entry to the industry would be raised. Another part of it was ideology. Privatization of a common resource was an idea that did not go down well with some people. Opposition came mainly from small-scale fishers who feared being bought out by large fishing companies. Also, restructuring of the industry was feared by some. Some regions of the nation might lose a critical mass of economic activity and finally be abandoned. Unwillingness to accept what I think is a necessary step to economic change prevented support for the move to ITQs. Finally, another factor was uncertainty whether this would be compatible with fisheries policy of the European Economic Union, which we have applied for but have not yet joined.

Ragnar Arnason.—I would like to mention some of the characteristics of ITQ systems. They work to increase economic efficiency. This has been the experience in Iceland, in New Zealand, and in some fisheries in Australia. Also, there are reports that experience with ITQs in Canada has been very good. So on both empirical and theoretical grounds, ITQs increase efficiency. And we are not talking about small figures. The total value of the world's fisheries is probably about US$70 or US$80 billion dollars per year. Based upon rent gains in fisheries we have studied, it might be possible to have pure rent gains worldwide on the order of US$30 to US$40 billion dollars. So it is pretty important to keep this in mind and not be overwhelmed by the problems of ITQ implementation experienced in various countries

Still, we must keep in mind the problems of ITQ fisheries. The ITQ system must be well enforced. But many fisheries around the globe are organized in ways that make it very difficult to enforce property rights regimes like the ITQ system. Certainly the volume of catches has to be very closely monitored. In those cases where you have many different landing stations, very high unit value of catch, and where you have small-scale operators, it seems to me it may not be feasible to operate the enforcement system needed for an ITQ regime. So this we have to keep in mind. Also, we have to keep in mind the social problems that are caused by ITQ systems, as mentioned by many commentators this morning. These problems can cause great controversy. Implementation of ITQs implies a reorganization of production relationships, which is generally resisted by the established groups. By the same token, ITQs imply a redistribution of income and wealth, they require new talents and new skills, and they imply a change in the distribution of political power and even social prestige. Finally, even to the individual operators, even though they/ITQs look good on paper, there is great uncertainty about how they are going to work. Even if they are successful in creating greater efficiency, is the increased income going to be expropriated by the government? So, it is not a surprise that ITQ systems are resisted by some portion of the population.

John Pope.—I think that I made clear earlier the importance of getting your industry members on board. This means that you have to develop a means of communicating with industry. One thing that struck me coming out of our own debate is that fishers actually see the world differently than economists or fisheries scientists or administrators. You may need to recognize that. I could tell a little story about the captain of one of our research ships. "I like my wife very much, John. But I don't like to lie alongside her for more that six nights in a row." Fishers are like that. They like going away to sea, and they like catching a lot of fish. Its the nature of the animal, and you have to recognize that. You have to ask yourself what your objectives are. In some fisheries it seems that economic efficiency is not necessarily the objective. For example, in Newfoundland the population exists mostly to fish; there is not much of any other reason for being there. If you implemented an ITQ system there, it seems to me pretty unlikely to end up in the hands of guys fishing in little boats out of out-ports. Yet they are the reason you have a population left in Newfoundland. So you have to decide what do you want of your population in general. Because is seems to me that all rent gets dissipated in the end; its just a question of where you establish your boundary. Now, a dilemma I always give to economists is a situation, like Newfoundland, where you have lots of little boats and where they dissipate the rent on building lots of boats locally. Or in comparison, you can have a big industry with ITQs, lots or profits, very rich fishers, and the rent all spent on wine, women and song. Now, Lee Anderson has a reason why that situation is better. I don't understand it, but he may give you an explanation later.

Bruce Rettig.—The next speaker, Lee Anderson, has played two important roles recently. He chairs the Mid-Atlantic Fisheries Management Council and he was instrumental in the development of the first ITQ system under the Magnuson Fisheries Conservation and Management Act. Lee also undertook a task for the National Marine Fisheries Service, visiting with people in every region of the country, trying to design what might be key elements of ITQs in various fisheries.

Lee Anderson.—I'll address John Pope's question first. I don't believe that the ITQs necessarily go to the biggest firms. They go to the more efficient folks, and those are the ones that can harvest efficiently. People with low opportunity cost of labor could end up with the ITQs. It could go either way. It is incorrect to say that ITQs always go to the larger industrial fisheries.

I want to address the question posed by our convenors: What is the most important aspect or consequence of ITQ management? I would say, "Do it right from the start, but be prepared to change if you have to." Do it right in two ways—the nature of the system and the implementation. The nature of the system has to do with the nature of the property right. Who can own it? Permanence, transferability, what are the types of regulation used in combination with the ITQ? What kinds of exceptions do you have, such as Ragnar Arnason mentioned (in Iceland there is a use-it-or-lose-it rule). Are you going to use taxes to collect rent? If you want the property right in order to get the advantages that it brings, you should have as clean and clear a property right as is possible.

Theory can tell how to design [property rights] from the start. This is what we did in the surf clam fishery. For example, if you want to have a tax system, set it up from the start. Do not leave a lot of uncertainty out there, because that will tend to lessen the advantages. Ragnar mentioned that Iceland did not have a perfect system from the start but that they altered the system over time. I agree that you can learn from experience under the ITQ system over time, but at the same time you may not be as lucky as Iceland was that the changes went in the right direction. I would mention that the new IFQ system for halibut and sablefish in Alaska is not a perfect system from the start. The limits on transferability between fleets do weaken the potential for gains from the system. It will be interesting to see what changes occur as the system moves along.

The second thing that you need to do right from the beginning is the implementation of the system. Barry Kaufmann talked about this very eloquently earlier in the program. I think you have to make sure the initial allocation is set up right. Then make sure the enforcement system does not have serious loopholes. We learned from the surf clam system that it is a lot harder to implement than you might think. You need to try to anticipate all the problems, but if you need to adapt the system later, be very careful. Every time you change some aspect of the ITQ, it weakens the system. NMFS has recently distributed a report on the surf clam system. I recommend that the audience looks at that report if it is interested about the details and problems of implementation. Look at the problems of monitoring and enforcement; fix the problems if they are really significant.

Bruce Rettig.—Before going on to our last panelist from the northern hemisphere, I would like to point out that we have not yet had a speaker talking about the Canadian ITQ systems. The Canadians have had some ITQ systems of long standing, some run by the Province of Ontario. There are a number of systems that have had terrible problems. Our next speaker, Jim Wilen, was on the faculty of the University of British Columbia and has followed the progress of IQ systems in Canada for a long time.

James Wilen.—What is the most significant consequence of ITQs? I want to take a big picture view. To me the most significant aspect of these systems is that they harness the power of rents. There is some nice symmetry here. In traditional fishery management systems—regulated open-access systems—rent is the source of the problem. Rent drives all the conflict between the regulators and the industry in open-access systems. In contrast, in ITQ systems rent becomes the motive for individuals to make conscious, foresightful, conservation-minded, rational decisions about how to use the fisheries. To me, the differences in the role of rent in the two systems is astounding and is the major difference between the two. In deciding whether it is worth shifting from an open-access system to an ITQ system, a common mistake is to think that the fishery will stay about the same if we don't

do something about it. That's false. It will always get worse. The pressure of rents is building in fisheries. Population growth drives prices up, and potential rents are always growing in traditional fisheries. So problems are bound to get worse in the future. When thinking about whether it is worth absorbing the transaction costs and overcoming the inertia of open access, we must think about what the system is going to be like a few years from now.

I want to give a quick example of how dynamic the system is. In 1980, the British Columbia halibut fishery was fishing over a period of about a month and a half. Because the government was worried about excessive entry, they implemented a limited entry system. That restricted the number of vessels to 435. What happened over the next 10 years was rather astounding. The fishers switched to circle hooks, which increased their productivity dramatically. As a result, the regulators had to start shutting the season earlier. In response to that, the fishers installed automatic baiting equipment and other things to increase their fishing power. As a result of that, the regulators had to crank down the season length even more. The fishery was backed into a 5-day season in 1990, even with a limited entry program in place. So, these systems are vulnerable and dynamic. There is always this force behind them which is exacerbating the management problem. Those of you who are in management should be thinking about this. Things are not going to get better.

You want to think about what is coming out of the systems where ITQs have been adopted. In the New Zealand snapper fishery ITQs are leasing for as much as NZ$6,5000 per metric ton (mt). That is equal to US$2.00 per pound. Groundfish fisheries off the USA are making about 30 cents per pound. So the ITQ system has the power to generate great rents. While many of our fisheries are not earning rents, the fishery is essentially valueless.

Bruce Rettig.—Thank you, Jim. Now it is time to take some questions from the audience.

Question #1 (Ray Hilborn).—My question is directed to both Lee Anderson and Jim Wilen. Lee said that he expects the catch per unit effort to increase in fisheries under ITQs, while Jim says that the catch per effort will decrease as fishers try to maximize the value of the quota they own. My experience is consistent with the latter. However, in cases where a fish stock is declining before the managers implement an ITQ, the stock should increase. You have to be very, very careful in how you measure catch and effort in an ITQ fishery.

Lee Anderson.—Yes, I agree.

Question #2 (Steve Pennoyer).—Lee, you are right that we have a lot of "bells and whistles" on our ITQ system in Alaska. This is in part due to the fact that Alaska is somewhat like Newfoundland, in that in many of our coastal communities people have little else to do. The question is for Arnason. You mentioned this morning that you have geographical allocation of individual fishing quotas, and you have a process, I think, by which communities vote on transfers of quota. Could you explain a little more about how the Icelandic community relations program works?

Ragnar Arnason.—There are geographical restrictions on the transferability of the quotas within the year. If operators want to transfer those quotas within regions, they will have to seek the agreement of the municipalities in the regions in question . If the municipalities object to this transfer, it is up to the Minister of Fisheries to decide. It turns out in Iceland that very few transfers are blocked by the municipalities and the Ministry. Somewhat unexpectedly, there has been very little geographic reallocation of the permanent quotas.

Question #3.—I have a question concerning the effects of ITQ systems on catch information. We have experience in Australia where, after the implementation of ITQs in the South East fishery, the *actual* catch for orange roughy (one of the most significant species in that fishery) was estimated to be somewhere between two and three times the *reported* catch. As someone involved in stock assessment, that is of concern to me. My second comment relates to the costs of the system. The increased cost in terms of administration, surveillance, and research sometimes engendered by these systems is particularly a problem in Australia where we have a number of relatively small and low-valued stocks. The cost of the implementation is more or less independent of the size of the fishery. So you have a significant economic cost. Since the ITQ seems to be an economic response to an economic problem, I wonder if there have been any serious analyses of costs and benefits after the introduction of these systems.

Lee Anderson.—I am currently doing an analysis of surf clams and have yet to come up with the numbers you may want.

Question #4.—Several of the panelists have mentioned the component of enforcement in ITQ systems. Ragnar noted this morning that there is a quota fee of 0.4 percent of the value of the quota. What do you get for this fee? Is it for enforcement?

Ragnar Arnason.—The fee is a contribution towards the cost of operating the management system. We estimate that roughly 1.2 to 1.4 percent of the value of the catches can be attributed to the management of the IQ systems. So the fee currently covered one-third of that.

Question #5.—What about the enforcement system in Iceland. Do you have observers aboard fishing vessels?

Ragnar Arnason.—The enforcement consists of a very extensive landings control system. Every landing is

measured by weight. There are also observers and landings controls in foreign countries where our fishing vessels might actually take their catches. We also have a secondary measure. We monitor the output of fish processing factories and compare these outputs with the reported inputs.

Question #6 (Francis Christy).—Clearly there should be access control for this microphone! I suggest an auction system. I am interested in the panel's response to a question concerning extraction of economic rents. This is a matter of some concern here in the United States regarding revision of the Magnuson Act. I would like to hear the views regarding whether rents should be extracted, and if so, at what level. Should it be sufficient to cover the cost of research, or research and enforcement? Should it cover a buy-back program or something like that? Or, at the other extreme, should the entire rent be extracted. Also, it would be interesting to hear what the panelists have to say about the system for extracting rent—should this be a user fee, a tax, a royalty on the catch, or an auction mechanisms? Should there be a property tax on the property right? What other kinds of mechanisms might be available for extracting those rents? What would be the consequences of that extraction of rents on the fishery? These questions should probably be the topic of another conference, but it would be nice to get your reactions on this. Thank you.

Rögnvaldur Hannesson.—First, I can understand the argument for letting rents remain in the industry. That is simply to make the system more entrenched, and to assure the participation of the industry in management of the fisheries. After all, I think the main benefit of the ITQ system is that the process becomes industry-driven, instead of being driven by public officials. And we all know what that system has resulted in. As I emphasized in my talk this morning, I tend to favor the extraction of rents for two reasons. First, I think that these resources should be regarded as being owned by the people living in the district of these resources. Secondly, the extraction of rents supports the legitimacy of the system. There is no doubt that in Iceland, for example, resistance to this system has developed because people have seen that some chosen few have been given assets for free, assets they have later been able to sell. That does not go down well with the general public. I think that the legitimacy of such systems could be enhanced by rent extraction.

There are various methods for doing this. One is to let some portion of the quotas disappear each year, so that people will have to buy them back in order to stay in business at the same level. As far as I can see, this would amount to a neutral tax on rents.

Lee Anderson.—I would disagree with Rögnvaldur on collection of rents for two reasons. One, if you say the system creating rents drives people to do the right thing, and then you turn around and have the government collect the rents, you take away the incentives to do the things that you want. This lead to a principle–agent problem. I would be very careful. I would say cost recovery, based on a tax of fish landings might be useful.

Question #7.—I note a problem that we are likely to face in European waters. That is that you only own a share of the total stock. So if you buy a share, you are buying a flat on the top of a building that is also owned by other European-flag boats. I wouldn't care to do that unless you could introduce the system on a European basis. But then you run into the problem that French and Danish interests in the fish meal industry for herring are completely different from the Dutch interests in the herring fishery for human consumption. Then you have to match those issues first before you try to impose a system like the ITQ system.

Question #8.—Yes, my question deals with the initial allocation when you first set up the ITQ system. We heard of several methods of doing that this morning. They range from allocating equal proportions of the quotas to each license holder, to allocating them in proportion to historical catches or in proportion to hold capacity. What would the panelists recommend for future allocation schemes? In particular, what methods would work best to obtain agreement among the group of people that you have recommended to actually do the initial allocation?

Ragnar Arnason.—I think the method of initial allocation is mainly a political question. There is no fixed rule here. First, from the viewpoint of economic efficiency, it doesn't really matter how the quotas are allocated as long as the quota market works fairly well. So the question boils down to the allocation of wealth and income associated with quota ownership. There you have to look at the particular situation. What is acceptable to the population at large? From my own viewpoint, I would like to see quotas distributed widely, preferably to every individual in the population. That would deal with the question of whether to appropriate the rents from the resource, since by and large the initial recipient of the quotas will get most of the rents generated.

Question #9 (Jim Easley). I have a two-part comment. First, many of our fisheries are participated in by both recreational and commercial fishers. Many of you may be aware that on the U.S. east coast and on the Gulf of Mexico, some of our fisheries have a larger recreational harvest than a commercial one. This may also be true in other parts of the world. My question is whether ITQs can be effective for those stocks that have significant recreational sectors. My second question is whether panel members have considered how to include the recreational sector in the initial allocation.

Lee Anderson.—The short answer is that you can set up a percentage of the annual quota to go to the commercial and recreational sectors. Then you have both sector quotas go up and down with the total allowable catch (TAC). The recreational sector could continue to be regulated with bag limits and other regulations. The commercial sector could be regulated within its share of the quota.

Andrew Branson.—In the New Zealand snapper fishery, we have a rebuilding program, which Ray Hilborn alluded to earlier. If you were to close the fishery for a large number of years, you could get something like a 10% increase in biomass. The problem is that if, whilst you are doing that, you don't restrict the recreational catch, then the recreational sector will continue to expand. In 10 years time, when the commercial industry reopens, all the rebuild will be allocated to the recreational catch.

Question #10.—I have a question concerning the danger of aggregation of quotas under an ITQ system. That danger would seem to be particularly severe where you have quotas allocated to a mixture of established industry and disadvantaged communities. Has there been experience with ITQs in such a mixed system, and if so, is there an effective and acceptable method to avoid the problem?

Jim Wilen.—The issue of aggregation and consolidation is a complicated one, and I don't think that you can predict whether there is a tendency to aggregate or not. I think that actually fisheries are too complicated to figure out whether consolidation will be a real concern or not. If it is, you can put on aggregation caps as is done in a number of fisheries. Another thing was done in the British Columbia (B.C.) halibut fishery. After the initial allocation is done, you have a period in which sales of quotas are not allowed. You have a period in which you watch the system until it settles down. Ragnar talked about asset markets for ITQs. They are volatile because at the beginning investors are trying to forecast something that no one has good information about. I think the B.C. system was very sensible. Put a lid on the system, and see what forces develop in it before you let it go. Then if it looks like there will be an undesirable degree of consolidation, you can put caps on it. Also, this gives the participants some experience in figuring out how the ITQ markets will equilibrate. If, during the trial period, you allow people to lease quotas, you can figure out from the lease market what is likely to happen to the asset market. This could be very important because at the beginning the market will be very uncertain and it will be unclear what level of asset price is likely to occur, or what forces are likely to cause consolidation and changes in industry structure.

Question #11 (Jake Rice).—First, I want to make an observation based upon the past few comments. It sounds like fisheries are run as a tool of economics in a country.

I have been dissatisfied by answers I have heard to the question concerning fisheries that have an important social role. It sounds that there are some real problems with IQs for which we are getting rather cavalier answers. The question I want to pose is whether there seems to be capital flowing into successful IQ systems, based upon the ability of these systems to attract capital. Then the fishery becomes more and more reliant on that capital to run the fishery. What many of us have to deal with is not how such a system would work in good times, but rather how such a system would react to a crisis. Watching any other commodity market faced with a poor crop prognosis you see money flow out of the market very quickly. Will a similar thing happen in these IQ systems. The first time we get a pessimistic stock forecast, and we need money for enforcement and the research most, will the capital flee elsewhere? This will make the system worse rather than better when the crisis hits.

Rögnvaldur Hannesson.—I don't think that the fishing industry is a very good instrument of social policy. Where this has been tried, it has ended up more often than not being a disaster. I think we have to face the fact that the fishing industry is primarily targeted towards producing products to be sold in highly competitive markets and creating income for its participants. To the extent that you try to realize a social objective at the expense of the economic viability of the fishing industry, you are making your industry less competitive in the international market for fish products. By emphasizing social goals and social policies we might be getting the industry trapped in poverty, an escalating circle of poverty. This doesn't let people compete in international markets with their products. It would be more sensible to let the industry operate along lines of economic efficiency. It would be better to produce an economic surplus, which could be invested in other industries for the benefit of the people in areas dependent upon fisheries.

Question #12 (Pope).—It is quite interesting to compare Iceland and Newfoundland. They have similar sorts of climates and resources, yet they are managed in such different ways. And I can't see that you are right. When fishers think of conservation, I think it means to them that their sons will be able to go fishing. That is the one concern I have. Yes, it will produce a viable industry, but will it be their sons that will go fishing? It reminds me of the enclosure movement in England that forced English peasantry off the land. So it may succeed, but will it succeed for the guys who are in it now?

Bruce Rettig.—All right, thank you for your comments. Now it is time to go south of the Equator. Harlan Lampe, it is your turn to make some comments.

Harlan Lampe.—I would like to direct my attention to the question posed by the organizers. In Chile, we have fisheries with two important characteristics: they are

mixed species fisheries, and they are integrated. Most fish plants own their own boats and produce at least 60% of their own inputs. So we have institutional quotas. We don't really care what boat produces the fish and what boat doesn't. We also have small-scale fleet independent operators, both industrial and artisanal boats. So we need to have quotas in several categories—institutional quotas for the integrated operations and vessel quotas for the small, independent vessels. We have more difficulty controlling the operations of the small vessels than the integrated operations. From the integrated operations, we have information flowing continuously on the fishing trip. One additional, problematic point. It is my estimate that we used to have 30–35% overcapacity in our fishing fleet, and 30–35% overcapacity for processing fish meal in Chile. Recently, an unusually high price of fish meal in relationship to soybean meal prevailed until the end of 1993. This encouraged investment. Add the fact that there is a surplus of fishing boats in the world, and we suddenly had fishing boats of 1,200 mt capacity searching for sardines and anchoveta. These boats are not particularly efficient. I would estimate that in some areas we have no less than 50% excess harvest capacity. Another factor is that in San Antonio, in the center of Chile, we have an investment in capacity to process fish meal of about 1,000 mt per hour. But there is no fish to be processed at this time. So we are in a crunch in the small-scale fisheries, where we would like to control capacity. We have such great overcapacity that the resistance to change is very great and we have had difficulty getting the industry to follow our lead on this. These are the factors that are particularly important to us.

In other countries, I think you also have integrated operations. You cannot talk about just fishing boats as though they are the only factor in the exploitation of fish. You have a large system and infrastructure which is also over-scaled. This makes the estimates of economic waste (that Francis Christy has made) much smaller than it might seem.

Andrew Branson.—I just want to tell a short and simple story. This predates the quota management system in New Zealand. But it led me and others involved in negotiation over the system to see the possible ITQ benefits. It occurs in the very far south of New Zealand, 47° south latitude, well into the "roaring forties," in the Fouveaux Strait oyster fishery. Rough water occurs there, and the fishery takes place in winter when there are high waves and cold winds. There is a fleet of 24 boats, all in the 70-foot class. They are dredging for oysters in depths of anywhere from 110 to 150 feet. There is a total quota, with competitive catch, of 115,000 sacks each season. The scene is set one dark and stormy morning. It is half past four in the middle of winter. The twenty-three boats are all lined up against the wharf, lights on, engines running, the crews onboard. The wind is rising and whistling through the rigging, and they all know that the seas are building on the outside. They are asking themselves: Shall we go to sea or shall we stay? All of the sudden the boat at the end of the wharf line casts off his lines and he heads for sea. Within 2 minutes the other 22 do the same. They have a rotten day at sea and burn a lot of fuel and break a lot of things and don't catch much. One morning exactly this thing happened. The first boat to cast off was the fastest boat in the fleet. The skipper cunningly headed out of the harbor, but then turned his lights off, killed the motor, and ducked into a little bay. He watched the other 22 going past, fighting their way through the chop. After they pass, he turned back, went home and returned to bed—very sensible. And the others had a miserable day, didn't catch much, and burned a lot of fuel.

Anyway, the fishers began to think about these sorts of things. In this fleet the very best boats were catching about twice the amount of the worst boat. But most boats were catching in the neighborhood of about 4 and one-half to 5 thousand sacks each. They suddenly had a bright idea. If you divide the quota of 115,000 sacks by 23, it comes out to exactly 5,000 sacks per boats. So we discussed with them, and eventually discussed with the minister of fisheries, why don't we have IQs? The Minister said, "Yes, for one year, without precedent, by way of trial, you can have a 5,000 sack quota each." So next winter, the first dark and cold morning with the wind rising and seas building up, not one single boat had its lights on, its engines on, its crew aboard. That sort of management has continued in that fishery ever since then. That experience was discussed at length among the industry as they planned and discussed and prepared for the change in culture for ITQs. The 23 boats can now plan their fishery. They plan when to go to sea. If it is a rude morning they don't have to go because no one is going to beat them to their catch. They have 5,000 sacks to catch when they want and where they want. Ray Hilborn is right, the CPUE is not indicative of the status of the stock because they go where they get the best oysters, not necessarily where they get the most oysters. But they can plan their investment and their fishing days. It used to be in this fishery that the majority of the catch was harvested at the start of the season. In March you had a glut, in August you had an under-supply. Now you can plan your distribution. The processing factories were short of oyster openers one day, they were laying off workers the next. No more. Now they can plan their oyster deliveries, they can plan labor requirements.

For me this was a good lesson. It tells me that in the pre-ITQ, as much as in the ITQ system, the most important factor was ability to plan your own destiny; to plan your own business and to generate these funny things that economists call efficiency and rents.

Phil Major.—What Andrew didn't tell you was that there was a lot of money in that fishery, and that it collapsed, and that all the people are busy working together to restore that fishery. They are doing a lot of research. That is what happens when a fishery declines like that. You get a community effort designed to reinforce what they have already achieved.

I want to elaborate on what I said in my address this morning. First, you have to understand there are two competing philosophies here in terms of property rights and other mechanisms for managing fisheries. You have the traditional way of regulating public property and you have private fisheries. You have an intervention arrangement with input controls, and a non-intervention arrangement with ITQs. You have a prescriptive arrangement without ITQs, and you have an ability to choose with ITQs. That is the fact that Andrew was describing in the oyster fishery. The ability of people to choose how to run their affairs. Isn't it amazing that we still think we can tell people how to run their own affairs.

Second, because of this differential philosophy, property rights in one guise or other are equally suitable for indigenous fisheries; they are suitable for high seas.

Third, who is going to manage fisheries in the future? I see here in this room this cozy arrangement in which there is a group of fishery management people, a group of fishers, a group of academic people. There may be some indigenous fishers, but very little environmentalists. Let me tell you this. World fisheries are 68% overfished. This is appalling. It is partly because of the cozy arrangement between fishery managers and fishers. There is vacuum here, and it will not be filled by fishery managers unless we seize some new initiative. It is going to be filled by conservation managers. They are coming down the road and they will seize control from you. And you will face a series of draconian fishery management measures. You have already seen this with the high seas driftnet fishery. New Zealand was one of the principle advocates of that. I was all for rational fishery management. I didn't think we needed a complete moratorium. But the Minister of Fisheries said, "Thanks for your advice, but we don't really need it." It had become, due to the conservation movement, a moral dilemma. You have seen that with turtles; and you have seen it in a different way with marine mammals. The reality is that conservation groups—these are people that have claims to the property right—are going to grab the high moral ground. They have every reason to do this, particularly when you look at the waste and gross overcapitalization in world fisheries. So we have got to make up our minds what we are going to do. If you do not grasp the opportunities that are here, the conservation groups will. And they will impose a range of controls on you that you find to be absolutely alien.

I have a couple of follow-up points. There was a question raised about economic rents and reasons why you want to have them. First, the conservationists have a good point. There should be some payment for damage or modification of the resource. Second, this is a public asset. Someone said that everyone in the country should be given a share of the rent; that would be one way of distributing the rent, I suppose. But the reality is that you have had people in fisheries for many years, and that rent has been used to support employment and excess capital. I think you should leave that rent there so that the industry can make its own adjustments. I don't think you can anticipate where and how the industry is going to change. In the New Zealand snapper industry, we have had changes towards smaller vessels going into longlining, employing more people, and getting higher value for their fish. So we have actually seen an increase in employment. I agree with Jim Wilen's earlier statement that you can't predict these changes. So don't get carried away with that particular issue.

One of the questioners claimed that we were giving cavalier answers. Unfortunately, in a private enterprise system you can't predict how people are going to shift their capital or what direction they will move. So you have to be a bit glib. But equally I think it is cavalier to say that we can make social decisions for other people. Certainly in an artisanal fishery, if you are worried that because of a lack of education or sophistication they will sell off their rights, you could make some rules to assure that that doesn't occur. But what you also have to understand is that those poor people have no rights now, and their rights are currently being imposed on by people entering their fishery. So if you give them some rights, at least they have something to protect. In the event of a fishery collapse, certainly you have to include measures to help people work together to come to a consensus. Andrew Branson mentioned earlier what we are doing in New Zealand to help encourage quota holder associations and stakeholder associations to assure that you don't have people trying to race each other for fish.

Bruce Rettig.—Thank you. Barry Kaufmann.

Barry Kaufmann.—It was nice to hear John Pope say that there was little doubt in his mind that there are significant economic benefits to be had from ITQs. I agree. Actually, the only fishery I have seen where it has gone badly was the South East trawl fishery in Australia. Implementation was the major problem there. One of the major benefits I have heard about from fishers is safety—not an economic factor. The fact that they don't have to go out in storms, that they can spend a weekend at home if they choose. Often you hear that fishers are not business people, that they just want to go out and fish. Yet every fisherman I know under an ITQ system says that they would never go back. If you ask them if it is per-

fect, they will say no. If you ask if they want to go back to input controls, there is no doubt about it, they do not want to go back.

This social policy question is a wild one for me. Take the Newfoundland situation and examine what you have done there. Basically, you had input controls. You had massive expenditures for fisheries biology. It has done little for the fishers. You had massive expenditures on fisheries management. It not clear to me who has benefited other than bureaucrats. What has been the outcome of all this? Basically, they have closed the entire east coast. So if you are concerned about ITQs and social considerations, what about input controls and social planning? It is not clear that it has gotten you anything but grief.

Martin Exel.—The main points I would like to make concerning ITQ implementation are four in number. First, take time, and get industry support. Second, change the framework; get bureaucrats out of management and get industry into it. Third, spend time to get your allocations right, because that is essential. Lastly, ensure your legislation is correct and your enforcement capacity is there from the very first.

Bruce Rettig.—Thank you very much Martin. Would people asking questions from the audience please queue up at the microphones.

Question #13 (J. Hastie).—My question is in the context of trawl fishery off the U.S. west coast of the USA, a fishery that is typified by vessels of less than 75 feet in length and where a high degree of observer coverage is not anticipated. The fishery pursues a number of different assemblages with a variety of species, of which some are currently under quotas and some are not, and the ratio of species in the catch can vary quite a bit from trip to trip. Do you see potential for applying an ITQ system to fisheries such as this without an observer program? If so, could you suggest what measures need to be stressed in designing the system.

Phil Major.—The answer is quite simply, "Yes." In your race for fish, people are not bothering about what they are catching, what the mix of species is or whatever. They just want to get out there and catch as much as they can, now. As noted by Andrew Branson, once you issue individual quotas, people will begin to figure out better ways of getting more value for their catch. In our oyster fishery, for example, fishers began figuring out how to get the best oysters instead of the most oysters. You actually put time into people hands to think about these issues, and they will start working out how to differentiate between the different bycatch species. We have seen this all in New Zealand. People say "Oh, we can't work out the bycatch, give us more quota." They are after the rent. So we stand hard on it and say we aren't going to give it to you. The fish are not as much together as the fishers would have you believe. I was just at a conference where people from the United States were showing how fish differentiate by temperature in the water column. People are actually able to target exactly what species they are going to catch by using temperature gear. They have been able to eliminate all the bycatch that they had previously experienced. Also, I don't think you should assume that they are going to continue trawling once they have ITQs. They are going to seek ways to use that resource more effectively to make more money. As another consequence of that, they will start looking at the bycatch—thinking, "How are we going to make money from this stuff?" Then the bycatch will begin to be used effectively too.

Question #14.—One of the important consequences of the catastrophe in the Atlantic groundfish fishery is that people start looking around for scapegoats. It would seem that partially implemented ITQs are being held up as a scapegoat in that particular catastrophe. I am wondering in the case I heard from Iceland and New Zealand, where we heard that in the initial evolution of ITQs there was a mix of effort and quotas in place, about the rhetoric we hear out of the Department of Fisheries and Oceans in Canada now being that we should have an ITQ system and input controls to go with it. So there could be catch limits per trip together with a seasonal ITQ. This seems to be at cross-purposes. In the case of Iceland and New Zealand, does it seem that that mix of regulations is feasible? And, if not, how can we persuade our administrators and scientists in particular how to do ITQs?

Andrew Branson.—One of the papers I alluded to earlier, authored by Mike Sissenwine, took a look at the New Zealand situation. It said something similar, that in the evolution of ITQs there should be an elimination of regulations controlling the inputs to the fishery. That paper describes how the input controls had not disappeared at that time. I can report that those regulations have still not disappeared. In terms of putting fishers in charge of management of their fisheries, they certainly like ITQs, but they also like being involved in the other elements of their fishery. And for a variety of reasons, in a variety of fisheries in different places and different times, they still have reasons for other kinds of control, of collective discipline about fishing. Such things as fishing in daylight, but not at night. In one fishery, for no apparent logical, economic reason they imposed upon themselves daily trip limits. The limit was related to what they could carry safely; it was related to what could be processed on shore during the day; it was a collective industry agreement. So, I believe that there is scope for a mix of quota management and a whole raft of other controls, if the industry wants it. If they see that is rational, they will accept it and they will do it.

Question #15.—In New Zealand the law is pretty settled that the quota share is a property right in perpetu-

ity and if it is ever revoked it is compensable by the government. In Australia and the United States, the position of the government is that the quota shares are revocable harvest privileges, and therefore arguably not compensable by the government if the regime is revoked.

My question is, "Where governments hold that the quota share is a revocable privilege, does this jeopardize, and to what degree, the economic benefits derived from ITQ programs?"

Phil Major.—Well, I cannot answer for the United States, but I think that once you allocate property rights, it doesn't matter what your legislation says about modifying them or taking them away. You will find it very, very hard to revoke them. That is why I say that you need to think very carefully about it before allocating the quota shares. In New Zealand we now know that a catch history will convert into property rights. We were issuing permits to harvest in a fishery with the stipulation that the permit would not establish a right to a catch history. The courts did not agree with us. It is very difficult to go back.

Question #16 (Jay Ginter).—I have a question for Mr. Major. In his presentation, he spoke of the policy–delivery split and the "corporatization of delivery." I wanted to know whether he could expand on that. Does this mean contracting for administration?

Phil Major.—Yes, it does imply contracting. The idea is to create an organization responsible for the functions of administration, enforcement, and research, which is separate from government. Why we call it "corporatized" rather than "privatized" is that it may be a body that is owned by government but will have directors from the industry and other interest groups appointed. So it is sort of a half-way house. Once it is corporatized, it could tender out or contract those services, and it would also be responsible to the Minister of Fisheries for the range of decisions that it makes. My colleagues here from industry would probably argue that that is not an optimal situation; they would rather see those functions privatized, although there would have to be an effort to include representatives of other interests groups, including recreationists, conservationists, and the Maori. And what government should do is retain some veto right over their actions or over the plans they intend to institute, and set a range of standards by which the privatized managers would operate.

Andrew Branson.—My colleague from the New Zealand government has stolen my thunder. He is dead right. From the industry viewpoint, we are in a "user pays" and so "user says" regime. We are not interested in dealing with monopolies, whether they be companies from Arkansas or government-owned enterprises. We don't care who does the job so long as it is done effectively, efficiently, and economically. We think that we should buy the services of administration, to the extent that we can, where the services are provided best. I quite understand that the government will retain their own policy unit, and if they do they can pay for that. But where we are spending our money in the future, we should get best value for money. We don't really mind who supplies the service.

Question #17 (Nina Mollet)—I have both a comment and a question for Phil Major. The comment is about corporations and companies. I think we have a little different situation in the USA, because we have corporations whose annual expenditures on lawyers and lobbyists probably exceed the annual GNP of New Zealand or Iceland. We have fishers in Alaska, for example, who are concerned about investments in fishing going to companies who have not been fishers but who have been chicken farmers or soup manufacturers. I think this is a real concern, and I share the concern that this distribution question has not been fully addressed. The question regards environmental concerns. In the North Pacific Fishery Management Council meetings, there have been quite a few environmental groups who have come forward with ideas for somehow rewarding "clean" fishers when we go to an ITQ system in Alaska. People who have been fishing fast and furiously should not necessarily be totally rewarded for that. Somehow we should come up with a system that rewards people who have already been conservationist minded. Do any of you have thoughts about that whether there is a reasonable way to do that, or is it "pie in the sky"?

Phil Major.—First, if you go to ITQs, in due course you will eventually get rid of most of the lawyers, because you will be doing things for yourself and you won't have other people trying to manipulate your property for you. The issue about the sharing of the resource is a very valid one. When we took this issue to the government in the first instance, and proposed the ITQ regime, they looked at it and said, "My God, the people that will benefit from this are the people who have been out there raping the resource. This doesn't seem quite right." This is what the conservationists have been saying, that there ought to be a charge for this program. But once everyone was in agreement regarding the direction to proceed, politicians really did not question that any further. And I think that it is quite legitimate to ask how you allocate these rights out. As I said earlier, if for the sake of efficiencies you don't want to charge a rent to individuals, you could retain some of the quota for the government instead of allocating out the full catch histories. That in itself is a way of penalizing those who may have been somewhat responsible for the excesses of the past. Of course, fishers don't ever believe that they were responsible. The reason that excesses occurred is that governments allowed them to do so and provided no incentives for them to act in a different way.

Question #18.—I am curious about the question of high-grading. I wonder how that has been dealt with by the various countries. Is there a mechanism in place to prevent that from becoming a big problem? It seems that once an ITQ is in place, the high-grading could become a big issue because it is so wasteful.

Andrew Branson.—Yes, there is a legitimate question regarding high-grading. The fishers will target the better fish and shape their season to obtain the better market price, but I understand that that is not what you are talking about. In my comments this morning, I tried to indicate that there were at least three fisheries in which high-grading was contributing very significantly to oversight and enforcement issues. In one of those fisheries, there are accusations of high-grading and denials of high-grading. This occurs in a fishery in which recreational interests and commercial interests conflict. The industry has provided funds for an observer corps and for various overflights of the fishery to establish daily positions of the fishing and to demonstrate the extent to which it does or does not occur. A lot of industry discipline is taking place. We think that the claims of high-grading are very excessive and that we will demonstrate exactly what takes place.

Question #19 (Alain Laurec)—I wonder if I am the official uneasy observer of this very interesting symposium. Because listening to this discussion of ITQs, I am just wondering whether the passion, talent, and enthusiasm of some success story teller could become counterproductive. You appear to be very hard-selling about this management tool. I wonder whether there are two risks following your discussion. First, the long-term objectives must be defined by politicians. One could wonder whether maximizing the rent is the objective for the politicians. I notice that in the European fisheries, for perfectly acceptable reasons, people try to maximize jobs at sea in specific areas in, for example, the next 10 years. This is a legitimate choice. Second, we are also choosing for the fishers. We seem quite convinced that fishers need to be included more actively in the process of management, but as pointed out by one of the speakers, maybe they would choose something else besides ITQs for their fishery. Also, I would like to quote the social issue. I can promise you that just indicating that the free market for individual fishing rights is the optimal approach, if put forward in Europe, it will be terribly counterproductive. A number of people are so scared about the unregulated market that if they have to choose between the unregulated free market and the bureaucrat, they will discover the beauty of bureaucrats. Just check what is happening in eastern Europe. This has to be considered seriously. This is why I would urge you to consider, at least for the scientists, what part could be played by the public authorities for regulating the market for individual rights. This has to be part of the debate.

Phil Major.—I think that the European Common Fisheries Policy is one of the great travesties of fisheries in our time. Frankly, the situation you have at present is a reflection of the inability of bureaucrats to come up with sensible arrangements to work things out. I don't have any doubt that if you sat down to work with all the countries involved in the common policy and worked with the industries there you might actually find that you come up with some solutions from a free market perspective. Sure, it is a political choice as to how you use the rent. And it is a choice as to whether you use the rent to create jobs or to let rent flow through to some other component of your economy where it will produce other goods and services. Yes, maybe there has to be a phase-out. But Laurec points out a problem with bureaucrats, and I am one of them: we hate not knowing what the outcome is going to be because we might be criticized, and we are very sensitive to criticism. The reality is that the free market will surprise you; it doesn't do what you think it will do. But usually it does better if given the opportunity. That is the risk you have to take.

Question #20 (Christy).—I would like to follow up on what Mr. Laurec said. This is supposed to be a global conference, but if you look around you will see that only half the globe is here. If you try to apply an ITQ system in the state of Kerala, on the east coast of India, where you have 600 km and 200 beach landing places, forget it! If you try to apply a limited entry system in that situation, forget it! It is great to hear about how these systems are perfect in developed countries where you have limited landing spots, and where you are not dealing with tropical fisheries where you have large range of species that come out of fishery. But how do you apply these systems in developing countries and Asia?

Phil Major.—I can't answer that directly. I was here at a conference on electricity here earlier this week. The president of Boeing was there. He said that there are three words they ban from their company: "We are different." What I hear Francis Christy saying is "It's different in Asia; it's different somewhere else." The president of Boeing said that they ban those three words because once you accept that you are different, you close your mind to the solutions. If I go to an engineering meeting, and I say "we're different," I have closed my mind to the solutions. Open your mind to the solutions! I can't tell you exactly the nature of property rights that will work in those regimes, but I can tell you that it will work if people apply their minds to finding the right range of solutions. I can also tell you that all the mechanisms that we have tried in the past are failures. Christy showed that himself. That is a testimony to what we have not done. So, don't think that we are different, everyone is the same.

Coffee Break 3:15 PM

Panel 2

Richard Marasco, Chair

Those of you involved in development of limited access systems are aware that fairness and equity become the most active issues. So in the interests of fairness, I want to start from the south and go north in this session.

Jim Mace.—I have thrown away my notes, because I was going to talk about the benefits of ITQ systems for operators of companies like mine in the fishery. I think the previous New Zealand speakers have covered that quite well. If I were to continue along that line, it might sound too much like preaching.

When I was preparing to come up here some of us discussed, half flippantly, whether it was really in our best interests to talk about ITQs in this part of the world. Because having a well managed and profitable fishery in this part of the world might not be in the interest of a small country like New Zealand that is trying to sell species like hoki, which is inextricably linked to the fortunes of pollock in the whitefish market. Having looked around here, I would say that the answer is that we should all work on these issues together. We are all working in the same industry around the world. Something that is very striking to me in Seattle is to see that the fish that is on the menu of first-class restaurants is largely halibut from Canada, fresh ITQ-caught halibut. The fish sold in the fish and chips stores is largely pulled out of cold storage, fish that was caught in one of the quick openings of a few months past. The message there is that under an open-access system the incentive is simply to get as much into the cold storage as possible and then get it out again as quickly as possible. All that does is drive down the prices. In New Zealand, we have the luxury of planning our catch carefully, so we can start on planning to add value to the product when we sell. I am firmly convinced that if we all had the time and luxury to do the job well, we would raise the earnings to all people in the business by improving the impression of fish in the market. It is clearly in the interest of New Zealand industry to come up here and encourage you to get away from a management system with the very short operational seasons you have at present.

I wanted to make a very brief point about the pursuit of social objectives. Perhaps because New Zealand is a small country we can afford to be a bit glib about this, but you need to think very carefully when you try to use fisheries policy to achieve social objectives. Our farmers in New Zealand would see their incomes as being depressed due to the European Community's common agricultural policy. So if you are going to have social objectives, you need to think about the effects they are going to have—often they are not what you expect.

The last point is with respect to the government taking a portion or all of the rents on fisheries. We have been through that debate very intensively in the last few months. I am pleased to say that the outcome of that debate has been reasonably satisfactory from our viewpoint. The government has decided they will take from the industry each year an amount of money that is sufficient to cover the cost of management and research and enforcement of the fisheries, but will not be taking a portion of the rents above that. From my perspective, rent-seeking is the engine that drives the capitalist economy. If you want to look at the opposite extreme just look at the command economy they had in the Russia where rent-seeking was illegal and the damage it did there. I think this has been salutary for us in New Zealand, because 10 years ago we had a government that was taking more and more of the rent from the people that were successful and our economy went backwards quickly. We have spent 10 years reversing that. As people are earning more rents, in whatever business, they are growing our economy. In my company last year we created 300 jobs. That was through a mixture of retained earnings and debt. Had we not had the retained earnings because they were appropriated by the government, then we would have created a lot fewer jobs. So from my viewpoint, the message is very clear: If you take rents, you must be aware that you are taking money that would have been used in growing the business and adding value.

George Clement.—In New Zealand it is like living in a laboratory. We have all kinds of experts and economists coming through. One came through the other day who was definitely against ITQs. His thoughts were as he walked through the airport "If ITQs were the answer, it must have been a silly question." He left two weeks later shaking his head and thinking "Well, they work in practice, I wonder if they work in theory." When I was growing up I always thought that the United States was the hallmark of capitalism. Its a little like stepping back 10 years. We've tried all these management techniques in New Zealand and they did not work. I'm not saying the ITQs are the only way to go, and I am not saying that they are perfect. We have spent most of the past ten years trying to improve it. But it is the best that anyone has come up with, and it works. And the reason that it works is it works for the fish. We have heard about managing people and managing profits, and that's fine. But that is all political stuff, and the fish don't have votes. If there are no fish in the sea, we don't have anything to manage. Everything we want to do on theory, or management, or profits comes to nothing. ITQs have worked for conservation in New Zealand. We have stocks re-building. When we put in some of the original quotas, they were a 75% reduction from the earlier catches. Those stocks are rebuilding. We have our detractors. We have had some

very healthy debates with groups from the conservation movement. Some of those groups you have here; one is Greenpeace. There have been strong debates from their view that we should have ecosystem management. I am happy to say that there has been an intensive dialog over the past few months, and that served to make them understand that in a multispecies fishery like ours we are dealing with multispecies management and it is working. We have significantly less bycatch and discard problems without the competitive race to catch fish. It was heartening to the industry to have one of the Greenpeace representatives to stand up and say they support ITQs. The conservation thing should not be overlooked.

The key to success is private property rights. No one respects government property. You have to give control to the people who can control those rights. We really moved on to the second generation of management issues, which is self-reliance. (Showing slide on overhead screen) Here is the picture of a fish, I haven't seen many of these at this conference so I thought I would show you what they look like. This is an orange roughy, a very important species to our fishery. There was a great deal of concern amongst the quota holders that the resource did not look as robust as we had hoped. We were not getting enough information from the scientists, simply because the didn't have enough resources. We have combined the quota holders to form a company (impossible to do in a non-ITQ system) to undertake a range of techniques aimed at exploratory fishing, fisheries research, and fisheries management.

What have we achieved? In the last 4 years we spent over $5 million in stock assessment, extra data collection, population modeling, stock structure analysis; we just surveyed the major productive areas using sidescan sonar to do an inventory of the habitat, like mapping the farm. We developed collectively, sharing the expense and the risks, exploratory fishing, seabed mapping, and surveying new stocks. We brought in expertise from the University of Washington and other companies and countries. We also operate a management system where we have seven areas within the existing area and have closed one area because the fishing in that area is too high.

So the system does work; it is not perfect, but it combines people cooperatively. If you have any doubts about ITQs, there is no better system devised, in our view.

Doug Hopkins.—First, I do not represent the entire environmental community by any means. The environmental community has a range of positions on the subject of ITQs, and I think you have already seen the environmental community becoming engaged in this issue. They are becoming engaged at a level of the debate that they have not been engaged in before. But there is one point on which there is consensus in the environmental community regarding ITQs, and that is that we want to see conservation benefits. We want to see the fish and the ecosystem better off as a result of the implementation of any management tool, including ITQs. There is a lot of skepticism that ITQ management will lead to conservation benefits. For example, and these are issues that have come up today, we are concerned about enforcement. Enforcement is critical to ITQ management working. Who is going to pay for it? Right now the Magnuson Act in the USA prevents the collection of user fees from the fishing industry. Will an ITQ system work in any complicated fishery in the USA without changing that? Probably not. We want to see any management scheme, including ITQs, lead to a reduction in bycatch. There are reasons why bycatch may or may not be reduced under ITQs. We are concerned about that. The list goes on: High-grading, cheating, potential for windfall profits flowing to entities that are not viewed very favorably by the environmental community right now. There is also the possibility of social and community disruption. There is the possibility that ITQ management may lead to a rash of takings claims when councils try to reduce TACs or try to terminate or change a system substantially. The example we have been hearing from New Zealand, where there was a need to go into the market and buy back ITQs using public dollars, is not an approach that would be supported by the conservation community.

What needs to happen in this country in the view of many of us in the conservation community is that the NMFS needs to change its role from an uncritical cheer leader for ITQs to a role as provider of information to councils and to industry about what works and doesn't work from the viewpoint of conservation management. Second, NMFS needs to take on the role of setting criteria and guidelines that the councils must follow if they want to set up ITQ management. The criteria would include characteristics of fisheries that would be appropriate for ITQ management, and the guidelines would be for designing ITQ schemes so that they maximize conservation benefits. ITQs seem to be a good idea to many people in this room. In theory and in practice there are many positive signs about the value of ITQs in fisheries management. But in this country, a major, highly public failure of an ITQ management program will not be brushed off by the conservation community, by Congress, or by many fishers. A high-profile failure like that could put this fishery management tool back in the tool box for a very long time.

Borrowing from Barry Kaufmann's presentation this morning, the most important principles to adhere to in converting to ITQ management are to go slowly and to make sure all the problems are identified and the loopholes are closed up front, before the plan is implemented. And be wary of the problem of theory versus reality. Finally, the Environmental Defense Fund has proposed that Congress create a pilot study to allow the entire USA to watch and learn from a limited number of ITQ management programs here in the USA. During this pilot study, while there is an intense focus on a limited number of fisheries which would include some fisheries that are already being managed by ITQs and some additional fisheries, there would be a moratorium on ITQs management for other fisheries.

Richard Marasco.—The bycatch issue has surfaced a lot during the session of the past several days. Our next speaker was included in the panel to shed some light on how bycatch is being addressed in the tuna fishery.

James Joseph.—I was just taken by the comment that Doug Hopkins just made, indicating that the non-government organizations are becoming much more active in fisheries, and some earlier comments that unless we did a better job, they may be in the driver's seat. Given the state of some of our fisheries and the way they are managed, maybe that is not such a bad idea.

The world is becoming increasingly more crowded. Within the professional lives of some of this audience, there will be 10 billion people on the earth. There will be more demands on all of our resources, water, land, and fish. In order to utilize fish in the future there are going to be more limitations on access to fish. Fishers are going to be more involved in managing fish I think, and there are going to be more responsibilities and requirements for fishers who want to harvest resources. I would imagine that in the near future most boats fishing on the high seas are going to be carrying observers to collect information not only for monitoring the catch but for scientific purposes. Observers and GPS systems and automated video systems will be used to monitor what is going on vessels to ensure that compliance is kept in order with management requirement. IQs are going to be a big part of that.

I want to talk about IQs from a different perspective: rather than IQs for target species, IQs for bycatch. IQs for bycatch have some benefits. It puts in the hands of fishers their own destiny, as an earlier speaker said. If there is a limit to what they can take as a bycatch, they are going to find innovative ways to minimize their bycatch and to maximize their catch. They will develop gear modifications to ensure that they target successfully and avoid bycatch species. They will develop strategies for time and area locations to fish that will minimize bycatch. I think that this kind of approach has a potential for helping. Obviously it is not the answer for bycatch in all fisheries. And bycatch is going to be the issue, in my opinion, that controls high seas fisheries over the next 10 years.

I want to mention the approach that has been taken in the tuna fishery in the eastern Pacific ocean with respect to bycatch. As many of you know, in that fishery there has been a very large bycatch of marine mammals. The fishers use the mammals to target the yellowfin tuna. Originally in this fishery, the mortality was very high, in the neighborhood of 200,000 animals a year. Programs were developed to reduce dolphin mortality—extension work, gear modification, etc.—and mortality came down relatively rapidly through time in the form of a learning curve. But it became progressively more difficult to reduce mortality further, so the governments actually adopted what the staff of the organization I work for recommended—that there be some mechanism established to limit the bycatch of dolphin. It took 3 years to convince the governments that this idea had merit, and then it became difficult to determine just how to implement it. Again, the staff's opinion was to assign individual limits to vessels.

The rationale was that if overall limits on mortality were set, there would be a race to take as much tuna as possible before the fishing was stopped. If limits were assigned to vessels individually, then the vessel captain himself would search for ways to reduce the mortality of dolphins. Indeed, this worked very successfully. In the first year of the program, dolphin mortality (which was already reduced to a low level) was reduced by more than a third. Dolphin mortality is continuing to go down under this scheme over time. Fishers have been extremely innovative in finding ways to reduce dolphin mortality because they know that if their dolphin mortality limit is filled they will be prohibited from fishing for tuna any further. I think this kind of concept can be applied to other fisheries where there is a significant bycatch problem. Not only will the concept of the limit be needed; it will have to be supplemented with research on the impact of bycatch on the stock and the strategies for gear modification, time/area strata, etc. But the program has worked beyond the wildest expectations of those who put it in place. A target was set for 1999 of less than 5,000 animals. That target was reached and surpassed in the first year of the program. In the second year it will be surpassed again.

Bill Robinson.—It seems like there is not a whole lot to say after a day of speakers covering the entire spectrum of issues. I will just make a couple of observations. I did have an opportunity to help develop the "nuts and bolts" implementation of the southern bluefin tuna quota system in southern Australia, working under a commonwealth form of government. And then I have had the opportunity to develop a license limitation and ITQ system here on the west coast of the United States.

The fact that you need the full participation of the fishing industry in developing the system cannot be overemphasized. Regardless of what form of government or what authorities you are operating under, you are doomed to failure if you do not have the support of the fishers.

So, this is the number one requirement. Regarding the southern bluefin tuna fishery, I recall Martin Exel's earlier description of the government's consultation with the industry. I would note that under the Australian form of government they had an advantage going into the process—the government set certain standards and priorities. There were three of these: first, the government said there will be ITQs; second, the objective of using ITQs is to move towards economic efficiency; third, the basis for initial allocation of the ITQs would be recognition of the commitment of tuna fishers to the fishery. Then the consultations with industry began. How to meet the three standards was worked out in a series of meetings. There was certainly an advantage to have the parameters established. Also, an advantage to that process was that once an acceptable agreement was reached with the industry, it could be swiftly implemented. Swift implementation of the system precludes the possibility of a number of problems that we have been running into here in developing a limited-access system on the Pacific coast.

In responding to a comment made by Doug Hopkins, our system here in the United States is different. It is more of a "bottom up" system of management. The Department of Commerce and the National Marine Fisheries Service have general oversight of fishery management activities, but the charge for developing policy in fishery management plans resides in eight regional fishery management councils. Those councils do not have the benefit of the kind of guidance that the Australian government provided in terms of what the basic parameters would be, and then developing a system to meet those parameters. The council must struggle with the job of developing their own goals. This leads us to management by the lowest common denominator. And that is reflected in the fact that to get a decision out of the council you have to satisfy a majority of voting members. To get that you often have to come up with goals that are conflicting or you have to compromise each goal to get acceptance.

In the ITQ system we are developing now there are three important factors in getting acceptance among members of the industry. The first is initial allocation. We have had a number of new entrants to the sablefish longline fishery during the development period of the ITQ system. The council is struggling with how to adequately recognize those new entrants and to recognize the historical participants. Windfall gain is another concern—that the initial allocation will result in windfall gains to a number of fishers. I find that a little ironic since we recently implemented a fully transferable license limitation program and there has been substantial influx of vessels during the limited permit period in the fishery. From the agency perspective, or at least my perspective, the windfall gains are not a great concern. It appears to me that the ITQ system generates capital in the industry, and this provides the capital basis for rationalization. In many cases, this precludes consideration of a taxpayer-funded buy-back system.

The other point I have is that in the term ITQ, the 'Q' means quota. But there are many other ways to skin the cat. We could have transferable units of fishing effort, fishing days, harvest capacity units, transferable fishing pot licenses, numbers of hooks. There is also the option of choosing "none of the above." In a multispecies fishery like the Pacific coast groundfish fishery—with some 80 different species, of which 15 or so are major target species—it is very difficult to take the plunge. There may be forms of management that lead up to a full-blown ITQ program and provide a kind of evolution. The Pacific groundfish fishery for a number of year has had a major objective of spreading the catch over the year to have fresh fish in the processing plants all year long. To do that, we have had trip limits. As capacity increased and harvest quotas came down, the original daily or single trip limits had to be expanded to become weekly. From there we added monthly trip limits to add flexibility for larger vessels to operate, and the trip limits eventually became so small that we created monthly cumulative trip limits. Since the license limitation overlaid that, this provides a kind of springboard for a quasi-IQ program which is now under consideration by the council. The idea is to allow the limited entry permits to be stacked on a vessel, and to assign a cumulative monthly catch limit to each permit. This essentially provides some of the benefits of an ITQ system to the fishers. So while this does not resemble a full ITQ system, it does build upon a set of management measures that the fishers are comfortable with, and have lived with for years. If the decision is ultimately to go to an ITQ system, that step would be like stepping into a lukewarm bath instead of into a cold shower.

Bruce Turris.—It is a real pleasure to listen to the panelists that have gone before me, and to realize that there are programs all over the world, and for us to get together to share information. It is also a pleasure to see Canadians and Americans in the same room talking about fish. I wonder whether Steve Pennoyer took the scenic inside passage on the way down from Juneau. You should know that they charged me $1500 to cross the border this morning, but the government has set up a fund to reimburse me.

I come here as a fishery manager. I have spent the last 6 years managing fisheries. Not just individual quota fisheries, I also manage competitive fisheries and I have listened to all of the arguments. I have been asked to answer the question, "What is the most important aspect or consequence of the ITQ system from my perspective?" That is difficult to answer because there are so many things that change when you go to ITQs, but perhaps the most important is the attitude of all the people involved in the program—the managers, the fishers, the scientists, the enforcement officers, the buyers of the product, the market. The attitude towards how they do their business changes considerably, and the result in my experience is that there is a net benefit to going to individual quotas. There is a change in the way of thinking about fishing. The change is from volume to value. The fishers now think about how to get the most for the fish they catch. Enforcement is important now to the industry. Before it was important only if they got caught. It is important now because it affects them. If there is cheating, it really affects their share of the allocation. In the halibut and sablefish fisheries I am involved with, the quota shareholders pay for enforcement, 100% of it. As an example, in halibut the industry used to pay a total of $4,000 per year. Last year they paid just under $1 million for the year. A lot of that went for enforcement.

The industry's concern for conservation has changed considerably. Under limited access they were not so concerned about conservation. Now they are not just fishing for today, they are fishing for the future. The asset value of their quota would fall if the fish stock falls. They are interested in protecting the value of that asset, especially in sablefish where they pay for all the science and research budget. The expenditure on research will be greater this year than it has ever been, because the industry has contributed to the effort and is asking for more scientific work, and they are involved in the science. They want to have a comfort level, or confidence in the science that is being done. That is not questioning the competence of our scientists; it is just a matter that, with the limited budgets we had, not enough science was being done. Their attitude towards planning has changed. Short-term planning is now secondary to long-term planning. Safety was brought up earlier by Barry Kaufmann. This is always a major issue, even in open-access fishing. It is just that the economic incentives don't necessarily promote safety under open access. The economics don't always promote safety with ITQs either. On the March 1 opening of our halibut fishery, 100 boats rush out there because the market price is the best they will experience all year. Everyone wants to get $4.50 or $5.00 per pound. So some boats are willing to go out there and take chances. The difference now is that we don't have 450 boats going out taking chances. They have a choice.

Regarding cost recovery, the industry now pays willingly because they realize the benefits of doing that. Their attitudes towards each other have changed. We now have many fishers who share information about where the best fishing spots are. Their attitude towards government has changed; they don't hate us as much anymore. There is a level of cooperation that is refreshing to see. Last, but definitely not least, is their attitude towards the future. With ITQs, the future is viewed as very good. This is reflected in the price of quota and the way the fishery is run. In the non-ITQ fisheries, generally we are running around trying to fix things all the time and wondering where we are going to be tomorrow.

Stuart Richey.—I think it is important to mention some of the down sides to ITQs since we are aware of the pain and problems that occurred in Australia's South East trawl fishery that Barry Kaufmann described earlier. The problem we had in the southeast fishery had all to do with the allocation process. The underlying reason for the introduction of ITQs was to rationalize the fleet and to remove at least 50% of the effort. This was to be done without any additional adjustment process. In my view, this made the implementation and allocation very difficult for the industry to accept. The social and economic dislocation coupled with the massive uncertainty and the needless grief caused by the prolonged allocation crisis was a major cause of the industry resistance to ITQs. The ITQ system was introduced into a fishery that had been managed by limited entry since 1985. It already had a system of limiting vessels by underdeck volume and engine power units. These were regarding as the currency in the fishery and were tradable. These units were defined in a major release by the minister of the day in 1987, and he indicated that that would be the basis of any future allocation of ITQs. A lot of money changed hands after that on the basis that people were buying a stake in the fishery.

However, to complicate the allocation process a little more, we had three distinct sectors in the South East trawl fishery. We had an inshore sector catching primarily market species, another sector fishing mainly eastern stripe, and a developing big-water fishery fishing mainly orange roughy, hoki, oreo dorries. These were all under the same license system, and license units were freely transferable between the sectors. So when we came to allocation, the allocation was deemed to be based upon your investment in the units in the fishery and your catch history. Unfortunately, the catch history had been acquired in many cases by boats fishing different areas and different depths. By the time we had allocation, we had an inappropriate species mix going to many of the boats. To adjust, the industry had to undergo an internal readjustment. The problems in the allocation process were exacerbated by large reduction in the orange roughy TAC at the same time. Also a zero TAC was set for the eastern gemfish fishery. These two species had been major target species in the southeastern fishery.

As a result of problems in the allocation process and the stated objectives of removing at least 50% of the effort in the industry, fishers themselves became totally fragmented; one fisher was pitted against another as each sought to obtain a viable allocation. This extended to the industry organization, the Southeast Trawl Fishing Industry Association, which had previously represented fishers and had been a successful lobbying organization on their behalf. As a result of the divisive nature of allocation, this organization was forced to agree to avoid all discussion of allocation issues. It took another 2 years to resolve the inequities that were inherent in the initial allocation. It also involved several adjustments to quota holders' allocations. These adjustments, as Barry Kaufmann described earlier, involved a review by Australian Fisheries Management Authority of the initial allocation process and a legal challenge in the federal court of Australia, which deemed that the original allocation formula was statistically flawed. A number of refinements were added after introduction of the management plan by the management advisory committee. During the period of changing these allocations, permanent transferability of the quotas was suspended, which meant that operators were prevented from restructuring their holdings to make a viable operation, and the non-viable operators could not sell out to leave the fishery.

So, aside from the uncertainty and instability, we had a process that was poorly implemented and poorly planned, with many unresolved issues before the plan was put into place. One of these problems was the issue of bycatch. Another issue that slipped up on us completely unexpectedly was the Australian capital gains tax system, which imposed a tax on anyone who swapped, sold, or traded quota. This has been a major hindrance in trying to restructure the fishery and get the ITQ system working. The uncertainties brought about by the unresolved issues initiated a period of great instability in the fishery and disillusionment on the part of many of the operators and a lack of confidence in the future of the fishery and of the benefits of ITQ management itself. Also, in this period financial institutions were reluctant to lend money on quota or to accept ITQ as security given the severe variations to the allocations.

So, in my view, the lessons to be learned from all this to anyone contemplating the introduction of ITQs is that ITQs should not be the sole method employed to rationalize the fishery, if the rationalization is to the extent required in the South East trawl fishery. Also, if the ITQs are to be introduced to a fishery that is under some form of limited entry or some form of input control, that control should continue to be the currency in the fishery and the basis of any future allocation within the fishery. The allocation method must be statistically, constitutionally, and legally sound, as proved not to be the case in southeast Australia. Not a lot of people will agree with me on this, but in a multispecies fishery consideration should be given to allocating only the target species. The issues of bycatch have not yet been fully resolved. Before any plan is introduced, the method of any future allocation of additional species should be decided before the initial allocation so the industry knows where they stand. Then we won't get this rush of people trying to establish catch history on paper and distorting the figures the scientists attempt to use in setting the TACs.

Given all the problems discussed up to now, the survivors of the last few years on the whole have accepted the change in culture, of changing from fish hunters to fish harvesters, and appreciated what we now consider the benefits of ITQ management, which is being able to rationalize our operations and increase our economic returns. And they realize they should have a secure investment in the fishery. This has resulted in their becoming more responsible and certainly more involved in fishery management issues.

Richard Marasco.—Thank you Stuart. I will now open the floor to questions.

Question #21 (Ray Hilborn).—Most ITQ fisheries that I know of had the good fortune to have a fish stock, or at least the value of the quota that is building over time, so that people that bought quota have usually done pretty well. Does anyone know of a case where the fishery has in some sense collapsed after introduction of IQs and a number of people have gone technically bankrupt? If so, how does that situation differ from the case where there is usually a massive government intervention and bailout?

George Clement.—The answer from New Zealand is no, we don't have such an experience. The attitude of the government is important here. The attitude in New Zealand is one of self-reliance. The ITQ is a private property; it is a risk they have to take and they are on their own. I would suggest that that is only a problem if you have social engineering in mind. Really, the overcapitalization that exists is between the vessel owner and his banker. Where there was restructuring in New Zealand, where we have reduced TACs after introduction of IQs, there has been no compensation. The costs fall where they may.

Jim Mace.—About 3 years ago my company bought another company that was actually larger than ours. When we did the financial analyses, we made our best estimates of what the likely scenarios would be with respect to ITQ adjustments. We factored in the likely reductions in orange roughy TAC. There is an expectation in New Zealand that those buying quota have to factor that in, and if they get it wrong, that is the risk they take. There is probably more of a risk in the single-species fishery than in the multispecies fisheries.

Bruce Turris.—Canada has a few examples of failures, and they are not just in ITQ fisheries. But they are not quantitative property rights. As in other countries, the political intervention in those failures depends upon

how that fishery plays in the political landscape. In some cases, there has been a lot of political intervention whether or not there are IQs, and in others, like the abalone fishery on the west coast, there were IQs and it has been shut down. There was no political intervention or attempt at compensation. The buyers of those quotas lost.

Question #22 (John Gauvin).—For Stuart Richey I have a question on capital gains taxes and how they affect share trades. As I understand it, in this country an ITQ transfer is considered a zero-based capital gain, and when a person sells one, they will pay on the full assessed gain. Was the problem in Australia that people did not know they were going to pay a capital gains tax? Is that what disrupted the market? I am also curious about the level of assessment in Australia compared to here where the maximum tax rate would be about 28%.

Stuart Richey.—I have a fair amount of experience with this, as I have been helping to prepare a submission to the Australian taxation office seeking to get "roll-over relief" for the industry from capital gains tax. In our case the capital gains tax rate is about 33%. Assuming you were a redfish quota holder, and I was a roughy quota holder, and we wanted to swap our quota holdings without paying cash or changing the value of our quota holdings, this is still deemed to be a sale of an asset. In the case of earlier entrants in the fishery, the base value of the asset is only about $20, so almost the entire value of the trade is a capital gain. To make it worse, we are moving to a system of statutory fishing rights, where each unit of quota is a statutory right on its own. So, if you hold 300,000 pounds of quota, you actually have 300,000 separate assets. If you are restructuring your quota holdings you are restructuring hundreds of thousands of assets. That has brought major quota trading to a halt at the moment.

Question #23 (Dan Huppert).—I am interested in the interaction of the bycatch problem and the IQs, so I am directing my question particularly to Jim Joseph and Jim Mace. I understand that the bycatch of the dolphins went down very quickly in the tuna fishery, and I am interested in how that happens. How were the fishers able to reduce their take of dolphins with IQs? For Jim Mace I am interested in whether there are multispecies bycatch problems in the trawl fishery in New Zealand, whether the IQs either exacerbate or ameliorate the problem, and what the mechanisms are for dealing with bycatch?

Jim Joseph.—Basically, it was a result of the fishers' will to reduce the bycatch. The way they did it was to modify some of their fishing practices, to be more careful in the "back-down" procedure, which is a way of letting the dolphins out of the net without letting the tuna out. It takes a longer time to do that. One of the important things they did was to actually pass up opportunities to catch yellowfin tuna when they calculate that there is a large probability of taking large numbers of dolphin. Very large schools of dolphin which carry large bodies of tuna with them were passed up. On many occasions they would let go of the whole complex that they had encircled when they saw some disaster or imminent mortality was developing. They used a variety of techniques. Another thing they do is to put themselves in danger, unfortunately, by putting people in the net with diving equipment (hooka gear) and walk the bottom of the net to make sure than any dolphins that might be sleeping or lying on the bottom can be released. There is a wide variety of methods they used, and it is virtually impossible to predict what fishers will do if given responsibility for their actions. My point is that we return a lot of the responsibility to them and they will find ways to reduce the mortality whether it is in a dolphin type situation or whether it is in a multispecies fishery.

Jim Mace.—In the New Zealand situation I think we have made rather substantial progress in dealing with bycatch issues, whether that be bycatch of commercially caught fish or bycatch of marine mammals. In both cases, there has been considerable work by the industry in developing codes of practice to change fishing practices to minimize the impact. For example, fur seal catch: In 1988, in the hoki fishery, the fur seals discovered that if they head in the direction of a trawl winch starting up, there was free feed to be had. Within 2 years the bycatch went up to 800 animals a year. We have now reduced that by more than 90 percent to 60 animals a year. We had similar results with our sea lion catch, and that has basically been due to the development of fishing practices.

With respect to finfish bycatch in the multispecies trawl fishery, there have been a number of developments, basically a matter of changing fishing patterns. One of the most common is to change fishing time or depth or the way the fishing gear is set, so that you are more likely to take your target species and less likely to catch the bycatch species. There has, of course, been some redistribution of quotas of those species so that operators in the multispecies fishery buy quota from others who would previously have targeted it. Therefore, although we still have some quota overruns, the problem is largely under control.

I would make one plea with respect to this issue. People are critical of ITQs, saying that you have bycatch problems. I think you have to look at the alternatives and see whether the problem will be better or worse under ITQs. I think that the program in New Zealand has been very successful.

Doug Hopkins.—Picking up on the last point, it is important to remember that the bycatch problem has many kinds of solutions. Even after conversion to ITQ management, the kinds of tools used before ITQs need to be kept available. There needs to be a variety of efforts to create incentives over and above the incentives that are created naturally through ITQ schemes for fishers to be innovative as quickly as possible to reduce bycatch. In addition to direct incentives to innovate and requirements for use of certain kinds of technology, that also may involve preferences in either reduced limitations on transfer or preferences in actual time at sea or even in initial allocations for the more selective fishing methods.

Question #24 (Paul MacGregor).—One of the vexing problems we have in the north Pacific is discards. Discards of prohibited species which are required by law and discards of target species that are required by law, but there is also a large component of economic discards—discards that are too small or not the right size and shape. This is becoming a highly visible public issue. The folks in the sector of the fishery that I represent attribute a large part of this discard problem to the race for fish and the fact that people don't have an opportunity to slow down their processing operations in a way to utilize fish that are odd size or something like that. I would like to know from any of the panelists who have experience with ITQ systems whether the system helps the fishers address that particular issue.

Bruce Turris.—By itself, no, but it provides you with a tool by which you can address other issues. In our halibut fishery we had a significant rockfish discard problem. We still have a rockfish discard problem. This is partly due to vessel size and partly due to shelf life. As you are probably well aware, rockfish don't keep as well as the halibut, but the mortality rate on the rockfish is basically 100%. So, before we had discard of rockfish, if they are not being used as bait. That still happens if fishers are on a long trip, but they have been given a 20% bycatch allowance by weight. So now we can quantify the bycatch, whereas under a competitive fishery you could never do that. So it has provided a vehicle. Fishers aren't necessarily keeping all the bycatch, but the incidence of bycatch retention is a lot higher than before. We now get the information for biological assessment purposes, and the fishers get the economic benefit of selling the fish. If they bring in anything more than the allowable retention, the government gets it.

Jim Joseph.—I want to make a couple of observations. Part of the motivation for developing the dolphin mortality limit, and the bycatch quota on dolphins, was to prevent the type of problems that were described by the questioner. In the eastern tropical Pacific, the purse seine fishery operates in three different modes. One mode catches tuna in association dolphins. The dolphin associated fish are large, sexually mature animals at nearly the optimal size in terms of yield per recruit. The other two modes, called "school fishing" and "log fishing," yield small tuna that are sexually immature and of sub-optimal size in terms of yield per recruit. School fishing and log fishing involve a reduction of 30% in yield compared with tuna caught in association with dolphin. In addition, there is a very large bycatch of other species, small unmarketable tuna as well as many other types of fish—mahi mahi, sharks, billfish and turtles. Although there was a very intense campaign to prohibit fishing on dolphins entirely, the IATTC member governments wanted to avoid forcing the fishery onto school and log fishing. That is one of the reasons they finally agreed to a dolphin mortality limit—in order to continue fishing on the larger fish, to eliminate the problem of growth over fishing, and to minimize the bycatch of other fish.

Question #25 (Phil Major).—I want to address a number of comments by Doug Hopkins. First, I want to make a correction, at the risk of sounding like an evangelist. ITQs are a philosophy, not a tool. Really, what we are dealing with here is a conflict between a controlled or directed economy and a free market. As you noted before, you can have other sorts of management tools along with ITQs, which is a reflection of the fact that it is an overriding philosophy and not a tool. Second, I invite you to come to New Zealand to study ITQs. Bring all the people you want. We may even be able to subsidize you. We would be delighted to have you come and study it because we have not had the time to stop and study it.

The next thing I want to say is don't procrastinate. Get on with the implementation of your ITQs now.

The last point is in the form of a question. If conservationists really think that they are better at managing the resource, they have concerns about the way the resource is managed, they want better balance in the ecosystem or they have social distribution concerns, why don't they support the implementation of ITQs? Then they can get in there and purchase them, and not fish them or redistribute them. This puts conservationists, or other people, who have other objectives, in the management seat on their own terms without telling other people, who might hold ITQs at the present time, how to deal with them. This is already done with wilderness areas. In fact, it seems that I myself have contributed to buying some of the Amazonian rainforest. It seems that this is one way that conservationists can really get into active fisheries management, and I wonder whether or not this is not a proposal that the Environmental Defense Fund and other conservation groups might not take up.

Doug Hopkins.—Let me take up two of your points. The first is your point about accepting a philosophy of "getting on with it." In this country there is a history of conflict when public resources have been transferred to private interests. The easy way to characterize this is that the long private control over public resources has led to substantial political power in the USA that led to decisions being made by Congress and administrative agencies that don't make a lot of sense to most people in this country, and yet are made year after year because a few people end up with a disproportionate amount of political clout. There is a real fear that a large-scale conversion of fishery management to ITQ systems will lead to a similar disproportionate amount of political clout. If the theory and evidence that ITQs lead to more conservation-oriented stewardship by quota holders is wrong or is wrong in some cases, then it will be very difficult to undo the system that is in place, even if one can get around the more limited concern about 5th amendment takings claims under our constitution.

The latter point—"Why don't conservation groups support ITQs and then buy up quota as quickly as possible?"—I should have mentioned earlier that the premise on which any environmental group would consider supporting an ITQ system is that, first and foremost, the TACs are set conservatively with a substantial degree of cushion built into them and that they are protective of entire ecosystems. Hence, the ITQ system is simply used to assure that the TAC is not exceeded. If the TAC is set properly, there should be no need to buy and retire ITQs.

Question #26 (Lee Anderson).—I have a question concerning the effect of ITQs on the crew share system. We all know that share systems are designed to distribute risk, and that the actual percentage set in any fishery depends upon the capital-to-labor ratio and the fish price. But with ITQs, the capital–labor ratio can change and the fish price can change. I will report that there has been a change in the share rates in the surf clam fishery. Before they changed some felt that labor was not getting their fair share. I am wondering whether other fisheries have experienced a change in the share rate.

Bruce Turris.—Yes, in all three of the IQ fisheries I have been involved with there have been changes in the ratio of crew share to boat share. There has been a reduction in the crew share settlement, but this is not across the board. There are some cases in which there is no change at all; in about half the halibut industry there was no change. In the half where there is quota leasing, the lease payment is figured into the crew settlement, causing a reduction in the crew's share. Also, there has been a reduction in the size of the crew. On average, our research shows that the individual crew share is larger.

Stuart Richey.—In Australia it has made no difference whatsoever. The share on the quota owned by the vessel remains the same. On quota that is leased, the cost of the lease is deducted first, and then the crew share is calculated on the normal basis.

Question #27 (Ellen Pikitch).—My question is directed to Doug Hopkins. The issue is the relationship between ITQ systems and conservation. I want to make my own observations on that, and then ask you again to address the position that Environmental Defense Fund has taken that we should have a few selected tests of the ITQ system, but have a moratorium on ITQ systems until we find out how those tests work out. It seems to me, on the face of it, that human nature says you take better care of your own property than when you rent someone else's. So I think there is a fundamental reason to believe we will have a better conservation ethic with ITQs. The other thing that was pointed out by Phil Major is that we now have a wealth of experience with these systems around the world. We can use experience with those systems to see whether conservation goals have been met or not. Personally, I have seen the Canadian cod fishery, with the enterprise allocation system, adopt the square mesh codends voluntarily. Those are known to reduce the catch of undersized fish; it made economic sense as well as conservation sense. There are other examples that have been given here. As a biologist who is primarily concerned with the conservation aspects, I know that there are instances in which conservation can be compromised. But I think we can identify those conditions and, instead of having a moratorium in ITQ system, follow your other suggestion and set some standards up front and insist that any new management measure put in place is at least as good as the existing management measures.

Doug Hopkins.—A decision by this country to convert any fisheries to ITQ management is a long-term decision that needs to be endorsed by the broader community, most of which is not represented here. Many of them would probably have a hard time understanding much of the discussion that went on here today. That community needs to be involved and brought in. There are plenty of members of Congress that don't understand ITQs. There are communities in New England that are deathly afraid of ITQs for a number of reasons. The message you should take away from the specific proposals I made are not the word "moratorium" but the words "go slow" and "guidelines and criteria." It would be difficult to develop criteria overnight. Councils are working on ITQ plans right now. It would be difficult for them to incorporate any of these guidelines and criteria from NMFS unless they are given some breathing room. It takes so long to get through the allocation battles in any ITQ program that calling for a "time out" shouldn't be threatening to anyone who is supportive of ITQs. For those opposing ITQs, the moratorium proposed by Environmental Defense Fund would appear inadequate.

Question #28 (Arni Thomson).—I just want to give a few brief comments based on the speakers I have heard over the last 3 days. Mike Sissenwine noted that Alaska accounts for 40% of the USA seafood production by value, and has potential for approximately 51% of the value. Barry Kaufmann and Philip Major have noted that we should not delay in setting up ITQ programs, but that we should get it right the first time. George Clement has noted that the USA is the world's center of capitalism, but we are a decade behind in fisheries management. Phil Major has noted that the traditional fisheries management is a travesty for the fish. Finally, I note that there is no entity that is more conspicuously absent at this conference on global trends than the state of Alaska, which controls over 40% of U.S. common property resources and is steadfast in blocking any meaningful analysis of ITQs in the North Pacific Fishery Management Council in favor of promoting the race for fish and the development of an anachronistic license limitation program.

Question #29.—I have a question for James Joseph about the incidental mortality of dolphins. If it has gone down so low, why are some countries still supporting the embargo of tuna? Do they think that the system is not working properly.

Jim Joseph.—Well, I didn't think I would be asked about embargoes here. I just came from Venezuela where I was deluged by questions because they are an embargoed country. What the question is referring to is the 1988 amendment to the Marine Mammal Protection Act, which calls for embargoes of imports from countries that don't conform to some standards set by the United States and which the USA requires of their own fleet. I must admit that the U.S. policy is kind of a moving target. It seems that every time a nation meets the standard that the USA requires, the standard is changed and the countries continue to be embargoed. At one time, there were 25 nations embargoed by the USA, primary embargoes and secondary embargoes, because there was not conformity on the part of producer countries with U.S. law. Dolphin mortality is very low now. As I said, it was 3,600 animals in 1993. That poses no threat whatsoever to the dolphin populations, which number 9.5 million animals. The mortality represents .04 percent of the population. But, nevertheless, the USA has these standards that it goes by. The newest amendments made to the Marine Mammal Protection Act prohibit fishing on some of the primary species and require other things of other nations. In many respects those nations can't meet those particular requirements, and so the embargo is sustained. In my opinion, the USA needs to reevaluate its policy with respect to marine mammals in the eastern Pacific ocean. Some of you won't like what I have to say, but the USA has a double standard with respect to how it treats marine mammals. On the one hand, the new proposed marine mammal legislation will permit U.S. fishers in U.S. waters to continue fishing on marine mammals even though those mammals are threatened or endangered. Whereas it will prohibit nations fishing on the high seas beyond the jurisdiction of the United States on populations that are not endangered or threatened from fishing. I think that has raises problems for many other nations and raises the kind of question that was just asked.

PANEL DISCUSSION 1
Changing Fishery Policy to Address Problems in World Fisheries

Tuesday, June 14, 1994

MODERATOR: EDWARD WOLFE, OCEANTRAWL, INC. SEATTLE

The panelists who participated in the ensuing discussion are as follows:

Dayton Lee Alverson, Natural Resources Consultants, Seattle, Washington

Bart Eaton, Trident Corporation, Seattle, Washington

Richard Gutting, National Fisheries Institute, Washington, D.C.

Joseph Sullivan, Mundt, MacGregor, Happel, Falconer, Zulauf & Hall, Seattle, Washington

Ed Wolfe.—We have an extremely well qualified and diverse set of panelists here today. They will be addressing the following questions: What is the biggest problem in world fisheries that prevents sustainability? And, what is the solution?

Lee Alverson.—What's wrong with fishery management on the global scale? I think that issue was addressed by almost every speaker in some degree this morning. We heard persistently about the issue of overcapitalization; we heard about lack of information and the quality of information. We certainly heard a great deal about uncertainty of the database, and there were hints about capacity to enforce management rules. An able enforcement regime can follow through with a punitive measure and catch the appropriate number of people. All of these ideas came to light this morning. I think the solution that I like the best, because it just seemed the easiest to me, was to turn the fishery over to the foreigners, because you can manage them any way that you want and the political ramifications of putting them out of business or setting a quota is not going to be debated to any degree at all. You've got them where you want them. That being unlikely, I think maybe a fair analysis of these opinions will be subjective in character.

I have the feeling that the biggest problem in terms of global fishery management is lack of political will at the national and international level. We have a whole host of regulatory regimes that everybody has committed to, and everybody has committed to conservation ever since the [Conference on the] Law of the Sea. There is a tremendous amount of literature on how to behave and what the objectives are, but none of us seems to be getting there very effectively. We have heard that there is undue political pressure on the part of the fishing industry, and that may have occurred and still may be occurring, but I still come back to the basic view that the responsibility is vested in national entities or international entities and that they are over-sensitive to such political pressures. The fault lies there, and there is where it needs to be resolved. Lack of political will, I think, is driven by the fact that we have gotten into such an overcapitalized position that disenfranchising a significant sector of the population is just not an easy task for a politician to face up to.

Beyond the overcapitalization issue and the problems it has generated, I would certainly look at a number of other factors (in no order of priority). First, inability to effectively monitor fisheries in many areas of the world, resulting I think in under-logging of catch (the so-called "black fish, gray fish") and all the problems associated with those who take advantage of the system. Second, I think there is a lack of data, certainly on certain stocks in terms of management. I do not consider this to be a significant factor in terms of mismanagement. But I think there is certainly a need for better data on a number of our stocks. Third, the inability to establish in the minds of the user groups that there is an advantage to taking short-term losses in order to achieve the longer term gains seems to be a tremendous obstacle in achieving conservation goals. Many users don't see themselves as the ultimate benefactor of the management process, doubt the fact that we are going to be able to rebuild the stock, and finally that if we do they will be the recipient. I think some manner of more effectively demonstrating that what you're trying to do will lead to more stable, economic viability for users is important. I suppose that with the collapse of the cod fishery in Newfoundland that message probably has gained significant importance in that part of the world.

Bart Eaton.—I can tell you that I haven't learned anything today that goes against what Oscar Dyson told me the first day I got on a boat. He said "Son, the fisherman that knows when to sell his boat, that's the one who makes

301

the most money." I'd also have to say that every time I'm with fisheries managers or a council meeting, whether they are public or private managers, I am reminded of the Jewish philosopher Philo who was from around the first century [AD]. His first claim to fame was that he was the first to articulate the basic difference between philosophy and religion. Basically, philosophy is based on reason, and religion is based on faith. It always seems to me when I am in a room of fisheries managers that there are a lot more preachers than there are philosophers. We really see that in the North Pacific.

To talk about fisheries problems in 5 minutes, you almost have to talk in bumper-sticker-style sentences because that is about all that can stick. That is probably all anyone can remember anyway. I'm not speaking about the problems of the world. I'm not much on giving advice to the world, but I think from what we've gone through, the main problems I've seen might apply in other places. The main problem is the confusion I know that exists in industry between biological possibility and economic probability. This was talked about earlier, but this really creates problems in an industry that tries to cope with fishery management as they try to do what they call "rent seeking"—trying to get ahead of management. It is especially a problem in distinguishing between what I call science and engineering. Now the second bumper sticker is so simplistic I am amazed that most people do not know it: no species can long exceed its carrying capacity in its environment. I will try to link that problem with economic development models: When you don't know where you're at, you can develop some false premises in your development models. I think it's very important that we be aware of that because it is true that the amount we can extract from the ocean is less than some of its parts. As long as we delude ourselves by building these separate models, we're always going to have problems.

The third bumper sticker, based on my experience, is that both state and local governments have been the main goading force toward creating fishery overcapacity. Industry doesn't develop in a vacuum. Industry develops by whatever is in its best interest of that day. When you have fishery loan guarantee programs, construction differential subsidy programs, capital construction funds, and maritime administration subsidies out there, people are going to take advantage of that and they are going to create overcapacity. What is very frustrating in industry is that some of our fishery managers require full use of capacity. Under the Magnuson Act, we had to build more harvest capacity before we could put the foreigners out. We had to build more boats before we could put the foreign joint ventures out. We have [sic] to build two domestic fleets. Or you have to extort capital investment on the beach if you want some concessions on the high

seas. But people in industry at the time are making logical decisions based on that time and place. How to solve it? Really, I don't know how to solve it, but I know any fisheries management should be coordinated with investment tax policy and with development policy. These should be coordinated because otherwise they are going to be at mixed purposes continually. That is what seems to put everything out of balance. In our case, very rarely do fisheries managers even know what programs are in existence for industry, or how they work, or what the implications are going to be on a decision that they make. So if there is one solution I can recommend, it is coordination of national polices so you're headed towards the same goal.

Dick Gutting.—The question that we got was what is "the" biggest problem? What is the single biggest problem that is preventing sustainability? Like Lee, I come to the same conclusions. It is the simple fact that we as human beings profoundly disagree with each other over the objectives of fishery management and why we are spending all of this money studying fish and going through hearing processes. I think this political problem, if you will, accounts for the decision loops that we saw described this morning and the paralysis in decision making. The evidence of this political problem is all too apparent. If you look at the way the initial allocation of individual fishing quota evolves, you will see the political problem in spades. You will see the disagreement over objectives in the discussion of community development quotas. The notion that somehow fisheries should be involved in social engineering and social reform. You will see this in the course of day-to-day technical attacks on fisheries biologists and in the need for more science, and in the kind of discussions about risks that we have had here.

I think that we are also beginning to see this political problem in the evolution of ocean use planning or zoning. The emergence of sanctuaries in ocean area management, which is replacing the resource management mechanism that we are all familiar with. The fact of the matter is that states and local governments and international agencies are beginning to set aside large areas of the ocean as sanctuaries. The root problem of all of this is a profound disagreement on what we want our fisheries to produce, on what the objective of management is.

I have to commend the scientists this morning. There were recognitions in passing that this was a problem. You heard phrases like "Fishery management is all about managing people, not resources." We heard references to political decisions and economic and social issues. But I have to say that I am disappointed in the scientific community, because it seems to separate itself from the political. It sort of doesn't want to get tainted. My solution is to bring on the psychiatrists, the anthropologists, and the political scientists, and to begin to study this political problem. I think we have a great opportunity to learn more about the nature of this problem. We have had 15 years of experience in the Magnuson Act and there have been success stories and there have been failures. What there hasn't been is an analysis, a scientific study of why things worked some places and didn't work in other places. We like to study fish and not study human beings. I think we need to get on with the task and address the political question head on. We do so indirectly every day in fishery management councils, Congress is touching on it a little bit, but from a scientific standpoint, I think there is a great opportunity to learn over the last 15 years.

My final addition is that in thinking about the political problem, we are well tooled and well equipped in our current fishery management structure to maintain stocks and keep stocks on a sustainable level. What we are not prepared politically to do, with the current management structure, is rebuild stocks. The evidence of that is when it comes to rebuilding everyone rushes to Congress for federal assistance or a bailout or a political decision. The current management structure is not strong enough, it doesn't have the tools. I think we need to think about rebuilding stocks and whether our fishery management system is strong enough for that.

Ed Wolfe.—Maybe we can ask Dick if the National Fisheries Institute were to fund this analysis that he is talking about, would that be helpful. Is it possible, Dick?

Dick Gutting.—Yes, we fund it every day with a lot of political pain. I think the industry would be willing to join with the academic community and with the government in setting up some honest guide to study what happened and why it happened, and I think it would be useful.

Joseph Sullivan.—I am going to strike off into entirely new territory here in terms of identifying what I think the biggest problems are concerning sustainability. I think the political tension is unresolved in fishery management process. Here, obviously, I am not in all new territory. I am right in sync with the viewpoint that Lee Alverson and Dick Gutting bring to the table. Nonetheless, I'd like to run through what I think are some essential themes in these issues. We have the tools for fixing specific management problems, such as overcapitalization and more sophisticated science. We're addressing habitat and ecosystem issues, but yet we're having an extremely difficult time employing those tools to solve the perceived problems. We understand, for example, that the "tragedy of the commons" is short-hand for saying that it is rational for individuals to overexploit a common property resource. We also understand that individual transferable quotas address that problem, perhaps as well as any possible fix could, yet we find it tremendously difficult to implement any sort of ITQ system, under any circumstances. I think the

tension stems from the diversity of interests and values that seek to be addressed in dealing with what we more and more consider to be a unitary resource complex. The allocation of wealth, as Francis Christy mentioned, is a political process to which values are brought and not from which the values are derived.

At the international level, I think the developing nations are hungry for capital, they're seeking at both local and national levels to bootstrap their economies, they're making an economic transition from agricultural and agrarian-based society that no longer supports their population to a more urban and to some extent industrial approach. Consequently, their national and local incentives are maybe to maximize their short-term yields rather than longer-term sustained yield. We see social dynamics in the international sphere that include protection of certain social sectors or that place high value on capturing control over certain types of highly valued resources, which lead to subsidies for industry sectors that may be either fully or even overcapitalized. On the U.S. national level, I think the tension might be most clearly demonstrated by the fact that we have eight regional councils that are applying a single set of national standards. I think, if it were intended that the standards be applied consistently on a national basis, we would have one council. No, we prefer to have this tension between what we consider a national policy and a reflection of local interests, local dynamics.

We have tensions within the national standards; I think the Magnuson Act national standards tell us that "optimum yield" is maximum sustained yield modified by appropriate economic, social, and environmental factors. These are often preceded as qualifiers that allow us to take into account local impacts of significance in the process of looking at what should be a national, maximum sustained yield. National standard number four says that there should be no discrimination between states, but how does one balance that against the incorporation of what are often local concerns in the economic, social, and environmental area? The tension, I think, has been evident throughout some very contentious North Pacific fishery management decisions, such as the inshore/offshore allocation of pollock quotas, which involved a state-of-the-art battle between experts, bolstering both sides with science being used essentially to drive what was more a political conflict. It brings to mind something that I found at the council meeting last week. This is a quote from the man to my right (Lee Alverson): "The human dimension in resource exploitation and management has become increasingly pervasive, and the use of facts and statistics in the public influence game has become an art to which science seemingly takes a back seat."

This political tension can be summarized as attempting to manage a natural resource in the best interest of the nation as a whole and a with concern for local characteristics of individual fisheries and local economics—to make sure they are not marginalized in the national management process. I think there are several real themes that are of some interest, themes that emerge when we look at how this tension has played itself out in the Pacific and North Pacific fisheries under U.S. management. I think one of the themes is a federalism conflict: national interest verses local economic development. This is the inshore/offshore conflict. I think in some respects it is the Community Development Quota program. Given that Alaska holds 6 of the 11 votes in the North Pacific Fisheries Management Council, it perceives that it has a license basically to manage the resource to benefit the state's interests. This political debate I don't think is going to be resolved easily. It runs back to the federalist debates, and I think that when we finally reach a conclusion on those we will reach a conclusion on how to balance the conflicting interests in this area.

Another theme that emerges is the conflict between artisanal and industrial fisheries. For example, look at the salmon bycatch issue in the midwater pollock fishery. The midwater pollock in the North Pacific is, if not the cleanest, certainly one of the cleanest fisheries in the world. However, what little bycatch it does have of salmon is tremendously controversial and the North Pacific Fishery Management Council, for about a year now, has hovered about the possibility of taking some very significant management actions that could impose tremendous costs on that midwater pollock fishery. The Alaska halibut and sablefish ITQ program and the Pacific coast limited entry permit formulas reflect an effort to provide some recognition to the importance of the local interests. The tensions are also connected to the efficiencies of operating scale for exploitation and management of certain resources. Pollock are processed in a large-scale, industrialized fishery, which is probably the most efficiently managed at that scale in terms of, for example, observation and enforcement. The fish catchers are small in number and large in size. There is tension and interaction between those large-scale fisheries and artisanal fisheries.

I think global competitiveness and national benefit have suggested to us that we should manage for efficiency, but local values such as employment, traditional community structures, and stability have been in the balance consistently. I think we also have what I would call a traditional values issue. I think that in our attempts to maximize for national benefit we often run into issues that are contentious such as economic discards, or bycatch. Assuming that the bycatch is accounted for as biological removal, there is no indication that the affected species is endangered by it. These become very significant factors in our concerns. I think those factors assume

importance because of what I will call a nostalgia for the artisanal as opposed for an actual artisanal value. In other words we are seeking to apply what we would consider to be subsistence values—such as "waste not, want not" or "keep what you catch"—in a context where from a market or economic perspective it is of questionable legitimacy to do that. It may be counterproductive if we increase biological removals or if we concentrate waste discharges in the process. So where does this put us?

Well, I have one recommendation in connection to the process. We need to recognize that politics is part and parcel of the process; it is foundational to the ability to resolve political tension and will often be foundational to the ability to implement an appropriate management regime. That leads us to a recommendation that we enter into and encourage dialogue between the significant players on either side of these tensions before we bring the issue of contest to the fishery management sphere. I think it is very important that the political dynamics be recognized and sorted out at that level rather than clothed in science and dealt with in a context where experts are put forward to provide a great deal of information, that ostensibly has to do with the fishery management implications of a decision, when the concern is actually the political ramifications of the decision. I think there have been some success stories when that has been done. I think the herring bycatch issue and the Bering Sea trawl fishery is in some respects a model for what I am suggesting. There is an opportunity for the fleet and the affected, or believed to be affected, fishers to get together and talk through these issues in advance of the issue being dealt with at the management level. A framework on which there is some consensus will lead to efficient and usually non-controversial implementation of the management regime. Attempting to deal with that contention within the management process doesn't seem to work well.

Ed Wolfe.—The panel today, as you know, was to be oriented towards global fisheries problems. But I would suggest to you, as Tip O'Neil said, "all politics is local." And when you get down to fisheries, most of our problems are local although we are dealing with them in a global arena with markets and we are fishing around the world. It all gets down to our regional areas, where we happen to be from. What I will do is comment on some remarks that were made and direct them to individual analysts or ask them to jump in, and then I will invite the audience to ask us any questions that you may have.

Getting back to the international issue, I think Lee made an excellent point, and I couldn't agree with him more. I assume he has been talking about the U.S.–Canada salmon issue, but if not I think that it is a pretty good example where there appears to be a lack of a political will to solve this problem. I would also suggest to you that there has been a lack of political will for a long time to resolve this problem for political reasons. Lee and I were involved in this process back in the mid- and early 1980s. We got a treaty, and the treaty has been going along and we have been putting Band-Aids on this treaty for a number of years. If you read the papers, you will know that there are major problems out there right now. I think for this issue to be resolved, there will have to be a very high level of intervention, and I think that is what will occur because of the overall bilateral relationship with Canada at this point. I would be very interested to hear Lee's point of view since he has been involved in this for over a decade.

Lee Alverson.—I had intended to try and deal a little more internationally, but I think there is an analog in the international setting for what is going on here. The North Pacific Salmon Treaty has a certain number of obligations that relate to both conservation and allocation. As you might suspect, each side perceives a different reason for why the treaty has broken down. One side concentrates largely on the allocation or equity issue—that each country of origin should receive salmon from its coastal waters where it spawns. The other side sees a diminishing number of resources that migrate up off Canada, probably not because the Canadians have been involved in intercepting those fish, but for a number of habitat reasons. The stocks are declining and we want to see a decrease in the interception. The fact that two sides can't get together seems to be based on the allocation and equity structure, and I guess it would be ungodly of me to take sides in this because I might have to switch citizenship. I would just say I do think it is an issue where the political will to get a resolution has to go beyond the local parties, which cannot find resolution because of their commitments to their local constituents. That sort of issue confronts people all over the world in terms of overcapitalization. Governments don't want to have to deal with the issue because it means disenfranchising people, economic difficulties for people, and there are trade-offs between achieving that disenfranchising. How do you protect all interests of the people who participated in the fishery? But I think we have a very strong commitment all over the world to the basic statutes of conservation as written under Law of the Sea Treaty, but we have a very frail commitment on the part of national leaders to really try to seek it.

Ed Wolfe.—I would just add to what Lee said that, during that period—during the negotiation and the final signing, having been there—we knew that this was a temporary fix. It was a Band-Aid at the time, and the Band-Aid stayed on a lot longer than we thought it would. It will be resolved shortly, and I think it will be resolved at the highest level, the Presidential level in our country.

The other point that I thought was well made today has to do with how we deal with fisheries in our country, and I will relate that to international issues. There are substantive problems that we deal with on a daily basis in fisheries. Obviously, there are legal problems, scientific problems, and policy problems that we deal with on a daily basis. But the people that I am looking at around the room—scientists and lawyers and policy people—would all agree with me, I think, that these very important scientific issues that are taught here and in other schools and the policy issues, they all come down to politics. That may sound terrible, and I don't even feel comfortable to say it, but I think it is a reality when I look back and I think about the international issues that some of us have been involved in: the salmon issue, the driftnet issue, the tuna issue in the South Pacific, and shrimp issues in Mexico. These are international issues. They all came down to regional pressures and regional interests. That is how they were resolved, with national assistance moving it forward.

Dick Gutting.—I think Bill Burke said something that we shouldn't lose sight of, because he's dead right. "We have a power vacuum. There is a hole in fisheries management authority beyond 200 miles and when you have a vacuum it won't stay there very long. People are going to move into that vacuum." Bill talked about coastal state expansion beyond 200 miles—unilateralism. There is another unilateralism going on to try to fill that vacuum, and that is using market power, using trade sanctions, using the strength of your import market to force other government agencies to do something out there in that vacuum. The fact is we are running out of fish, the demand is going up, that vacuum is becoming more and more serious, and we have to deal with it. I think resolving it at the international level is by far the biggest challenge that we have, and has got to be the focal point of our efforts right now.

Ed Wolfe.—My final point before we move onto questions would be just to pose a question to you. Is the problem we are talking about, the resource problem, an economic problem or is it a scientific problem or is it both? From our perspective in the Pacific Northwest it is a regional problem in that there are just too many boats, and I think this is an international problem also— there are too many boats in the fishery. We need to limit the number of vessels that are operating. I think our scientists have done a very good job of managing the science. For the most part, our stocks in this part of the world are sound and probably as healthy, if not healthier than any other stocks in the world. And we are proud of that. So I would promote limiting the number of vessels. That will go a long way towards resolving many of the problems that we have in the Pacific Northwest, hopefully through some sort of comprehensive rationalization program, whatever that may be. But I think it will have some limitations on vessels within our fishery in this part of the world, and it is a regional problem, yet it is an international problem and it effects everybody in other fisheries around the world.

Lee Alverson.—I would just like to make one final comment. Everybody today talked about the problems facing world fisheries. And there have been a number of items that have been brought to everyone's attention. The people talk about the lack of data, the lack of enforcement capability, overcapitalization, a half-dozen or more additional items. Yet I think Dick Gutting is right in the fact that we are trying to write a prescription when nobody has sat down and done a detailed diagnostic look at success and failures around the world. I think before we do that, it would probably be nice to have someone sit down and say why are some fisheries in an overfished position, and why are some other fisheries essentially well managed and productive over a long time period?

Ed Wolfe.—That concludes our presentation. We would be happy to entertain and attempt to answer questions from the audience.

Question #1.—I might have missed it this morning, but I was just wondering if there is a global estimate on the extent of overcapitalization?

Francis Christy.—We did make an estimate in the FAO analysis indicating that the total global revenues from fishing were $70 billion in 1989. The total operating cost of the global fishing fleet was $92 billion and the total capital in operating costs combined was $124 billion dollars. So this in essence shows overcapitalization. Beyond that, there is an estimate of the economic rent that is being dissipated of $30 billion. So you have an estimate maybe on the order of $60 billion dollars a year being wasted globally because of overcapitalization.

Now I would like to take this opportunity to address another point. I found the discussion this afternoon to be very refreshing. I think we are talking about the question of political will. This morning I tried to point out the reason we don't have the management is because we haven't made the political decisions. It is not so much a matter of political will as it is a matter of making decisions to allocate wealth, or redistribute wealth. That has to be done, and as long as we have the open-access condition, this can only be done by arbitrary means. We don't have this in any other industry to speak of. In other industries, the access right is allocated through the market mechanism; this is what we should be striving for. To get to that point, we have to make the initial decision on allocating access rights through whatever system we use, whether it is an ITQ system or entry limits scheme or territory use right or whatever. That is the basic impediment to adopting a better management measure.

However, once those decisions have been made, then

the role of government and the role of scientist change considerably. Then you have created, on one hand, a real demand for a lot of the information because scientific information can be used. At the moment, the scientific information to some extent is useless. You can come up with the best models, but if you can't make the decisions to manage the fishery to control access, it is of no real value to mankind. It may be of value to the fish, but it is of no real value to mankind. Once you have established a real system where property rights exist instead of the open access condition, then that information will appreciate very greatly in value. I am reminded of the situation in the northern Australian prawn fishery where they established a license limit scheme, and the fishers found it in their own interest to postpone their harvest to catch the larger prawns rather than racing because they had an exclusive right. They also found it in their best interest to employ their own scientists to tell them when to move from one stock to another stock. What I am getting at is I think there are two phases. The first phase is the political decision that has to be made to distribute the wealth through some sort of closed-access system. Once that decision is made and the property rights are created, then the allocation decisions are made automatically through the market place.

Dick Gutting.—Thank you. I have to complain here. All of us have been reading reports about the ITQs and, before that, limited entry. We are well aware of the grand scheme and the beauty of the concept. What I am appealing for is a little understanding that the initial allocation of quotas is not just a mere problem that one has to work through and the whole system comes into play. That initial allocation of resources is why we are spending 10 or 15 years trying to work out the system. If you believe in these systems, as academics, if you really think that we will optimize our benefits, then my appeal to you is help us get through that initial decision. That is the killer decision. That is where the bottleneck is. That is where I don't quite frankly see the professional academic focus. How do you get 100 people or 1,000 people through that bottleneck? And how has it been done in XYZ fishery, and why hasn't it been done in other areas? I have large holes in my back as a result of participating in the process. I don't like holes in my back. Please give us some insight. Give us some real examples that we can point to and say, "See they did it here or they didn't do it there."

Bart Eaton.—I just wanted to address overcapitalization. I don't think that overcapitalization is a problem. The capital is going away. The bankruptcy judges are taking it away every day. The fishing power is staying. We are going to have a much lower capitalized fleet here in 3 or 4 years, but we are still going to have the same amount of power.

Joe Sullivan.—One of the points I would like to make echoes the same themes that we have been hearing. One of the difficulties I see is that we have a fisheries management process that develops our policy (in the United States context anyhow) and provides that policy to the management administration and then asks them to implement it. We are asking that process to address the first phase of quota allocation. That is a political issue. In my mind it is questionable whether it should even be addressed within the fishery management policy development process, whether it is efficient to do so, or maybe that process should take place outside of the context of fishery management decision making. Rather, should this be an issue addressed at the national political level or inter-local political level, resolved there, and then to some extent brought forward from there to the management process. Then from that point, the management plan can be implemented.

Question #2 (Bert Allsopp).—Sorry to intrude, but I come from a country where we have been frustrated by the advice of all the developed nations who have been suggesting maximum sustainable yield. But in the North Atlantic or the North Pacific or in the Indian Ocean, I've seen the stocks collapse. I have worked with the United Nations Food and Agricultural Organization, and I have seen the advice being transferred about things that you advocate and that have failed. I have seen places like Iceland tenaciously hold to the management principles without any interruptions or political accommodation from various parts of the country, locally or otherwise, and they have succeeded. I have worked in countries where they have maintained that they must manage their spiny lobster stocks in Belize, while you have been vacillating about it in Florida, and lobster stocks have been vacillating in other countries. I have seen places where the fish bycatch has been overexploited and these issues have never been addressed. I've seen the South Pacific Fisheries Commission tenaciously hold to the management of their tuna stocks, Seychelles holding to their management, and the Maldives, and so forth.

You lack the political will. The academics have been unable to influence the politicians. There have been changes in administrations either in states or in provinces in Canada, which have not coincided with the breeding cycles of the salmon, or the cod, or whatever. In fact, the issue is, "Has the developed world taught the developing world anything about fisheries management?" This morning you have had the issue presented by Daniel Pauley, indicating very clearly, lucidly, that in fact when you overcapitalize, you lose money, and in the artisanal fisheries social and economic gains (as food brought in) are greater per person. You've seen the maximum bycatch discards, etc., which do not end up being food for people. To the question addressed by the panel "What is

the biggest problem in world fisheries sustainability?" my answer is greed. Thank you.

Dick Gutting.—I think that there's plenty of greed out there, but I also think that we ask our fisheries to do many different things. It may not be a major issue in the Pacific Northwest, but I assure you in the Gulf of Mexico and along the South Atlantic, trying to balance recreation opportunity with food, you can call it greed if you will. Still, we're asking our fisheries to produce both recreation and food. That's a two-edged demand which is often inconsistent because one demands efficiency and the other demands inefficiency. We demand that our fisheries provide employment at the same time we demand that they provide food, and again, there are these conflicts. I think that "greed" has a nice ring to it, but I think that it's a little more complicated than that.

Lee Alverson.—I'd like to jump in and comment just a little on the greed issue, because I think that it's a very easy thing to say, and everybody raises their hand and you get the audience to clap. Fact: Greed probably exists in all sectors of society, whether it's fisheries, whether it's industry, whether it's academia. A lot of rich people are sitting on top of good contracts to be consultants to high-level exploiters. Society has to develop the boundaries in which people can operate, and those are the boundaries that control greed. And in fisheries, if greed is out of line, the social structures and institutional arrangements need to be refurbished.

Joe Sullivan.—I'd like to respond by saying that I think that there is indeed an appropriate role for academicians in the process. But I guess what I see is that currently the step in the process in which academicians are typically brought in may be inappropriate. The point that I was trying to make in my initial presentation is that I think that there are certain basic political tensions that will condition how an analysis is undertaken and what types of recommendations will stem from that analysis. Political issues of a federalism sort determine whether you have a certain amount of local control, local employment, local economic development given a priority, or whether you emphasize national efficiency. Those decisions, I think, really are value choices that need to be made as part of the political process; to ask a scientist to assign relative weight to those values I think asks them to do something that they're not well equipped to do. But when you get to the point where some of the basic premises have been addressed and where some of the basic priorities have been established, I think it makes perfect sense in the context of negotiating a resolution between conflicting user groups to bring in people with academic expertise who can measure value, establish value, can show how value can be traded or maximized in that process. I think that can be of great assistance in resolving some of the second-level tensions associated with the impact of decisions. But it is a problem, I think, to try to employ academics too early in the process, which really frustrates [sic] both the process and the academic.

Ed Wolfe.—Maybe just one final point by the panelists. I would also add that the services this University, and I'm sure other universities in other regions, provide to the regional fishery body, the regional councils, on the scientific committees and on the advisory panels is a very valuable resource.

Dick Gutting.—I have two suggestions: Maintain the discipline that is evident here, but which was not evident 15 years ago. The discipline of distinguishing political decisions from scientific ones. I think that a number of the scientists came forward and emphasized the importance of keeping clear the difference between those two. That was not happening in the beginning of the Magnuson Act, and I'm delighted to see the scientific community beginning to come forward on that. The second suggestion I have is, you can think of the fishery management process as a human process, a group of individuals trying to resolve conflicts. When we have taxes and it's complicated, we go to an expert. I think all of us in the process think we're experts in the process and we're not. I really think that there is a role in the academic community to study the system, the process, conflict resolution, and draw upon success stories, perhaps in other resource management areas, and bring new ideas as to how we can structure conflict resolution in fishery management. I think that other agencies and other programs are going through conflict resolution processes that hold a lot of promise.

Lee Alverson.—Well, I want to help Ellen where I can. To begin, I didn't hear any statements that I thought were implying that the academic institutions or the academicians did not have a significant role. I think the point being made is that the decision-making process in fisheries, both nationally and on a global scale, is driven significantly by a political process. And certainly, academia can and is frequently a part of the political process. They certainly can be the arbitrator of facts. They have an important role in developing new concepts and methodology, looking at the consequences of our activities both in the social, scientific, and economic arena. They've done that in the past. I think that they need to continue that process. Certainly one thing, and I've made this comment before, is that the news media today is full of absolute nonsense about what's going on in fisheries. Factually incorrect consistently. Certainly academia can play a role in sorting out the factual basis of what I consider to be misleading the public continually in what's going on in fisheries. So I think they have a very important role in developing new hypotheses and ways to manage—looking at better ways to analyze data, to collect and interpret the facts quicker, and I think that they have and continue to have a significant role in that arena.

Question #3 (Brian Pierce).—I am speaking from a non-academic background, if you like. Firstly, worldwide when I've traveled around, I have yet to meet anybody that doesn't support wise stewardship of fish resources. And yet on the ground, in reality, when I present biological advice, I lose about 80% of my battles. I can usually predict that when somebody's got a vested interested and can make money out of doing the opposite of whatever we biologically suggest is best for society and for its resource, I tend to lose. I've heard from Dick that we should do some research on that, and I agree it's a really interesting series of experiments that are out there that we've actually performed ourselves. But I don't see that research in itself has actually solved a lot of problems worldwide; it still returns to the political issue that you guys raised. My question, my reason for being here, is to ask you guys, how managers and how the real resource owners can have a more effective input into the political process, because certainly the folks who can make a few thousand dollars out of the resource are exceptionally good at it.

Dick Gutting.—I totally agree, research is not going to get us there; it may help. When you devote your life to fisheries, and I've been in this game for 25 years, you tend to think that we're damn important. You forget that we're a very small percentage of the GNP. I think you've heard some of the panelists suggest that maybe within the fishery community, because of the self-interest, whether we're an elected official sitting in the middle of a fishing community, or someone in that community itself, we may lack the intestinal fortitude to make the kind of decisions that you suggest are needed. That means that our problems or our issues have to be elevated out of our own little backyard and onto a larger stage. I think that was the suggestion on salmon, and I think that it applies in other areas as well. How one elevates decisions out of the fishing community, when the fishing community isn't always too pleased that that occurs, is a difficult question. I think a lot has to do with what Lee Alverson just said, and that's speaking out at a national, educational level, and responding when you see the front page of Time magazine. Getting the word out, as best you can, to a broader constituency is the only answer I can think of. I think the environmental community in the last 2 or 3 years has begun to serve that function; however, like any other special interest group, they have an interest which I don't think is necessarily coextensive with the public's interest. They need to raise money, they need to have certain problems and crises, and they present things a certain way to raise money. I think public education is a function that you can perform.

Question #4.—It's not really a question, but, frankly speaking, coming from Europe, I must indicate that I don't feel very easy in the discussion we had the second half of this afternoon. First point: If this is world fisheries, I do not recognize myself in a world debate, frankly speaking. I don't have the feeling that the question is to be for or against ITQs. It seems to me a very American debate now. I don't want to enter this debate, but I want to make it clear that it's not a world-scale debate, up to now. I would like also to add a comment on the political will. I am involved now in some political decision, and I am very scared by this idea of the lack of political will. What does it mean? I am afraid that many scientists have kind of a dream of the enlightened despot. They would know what is good to decide. A good politician will be a politician who will decide exactly what was advised. I do follow the point of view according to which scientists should try to understand what went wrong in the past. Because if the political will did not appear, maybe it's also because the parts have not been clearly defined between the scientist and the politician. If we are all of us within democracies, or most of us, if the political will doesn't exist, it's maybe because the debate was not organized in the proper way. And it's also part of the responsibility of the scientific community.

Ed Wolfe.—Thank you. I thank the panelists. I thank Dan Huppert and Ellen Pikitch for inviting us, and I thank you for being attentive this afternoon. Good day.

PANEL DISCUSSION 2
Managing for Sustainability

Wednesday, June 15, 1994

MODERATOR: CLARENCE PAUTZKE, EXECUTIVE DIRECTOR,
NORTH PACIFIC FISHERY MANAGEMENT COUNCIL, ANCHORAGE, ALASKA

The following panelists participated in this discussion:

Suzanne Iudicello, Vice President for Programs and General Counsel, Center for Marine Conservation, Washington, D.C.

Don McCaughran, Executive Director, International Pacific Halibut Commission, Seattle, Washington

Ole Mathisen, Professor, School of Fisheries & Ocean Studies, University of Alaska, Juneau, Alaska

Marc Miller, Professor, School of Marine Affairs, University of Washington, Seattle, Washington

Jake Rice, Head of the Marine Fish Division, Pacific Biological Station, Nanaimo, British Columbia

Clarence Pautzke.—Two questions have been given to our panelists: How should we manage fisheries given the change in ocean environment? And how should we regulate artificial enhancement programs for long-term sustainability? I put those together into the question: When we've managed all the stocks down to their lowest sustainability, will the aquaculturist be there to bail us out? I think we have five good panelists here who have been through varied backgrounds and experiences as far as different fisheries. The first one is Suzanne Iudicello.

Suzanne Iudicello.—Good afternoon. The first thing I'd like to do is answer the second question with a very brief, U.S.-specific response. How should we regulate artificial enhancement programs for long-term sustainability? I would suggest that if those federal permitting agencies who are called upon to allow artificial enhancement programs within U.S. waters would simply comply with the National Environmental Policy Act, the proponents of such facilities would have more predictability on alternatives and consequences and costs in their operations. The public would have more information on consequences to the environment and human health; and the decision makers would be better informed in terms of other environmental consequences and off-shoots, and we would probably have a better regulatory regime. So that's a very simple answer.

The first question is the one that's more fun, and I would like to use my time to respond to it. How should

we manage fisheries given a changing ocean environment? I think the second part of question is what is most important. There's kind of a fatalist assumption there: "given a changing ocean environment." It brings to mind the saying that you can't see the forest for the trees. And if the practice of not seeing the forest for the trees has brought us to our present state of affairs in the Northwest logging industry, then our inability to see the ocean for the fish, because we see fish as a product, I would suggest has brought us to the present state of affairs in our global fisheries management. So, what do we need to do to recognize that second element in the equation, recognizing a changing ocean environment? How is the ocean changing? Global climate change has brought about phenomena from coral bleaching to sea-level rise to pollution. We've heard about it in the context of aquaculture facilities and in the context of salmon habitat. It has been predicted that by the year 2010, three-fourths of the population of the United States will live within 1 hour's drive from the coast. What is that going to do to alter the physical capacity of the ocean? Transportation of alien species, they come in ballast water tanks, they come in the bottom of vessels, they're moved from ocean to ocean, ocean to river, river to lake, to inland places. What today is a zebra mussel problem in the Great Lakes in the United States may tomorrow be, who knows, one of these shrimp-borne diseases that might be transported. Further, bycatch and incidental catch in directed fisheries. Finally and probably most important, every single one of these human-caused changes in the ocean environment is driven by the most important human consideration to the ocean, and that is population growth.

So, all of those are human-caused changes, but only one of those things is even susceptible to solutions that can be derived from fisheries management. So then we get back to the first part of the question: How should we manage fisheries? Well, given that whole list of threats, it's a pretty dismal picture. You're talking about the tiniest portion of the equation, where you're managing human behavior in one particular activity, but you've got all this other human activity over which fishery managers have little if any control. So what do we do? Well, let's try looking at the forest. Let's look at the whole system; let's look at the ocean. Let's look at fish not just as a product but as a predator, as prey, as the object of recreation, as a source of tourism, maybe even as a source of aesthetic inspiration for some people. Let's not forget the human element of the system, not just those who make their living from catching fish, but those who affect the ocean: the people who are moving to the coast, the people who eat fish, the people who want an ocean view. So what do we do about those people? We have to engage them in the process.

Right now, most of fishery management and certainly most of the fishing industry view the public either as just consumers or folks like those who I represent, the "greenies" (we like to call ourselves the "blue-ies" because we deal with the ocean), but the greenies and the environmentalists are who are causing you all these problems. Well, why don't you bring us into the process? Because openness, whether it's transparency in the international context for non-government organizations or participation in the regional councils in the United States, makes us an informed public who can help to advance the case to fix some of those other human-caused problems, such as habitat degradation, pollution, and so forth. You can't place the entire burden of those issues on people who catch fish for a living. So, in conclusion, I would say that if we are going to manage fisheries sustainably in the face of a changing ocean environment, you have to look at the public not just as consumers, and not just as targets for new markets, but as a potential informed group of opinion leaders, of advocates for sustainable fishing in the future. Because with that constituent group, we could possibly build the political will today to conserve fish for the future. And then we wouldn't have to spend our marketing dollars and our fishery research dollars figuring out ways to convince people that they really want to eat mackerel hot-dogs and krill patties. Instead, if we had fish for the future, they could be eating the haddock and the salmon and the swordfish that they really want to eat. Thank you.

Ole Mathisen.—At one point of my education, I was made to read the dialogues of Plato. I do not remember much except his discussion of the ideal world and the real world. The ideal world is universal and [has] changeless concepts. And it was a shadow or a projection of the real world. I'd like to submit that the management decisions today should be more experimental in nature. There can be controlled experiments, instead of all this presenting of indisputable knowledge. And, of course, the reason for saying this is that we have the annual variability, and then we have the long-term changes in productivity. Furthermore, when I look at the decline of so many stocks of fish all over the world, I sit back and want to manage for sustainability. You have to introduce

the concept of threshold limits. In other words, the population size below which you never go. And again, in order to do so, you need to be able to assess the stock. Then of course, you have this question of what is the upper limit of the harvest level. Well, again, I know that you can calculate many yields; I do not think there are overestimates until you mix it with population genetics, which is happening these days.

Furthermore, I have never had much sympathy for people talking about too large a spawning biomass, or overescapement of salmon. As we are all getting to know more about the behavior of these animals, we can see that large, excessive escapements of a large spawning biomass has a function. It might be to fill out niches which are lacking. However, aside from this, there are long-term changes and they can come rather abruptly, as witnessed in the late 1940s, when the salmon populations on both sides of the North Pacific Ocean declined. It came too quickly for the industry to adjust, and it caused a drastic economic restructuring of the industry. Now, I do think that we are on the verge of being able to predict some of these long-term changes, using indicators like earth surface temperature anomalies, the atmospheric circulation index, and the earth rotation velocity index. They have all shown close relationships to long-term changes in the abundance of salmon, herring, pollock, and even marine mammals. Stability, or sustainability in the future in the capture fishing industry will, to a large extent, hinge on the extent [management] can make predictions based on these recommended studies in this area.

Now, you have the second question: How should we regulate the fishing enhancement programs? We are told not to be parochial, but let me say that at least in Alaska, the justification for nonprofit hatcheries was to buffer the low years, to fill out the valleys. And of course there are simple fallacies here. You don't build a large hatchery or construction without using it every year. Second, although there may be some advantages in the freshwater stage, in the long run the production from these hatcheries has shown a tendency to follow the production fluctuation of the wild stocks. What probably happens is their initial gain achieved during the freshwater stage is lost by predation, accumulation of predation at time of release of fry that have been hand-fed up to this point. Of course there is another side, and that is the loss of genetic diversity, which in the Northern Hemisphere has been acquired during the last 12,000–15,000 years, since the last glacier period. It's easy to say that we must conserve genetic diversity; it's much more difficult to see how it can be done. In a large-scale hatchery operation, it's very difficult to maintain your genetic diversity. Sustainable harvest is ultimately linked to increased knowledge, and I'd like to submit that the knowledge has to come from directed questions. Eventually these directed questions will lead to basic research. I believe that institutions like the School of Fisheries here in Washington and elsewhere were created for this purpose. They can only justify their existence on this ground. Thank you.

Clarence Pautzke.—Next, we have Jake Rice who is now from British Columbia, but who also had experience off Newfoundland with the cod fisheries. He tells me that the cod are not really gone from the east coast, they're just on their way around to the west coast.

Jake Rice.—I'm also only going to deal with the second question. Five minutes isn't a lot of time, and I don't know much about the first question. I don't know much about aquaculture, so I'll take this opportunity to pontificate about things I do know about. Now, to me, looking at how we mange fisheries in a changing environment, or any environment at all, there's five points we need to really focus on. A number of them came out of the talks we've had in the last couple of days. If we put those six points together, I think we'll find a guide to how we manage fisheries in a changing environment.

First, we need to focus on forecasting. Lots of people in this room have contributed to the really rich fisheries literature in backwards-looking types of assessment methods. Forecasting is a mathematical discipline, and not enough of us, and I include myself in this group, pay enough attention to forecasting as a tool. We're going to have to do a lot more forecasting as we try to deal more realistically with evaluating risk, dealing with the possibility that the environment is going to be different in the near future, whether it's a point regime shift, a gradient, whatever. So forecasting is one thing we really need to deal with seriously.

Another important thing we need to consider is inertia. And I mean inertia in a lot of ways. Fortunately, although the environment is changing, it does have some inertia. Today isn't a perfect guide for tomorrow, but it's a better guide than lots of other things we could be using. And in this forecasting sense, the inertia that the ecosystem system has is a real ally. We need to use it. But there's another kind of inertia, and that's the decision-making inertia. Even after you've diagnosed that something has changed in the ecosystem, and some action is necessary, there's an awful lot of institutional inertia that makes fisheries management systems respond way too slowly to the need for change. And we have to deal with that institutional inertia effectively. This is something different than the lack of political will. It's the real reluctance we have to respond quickly to changes. And that's an important thing to deal with when we're trying to manage fisheries in a changing environment. Postponing reaction to a change can really increase the pain you have to take when you finally do react.

Next, regarding data, I firmly believe that modest amounts of really reliable, really relevant data are much more valuable than large amounts of amorphous data of questionable reliability. Fisheries in general and ecosystem-oriented research is notorious for collecting very large but inconclusive data sets. We have to cure ourselves of that very attractive fallacy if we're going to manage fisheries in a changing environment. I firmly believe, and we can discuss this later, that even if you choose less than the ideal thing to measure, measuring a few things well is better than measuring many things poorly. Also under data, I'll touch on the infatuation that I personally have with databased, rather than model-based ways to analyze forecasts. But I couldn't cover that topic in 5 minutes here.

Regarding diagnostics, I believe we have something to learn from medicine. Medicine for hundreds of years had a pretty good track record; long before it understood the full pathology of a disease. Important diseases had symptoms that could be recognized and there were treatments that could be applied for the benefit of the patient. This is not a perfect model for fisheries management, but it's a tool we don't use. There's a paper that I was very impressed with by Chris Hopkins on the Barents Sea capelin, where there is a measure that can be taken that can really highlight when the Barents Sea capelin stock is in trouble. And it can give you that information in time to react to it. I have more faith in that as a guide to management in a changing environment than much of the work done on the huge Barents Sea ecosystem modeling activity that's going on. And I think that there's lots of opportunities to look for diagnostics of change and react to them, without having to do a full ecosystem system study before we get into the issue of managing the ecosystem.

Less—sometimes, if not always, we will have to use less to manage fisheries in a changing environment. And finally, we need to deal with effects of our management. I think Keith Sainsbury is the only person who mentioned this in the last two days: The fact that after we've done everything else right and prescribe a reaction, the fishers react to our reaction. That's a tractable problem; we should be studying how that thing that we manage, the fishing fleet, responds to our management initiatives. It's called implementation uncertainty; we can do something with it. Now, if we combine these five points, it tells us how we should manage fisheries in the future. We should fiddle with them. Thank you.

Don McCaughran.—Let me answer the second question first, because that's rather easy for me to answer. How should we regulate artificial enhancement programs for long-term sustainability? Well, I would say, "carefully." We've heard that this afternoon; there's good evidence for that. And the other answer, and probably the more important one is to talk to somebody like Ole Mathisen, because I don't know a thing about it. So I will concentrate on the first question, which I know a little bit more about, and that's how we should manage fisheries given a change in ocean environment.

Well, Jake said we should do some forecasting. Being a statistician, and a very conservative kind of statistician, I don't think a lot about forecasting, and the older I get, and this is in terms of months, the more I'm beginning to agree with Ray Hilborn. About the only thing that scares me is that by the time I finish my career, maybe I'll agree with him 100%. However, there's a lot of uncertainty, and you know that when you go beyond the range of your data, your confidence bounds expand extremely rapidly, and you'd quickly get to the point where you shouldn't make any decisions based on that kind of uncertainty. Now, we're supposed to talk globally, but I would like to talk about Pacific halibut because there's an awful lot of data there and it's an old, managed fishery, and it does occur in both the Pacific and the Atlantic. And perhaps under the Arctic Sea as well. So in some ways, it's global. We have about a 20-year cycle in recruitment. Roughly, 1945 was a peak, and 1965 and 1985 were also peaks of recruitment, and right now at 1995 we expect a lull. But what do we make of that? How many fisheries have data going back that far? In the North Pacific, very few, if any. And what am I supposed to think of this? Is this the effect of the environment?

There's some interesting variability in the stock–recruitment relationship for halibut. You can fit most any kind of model through the data, like you can through most stock–recruitment relationships, and the reason for that, I think, is the effect of the environment on halibut stocks. The environment has a huge effect on halibut stocks. Now we saw that 20-year cycle in recruitment; we don't know if that's environment, or something intrinsic, or what it is, but we know it's there. We see a tremendous amount of scatter in the stock–recruitment data like this. We've seen very high spawning biomasses produce very good recruit populations, and of course, very low spawning biomasses produce very good recruits. In general, there is some relationship that would show it decreasing, but that's probably the effect of the environment. So in halibut management, we just say there's a lot of environment going on. We can't forecast, nobody can forecast worth a damn. We can't even tell you what environmental things are important, except perhaps ocean current drifting halibut onto shallow grounds when they're ready to settle out. There might be some mechanism we can speculate about, but we don't really know. So what we do is this: Annually, we do the best job we can of estimating the standing biomass, the exploitable biomass (the 8-year-olds to 20-year-olds). So we get a stock assessment every year and we take a constant proportion. In other words, we have a constant exploitation rate.

You have to get good estimates, or consistent estimates over time of spawning biomass, and that will fluctuate with the environment—so you don't care about the environment, you just have to get a good estimate. And then you take a consistent proportion of that in your fishery every year. There's two tricks then. The first trick is getting consistent biomass estimates; the other trick is what proportion should that be? How do we figure that out? The only tools we have available, and we all use them, and most of you are familiar with this, are models. We build computer models, we build elaborate simulations with all sorts of strategies. And we do a lot of computer simulation, and you choose different exploitation rates and you see what happens under a lot of different scenarios. And you try to protect yourself. But what do you protect ourselves from? That's the third question. Well, what we've decided to do is say we're going to protect the spawning biomass. We don't want it to drop below a certain value. And so we choose a strategy, in other words an exploitation rate, that in the long term, using simulation, will not cause the spawning level to drop below some historical level. That's one approach, and it's the approach we've done.

What we're really saying here—and this applies to a lot of other species besides halibut—is the environment is going to change; it's going to change your fish stocks. So you have to develop a method in which you don't have to rely on forecasting. You have to estimate what you've got and take a certain proportion. And then you have to think a great deal about what that proportion should be, and you have to think a great deal about how good a biomass estimate you can get. I think that kind of a strategy is probably better than trying to develop some hypothetical forecasting where the uncertainty is very great. Not to say there isn't a lot of uncertainty here, but you can remove a lot of the uncertainty in terms of causes and so forth by just operating with what you've got. The problem becomes what those exploitation rates should be, and of course that requires a lot of modeling and you just hope that it works. How do you know if it works? Well, if your estimates look like what the fishers are telling you, if they look like the survey work that you're running at the same time, if things are kind of consistent with anecdotal data plus survey data, then you must be on the right track. But there's really no other measure of how good you're doing, except if in the long term, it works. And in Pacific halibut, it seems to work. So, that's one way we've thought in managing fish with changing environment. We've tried to avoid the forecasting. We've tried to do something that works today, and back this up with a tremendous amount of computer simulations and modeling and so forth in choosing exploitation rates. I know I was kind of parochial in terms of talking about halibut, but it's a nice species to talk about because we've managed it for a long time, and it does occur in both oceans.

Clarence Pautzke.—I saved Marc Miller until the end because when the stock collapses, or if something bad happens to the fishery, I can still get in my car and drive home. I think that most of you in the audience who are biologists and fisheries managers probably will still have your pension funds. But if we don't make the right decision and the fishery is having problems, then it's the people out there who are actually fishing on the stocks who have to tie up their boats. Marc Miller's experience and background are in the social sciences. He is the social scientist who is on our Scientific and Statistical Committee; in fact, he's the only one we've ever had in the North Pacific.

Marc Miller.—It's my pleasure to be here. Let me say at the outset that at the University of Washington, the Schools of Fisheries and of Marine Affairs have a Joint Program in Fishery Management, and we're reformist in our view. We're trying to encourage students to understand the management side as well as the science side, and essentially we see fishes behaving in the constraint of the ocean, and people behaving in the constraint of something of what you might call society or institutions.

I have taken some liberties with the question set here before us. I've looked at it and noted that the panel deals with managing for sustainability, so then I wonder, "Well, sustainability of what?" Then I look at the first question. Well, I'm going to amend that question to fit my disciplines (I'm a cultural anthropologist), and I'm going to substitute for "ocean environment"— the constraint on fish—to "institutional, or ethical environment"—and that's the constraint on what it is that people do. Essentially, what usually happens is we take oceans and institutions as given in sort of a parametric sense and then we try to anticipate what it is that folks will do.

Every now and then the real world surprises us and the environment itself changes. El Niño is an example of that, and global warming is an example of that, and on and on. The same thing happens on the human side. From time to time, humans redesign their values, their ethical postures, and institutions themselves change. And in a sense, if people define situations as real (this is well known in sociology), they are necessarily real in their consequences. So it's important to understand how people approach what we routinely call fishery problems, and to understand how an ethical problem to one person is only a technical or logistic problem to someone else.

First, I ask, "What is a fishery from a sociological point of view?" For the "fishing industry" (I'm using the term to include commercial, recreational, charter, and subsistence people), it turns out that we know very little about how people are committed to those different activities. Or we don't even know how people move in and out of fishing as an occupation or as a way of life. The "public element" has to do with diverse social movements and special interest groups who feel, increasingly, that they have a direct connection to what we call fishery resources, via an emotional tie. And then we've got the "management" element here at the federal, state, and traditional sectors, and it turns out that from a sociological point of view, these people are definitely a part of the system under inquiry. So health of a system, viability of a fishery system, has to do not just with the habits and behaviors of fishers, but with how they communicate or do not communicate with their management sector and the other public element constituencies.

Let me talk just a little bit about values here, and how values drive this whole thing. This is a well known statement of Aldo Leopold in 1949: "A thing is right when it tends to preserve the integrity, stability, and beauty of the *biotic* community. It is wrong when it tends otherwise." You can equally well apply that to the ecology of humans. And you can decide for yourselves, "What is social integrity? What is social beauty? What is social stability?" Expanding the notions of values over the last 100 years, we've had what's sort of called an "experiment in conservation." I think that it's important for us to realize that conservation is a double-edged ethic. I refer to the extractive conservation associated with the "wise use" thinking of Gifford Pinchot, and I contrast that with the aesthetic conservation of John Muir. I've identified two executive agencies here: the Forest Service and the Park Service, to illustrate different fundamental orientations toward the relationship between humankind and nature. Importantly, both of these can be and are successful, and they are both improvements over the alternative, which is sort of a laissez faire, cavalier use of resources. Sort of from a clearcutting, "do as you will," to the extractive or aesthetic variants here, both stressing sustainability.

You might wonder what we have as an ideal for better management down the road. There has been an influential report, the well-known Brundtland report, which offers one variation of what sustainable development can be if it's not oxymoronic in the first place. That's not exactly my favorite formulation. Bob Francis and I, in fact, prefer this one by Gregory Bateson from his fabulous article, "On Steps to the Ecology of Mind." Gregory Bateson takes his proposition and then identifies, carefully, each word in it. Essentially, it's the same sort of ideal. In the last years in different resource management settings, different practitioners and scientists have called for a new kind of an order, a new kind of a paradigm. Daniel Botkin articulates that there's a difference between new management and the old management. I think that it's important for people in fisheries in a sense to not refuse to see the "forest" for the fish. There's been a dramatic change in how forests have been managed, wildernesses have been managed, parks have been managed over the last 25 years in this country. Fishery managers can not be indifferent to that history.

Then you wonder, well, what might be a solution, given that challenge? Essentially, I would be calling for a new fisheries in the pattern of a new forestry. Here's one variation, from *Wilderness Experience* recently: "Ecosystem management is regulating internal ecosystem structure and function plus inputs and outputs to achieve socially desirable conditions. Thinking like an ecosystem here it's important to reflect attention to environmental and socioeconomic concerns." In the forest setting, ecosystem management emerged in this way: it's a "strategically planned and managed ecosystem to provide for all associated organisms as opposed to strategy or plan for many individual species." So, you say what is the status quo? What is the orthodoxy? Consider that to be single-species management. Essentially, ecosystem management is asking a larger question. That given, I wrote a paper with a colleague recently dealing with the reauthorization of the Magnuson Act, the Endangered Species Act, and the Marine Mammal Protection Act. It turns out that you have to be conversant about all three of these acts if one is to do a responsible job in fisheries management alone. Yet I know too many people who know one of these acts well and have never seen first-hand the other two.

You can discuss sort of an institutional ecology of laws; these laws do reflect, in legal text, a reflection of the values of society. Notice that they have slightly different objectives across the three Acts. Optimum yield, fishery impact statements, optimum sustainable population, critical habitat—it depends on different kinds of science. In a sense, however, they are all better questions than the standard, single-species management approach. Note that, in the Magnuson case, it's incumbent on researchers to formally study—I don't mean guess; I'm not talking folk science—formally study sociological conditions, cultural conditions, and how people are involved in fisheries, in addition to the economic studies. That's to balance along with ecological information to adjust maximum sustainable yield into an optimum yield. In my view, that's a better question. In the other two instances, it's an adjustment from a single-species approach to one showing concern for people and for environmental and ecological attributes. All three of those in a sense are illustrations of an ecosystem management approach, given the fact that you're dealing in the first place with something called a fishery. I think it's important to understand, for example, that a forest, a park, and a wilderness have connotations. They are managed for different objectives. And so you could manage, from some people's point of view, a forest as a failed park, and a park as a failed wilderness.

The best I can do then, is argue that ecosystem management is an adaptive departure from the status quo. It asks a broader and more responsible question than the orthodox doctrine it displaces. It's a process of stewardship in which goals are problematic. By that, they are emerging through the process of representative government. Thank you.

Clarence Pautzke.—First I'm going to turn to the audience to see if anything here has stimulated you to ask some questions. I know that I'm very interested in how many people out here have had to actually shut down a whole industry—I know one of the panelists here has—and whether they were willing to go on very scant data, or indicators of the fishery, or did they always want more data as they sweated the decision on what to say on what the stocks were going to yield for the next year.

Question #1(Ray Hilborn).—All right, the title of this symposium is "Global Trends," and one trend that hasn't been discussed much—Marc Miller probably came as close as anyone—is what one might call the Animal Rights Movement. We've already seen whale fisheries pretty well ruled out as a commercial proposition for the future. I have a suspicion that bluefin tuna, and some of the larger fishes, may be coming up next. I was wondering if any of the panelists want to comment on how they see this changing socioeconomic trend in terms of fisheries management?

Suzanne Iudicello.—Well, first of all, neither I nor the organization that I represent would characterize ourselves as Animal Rights representatives. I think in terms of Marc's spectrum—with Pinchot on the right and Muir on the left—we are somewhere more toward the Muir side, but we're not to the John Muir end of the spectrum. I think that the way you have to account for the public opinion on the Animal Rights side is to try to talk to folks about conservation. There's a difference, I think, between conservation and protectionism, between a choice about values that says we can find a sustainable use—that means we can use ocean creatures and systems for the future and for the present—which is different from a point of view that says, "I'm not going to eat fish. I'm not going to eat veal. I'm not going to wear a leather belt. I'm not going to wear leather shoes." There's a whole range of opinions, and various groups fall in various points on a spectrum.

With regard to the bluefin, I think the bluefin has definitely become a poster child species. It is a charismatic megafauna. It's the closest thing that we have in the fish world to a whale. Maybe sharks will be the next "featured" species, if you will. But I think the groups that are using these images of the apex predators to try to get people interested in fish are doing so, not to stop fisheries on them, but to engage people in the whole discussion of fishery management. There are no "hug-a-halibut" posters out there. We tried the idea, but nobody wanted to buy one. The whole idea of trying to engage the public in these complex and very complicated mathematical issues of fishery management, not to mention the economics and sociology of it, is not easy. It's not as simplistic as "Save the Whales." So, you're going to see in the coming months and years more creative ways to try to engage folks in the kinds of issues you've been talking about—we've been talking about—for the past couple of days.

Question #2 (Ray Hilborn).—Could I just mention that Greenpeace in New Zealand has the Ministry of Fisheries in court over the mathematics of one of their stock assessments?

Jake Rice.—Yes, I've got a somewhat different view of this, having lived through the demise of the Newfoundland seal fishery due to slander and misrepresentation of the facts associated with quite a sustainable harvest of a renewable resource. Society has the right to set values and society has the right to change values. If society, in its infinite wisdom, chooses not to allow certain organisms to be harvested, so be it. It is our job as resource mangers to represent the will of our society as best we can. Some of us are still naive enough to think that we can educate the public and have them make only wise ethical decisions, but some of us are a little more pessimistic than that. I don't think very many people bothered to do any calculations on what the future impact of all those seals that were saved by the end of the seal fishery was going to be on the charismatic outport fishers of Newfoundland. And yet many people believe that there has been an impact of those cute, cuddly seals that haven't been killed on hard working fishers. We're going to face a lot more societal values that are somewhat different than we're used to dealing with in the future. I don't think that we should be building some way to resist them. We're going to have to accept them, and maybe do a better job of trying to educate the public about what the consequences of some of our ethical choices are. None of us have been doing very good about foreseeing them. That's the future, let's deal with it. People's opinions change just like climates do.

Marc Miller.—I like your question, Ray. It's interesting; it sort of illustrates a point I was trying to make about the difference between how something might be coded as a moral issue by one person and a technical or scientific question, or a straightforward routine issue by someone else. It turns out that population thinning is a good illustration of that. And it doesn't matter whether you're talking marine mammals or human populations, whether you're talking whales and fur seals, or you're talking birth control and infanticide and abortion. If these are coded as ethical or moral issues by people, then they are issues about which people have no intention of com-

promising at any time. And oftentimes as managers, we sort of assume that reasonable people together will come to the same position. Now that's not entirely clear when things are cast in ethical terms. I tell people that natural resources and environment as terms are competing symbols. One person's natural resource is someone else's element of the environment. If you go to other cultures, while you don't find bumper-stickers saying "hug a halibut," you most definitely find people who say that they are related to halibut, or related to sharks. So these are fundamental orientations that need to be explored formally, rather than anticipated or presumed away.

Question #3 (Doug Butterworth).—Two questions: First, I'm very disappointed that Don McCaughran isn't fighting with Ray Hilborn any longer, so I'm going to try to give him someone else to fight with. Basically, I've thought Don McCaughran gave the best answer to the first question before the panel. Essentially what I think you were asking about was management strategy. The question I want you to fight with me about was, I get the impression, looking at what you put up there, that you're underexploiting your resource, and I want to know, "Why are you doing that, and do you feel happy with that?" And a question to Jake Rice as well: He raised a question of diagnostics and he gave the medical analogy. I'm not very happy with that because I'm perfectly happy with doctors having had a very large sample size to develop their diagnostics. I'm not sure in fisheries we do have a large enough sample size to be confident about diagnostics to use them in that way yet.

Don McCaughran.—Well, one might look at the bottom line for Pacific Halibut. The fishery started in about 1883 or something like that, and it's been managed under a group that's been put up to manage it since 1923, and by God, we've still got a few halibut around. So maybe it's not bad. The International Pacific Halibut Commission has always been conservative, except for one point in it's history, and that point was a bad point because there was this concept called maximum sustainable yield. Way back when, people thought that was a good thing to test, so let's push the halibut yield to test the MSY principle, and that was actively done by the Commission, but it happened to coincide with the advent of the foreign fleets arriving in the North Pacific and beating the hell out of halibut with bycatch. Both things coincided and the stocks collapsed. That was the dark side of the Halibut Commission's management. Since then and before then too, we've always been very, very conservative. The bottom line is we still have halibut, so maybe we should be conservative.

Jake Rice.—I actually agree with Doug. I'm not particularly comfortable with the medical diagnostic analogy. I'm trying to capture something somewhat more complex than that in one-sixth of a 5-minute talk, and that was a convenient shorthand. I will say that as we move to more ecosystem management, what we're encountering is the need for some management tools. If the choice is, "Well gee, we're going to monitor the ecosystem, study the ecosystem, and build an ecosystem model that consists of something more than a bunch of made-up functional relationships unconstrained by any data," I have more confidence in a diagnostic approach if we do have some reliable time-series than in those ecosystem models as a source of advice. The other thing I will say in defense of the analogy is the fact that it is much quicker and much cheaper than waiting until we have an ecosystem model that we believe and that we've tested adequately. If we've got some diagnostics, even though they may be faulty, if they give us a message, I think it's worth paying attention to. I could be wrong, but I wouldn't want to ignore it. Nor would I want to wait 20 years for the ecosystem model before I do something about it. It would be only in the sense that it's a tool, it's a thing we can use more than we are. It's certainly not an endpoint.

Clarence Pautzke.—When we were talking about halibut and diagnostics, we were looking at what, a 20-year cycle? Is that what you showed up there? When you're managing those fisheries, do you really think you're helping to produce those cycles or are you just hanging on to the cycle and holding your breath and hoping it goes out of the trough and back up to the next peak?

Don McCaughran.—Well, we have three peaks, but is this a cycle, or is this just a phenomenon?

Clarence Pautzke.—Do you think you can manage it? Is there a particular diagnostic that you can look at, and are you comfortable enough that you know what's going on out there with oceanic change that you're able to adjust your management?

Don McCaughran.—Well, we're trying to remove the oceanic change from it by looking at standing stock and taking some conservative portion of that standing stock and then looking at a lot of different possibilities through modeling to see if that's not a bad strategy. That's what we've done, and that's the state of the art right now. We hope that we're doing right.

Clarence Pautzke.—And Suzanne, when we're talking about sustained fisheries and sustainability, would you consider the Halibut Commission a success?

Suzanne Iudicello.—Based on the record, yes.

Question #4 (Bart Eaton).—We're talking about sustainability here, and you touched on it there with your last line of questioning. The question I really have is: Are we talking about sustainability of the resource, like in the halibut fishery? Are we talking about sustainability for the fishing industry or the fishery, or is it sustainability of biodiversity of a given ecosystem? To me, there are three separate goals there. How are we combining those?

Or are we going to manage for a separate one? I guess I'd like to hear the panel's thoughts on that.

Suzanne Iudicello.—I'll take a short whack here. I think when Marc Miller was talking about looking at the system—and a lot of people have talked about ecosystems over the last 2 days—we may look at the Halibut Commission as a successful experiment in a sustainable fishery. That is, there are persons catching halibut, making money to some extent, and it's been going on for a number of years. Whether the Bering Sea and Gulf of Alaska ecosystem has been sustained in terms of its diversity and integrity is another question. I don't think any of us has sought to answer that yet. In terms of the whole system question, we still regulate and manage and consider certain types of bycatch not even as bycatch. It isn't even listed in the list of prohibited discard species. There are a lot of organisms that come up in nets, for example, that are just discounted entirely. There's no regulatory regime, there's no accounting for them, there's no measuring for them, they're just trash. Well, somebody's jellyball or trash fish may be food or nutrient to someone else. I think that we're just at the beginning of trying to understand that. There has been the development of a conservation biology on the terrestrial side, and I think on the marine side, we're still not there yet. We're just beginning.

Question #5 (Ellen Pikitch).—I want to explore the medical analogy. I think that in a way it's a good one. One of the things that you've been asked to address is the sustainability question. I think one of the observations that I have is that we have seen "failures" in fishery management where we have not sustainably managed resources. We thought that we failed because we didn't know enough and that if we just collected more data, then we would know more and that wouldn't happen. I think we're coming to the realization that we can collect all the data we want, and there are lots of things outside of our control that will cause fisheries to not fare well for certain periods of time. I think there's a growing awareness of the level of uncertainty in fisheries and in fisheries management, but I'm not sure we've yet embraced the fact that stocks not prospering, will be a continuing fact of life, and that all the knowledge in the world, and all the management techniques in the world aren't going to prevent those things from happening. I think that there's a large element of that, although it's a very complicated story, in the Northern cod case.

And here's where I'd like to expand a bit on the medical analogy. When you're trying to live a healthy life, live a long life, not get cancer, you don't smoke cigarettes, you exercise all you can, you eat the right foods, and low and behold, some people develop cancer anyway. They go to the doctor and the doctor, if it's a good doctor, uses the latest techniques, and may or not diagnose the problem well, and let's say you do have cancer. They diagnose it, and they set you on a course of treatment. They know that a certain percentage of the people will survive and go into remission, and others will die. I think, perhaps, that thinking that way about fisheries management can help us. Because if you look medical science over time, we have imperfect information, but we diagnose people anyway. We treat them anyway, and what we see is an improvement over time in the treatment of most diseases and in life expectancy. I guess that I just wanted to elaborate a little bit on those thoughts and say maybe we should be looking more at fisheries management that way. I like Jake's suggestion.

Clarence Pautzke.—I would respond, and maybe Marc wants to respond. One of the problems with a patient with cancer is that you have to prepare them properly for the bad news. I think that many times we're not prepared as fisheries managers to deliver the bad news, and so we procrastinate as the stocks get worse possibly, until they get to that slippery slope, when then it's time to move to the Pacific coast.

Question #6 (Loh Lee Low).—Well I thought I had a friendly panel, so I gathered some guts to come up here and say a few words. The main theme here is: How should we manage fisheries given a changing ocean environment? How should we regulate artificial enhancement programs for long-term sustainability? I would like to say that, although a lot of the talks here have been focused on the technical and science side, maybe that's not a problem. It's basically managing with the best available sources of information and with accountability—accountability in the sense that the process has to be very, very transparent to everybody. Because when that process is public, such as the North Pacific Fishery Council process, any input should be very responsible. We may have to develop firmer set of rules on how that input is to be incorporated. You know in the technical game there's only so many ways you can score two points at basketball. In the fisheries field, we don't have that. We don't have a firm set of rules. I think we need to tighten them a little better so that the proponents, opponents, and people who sit on the fence have a clear input through that process. Since that input is so clear, they would have to act fairly responsibly. I wouldn't come out here and utter clearly erroneous material if I knew that I was going to be accountable for it. And that's a process which is actually taking place in the North American fisheries management process. We may eventually get to the point where we'll be all managing these fisheries with information, with accountability, and with responsibility.

Question #7.—I sometimes get the feeling that the scientists in this era become more or less paralyzed by the idea that all the data they have are uncertain. There's just so much uncertainty everywhere. As a scientist, I

would also highlight the things we are quite certain about: that is, the number of stocks in the world which are over-exploited. We are quite sure that the cause is that the fleets are too large, there's an overcapacity. If we use the fleets, we can use some of the issues that Jake was bringing up, like less fishing, [which] gives you more inertia in the system because you have more year classes around, more biomass, so you can respond much more efficiently to the developments in the system. I think actually that we know a lot about the effects of fishing on fish stocks, and we can use that information in order to improve management. We shouldn't hide ourselves behind our uncertainty about everything.

Clarence Pautzke.—But then I don't think we should hide ourselves behind the premise that it's large fleets that are causing all the problems. I think that the managers have the tools they need, at least under the Magnuson Act, to control the amount of catch out there, regardless of how big the fleets are. It's a matter of whether they use those tools that are available to them to shut down a large fleet in the face of the political realities?

Marc Miller.—I don't think it's a good idea to get too romantic about the medical analogy here, because in a sense it illustrates the opportunity for a problem of hubris. What happens in health care is we have people who are technically very well trained, keen on technology and science, with a poor understanding in fact of their patients. And we've had problems with people who have assumed—doctors in particular—that life is necessarily good and should be sustained at any cost. And human constituencies are arguing for different endings. And so the lesson here I think is for managers in science, as well, to take more seriously the variation in the cultural orientation of their constituencies and to examine that.

Question #7 (Jim Beckett).—Yes, I think we've touched upon it a couple of times and I think actually Doug Butterworth raised it when he suggested that halibut had been underutilized. I think we have to accept that we are going to underutilize the resources because we're going to have to be cautious and we're going to have to be flexible. And that means we cannot push the resources to the maximum. And I would say with the involvement that Jake and I have had on the Atlantic fishery in Canada, we just cannot try and push. We thought we were being conservative and going at the $F_{.0.1}$ level. It has been quite apparent that because of the way that the fishery's been undertaken and the data that we've had, we've not been very conservative. So we just have to accept that we cannot get that last piece of fish out of the ocean. And that is how we're going to be sustainable. Thank you.

Clarence Pautzke.—Can I ask one question of you before you leave? I assume there wasn't a catastrophic drop in the stocks back there that just happened overnight. What signals or diagnostics did you see and why did the scientific community or the managers not react quickly enough?

Jim Beckett.—Let's see. Do we have 2 hours? Actually, there may have been diagnostics and slow change over time but the actual event was pretty dramatic. We appear to have lost 600,000 metric tons of spawning fish over the space of a few months. We do not know where they went or what happened. But we do seem to have had a pretty dramatic effect that actually did precipitate this. It certainly overcame the political inertia in a hurry.

Clarence Pautzke.—I call that catastrophic. I want to thank our panel very much for being with us. Thank you.

Index